NATURE ACROSS CULTURES

SCIENCE ACROSS CULTURES:
THE HISTORY OF NON-WESTERN SCIENCE

VOLUME 4
NATURE ACROSS CULTURES

Editor:

HELAINE SELIN, *Hampshire College, Amherst, Massachusetts USA*

A list of titles in this series may be found at the end of this volume

NATURE ACROSS CULTURES

Views of Nature and the Environment in Non-Western Cultures

Editor

HELAINE SELIN

Hampshire College, Amherst, Massachusetts, USA

Advisory Editor

ARNE KALLAND

University of Oslo, Oslo, Norway

KLUWER ACADEMIC PUBLISHERS

DORDRECHT/BOSTON/LONDON

Library of Congress Cataloging-in-Publication Data

Nature across cultures : views of nature and the environment in non-western cultures /
   Helaine Selin, editor; Arne Kalland, advisory editor.
      p. cm. – (Science across cultures ; v. 4)
   Includes bibliographical references and index.
   ISBN 1-4020-1235-7 (cloth : alk. paper)
      1. Human ecology–Cross cultural studies. 2. Philosophy of nature–Cross-cultural studies.
   3. Indigenous peoples. 4. Nature–Effect of human beings on. 5. Environmental protection.
   I. Selin, Helaine, 1946- II. Kalland, Arne. III. Series.

ISBN 1-4020-1235-7

Published by Kluwer Academic Publishers,
PO Box 17, 3300 AA Dordrecht, The Netherlands.

Sold and distributed in North, Central and South America
by Kluwer Academic Publishers,
101 Philip Drive, Norwell, MA 02018, USA

In all other countries, sold and distributed
by Kluwer Academic Publishers,
PO Box 322, 3300 AH Dordrecht, The Netherlands

Printed on acid-free paper

Printed and bound in Great Britain by Antony Rowe Ltd.

INTRODUCTION TO THE SERIES

SCIENCE ACROSS CULTURES:
THE HISTORY OF NON-WESTERN SCIENCE

In 1997, Kluwer Academic Publishers published the *Encyclopaedia of the History of Science, Technology, and Medicine in Non-Western Cultures*. The encyclopedia, a collection of almost 600 articles by almost 300 contributors, covered a range of topics from Aztec science and Chinese medicine to Tibetan astronomy and Indian ethnobotany. For some cultures, specific individuals could be identified, and their biographies were included. Since the study of non-Western science is not just a study of facts, but a study of culture and philosophy, we included essays on subjects such as Colonialism and Science, Magic and Science, The Transmission of Knowledge from East to West, Technology and Culture, Science as a Western Phenomenon, Values and Science, and Rationality, Objectivity, and Method.

Because the encyclopedia was received with critical acclaim, and because the nature of an encyclopedia is such that articles must be concise and compact, the editors at Kluwer and I felt that there was a need to expand on its success. We thought that the breadth of the encyclopedia could be complemented by a series of books that explored the topics in greater depth. We had an opportunity, without such space limitations, to include more illustrations and much longer bibliographies. We shifted the focus from the general educated audience that the encyclopedia targeted to a more scholarly one, although we have been careful to keep the articles readable and keep jargon to a minimum.

Before we can talk about the field of non-Western science, we have to define both non-Western and science. The term non-Western is not a geographical designation; it is a cultural one. We use it to describe people outside of the Euro-American sphere, including the native cultures of the Americas. The power of European and American colonialism is evident in the fact that the majority of the world's population is defined by what they are not. And in fact, for most of our recorded history the flow of knowledge, art, and power went the other way. In this series, we hope to rectify the lack of scholarly attention paid to most of the world's science.

As for defining science, if we wish to study science in non-Western cultures, we need to take several intellectual steps. First, we must accept that every culture has a science, that is, a way of defining, controlling, and predicting events in the natural world. Then we must accept that every science is legitimate in terms of the culture from which it grew. The transformation of the word science as a distinct rationality valued above magic is uniquely European. It is not common to most non-Western societies, where magic and science and religion can easily co-exist. The empirical, scientific realm of understanding and inquiry is not readily separable from a more abstract, religious realm.

Nature Across Cultures is the fourth book in the series. It includes 23 chapters. Most deal with views of the environment as they are perceived by different cultures: Australian Aboriginal people, Native Americans, Polynesians, Indians, etc. The book also contains a variety of essays on broader issues, such as Images of the Other, Traditional Ecological Knowledge and Indigenous Knowledge, and The Global Mobilization of Environmental Concepts: Re-Thinking the Western/Non-Western Divide. The final section contains articles on views of nature and the environment from the points of view of Buddhism, Confucianism, Daoism, Hinduism, Islam, and Judaism.

We hope the series will be used to provide both factual information about the practices and practitioners of the sciences as well as insights into the worldviews and philosophies of the cultures that produced them. We hope that readers will achieve a new respect for the accomplishments of ancient civilizations and a deeper understanding of the relationship between science and culture.

TABLE OF CONTENTS

ACKNOWLEDGMENTS

I would like to thank the contributors to this volume; I am so impressed with how committed they are to their subjects while at the same time keeping their academic distance. It was a pleasure to read and work with their writing. Thanks especially to Arne Kalland, the Advisory Editor, who read and commented on all the articles in addition to writing his own piece. He is the best editor I have worked with; you can send him an article on Monday and have it back by Wednesday with ten pages of comments and a bibliography. Thanks to my Kluwer family: Maja de Keijzer and Andrea Janga in Dordrecht, and Phil Johnstone in England, who does such a lovely job of making the books beautiful. And thanks, always and again, to my loving family, Bob and Lisa and Lisa and Tim.

ABOUT THE AUTHORS

HELAINE SELIN (Editor) is the editor of the *Encyclopaedia of the History of Science, Technology, and Medicine in Non-Western Cultures* (Kluwer Academic Publishers, 1997) and Science Librarian and Faculty Associate at Hampshire College in Amherst, Massachusetts, USA. In addition to editing the new series, *Science Across Cultures*, she has been teaching a course on the Science and History of Alternative and Complementary Medicine. She has also begun work on a revised electronic version of the encyclopaedia.

ARNE KALLAND (Advisory Editor, Environmentalism and Images of the Other) is professor in the Department of Social Anthropology, University of Oslo, Norway. His main fields of interest are maritime resource management, people's perceptions of nature, and the environmental movement, with a particular focus on Japan. Among his books are *Rice or Opium? An Introduction to the Golden Triangle* (in Norwegian, Universitetsforlaget, 1985), *Fishing Villages in Tokugawa Japan* (University of Hawaii, 1995) and *Marine Mammals in Northern Cultures* (with Frank Sejersen, in press). He is the co-editor of *Asian Perceptions of Nature: A Critical Approach* (with Ole Bruun, Curzon, 1995), *Japanese Images of Nature: Cultural Perspectives* (with Pamela Asquith, Curzon, 1997), and *Environmental Movements in Asia* (with Gerard Persoon, Curzon, 1998).

WILLIAM BALÉE (Native Views of the Environment in Amazonia) is Professor of Anthropology at Tulane University, New Orleans, Louisiana, U.S.A. He chaired the department from 1998–2001. He is the author of *Footprints of the Forest: Ka'apor Ethnobotany* (Columbia University Press, 1994), which won the Klinger Award from the Society for Economic Botany. He edited *Advances in Historical Ecology* (Columbia University Press, 1998) and was coeditor of *Resource Management in Amazonia: Indigenous and Folk Strategies* (New York Botanical Garden, 1989). He edited the *Journal of Ethnobiology* from 1999–2002. His current research focuses on the historical ecology of Amazonian landscapes that are occupied and utilized today by diverse societies associated with the Tupi-Guarani branch of the Tupi language family.

FIKRET BERKES (Local Understandings of the Land: Traditional Ecological Knowledge and Indigenous Knowledge) is Professor of Natural Resources at the University of Manitoba, Winnipeg, Canada. He holds a Ph.D. degree from McGill University, Montreal, in applied ecology, and he has done postdoctoral studies in anthropology. Working at the interface of natural and social sciences, Dr. Berkes has devoted most of his professional life to investigating the interrelations between societies and their resources. His main area of expertise is common-property resources, community-based resource management, and traditional ecological knowledge. He teaches in these fields, contributes to theory, and applies his experience in a range of geographical areas. His recent publications include three books: *Linking Social and Ecological Systems* (co-edited with Carl Folke; Cambridge University Press, 1998), *Sacred Ecology* (Taylor & Francis, 1999), and *Managing Small-Scale Fisheries* (co-authored with R. Mahon, P. McConney, R. Pollnac and R. Pomeroy; International Development Research Centre, 2001).

JOHN BERTHRONG (Confucian Views of Nature) educated in sinology at the University of Chicago, has been the Associate Dean for Academic and Administrative Affairs at the Boston University School of Theology since 1989 and Director, Institute for Dialogue Among Religious Traditions since 1990. Active in interfaith dialogue projects and programs, his teaching and research interests are in the areas of interreligious dialogue, Chinese religions, and comparative theology. His most recent books are *All Under Heaven: Transforming Paradigms in Confucian-Christian Dialogue* (State University of New York Press, 1994), *The Transformations of the Confucian Way* (Westview Press, 1998), and *Concerning Creativity: A Comparison of Chu Hsi, Whitehead, and Neville* (State University of New York Press, 1998). He is co-editor with Mary Evelyn Tucker of *Confucianism and Ecology: The Interrelation of Heaven, Earth, and Humans* published by Harvard University Press in 1998. In 1999 he published *The Divine Deli* (Orbis Books), a study of religious pluralism and multiple religious participation in North America. Mostly recently he collaborated with Evelyn Nagai Berthrong on *Confucianism: A Short Introduction* (Oneworld Press, 2000).

ELLEN BIELAWSKI ("Nature Doesn't Come as Clean as We Can Think It": Dene, Inuit, Scientists, Nature and Environment in the Canadian North), born in Alaska, has worked with Inuit and Dene since 1975. She received her doctorate in 1981. She is a free-lance scholar and writer and a Research Associate of the Arctic Institute of North America, University of Calgary, Alberta, Canada. She divides her time between Edmonton, Alberta, where her sons attend school, and the North.

ANNIE L. BOOTH (We Are the Land: Native American Views of Nature) is an Associate Professor in the Environmental Studies Program at the University of Northern British Columbia, Canada. Her current research interests include Native American Natural Resource Management and views of the environment, resource dependent communities, public participation in resource planning and environmental ethics.

DAVID L. BROWMAN (Central Andean Views of Nature and the Environment) received his Ph.D. in anthropology from Harvard University in 1970, and has spent the last 35 years involved in research projects on various Central Andean topics ranging from geophagy to pastoralism, resulting in more than 120 articles, chapters, edited books and other publications. He has a particular interest in the origins of agriculture and complex society, and the impact which environmental parameters have on the development of cultural traditions in this region. He currently serves as Professor of Anthropology and Chairman, Interdisciplinary Program in Archaeology at Washington University in St. Louis, Missouri, U.S.A.

D. P. CHATTOPADHYAYA (Indian Perspectives on Naturalism) received his LL.B., M.A., and Ph.D. degrees from Calcutta and the London School of Economics. He was Professor at Jadavpur University in Calcutta and Founder and Chairman of the Indian Council of Philosophical Research. He was Chairman of the Raja Rammohan Roy Library Foundation in Calcutta and President of the Indian Institute of Advanced Study in Shimla. He also served as Union Minister of Commerce and Governor of the State of Rajasthan. At present he is the Chairman of the Indian Philosophical Congress, the Centre for Studies in Civilizations, and Director and General Editor of the Project of History of Science, Philosophy and Culture in Indian Civilization (50 Vols.). Among the 30 books he has authored and edited are *Individuals and Societies* (Allied Publishers, 1967), *Environment, Evolution and Value* (South Asian Publishers, 1982), *Induction, Probability and Skepticism* (State University of New York Press, 1991), and *Societies, Cultures and Ideologies* (2001).

HAROLD COWARD (Hindu Views of Nature and the Environment) is a Professor of History and the Director of the Centre for Studies in Religion and Society at the University of Victoria, Canada. He received his Ph.D. from McMaster University. With a teaching focus on India, comparative religion, and ecology, he has supervised numerous students for their MA and PhD degrees both at the University of Victoria and at the University of Calgary. He is a Fellow of the Royal Society of Canada. He has been the recipient of numerous research grants from the Social Sciences and Humanities Research Council and the Ford Foundation. He has been a Visiting Fellow at Banaras Hindu University and the Institute for Advanced Studies in the Humanities, Edinburgh University. He has written sixty-two articles and is author/editor of thirty-two books, including: *Hindu Ethics* (State University of New York Press, 1988); *The Philosophy of the Grammarians* (Princeton University Press, 1990); *Derrida and Indian Philosophy* (State University of New York Press, 1990); *Mantra: Hearing the Divine in India* (Anima, 1991); and *Population, Consumption, and the Environment* (State University of New York Press, 1995). His current research is focused on medical and environmental ethics.

SUSAN M. DARLINGTON (The Spirit(s) of Conservation in Buddhist Thailand) is an Associate Professor of Anthropology and Asian Studies at Hampshire College, Amherst, Massachusetts, USA. She has been engaged in research in Thailand with monks doing rural development and environmental conservation since 1986. She is currently working on a book on environmentalist monks and the environmental movement in Thailand.

MICHAEL R. DOVE (The Global Mobilization of Environmental Concepts: Re-Thinking the Western/Non-Western Divide) is a Professor of Social Ecology, Professor of Anthropology, and Chair of the Council on Southeast Asian Studies at Yale University. He holds a B.A. from Northwestern University and M.A. and Ph.D. degrees from Stanford University. Professor Dove's research focuses on the interaction between local communities, national governments, and global agencies concerning the use of natural resources. He spent two years in a tribal longhouse in Borneo studying swidden agriculture, six years as a research advisor in Java studying the formation of government resource policy, and four years in Pakistan advising its Forest Service on social forestry policies. Recent research, funded by the MacArthur Foundation, examines the impact of supra-community, institutional factors on biodiversity. Other research and teaching interests include: the theory of sustainable development and resource use; contemporary and historical environmental relations in South and Southeast Asia; the history of market linkages in the tropical forest; the study of developmental and environmental institutions, discourses, and movements; and the sociology of resource-related sciences.

ROY C. DUDGEON (Local Understandings of the Land: Traditional Ecological Knowledge and Indigenous Knowledge) is a doctoral candidate in cultural anthropology at the University of Manitoba, Winnipeg, Canada. His doctoral dissertation, in preparation, examines the many similarities between ecological philosophy and Native American philosophies, as expressed in their own literature. His research interests include traditional ecological knowledge, Native American history, culture, politics and literature, ecological philosophy and anthropological theory. He is currently teaching a course on Ecology, Technology and Society as a sessional instructor at the University of Manitoba, and another concerning Cultural Perspectives on Global Processes at the University of Winnipeg. He is also a founding member of the Green Party of Manitoba.

ROY ELLEN (Variation and Uniformity in the Construction of Biological Knowledge Across Cultures) was educated at the London School of Economics and has taught for the last 30 years at the University of Kent at Canterbury, U.K., where he is presently Professor of Anthropology and Human Ecology. His current research interests are environmental anthropology, ethnobiology and the social and ecological organisation of regional trading systems. He has conducted fieldwork in Indonesia (the Moluccas, Sulawesi and Java) and in Brunei. He is the author of *Nuaulu Settlement and Ecology* (Nijhoff, 1978), *Environment, Subsistence and System* (Cambridge University Press, 1982), and *The Cultural Relations of Classification* (Cambridge University Press, 1993). Amongst the works he has edited, *Redefining Nature* (with Katsuyoshi Fukui,

Berg, 1996) and *Indigenous Environmental Knowledge and its Transformations* (with Peter Parkes and Alan Bicker, Harwood, 2000), are the most recent.

JEANNE KAY GUELKE (Judaism, Israel, and Natural Resources: Models and Practices) is Professor of Geography at the University of Waterloo in Ontario, Canada, and former Dean of its Faculty of Environmental Studies. Her articles on interfaces between religion and environment and in North American historical geography have appeared in a variety of scholarly geography and interdisciplinary journals and edited volumes.

EDVARD HVIDING (Both Sides of the Beach: Knowledges of Nature in Oceania) is professor of social anthropology at the University of Bergen, Norway. He has carried out several years of field research in the New Georgia Group of the Solomon Islands, as well as brief work in other parts of the South Pacific as adviser to regional organizations. He has published widely on the human ecology of coral reef and rainforest, ethnobiology and ethnobotany, resource managament issues, cultural history, seafaring, kinship, and cosmology. Among his books are *Of Reef and Rainforest: A Dictionary of Environment and Resources in Marovo Lagoon* (University of Bergen, 1995), *Guardians of Marovo Lagoon: Practice, Place, and Politics in Maritime Melanesia* (University of Hawaii Press, 1996), and *Islands of Rainforest: Agroforestry, Logging, and Ecotourism in Solomon Islands* (Ashgate, 2000, with T. Bayliss-Smith). He currently works on the comparative history of interisland exchange in the Pacific.

JOHN KESBY (The Perception of Nature and the Environment in Sub-Saharan Africa) is attached to the Anthropology Department at the University of Kent at Canterbury, England, where he continues to teach a little after his retirement in 1998. At present he is particularly concerned with the human experience of animal symbols and with classifying the cultural areas of the world during the modern period. Among his publications that are particularly relevant to this volume are the books *Rangi Natural History: The Taxonomic Procedures of an African People* (Human Relations Area Files, 1986) and *The Cultural Regions of East Africa* (Academic, 1977). Apart from his interest in the peoples and natural history of Africa, he also has a long-running preoccupation with the ecology and conservation of the chalk downs of southern England.

JAMES L. KOHEN (Knowing Country: Indigenous Australians and the Land) is a Senior Lecturer in the Department of Biological Sciences at Macquarie University, Australia, where he teaches courses in Aboriginal Environmental Impacts and Aboriginal Bioresources. He also teaches two courses on Aboriginal Prehistory on secondment to the Department of Indigenous Studies – Warawara, and contributes to the Advanced Diploma of Community Management for Aboriginal and Torres Strait Islander students. His research interests extend across the disciplines of environmental science, archaeology, Aboriginal history, Native Title and Aboriginal ecotourism. He has worked extensively with Aboriginal communities in the Sydney area, particularly the

Darug and Gundungurra people, and with communities in the East Kimberley region of Western Australia. His publications include *Aboriginal Environmental Impacts* (UNSW Press, 1995) and "First and last people: Aboriginal Sydney" in J. Connell, ed. *Sydney: The Emergence of a World City* (Oxford University Press, 2000, pp. 76–95).

S. PARVEZ MANZOOR (Nature and Culture: An Islamic Perspective) is a critic who actively contributes to the Islamic debate in the West. Besides various academic roles, he was the editor of the Muslim journal *Afkar-Inquiry* and is currently editor of the *Muslim World Book Review*. His writings have appeared in various books and journals and have been translated into Arabic, Turkish, Malay and Urdu. He is currently preparing a study entitled *Faith and Existence: The Problem of Norm, History and Utopia in Islamic Thought*.

JAMES MILLER (Daoism and Nature) studied in Durham, Cambridge, Beijing and Boston, before taking up his current position as Assistant Professor of East Asian traditions at Queen's University, Canada. He is editor of www.daoiststudies.org and co-editor of *Daoism and Ecology* (Harvard University Press, 2001).

PORANEE NATADECHA-SPONSEL (Buddhist Views of Nature and the Environment) has an M.A. in philosophy from Ohio University and an Ed.D. in educational foundations from the University of Hawaii. She is an academic officer and adjunct professor at Chaminade University in Hawaii, where she directs the Gender Studies Certificate and also teaches courses in philosophy, religion, and sociology. She collaborates with Leslie E. Sponsel in research on the relationships between religions and environments in southern Thailand, especially Buddhist ecology, sacred places and biodiversity conservation.

GRAHAM PARKES (Winds, Waters, and Earth Energies: *Fengshui* and Awareness of Place) is Professor of Philosophy at the University of Hawaii at Manoa. Among the books he has edited, translated, or authored are: *Heidegger and Asian Thought* (University of Hawaii Press, 1987), *Nietzsche and Asian Thought* (University of Chicago Press, 1991), Nishitani Keiji's *The Self-Overcoming of Nihilism* (State University of New York Press, 1990), *Composing the Soul: Reaches of Nietzsche's Psychology* (University of Chicago Press, 1994), Reinhard May's *Heidegger's Hidden Sources: East-Asian Influences on his Work* (Routledge, 1996), and François Berthier's *Reading Zen in the Rocks: The Japanese Dry Landscape Garden* (University of Chicago Press, 2000). He is currently finishing up a new translation of Nietzsche's *Also Sprach Zarathustra* for Stanford University Press and is also working on two video projects: *Nietzsche's Thinking Places: from the Alps to the Mediterranean*, and *The Role of Stone in the Zen Rock Garden*.

LESLIE E. SPONSEL (Buddhist Views of Nature and the Environment) earned the B.A. in geology from Indiana University and the M.A. and Ph.D. in anthropology from Cornell University. He is a professor of anthropology at the University of Hawaii, where he directs the Ecological Anthropology Program and teaches courses on historical, cultural, and spiritual aspects of human ecology as well as peace studies and human rights. Among the publications he edited are *Indigenous Peoples and the Future of Amazonia: An Ecological Anthropology of an Endangered World* (University of Arizona Press, 1995), and *Endangered Peoples of Southeast and East Asia: Struggles to Survive and Thrive* (Greenwood, 2000). Sponsel is currently editing the forthcoming book *Sanctuaries of Culture and Nature: Sacred Places and Biodiversity Conservation.* Since 1986 he has been visiting southern Thailand to collaborate with biology colleagues at Prince of Songkhla University in Pattani to explore various aspects of the relationship between religion and ecology, especially Buddhist ecology, sacred places, and biodiversity conservation. Poranee Natadecha-Sponsel collaborates in this research.

RICHARD W. STOFFLE (Landscape, Nature, and Culture: A Diachronic Model of Human-Nature Adaptations) is a senior cultural anthropologist at BARA (Bureau of Applied Research in Anthropology, University of Arizona). Since 1976, he has worked with more than 80 American Indian tribes and many federal agencies to identify and present Indian environmental views and concerns in land management decisions. Stoffle has worked also with several fishing communities in the Caribbean and Great Lakes region addressing similar management issues of environmental and resource use. His more recent publications include American Indian histories with the Nevada Test Site and with Nellis Air Force Base and articles on traditional environmental knowledge in *Human Organization, American Indian Quarterly,* and *Current Anthropology.*

REBECCA S. TOUPAL (Landscape, Nature, and Culture: A Diachronic Model of Human-Nature Adaptations) is a research specialist with the Bureau of Applied Research in Anthropology, University of Arizona, USA. Her aca demic background in range management, landscape architecture, natural resource policy, and cultural anthropology supports her research interests in ecological change, human-nature relationships, and cultural landscapes. She has applied this background to natural resource management issues working with landowners, agencies, partnerships, and tribal groups in the western United States. Her publications include a comparison of institutional and grassroots definitions of successful conservation partnerships in *High Plains Applied Anthropologist* (20(1): 53–66), and an article in *Environmental Science and Policy* (4 (2001): 171–184) on the use of ethnography with geographic information systems to identify and protect cultural landscapes.

JOHN A. TUCKER (Japanese Views of Nature and the Environment) completed his Ph.D. at Columbia University in 1990, through the Department of East Asian Languages and Cultures. His dissertation, *Ch'en Pei-hsi's Hsing-li tzu-yi and Early Tokugawa Philosophical Lexicography,* traced the impact of Pei-hsi's *Tzu-yi,* a late-Song dynasty Neo-Confucian text, on a host of seventeenth-century Japanese works. Tucker is also the author of *Ito Jinsai's Gomo jigi and the Philosophical Definition of Early Modern Japan* (E. J. Brill, 1998). His specialty is in East Asian thought, specifically Tokugawa Neo-Confucianism. Tucker is also past editor of *Japan Studies Review*, an annual publication of the Southern Japan Seminar.

MARY EVELYN TUCKER (Worldviews and Ecology) is a professor of religion at Bucknell University in Lewisburg, Pennsylvania, USA, where she teaches courses in world religions, Asian religions, and religion and ecology. She received her Ph.D. from Columbia University in the history of religions, specializing in Confucianism in Japan. She has published *Moral and Spiritual Cultivation in Japanese Neo-Confucianism* (State University of New York Press, 1989) and co-edited *Worldviews and Ecology* (Orbis, 1994), *Buddhism and Ecology* (Harvard, 1997), *Confucianism and Ecology* (Harvard, 1998), and *Hinduism and Ecology* (Harvard 2001). She and her husband, John Grim, directed a series of ten conferences on World Religions and Ecology at the Harvard University Center for the Study of World Religions from 1996–1998. They are the series editors for the ten volumes which are being published from the conferences by the Center and Harvard University Press. They are also editors of a book series on Ecology and Justice from Orbis Press. In addition, they are now coordinating an ongoing Forum on Religion and Ecology (FORE). Mary Evelyn has been a committee member of the Interfaith Partnership for the Environment at the United Nations Environment Programme (UNEP) since 1986 and is Vice President of the American Teilhard Association.

M. NIEVES ZEDEÑO (Landscape, Nature, and Culture: A Diachronic Model of Human-Nature Adaptations) is an archaeologist with a specialization in Southwestern archaeology, ethnohistory, and the contemporary ethnography of cultural resources. Dr. Zedeño has worked for seven years on environmental and cultural assessment projects involving American Indians and Federal agencies and has been the senior researcher on a number of them. Zedeño has published a monograph and a number of articles on American Indian cultural landscapes (*Journal of Archaeological Method and Theory, Culture and Agriculture,* and a chapter in *Explorations in Social Theory*), archaeology (*Anthropological Papers # 58, Journal of Anthropological Research, Kiva*), and the history of anthropology (special issue of *Journal of the Southwest*).

INTRODUCTION

Nature Across Cultures: Views of Nature and the Environment in Non-Western Cultures explores the beliefs and practices of cultures around the world. Ideas about land and nature are central to every culture. There are no universals regarding what it means to live in your environment. Environmental knowledge and accompanying practices in all societies are closely associated with other widely held values about how people understand the world and their place in it. Even though these values change with new knowledge and new technologies, we can still speak with confidence about culturally specific systems. A society's views on nature and the environment arise from and reflect its cultural beliefs and customs. At the same time, for centuries if not millennia there have been exchanges and cross-fertilization among environmental systems around the world.

All people everywhere transform nature. This is the case with huge industrial societies and with small hunting-gathering ones. The scales may be different, but people need to make use of the land and water to survive. Hunter-gatherers burned the forest to make it easier to find game and to encourage certain species. Some Native American cultures killed more buffalo than they could possibly eat or use. Early agricultural societies dammed rivers and irrigated. These practices were not necessarily harmful to the land – much of the latest evidence shows that controlled burning in many societies, such as in Aboriginal Australia, encouraged biodiversity and prevented large uncontrolled burns by keeping the understory low and limiting the fuel available to burn. At the same time, it is clear that practices like these are what we now call ecologically sound, even though the cultures using them did not always articulate the practices in anything like philosophical terms.

Why should modern Westerners be interested in the environmental practices and beliefs of other cultures? Do they offer something that might reverse some of the environmental degradation of the past 100 years, or can we find something in their beliefs that might make the planet more sustainable now? This is probably not always the case, and yet we are drawn to what we see as a gentler way of dealing with the environment. There are a few caveats and problems with this. First, after years of disparaging the scientific and environmental ideas of non-Western cultures, the pendulum is swinging the other way, and there is often now a tendency to over-romanticize and imagine a golden age of environmental harmony. This presents a rather simplistic notion of other cultures as monolithic and untainted. We have tried in this book to emphasize analysis over advocacy, although both do appear.

A potential danger is that by romanticizing other non-industrial cultures, we

xix

H. Selin (ed.), Nature Across Cultures: Views of Nature and the Environment in Non-Western Cultures, xix–xxiii.
© 2003 *Kluwer Academic Publishers. Printed in Great Britain.*

sometimes prevent them from reaping some of the benefits of modernization. Arne Kalland talks about this in his chapter on Environmentalism and Images of the Other. To many people, whaling is acceptable only if it is for subsistence purposes, but he argues that to make it illegal for them to sell parts of their catch not only deprives them of potential benefits of the market but also denies them their rights of self-determination. Also, if we persist in using the romantic images of other cultures, we run the risk of belittling them, making it clear that they are so different from us that they are not welcome to share in our riches.

Another distinction readers must make is between the modern and the historical. When some writers discuss native perspectives of nature, there is a tendency to lump people together from across the recorded historical period. As Annie Booth says in her article on Native Americans, "The problems with this approach should be obvious: none of us lives like our ancestors did 200 years ago." We face different challenges from those of our ancestors, and many of those challenges affect how we relate to the land. We must not conflate what is known about the beliefs and behaviors of historical cultures and today's people. "While they inherit their past, they must live in the present."

Another theme that runs through this book is the interconnection between religion and nature. These are considered very separate spheres in the West, at least in recent times. *Nature Across Cultures* contains chapters on Judaism, Islam, Buddhism, Daoism, and Confucianism, as well as Mary Evelyn Tucker's essay on Worldviews and Ecology, but in fact what Westerners call religion is present in all the cultures discussed in this volume. Non-Western cultures, and not just tribal cultures, do not necessarily see people and nature as separate entities; they know that we are affected by our surroundings and we affect them. No great intellectual leap is needed to realize that where we are affects who we are.

Buddhism gets a lot of coverage in *Nature Across Cultures*. Sponsel and Natadecha-Sponsel discuss Buddhist views of nature and the environment, but Buddhism plays an important part in Tucker's article on Japan, in Kalland's Environmentalism and Images of the Other, in Chattopadhyaya's chapter on India, and in Darlington's on Thailand. In some of these, the authors argue that following Buddhist precepts can effect serious environmental change for the good, while Darlington shows how the Thai government used Buddhism and Buddhist monks to promote both development and conservation at different times. Kalland points out that we have to be careful when we try to induce ecological practices from philosophical traditions. That is really at the root of what we hope to do in this book: we present many different viewpoints and ideas from many societies, but we do not think that they necessarily offer a solution to the world's current environmental practices. Certainly if we followed the precepts of the world's religions, we ought to live in peace. But it is clear we are far from that. Similarly, no one can say that the historical Buddha (Siddhattha Gotama [or Siddhartha Gautama in Sanskrit], who lived on the South Asian subcontinent about 2500 years ago), for example, was an ecologist, but perhaps some of his teachings can be used today to promote a healthy environmental ethic.

Nature Across Cultures is loosely divided into three sections. The first six chapters deal generally with notions of nature and environment, although all use specific examples from many societies. Arne Kalland's introductory essay, "Environmentalism and Images of the Other", sets the stage for the rest of the volume, as he describes the concepts of both the Noble Savage and the Noble Oriental and how those images have shaped our views of and dealings with the rest of the world. Michael Dove *et al.*, in "The Global Mobilization of Environmental Concepts: Re-Thinking the Western/Non-Western Divide", use case studies from Amazonia, Borneo, Nepal, and the Northern United States, among others, to illustrate how systems of environmental knowledge and practice in even the most marginal communities are influenced by ideas from other parts of the globe. Roy Ellen's paper examines the extent to which knowledge of biological entities and processes varies according to different life experiences and cultural traditions. He shows how knowledge is culturally embedded and characterizes the relationship between biological science, folk knowledge, and scholarly knowledge. Roy C. Dudgeon and Fikret Berkes discuss Indigenous Knowledge and Traditional Ecological Knowledge, showing how an explicitly ecological approach to the study of traditional knowledge suggests that there is much more to be learned from traditional peoples than simply techniques for development. Richard Stoffle, Rebecca Toupal, and Nieves Zedeño consider ways in which human–nature relationships change through time and contribute to the development of local knowledge, environmental values, place attachments, and cultural landscapes. Mary Evelyn Tucker explores the connections between worldviews and ecology, using examples from many religions to explain attitudes and ethics towards nature.

The central core of the book explores nature and the environment in eleven different places, from Native America to Aboriginal Australia. Susan M. Darlington, in "The Spirit(s) of Conservation in Buddhist Thailand", shows how the Thai government was able to use Buddhism both to promote develop ment and later to promote conservation, when the deleterious effects of development began to emerge. She also illustrates the mix of spirit belief and Buddhism, a hybrid form of religious belief common to many cultures. D.P. Chattopadhyaya talks about the long Indian tradition of naturalism in India. He draws a distinction between naturalism in terms of materialism, within the framework of naturalism without God, and as a complement to spiritualism. John A. Tucker provides an exploration of the diversity in understandings, positive and negative, of nature and the environment in works of Japanese literature, religion, philosophy, and political and legal thought, from earliest time to the present. Graham Parkes does much the same with Fengshui, the Chinese art of siting graves and houses and creating a peaceful space – a set of sensible recommendations grounded in sensitivity to the natural environment. John Kesby takes on sub-Saharan Africa in his essay, condensing the views of a vast continent. J.L. Kohen discusses Australian Aboriginal people, focusing especially on uses of fire and the negative results of banning controlled burns. In his overview of the vast region of Oceania, Edvard Hviding turns our attention to how the people of coral atolls and volcanic islands view, know, use and manage their tropical environments. Hviding shows how Pacific islanders' worldviews emphasize connections of ecological character, between land and sea, and of social and cultural character.

There are four articles on the American part of the non-Western world. William Balée's entry on "Native Views of the Environment in Amazonia" explores how local peoples from across a cultural and linguistic spectrum recognize, name, classify, and manipulate the biotic and environmental diversity of the Amazon region. David Browman looks at Andean folklore regarding the weather and planting, and shows how modern scientific techniques validate the folk beliefs. He focuses particularly on the Lake Titicaca basin, and the use of stars, winds, and water colors and how they accurately predict future climatic events. Ellen Bielawski, in "'Nature Doesn't Come as Clean as We Can Think It': Dene, Inuit, Scientists, Nature and Environment in the Canadian North", shows how native knowledge differs from western science and how research based on combining indigenous knowledge and science might yield rich results. Annie L. Booth talks about Native North America south of the area covered by Bielawski, separating myth from reality and presenting a clear picture of native relations to the land.

The final section of *Nature Across Cultures* is devoted to the study of different religions and how they view nature. Of course, religious beliefs are not ecological ones, but there are connections between the two that may be used to encourage sensible, sustainable policies. Leslie Sponsel and Poranee Natadecha-Sponsel provide a systematic survey of the relationships between Buddhism and nature through their discussion of the life and teachings of the Buddha, the monastic and lay communities, Buddhism in the West, problems and limitations, and the future. John Berthrong traces the historical development of Confucian theories of nature by showing how various authoritative texts and individuals addressed the question of understanding how human beings live in the context of the cosmos. He treats Confucian theories of nature as part of a living, changing discourse of natural philosophy that stretches from the time of Master Kong (Confucius) to modern Confucian intellectuals today. James Miller examines classic Daoist attitudes towards nature, focusing on the sky, the earth and the body as the three important fields in which Daoism operates. Harold Coward discusses how Hindu principles and practices keep humans and their environment in harmony. S. Parvez Manzoor shows how Islam's vision of nature and culture emanates from its belief in the divine transcendence. The article describes the Islamic perception of nature as a symbolic phenomenon rather than an autonomous and self-subsistent reality. Finally, Jeanne Kay Guelke addresses the themes of water resources, crop production, climate, range management, and heritage resource preservation in both biblical and modern Israel.

Let us go back to the notion that other societies had and have a better way to live with nature. I can find no better way of ending these introductory remarks than to quote Michael Dove *et al.* As they say in their chapter,

The purported divide between Western environmental science and non-Western systems of environmental knowledge, although continually represented as an important boundary marker heavy with symbolic meaning, is problematic. The two systems are historically inter-mingled. And there is also inter-mingling within each system. Neither is monolithic; both encompass multiple, diverse, and sometimes conflicting paradigms. The division between East and West or South and North, when discussing environmental knowledge and practice, is thus an essentialist fallacy. *The linkages are more compelling than the divide.* The boundary between East and West is more meaningful as a metaphor than as a geographic fact.

Helaine Selin
Amherst, Massachusetts, U.S.A.
January 2003

ARNE KALLAND

ENVIRONMENTALISM AND IMAGES OF THE OTHER

Human beings share with many other social animals the ability to discriminate between one's own and other groups, but to legitimize this distinction in terms of moral evaluation is probably uniquely human. The Other has always been important in order not only to define ourselves as human beings – whether the Other is defined biologically (humanity vs. animality), socially (e.g., sex, age, clan), spatially (e.g., community, tribe, nation) or temporally (past, present, future) – but also to put forward claims of a moral order. In a chapter for a book entitled *Nature Across Cultures*, the main focus will be on the spatial, contemporary situation.

The Other is frequently portrayed in negative ways in order both to build a positive self-identity and to legitimize one's own superiority and even conquest and domination of the less "civilized". From this comes the widespread idea of the savage or barbaric outsider or stranger.[1] There has been a tendency to describe an "Ignoble" Other in terms of what they lack – depicting them as filthy and ignorant people hardly living better lives than beasts. They were allegedly steeped in superstition and were under the tyranny of witch doctors and medicine men, and their cruelty included human sacrifice and cannibalism. Such a view has legitimized both physical extermination as well as military conquest and attempts to "civilize" the Other. But the Other may also be portrayed as innocent and uncorrupted by civilization and market forces. This positive image of a "Noble" Other has served as a powerful, internal cultural critique, not least in the industrialized West. It has even been claimed that such images are fundamental to the radical environmentalist critique of industrialism (Milton, 1996: 109).

Inspiration for formulating new concepts and perceptions of human-nature relations comes from a wide variety of sources. Two sources have in particularly been harnessed to construct these images, i.e. the noble savage – first of all represented by Native Americans – and the noble Oriental. Whereas mostly small populations of technologically simple hunter-gatherers and rainforest peoples living in marginal areas have given rise to the image of noble savage (Ellen, 1986), technologically advanced and highly urban cultures in East and South Asia have given rise to the noble Oriental counterpart. Although both

1

H. Selin (ed.), Nature Across Cultures: Views of Nature and the Environment in Non-Western Cultures, 1–17.
© *2003 Kluwer Academic Publishers. Printed in Great Britain.*

these images have served important roles in western cultural critiques and thus share some underlying characteristics, there are also, as we will see, important differences between them, not least in how they relate to practice.

Not surprisingly, this positive interest for the ecologically Noble Other has not been left uncontested. A wide spectre of criticism has been raised against what has been claimed to be an unjustified romantic view of the Other, ranging from uncovering empirical evidence of environmental destruction caused by these allegedly ecologically wise peoples to theoretical discussions about the relationship between perceptions and praxis. This criticism has often been confused with the image of the Ignoble Other but should be recognized for what it is: i.e., a third position that tries to humanize the Other by refuting the relevance of moral evaluation, accounting for both cultural pluralism and moral relativism. It is the aim of this chapter to summarize how images of the Other have been used, mostly in contemporary western environmental writings.

THE NOBLE OTHER

Within the environmental movement there is a widely held notion that one of the roots for current environmental problems is the inadequacies of the Judeao-Christian and Cartesian worldviews (e.g., White, 1967), which are blamed for having alienated modern man from nature. A new ecological paradigm is therefore frequently called for, a paradigm where "man" and "environment" no longer are seen as separate and opposite entities but where "organisms and environment form part of one another" (Dickens, 1992: 15). Some people (e.g., White, 1967; Passmore, 1980; Kalupahana, 1989; Booth and Jacobs, 1990) suggest that the solution to the environmental crisis may be sought in re-interpretations of western concepts and perspectives, whereas others – both scientists and laymen alike – have searched for new inspiration to correct these ills from outside western traditions. A large body of scholarly as well as popular literature offers alternative worldviews to the prevailing western ones – usually depicting man as an integral part of nature instead of being separated from it and trying to dominate it. They portray man and environment as a harmonious unity of mutual respect, complementarity and symbiosis built on profound ecological wisdom; their views are holistic-organic rather than atomistic-mechanistic as in the industrial West (Callicott and Ames, 1989: 5).

The notion of the noble savage, a term mistakenly attributed to Jean-Jacques Rousseau (Fairchild, 1961; Ellingson, 2001), is by no means new. It predates Romanticism by more than a millennium, and goes back at least to the Roman historian and praetor Tacitus (*ca.* 56–120), who described the Germanic tribes to the north of the Roman empire as environmentally friendly and living in harmony with their forest environments (Olwig, 1995). But it was during the late Renaissance that the notion of the noble savage became important as a cultural critique, first in France not least through the writings of Michel de Montaigne in the 16th century (Berkhofer, 1978: 75), although Marc Lescarbot allegedly coined the term itself in 1609 (Ellingson, 2001: 21–22). The savages are here depicted as living closer to nature without being imprisoned in artificial cultures.[2]

With the growing concern, particularly in industrialized countries, about the present rate of degradation of the physical environment and a perceived looming global environmental crisis, the ecological Other has in recent decades got new meanings, not least reinforced by the notion of sustainability (WCED, 1987). Typically, indigenous peoples are depicted as living in harmony with nature, in fact so close that humans and other forms of life constitute a single society, the whole world being one's extended family (e.g., Nash, 1989: 117; Posey, 1999: 5). It is precisely the assumption that non-industrial indigenous societies live sustainably in their environments that makes them a powerful argument that industrialization is the cause of environmental destruction (Milton, 1996: 109). The notion of sustainability, however, resembles models developed within rather outdated functionalist and neo-functionalist schools of anthropology, depicting man and nature co-existing in some kind of homoeostatic equilibrium, thus giving support to the functionalist trap of equating perceptions with practice.

It is common both in the environmental and anthropological discourses to read conservational motives into the practices and worldviews of local peoples. Orlove and Brush (1996: 335) have suggested four reasons why indigenous peoples often are depicted as living in harmony with nature: (a) they have a long history without apparently disrupting ecosystems, (b) they have a rich inventory of traditional environmental knowledge, (c) they have specific management practices based on such knowledge about nature, and (d) they have religious beliefs about ritual uses of animals and plants that safeguard their sustainable use. The interest in what has variously been labelled "indigenous knowledge" (IK) or "traditional ecological knowledge" (TEK) – i.e. (b) above – has been particularly intense. [Editor's note: See the article by Dudgeon and Berkes in this volume.] This interest is more than academic. Indigenous knowledge is often seen by non-government organizations (NGOs), media and the public as an alternative to scientific knowledge, an alternative better able to address urgent problems of resource management. But, as pointed out by Peter Brosius (1997), indigenous knowledge tends to go through a transformation in environmental discourse. Analysing the discourse on the Penan in Sarawak (East Malaysia) he noticed that rhetoric from an Amazonian context had informed the discourse on the Penan. Doing fieldwork in the 1980s, Brosius had been surprised that the Penan paid so little attention to medicinal plants. But the narrator in the film *The Penan: A Disappearing Civilization in Borneo* (1989) nonetheless claimed that "with more than 40,000 years of experimentation and observation, the Penan have enormous medical knowledge which Western science cannot duplicate" (quoted in Brosius, 1997: 61). Brosius also noticed that their rather down-to-earth knowledge of their environment was reduced to a question of the sacred or ineffable. More generally he claims that, "one is imposing a falsely universalized quality on a range of peoples, and thereby collapsing precisely the diversity that defines them. The Penan are transformed into a homogenous 'indigenous people', or 'forest people'" (Brosius, 1997: 65). Some anthropologists (but by no means all) have contributed to this transformation and thereby to the stereotyping of the ecologically noble savage. It is ironic that popular descriptions of indigenous peoples make them remarkably similar and thus invalidate these peoples' main argument for their rights, i.e. cultural diversity.

Besides its ahistoricity and reductionism, the culture-ecological approach – which sees human populations and their culturally patterned behaviour as parts of ecosystems – tends to confuse the effects of something with its cause (Vayda, 1986). The very fact that both indigenous peoples and wildlife have survived is taken as a proof that they are conservationists. Even the most bizarre customs are seen as having an ecological function in balancing human populations and activities with the natural environment (Headland, 1997). A classic example is Roy Rappaport's *Pigs for the Ancestors* from 1968, where the complex ritual cycle and warfare among the Maring Tsembaga – a group of shifting horticulturalists in the highlands of Papua New Guinea – were analysed in such terms. The Maring raise pigs in large numbers, but as the herds grow the pigs take up more of the women's time and energy until their numbers reach such high levels that they become important competitors to the human population itself. At that time pigs are slaughtered, a sacred scrub *rumbin* planted some years previously is uprooted and a pig festival (*kaiko*) is held. During the festival, which may continue for a year, alliances are made for the approaching warfare. Fattened on pigs, the warriors gain strength for the battle against other groups. The planting of a new *rumbin* after peace has been restored symbolizes that the group has managed to retain its territory; failure to do so indicates that the territory has been abandoned. Rappaport argues that rituals and warfare re-establish an ecological balance between humans, pigs and the environs by disseminating the human population and regulating the number of pigs.

Rarely did authors like Roy Rappaport and Marvin Harris – who both see culture in terms of ecological adaptation – ask whether the goals could have been achieved in other, less bizarre, ways. "Primitive" people are seen as functionally adapted to the environment through cybernetic loops and homoeostases (*cf.* Rappaport, 1967, 1968). In this equilibrium model human beings are assumed to behave according to pre-programmed genes or, more commonly, to a culture which is seen as the outcome of some sort of a Social Darwinian selection of behaviourial traits which are ecologically adaptive. Culture becomes the mechanism which balances population size with the resource base. The ecosystem controls human beings and deprives them of their free will as autonomous agents (Vayda, 1986). Implicit in this view is the notion that people live in harmony with nature. With this romantic idealization of the natural, it is not surprising that it is the peoples that apparently have the least impact on the environment and use the land as a subject of labour and not as an instrument of production (Meillassoux, 1972) that most commonly figure as the "noble savages" (Ellen, 1986).

A similar idea exists, however, about advanced agricultural societies in Asia – the ecologically noble Orientals (Bruun and Kalland, 1995). Asia has long figured as the Other to Europeans (Said, 1979). In the 16th century St. Francis Xavier praised the Japanese as "the best race yet discovered" and did not believe any match would be found "among the pagan nations". In one respect they were even superior to the Europeans: "however poor a noble may be, they pay him as much honour as if he were rich" (quoted in Cooper, 1965: 60). During the 18th century there was in Europe a craze for everything Chinese, and to the intellectuals of the Enlightenment, China became a Utopia, a country ruled according to Confucian morality and reason and not given over to metaphysics and religion. She was seen as a meritocratic society where it was possible to work one's way up through the bureaucracy, and hence it became a model for those who fought for a more democratic Europe. These positive images gave way to contempt after industrialization in Europe and the demo-cratic revolutions in the United States and France. Western self-confidence was further enhanced by the Social Darwinism that followed in the wake of Charles Darwin's *On the Origin of Species* (1859).

Nonetheless, it was in the second half of the 19th century that the image of an environmentally noble Japanese emerged. With the opening of the country to westerners in 1854, visitors started to report their impressions. In 1878 the missionary Isabella Bird observed that "[the town Niigata] is so beautifully clean that.. I should feel reluctant to walk upon its well-swept streets in muddy boots" (Bird, 1973: 116). Our image of late 19th century Japan owes more to Lafcadio Hearn than anybody else. Fleeing western materialism he settled in Japan in 1890, married a local girl and became a Japanese citizen. And he sets the tone for much of the subsequent writing on the relationship between the Japanese and nature. Learning the art of *ikebana*, Japanese flower arrangement, he remarks,

> I cannot think now of what we Occidentals call a 'bouquet' as anything but a vulgar murdering of flowers ... Somewhat in the same way, ... having learnt what an old Japanese garden is, I can remember our costliest gardens at home only as ignorant displays of what wealth can accomplish in the creation of incongruities that violate nature ... Until you can feel, and keenly feel, that stones have character, that stones have tones and values, the whole artistic meaning of a Japanese garden cannot be revealed to you ... [T]he soul of the [Japanese] race compre-hends Nature infinitely better than we do (Hearn, 1976: 345–346).

Others have connected this aesthetic appreciation of nature with religion. To H. Byron Earhart (1970: 4), the "appreciation of nature is a mixture of aesthetic and religious appreciation of the countryside". Daisetz Suzuki, the scholar who probably has done more than anybody else to introduce the ideas of Eastern religions to the American popular world (Nash, 1989: 114), writes in his celebrated *Zen and Japanese Culture* that, "the appreciation of the beautiful is at bottom religious, for without being religious one cannot detect and enjoy what is genuinely beautiful" (Suzuki, 1988: 363). It is often assumed that this religious aestheticism is translated – at least before the dawn of westernization – into behaviour. "The Japanese lives too close to nature for him to antagonize her, the benign mother of mankind," writes another authority on the subject (Anesaki, 1973: 6). From this followed the widely held notion that the Japanese both love and live in harmony with nature.[3] Generalising for the whole of the Orient, Suzuki writes that in "the East ... this [Western] idea of subjecting Nature to the commands or service of man according to his selfish desires has never been cherished ... Nature has been our constant friend and companion, who is to be absolutely trusted ..." (1988: 334).

Religion plays the same role in the myth of the ecologically noble Oriental that TEK plays in the myth of the ecologically noble savage.[4] Like the worldviews of North American Indians, but unlike anthropocentric Christianity, Eastern religions have been pictured in eco- or biocentric terms, thus meeting the demand for an ecological paradigm that unites man and environment as parts of one another.[5] "Ancient Eastern ideas closely paralleled the new assumptions of ecology", writes Roderick Nash (1989: 113), thus echoing ideas expressed by Fritjof Capra in his *The Tao of Physics* (1975). Therefore, Eastern religions – and particularly Daoism and Zen – profoundly influenced the Deep Ecology movement (Devall and Sessions, 1985: 100–101).

Since the publication of Lynn White's seminal paper, "The historical roots of our present ecological crisis", in 1967, there has been a proliferation of projects that have been informed by the possible relevance of religions to biological diversity and sustainability. The 1986 meeting in Assisi to mark the 25th anniversary of the World Wildlife Fund (alias Worldwide Fund for Nature, WWF) marked an important step in that direction, with representatives from five "world religions" (Buddhism, Christianity, Hinduism, Islam, Judaism) coming together to discuss the ecological insight and relevance of their creeds (WWF, 1986). Since then a series of conferences on "religion and ecology" has been held at Harvard University and elsewhere (some of which have resulted in publications), the associated *Forum on Religion and Ecology* has been formed, and a large number of books have been published on the subject (e.g., Callicott and Ames, 1989; Badiner, 1990; Chapple, 1993; Hamilton, 1993; Nelson, 1998). It has become fashionable to read ecological insight into religious dogmas, an approach Poul Pedersen (1995) has termed the "religious environmentalist paradigm".[6] [Editor's note: See Mary Evelyn Tucker on Worldviews and Ecology for her views on this topic.]

EMBRACING THE RELIGIOUS ENVIRONMENTALIST PARADIGM

Pedersen asks why the concern for the environment frequently is expressed in religious terms (1995: 258). To him the concept of ecology is intimately con-nected with modernity. It is the interpretation of science and has a global validity – i.e., it is universal and decontextualized. Religion, on the other hand, is contextual and embedded in local communities. This does not mean that people do not have ecological insight, but this is something which must be empirically investigated – not taken for granted – and the most profitable place to look is not necessarily religion. Nor does it mean that religious practice may not have positive ecological effects, but we should be careful not to confuse positive side effects with causes (see below).

Why then do people take the trouble to try to read ecological insight into their religious creeds? Pedersen suggests an answer which is radically different from those suggested by some contributors to this volume:

> By offering to the world what they hold to be their traditional, religious values, local peoples acquire cultural significance. When they speak about nature, they speak about themselves. They

demonstrate to themselves and to the world that their traditions, far from being obsolete and out of touch with modern reality, express a truth of urgent relevance for the future of the Earth. This achievement, with its foundation in appeals to imagined, traditional religious values, represents a forceful cultural creativity which would not have worked by the invocation of 'pure' ecology or environmentalism (1995: 272).

Pedersen takes his examples from Asian religious writers of Hindu, Islamic, and Tibetan Buddhist texts. Let me take an example from the anthropological literature, which shows that both native writers and their sympathetic external advocates embrace the religious environmentalist paradigm. Citing Johan Galtung (1988), who has claimed that Buddhism is closer to nature than any of the other world religions, Leslie E. Sponsel and Poranee Natadecha-Sponsel (1993) are among the many who have suggested that Buddhism may be used to solve the problems of loss of biodiversity. [Editor's note: See their article in this volume for a full explication of their views.] They claim that environmental ethics is an inherent part of Buddhism most clearly expressed, perhaps, through the notion of the Middle Way, the elimination of material desires, *karma*, reincarnation and a holistic perspective. I will not go into detail here but limit myself to illustrating how they juxtapose Buddhist and Western worldviews in the following set of (typically western?) oppositions (1993: 87, Figure 4.1):

Table 1 Buddhist worldviews and Western worldviews (from Sponsel and Sponsel, 1993).

Buddhist worldview	Western worldview
ecocentric	anthropocentric
nonviolence	violence
mental control	technological control
need and being	greed and possession (consumerism)
spiritual development	economic development ("growth mania")
ego extinction	species extinction

What the authors seem to compare is a Buddhist ideal and what is best termed a caricature of western practices. It is based on a selective reading of Buddhist dogma and an equally selective reading of western practice. In their analysis Thailand remained in a "dynamic ecological equilibrium for millennia" (p. 86) until the introduction of western modernization about 200 years ago. This allegedly weakened their "adherence to traditional Buddhism and Thai culture" (p. 87), causing moral collapse and environmental degradation. Typically, western influence is blamed for all ills (cf. Edgerton, 1992; Krech, 1999: 152), and modernization is hardly given any merit whatsoever. They are not alone in this view (e.g., Callicott and Ames, 1989; Hargrove, 1989). A corollary to this is that the Other should be allowed to continue its blissful, Edenic life of innocence and be protected from vicious influences from the modern world. Based on a visit to Japan in 1889, Rudyard Kipling observed,

It would pay us to establish an international suzerainty over Japan: to take away any fear of invasion and annexation, and pay the country as much as ever it chose, on condition that it simply sat still and went on making beautiful things while our learned men learned. It would pay us to put the whole Empire in a glass case and mark it *Hors Concours*, Exhibit A (Kipling, *From Sea to Sea*, quoted in Lehmann, 1978: 25).

Kipling's suggestion may sound ridiculous, but even today there is a strong desire to place exotic cultures in museums and deny them the advantages of modern societies. Although many anthropologists and others have repeatedly stressed that traditional knowledge must not be understood as lack of change but in terms of how knowledge is acquired, accumulated and passed on to following generations (e.g., Posey, 1999: 4), Western environmentalists have used technological progress or integration into the world market as arguments against, e.g., Makah whaling (Ellingson, 2001), Sáami use of snowmobiles (Beach, 1993), and Kayapó forestry (Conklin and Graham, 1995). By using images of the Other in social critiques of modernity, it becomes imperative to stress what the Other is not, namely modern. Modernization apparently plays havoc with the authenticity of these peoples. Only by being "authentic" – that is "uncontaminated" by modern ways – are they noble and worth our consideration. Corrupted by modern ways they become "fallen angels" (Berkes, 1999). The notion of the ecologically noble Other thus serves to lock them in an "ethnographic present" of more idyllic premodern days (cf. Berkhofer, 1978: 29).

Many scholars have stressed the ambivalent position money has in Western culture: on the one hand it is seen as a liberating force, on the other it is seen as a dangerous, corrupting agent. Whereas the former dominates neo-classic economic theory, the latter is almost equally dominant within the environmental movement. In this negative view, which goes back through Marx and Aquinas to Aristotle, money is seen as corrupting social relations as well as morality (Parry and Bloch, 1989). Vandana Shiva (1992) is one of the many who makes an explicit distinction between traditional, subsistence societies existing in harmony with their environment (actually there is no separation between the two; they are one and the same) and modern societies based on the market economy. To her a market economy is incompatible with sustainability. Anthropologists also have lamented that indigenous peoples are losing their innocence by being drawn into the forces of the market economy (e.g., Rappaport, 1979). Money has no role to play in the Garden of Eden. By demanding authenticity and denying indigenous peoples access to the world market, it can be argued that the notion of the Noble Other has become a powerful weapon in the hands of neo-imperialism (e.g., Conklin and Graham, 1995; Kalland, 1999).

HUMANIZING THE OTHER

Many observers have recently questioned the truth of the myths of the ecologically noble Other, whether they are savages or Orientals. One strategy has been to document *practices* that are not necessarily benign to the environment. It has been reported that the G/wi Bushmen in the Kalahari, for example, poison waterholes and kill a number of animals they do not use, despite the belief that all life forms belong to the God Nlodima and nothing should be hunted without reason (Broch, 1977). The Mbuti pygmies in Zaïre regard the forest as their parent but nevertheless burn it down in order to drive out game (Turnbull, 1961). Historical ecology has indicated that a large number of endemic species of birds, including the giant moa in New Zealand, were exterminated by indigenous peoples in Polynesia before the arrival of the Europeans to the Pacific. More than half the endemic birds in Hawaii may have been exterminated (Kirch, 1982; Olson and James, 1984; Krech, 1999). Japan experienced serious deforestation about 1200 years ago and again about 400 years ago, long before her culture was infiltrated by western ideas (Totman, 1989) and despite Buddhism's allegedly protective attitude toward trees. The Japanese managed to correct the situation in time, whereas the prehistoric Maya and Indus civilizations seem to have been unable to halt depletion of their forests. Cree and other native peoples in North America have been reported to kill indiscrimately (Brightman, 1987), although their environmental values are said to be based on humanistic notions and morality towards nature where animals have intrinsic value (Berkes, 1988). Reviewing the environmental history of the North American Indians, Shepard Krech III (1999) argues that several species of prey at times were severely depleted due to wasteful killing, where meat was left to rot in large quantities. There are even reports that some Native Americans torture and kill small animals for fun (Broch, 1977), and the Nuaulu in Indonesia and the Kwaio in the Solomon Islands deliberately cut all kinds of vegetation as they pass through the forests for the same reason (Ellen, 1986: 11; Keesing, 1976: 116). Headland was shocked by the lack of a conservation ethos among the Agta in the Philippines: they cut down trees to harvest the fruit and "pollute the air they breathe far more than most industrial nations" (1997: 607).[7]

Reviewing literature on conservation and sustainability in small-scale societies, Eric A. Smith and Mark Wishnie recently concluded that, "on balance, the evidence on faunal impacts on small-scale societies indicates that conservation is absent and depletion is sometimes a consequence" (2000: 509). What little evidence there is for conservation involves plant resources. They found that a crucial deciding factor is whether the society in question controls access to territories (ibid.). To appreciate this conclusion a distinction must be made between sustainability and conservation. Sustainability may be taken to mean an extraction of natural resources at a level that does not lead to overharvesting of species or to habitat degradation, a situation that for various reasons can prevail regardless of people's knowledge or intentions. Conservation, on the other hand, implies a degree of design. To Smith and Wishnie conservation must not only prevent or mitigate resource depletion, species extermination or habitat degradation, but must also be designed to do so (2000: 501). Otherwise sustainability might be a mere side effect of external factors. When these peoples do not do more harm to their environments, it might be because of low population density, relative isolation with no external pressure, few market outlets, limited technologies or unattractive resources or habitats. Hunn (1982) has called this situation "epiphenomenal conservation" in contrast to "genuine conservation" that is specially designed to lower the level of exploitation that otherwise would have been unsustainable (cf. Ellingson, 2001: 351). In other words, a society can be sustainable without conserving or conserve without achieving sustainability. The great majority of small-scale societies may live sustainably without being conservationists.[8]

Why, then, do so many anthropologists continue to support images of the noble others?[9] Robert B. Edgerton suggests that many anthropologists have chosen not to write about misadaptation, suffering and discontent in order not to offend the people being described or discredit them in the eyes of others. Therefore, "some anthropological monographs ... are idealized, even romanticized portraits" (Edgerton, 1992: 5). But revisionism does not necessarily imply that the Other turns ignoble, as Ellingson (2001) seems to argue.[10] One should not confuse the denial of the Noble Other with the notion of the Ignoble Other. There is, as Conklin and Graham (1995) forcefully point out, a middle ground which leads to a humanization of the Other. S/he is neither a saint nor a demon, but simply a human being with all her or his faults and virtues.

In a debate in *Current Anthropology*, Alice R. Ingerson believes that anthropologists do not cultivate the Noble Savage myth merely as a courteous gesture to protect one's informants' feelings but for political reasons (Ingerson, 1997: 615):

> [M]any anthropologists have used ahistorical romanticism as a way of encouraging environmentalists to see forest peoples as 'adapted to' or 'defenders of' supposedly stable ecosystems. This tactic clearly worked in the short term ... It almost doesn't matter whether this tactic was cynical – a way of manipulating the environmentalists – or reflected profound romanticism among the anthropologists themselves.

In the same volume, Leslie E. Sponsel expresses concern for the political consequences of ecological revisionism that questions notions of the ecologically noble savage, warning that this may provide material that can be used to undermine the interests of indigenous people. He is afraid that when scientific evidence suggests that, for example, native Hawaiians were destructive of their environment, this might be one factor jeopardizing their rightful claims to land and water (Sponsel, 1997: 621–622).

Obviously, such concerns should not be dismissed lightly. The notion of the noble ecological Other implies that culture has an important role to play in securing sustainable use of natural resources. With such a perspective it is no wonder that revisionism may be seen as a threat. This despair is well expressed by one environmentalist at a meeting of the American Anthropological Association:

> [I]f we can't count on culture to make indigenous peoples conserve the rain forests, then turning the forests over to them would be like turning the forests over to a random assortment of people ... (quoted in Ingerson, 1997: 615).

However, today it is a widely held notion that natural resources are best regulated if local communities, which depend on these resources for their nutritional, economic, social, and cultural needs, are brought into active participation, a principle incorporated into the IUCN/UNEP/WWF report *Caring for the Earth: A Strategy for Sustainable Living* as well as in *Agenda 21*. There are therefore good reasons to believe that it makes a big difference whether the management of rain forests is handed over to indigenous people living in the forest or to a random set of people seen at an airport in the United States.

More important than designing strategies for sustainable management of natural resources is the question of human rights. Human rights are unconditional and do not rest on whether people's environmental ethics are pleasing to external observers or not. Indigenous peoples' rights to resources are not conditioned on culture or on how "authentic" they might be. They have rights because they are people, not because they are saints (e.g. Ward, 1993). Over the years indigenous peoples around the world have slowly gained recognition, not least because they have skilfully appropriated two important global discourses – those on human rights and environmentalism. At times these two discourses have pulled in the same direction, as when protection of rainforests is a means to secure the rights of those living therein, making alliances between indigenous peoples and environmentalists possible. At other times the discourses are at odds, as when it comes to the question of whaling. Here the clash between indigenous peoples and environmentalists has at times been violent; the most recent example is the resumption of whaling by the Makah Indians in the State of Washington in 1999.

When the International Whaling Commission (IWC) exempted aboriginal subsistence whaling (ASW), defined as "whaling for purposes of local aboriginal consumption carried out by or on behalf of aboriginal, indigenous or native peoples who share strong community, familial, social and cultural ties related to a continuing traditional dependence on whaling and the use of whales" (IWC, 1981) from the moratorium on whaling, preferential treatment was given to indigenous peoples when it came to the exploitation of renewable natural resources. Considering the outrageous injustices inflicted upon aboriginals in the past, few voiced any objections to what seemed to be an attempt to put things right. But concessions are seldom given out of altruism, and the rights to catch whales were only obtained at considerable cost. Every year the Iñupiat of Alaska must testify that they have not traded in whale products, unless they are prepared to lose their authenticity as an indigenous people. This privilege, then, has become a double-edged sword, because it implies a static view of a people and its culture and can be used to deny indigenous peoples their obvious right to develop on their own terms. To many IWC members whaling should apparently only be allowed as long as it is conducted by small ethnic minorities who are perceived as lacking unifying political institutions, use "simple" technologies, and whose economic exchanges are believed to exist within the confinement of a non-commercial economy. It can therefore be argued that they have been deprived of their rights to self-determination and that the concept of ASW has become a powerful weapon in the service of imperialism (Kalland, 1999). The Makah are only the last victims of this policy. Indigenous sealers have experienced the same kind of extortions (Wenzel, 1991). ILO 169 and other international conventions recognize the rights of indigenous peoples to their natural resources and way of life (Ward, 1993), and it should therefore be unnecessary to legitimize such rights by giving voice to the myths of their nobility.

* * *

Krech (1999: 26) reminds us that, "narratives about Native North Americans are contingent on the times in which they were created." Our images change over time, not necessarily because the Other changes – although that may be one factor – but because we change. The Other is first of all understood as an antithesis to oneself, and our images may tell more about ourselves than about the Other. The image of the Other is therefore highly complex and has been used in many different contexts (Said, 1979) and for many different purposes.

The notion of the ecologically Noble Other has several important implications, each of which casts serious doubts on its soundness. First, as stated by Brosius (1997), the image of the ecologically noble savage creates a picture of a homogenous "indigenous people" by conflating the very cultural diversity which is so vital to the global discourse on the rights of indigenous peoples. Second, as Guha (1989) argues, the myth of the ecologically noble Oriental reproduces, as its noble savage counterpart, a false dichotomy between the rational and science-oriented Occidentals and the spiritual and emotional Orientals. Third, as stressed by Conklin and Graham (1995), the image demands that people meet claims for authenticity or run the risk of being treated as fallen angels. Rights bestowed on indigenous peoples because they have, as seen by some outsiders, a benign relationship with nature, can easily have both positive and negative consequences. Therefore many indigenous people themselves reject an "ecologically noble savage" approach that justifies rights in terms of a romantic view of their relationship with nature (Posey, 1999: 7). As stated by Aqqaluc Lynge, president of the Inuit Circumpolar Conference: "We are living in the modern world and have the same economic needs as everyone else. We are not here to live out the fantasies of white people about Eskimoes" ("Eskimoes surrender to lure of oil", *The Independent*, June 20, 1998).

It would be foolish to deny that TEK (or any worldview) may play a role in securing sustainable use of natural resources, and I have elsewhere argued that such knowledge is essential (Kalland, 2000). But two points should be made. First, experience from North America and Asia, and elsewhere for that matter, clearly tells us that neither profound knowledge about the environment nor sound environmental ethics is sufficient to prevent degradation of natural resources. We therefore need to tread cautiously when inducing ecological practices from philosophical traditions. Discrepancies between theory and practice should not surprise us (see Holy and Stuchlik, 1983). We cannot *a priori* assume that people's perceptions and norms are mirrored in their actual behaviour. If such a connection is present this is not necessarily a result of ecological understanding and conscious conservation but might be a coincidental side effect of something else. Moreover, there exist conflicting values in all cosmologies and knowledge systems, and a religious creed which is believed to encourage people to conserve natural resources by giving moral support to certain norms may also provide people with the means to circumvent the same norms. It is tempting to suggest that any religion is likely to support values that inhibit over-exploitation of natural resources as well as values that facilitate or legitimate such behaviour. Explaining behaviour from ideologies may rest on selective reading of evidence. Rather than norms determining behaviour we find that goal-oriented behaviour is legitimized by appealing to certain norms.

Second, although alternative knowledge systems can provide us with important new insight which can stimulate us not only to reflect on our own relationship with nature but also to actively construct new understandings through a process of syncretism, a word of warning is in place. Incorporating elements from one culture into another locates these elements in a totally new context. This might give very different and unexpected outcomes. One particularly relevant example is how ideas from Native North American and Oriental philosophies have been appropriated and merged with a Western tradition of absolutism. This has produced environmentalist and animal rights discourses which are quite alien to the donor cosmologies. Not only has the animal rights discourse – and to a lesser extent the environmental discourse – transformed respect for game animals into "intrinsic value", but their missionary zeal also stands in sharp contrast to the contextual approaches of many local peoples. Ironically then, the environmental and animal rights discourses may pose a threat to the lifestyles of local people who depend on the utilization of natural resources.

NOTES

[1] The stranger can also be a welcome provider of good luck and riches and at times even be seen as a God, as in Japan (Yoshida, 1981).

[2] Berkhofer (1978: 77–79) makes a distinction between an early rational and sensical "enlightened savage" and a more emotional and sensitive "romantic savage" which became more pronounced by the end of the 18th century.

[3] Elsewhere I have argued that this is a naïve and romantic interpretation of Japanese attitudes to nature (Kalland, 1995, 2000; Kalland and Asquith, 1997).

[4] I here take the term "myth" to mean popular ideas and beliefs without implying that myths are false or illusory.

[5] For a discussion of these terms, see e g , Shaner, 1989. Ramachandra Guha (1989) objects to the attempts to turn Oriental religions, such as Buddhism, Hinduism and Daoism, into ecocentric religions. He looks upon this, what he calls appropriation of Oriental religions, as yet another expression of westerners' need to universalize their messages and to uphold a false dichotomy between the rational and science-oriented Occidentals and the spiritual and emotional Orientals.

[6] In a similar vein, Guha (1998) has objected to the tendency to read ecological wisdom into the writings of Mahatma Gandhi, arguing that Gandhi was far from the ecological saint he is claimed to have been.

[7] See also Redford, 1991; Edgerton, 1992; Redford and Stearman, 1993; Stearman, 1994; Milton, 1996.

[8] Milton puts it this way: "Non-industrial peoples do not think like environmentalists. ... [T]he practices in which some non-industrial peoples engage may be environmentally benign, but their cultures, their ways of understanding the world, are not" (Milton, 1996: 113–114).

[9] Ellingson (2001: 353) argues that the popularity of the noble savage myth is in itself a myth, constructed by the revisionists. He quotes John H. Bodley (1997: 611) who claims that, "the problem with ecological revisionism is its tendency to exaggerate and misattribute the myths it seeks to demolish."

[10] The image of the ignoble Japanese that emerged in the West during the 1970s and 80s must be seen in light of Japanese economic successes, although fuelled by several environmental disasters. By linking economic success to pollution at home and depletion of forests and oceans abroad, it became possible for Westerners to regain the moral upper hand. This image, which became an

important ingredient of Japan-bashing, was promoted by Western journalists, environmentalists and politicians but seldom by scholars.

BIBLIOGRAPHY

Anesaki, Masaharu. *Art, Life, and Nature in Japan*. Tokyo: Charles E. Tuttle, 1973.

Badiner, Allan Hunt, ed. *Dharma Gaia. A Harvest of Essays in Buddhism and Ecology*. Berkeley: Parallax Press, 1990.

Beach, Hugh. 'Straining at gnats and swallowing reindeer: The politics of ethnicity and environmentalism in Northern Sweden.' In *Green Arguments and Local Subsistence*, G. Dahl, ed. Stockholm: Stockholm Studies in Social Anthropology, 1993, pp. 93–116.

Berkes, Fikret. 'Environmental philosophy of the Chisasibi Cree people of James Bay.' In *Traditional Knowledge and Renewable Resource Management in Northern Regions*, M.M.R. Freeman and L.N. Carbyn, eds. Edmonton: Boreal Institute for Northern Studies, 1988, pp. 7–21.

Berkes, Fikret. *Spiritual Ecology. Traditional Ecological Knowledge and Resource Management*. Philadelphia: Taylor and Francis, 1999.

Berkhofer, Robert F., Jr. *The White Man's Indian. Images of the American Indian from Columbus to the Present*. New York: Alfred A. Knopf, 1978.

Bird, Isabella L. *Unbeaten Tracks in Japan*. Rutland, Vermont and Tokyo: Charles E. Tuttle Company, 1973 (1880).

Bodley, John H. 'Revisionism in ecological anthropology.' *Current Anthropology* 38(4): 611–613, 1997.

Booth, Annie L. and Harvey M. Jacobs. 'Ties that bind: Native American beliefs as a foundation for environmental consciousness.' *Environmental Ethics* 12(1): 27–44, 1990.

Brightman, Robert A. 'Conservation and resource depletion. The case of the boreal forest Algonquians.' In *The Question of the Commons. The Culture and Ecology of Communal Resources*, B.M. McCay and J.M. Acheson, eds. Tucson: University of Arizona Press, 1987, pp. 121–141.

Broch, Harald Beyer. 'Den økologiske 'harmonimodell' sett i lyset av jegere og sankere, eller de såkalte naturfolk.' *Naturen* 3: 243–247, 1977.

Brosius, J. Peter. 'Endangered forest, endangered people: Environmentalist representations of indigenous knowledge.' *Human Ecology* 25(1): 47–70, 1997.

Bruun, Ole and Arne Kalland. 'Images of nature. An introduction to the study of man-environment relations in Asia.' In *Asian Perception of Nature: A Critical Approach*, O. Bruun and A. Kalland, eds. London: Curzon Press, 1995, pp. 1–24.

Callicott, J. Baird and Roger T. Ames. 'Introduction: The Asian traditions as a conceptual resource for environmental philosophy.' In *Nature in Asian Traditions of Thought: Essays in Environmental Philosophy*, J.B. Callicott and R.T. Ames, eds. Albany: State University of New York Press, 1989a, pp. 1–21.

Callicott, J. Baird and Roger T. Ames, eds. *Nature in Asian Traditions of Thought: Essays in Environmental Philosophy*. Albany: State University of New York Press, 1989b.

Capra, Fritjof. *The Tao of Physics*. Suffolk: Fontana, 1976.

Chapple, Christopher Key. *Nonviolence to Animals, Earth and Self in Asian Traditions*. Albany: State University of New York Press, 1993.

Conklin, Beth A. and Laura R. Graham. 'The shifting middle ground: Amazonian Indians and eco-politics.' *American Anthropologist* 97(4): 695–710, 1995.

Cooper, Michael, ed. *They Came to Japan. An Anthology of European Reports on Japan, 1543–1640*. Berkeley, Los Angeles and London: University of California Press, 1965.

Devall, Bill and George Sessions. *Deep Ecology. Living as if Nature Mattered*. Salt Lake City, Utah: Gibbs Smith, 1985.

Dickens, Peter. *Society and Nature. Towards a Green Social Theory*. London: Harvester Wheatsheaf, 1992.

Earhart, H. Byron. *A Religious Study of the Mount Haguro Sect of Shugendō. An Example of a Japanese Mountain Religion*. Tokyo: Sophia University, 1970.

Edgerton, Robert B. *Sick Societies. Challenging the Myth of Primitive Harmony*. New York: Free Press, 1992.

Ellen, Roy F. 'What Black Elk left unsaid: On illusory images of Green primitivism.' *Anthropology Today* 2(4): 8–12, 1986.

Ellingson, Ter. *The Myth of the Noble Savage*. Berkeley: University of California Press, 2001.

Fairchild, H.N. *The Noble Savage: A Study of Romantic Naturalism*. New York: Russel and Russel, 1961 (1928).

Galtung, Johan. *Buddhism: A Quest for Unity and Peace*. Honolulu: Dae Won Sa Buddhist Temple of Hawai'i, 1988.

Guha, Ramachandra. 'Radical American environmentalism and wilderness preservation: A Third World critique.' *Environmental Ethics* 11: 71–83, 1989.

Guha, Ramachandra. 'Mahatma Gandhi and the environmental movement in India.' In *Environmental Movements in Asia*, A. Kalland and G.A. Persoon, eds. Richmond, UK: Curzon Press, 1998, pp. 65–82.

Hamilton, Lawrence S., ed. *Ethics, Religion and Biodiversity: Relations Between Conservation and Cultural Values*. Cambridge: The White Horse Press, 1993.

Hargrove, Eugene C. 'Foreword.' In *Nature in Asian Traditions of Thought: Essays in Environmental Philosophy*, J.B. Callicott and R.T. Ames, eds. Albany: State University of New York Press, 1989, pp. xiii–xxi.

Headland, Thomas N. 'Revisionism in ecological anthropology.' *Current Anthropology* 38(4): 605–630, 1997.

Hearn, Lafcadio. *Glimpses of Unfamiliar Japan*. Rutland, Vermont and Tokyo: Charles E. Tuttle Company, 1976 (1894).

Holy, L. and M. Stuchlik. *Actions, Norms and Representations*. Cambridge: Cambridge University Press, 1983.

Hunn, E.S. 'Mobility as a factor limiting resource use in the Columbian Plateau of North America.' In *Resource Managers: North American and Australian Hunter-Gatherers*, N.M. Williams and E.S. Hunn, eds. Boulder, Colorado: Westview, 1982, pp. 17–43.

Ingerson, Alice E. 'Revisionism in ecological anthropology.' *Current Anthropology* 38(4): 615–616, 1997.

IUCN/UNEP/WWF. *Caring for the Earth: A Strategy for Sustainable Living*. Gland, Switzerland: IUCN/UNEP/WWF, 1991.

IWC (International Whaling Commission). *Report of the Ad Hoc Technical Committee Working Group on Development of Management Principles and Guidelines for Subsistence Catches of Whales by Indigenous (Aboriginal) Peoples*. Cambridge: IWC, 1981.

Kalland, Arne. 'Management by totemization: Whale symbolism and the anti-whaling campaign.' *Arctic* 46(2): 124–133, 1993.

Kalland, Arne. 'Culture in Japanese nature.' In *Asian Perception of Nature: A Critical Approach*, O. Bruun and A. Kalland, eds. London: Curzon Press, 1995, pp. 243–257.

Kalland, Arne. 'Aboriginal subsistence whaling: A concept in the service of imperialism.' *Dark Night Field Notes* 14: 23–25, 1999 (1992).

Kalland, Arne. 'Indigenous knowledge: Prospects and limitations.' In *Indigenous Environmental Knowledge and its Transformations, Critical Anthropological Approaches*, R. Ellen, P. Parkes and A. Bicker, eds. London: Harwood Press, 2000, pp. 319–335.

Kalland, Arne. 'Holism and sustainability: lessons from Japan.' *Worldviews: Environment, Culture, Religion* 6(2): 145–158, 2002.

Kalland, Arne and Pamela Asquith. 'Japanese perceptions of nature: Ideals and illusions.' In *Japanese Images of Nature. Cultural Perspectives*, P. Asquith and A. Kalland, eds. Richmond, UK: Curzon Press, 1997, pp. 1–35.

Kalupahana, David J. 'Toward a middle path of survival.' In *Nature in Asian Traditions of Thought: Essays in Environmental Philosophy*, J.B. Callicott and R.T. Ames, eds. Albany: State University of New York Press, 1989, pp. 247–256.

Keesing, Roger M. *Cultural Anthropology: A Contemporary Perspective*. New York: Holt, Rinehart and Winston, 1976.

Kirch, Patrick V. 'The impact of the prehistoric Polynesians on the Hawaiian ecosystem.' *Pacific Science* 36: 1–14, 1982.

Krech, Shepard III. *The Ecological Indian. Myth and History*. New York and London: W.W. Norton and Company, 1999.

Lehmann, Jean-Pierre. *The Image of Japan. From Feudal Isolation to World Power 1850–1905*. London: George Allen and Unwin, 1978.

Meillassoux, Claude. 'From reproduction to production. A Marxist approach to economic anthropology.' *Economy and Society* 1: 93–105, 1972.

Milton, Kay. *Environmentalism and Cultural Theory. Exploring the Role of Anthropology in Environmental Discourse*. London: Routledge, 1996.

Nash, Roderick Frazier. *The Rights of Nature: A History of Environmental Ethics*. Madison: The University of Wisconsin Press, 1989.

Nelson, Lance E., ed. *Purifying the Earthly Body of God: Religion and Ecology in Hindu India*. Albany: State University of New York Press, 1998.

Olson, Storres L. and Helen F. James. 'The role of Polynesians in the extinction of the avifauna of the Hawaiian Islands.' In *Quaternary Extinction: A Prehistoric Revolution*, P.S. Martin and R.G. Klein, eds. Tucson: University of Arizona Press, 1984, pp. 768–780.

Olwig, Kenneth R. 'A British Italy in the North: Landscape, *landskap*, and the body.' In *Nordic Landscapes: Cultural Studies of Place*, A. Linde-Laursen and J.O. Nilsson, eds. Copenhagen: Nordic Council of Ministers, 1995, pp. 154–169.

Orlove, Benjamin S. and Stephen B. Brush. 'Anthropology and the conservation of biodiversity.' *Annual Review of Anthropology* 25: 329–352, 1996.

Parry, Johnny and Maurice Bloch. 'Introduction: Money and the morality of exchange.' In *Money and the Morality of Exchange*, J. Parry and M. Bloch, eds. Cambridge: Cambridge University Press, 1989, pp. 1–32.

Passmore, John. *Man's Responsibility for Nature*, 2nd ed. London: Duckworth, 1980 (1974).

Pedersen, Poul. 'Nature, religion and cultural identity: The religious environmentalist paradigm.' In *Asian Perception of Nature: A Critical Approach*, O. Bruun and A. Kalland, eds. London: Curzon Press, 1995, pp. 258–276.

Posey, Darrell Addison. 'Introduction: Culture and nature – the inextractable link.' In *Cultural and Spiritual Values of Biodiversity*, D.A. Posey, ed. London: Intermediate Technology Publication/UNEP, 1999, pp. 3–18.

Rappaport, Roy A. 'Ritual regulation of environmental relations among a New Guinea people.' *Ethnology* 6: 17–30, 1967.

Rappaport, Roy A. *Pigs for the Ancestors: Rituals in the Ecology of a New Guinea People*. New Haven, Connecticut: Yale University Press, 1968.

Rappaport, Roy A. *Ecology, Meaning and Religion*. Richmond, California: North Atlantic, 1979.

Redford, Kent H. 'The ecologically noble savage.' *Cultural Survival Quaterly* 15(1): 46–48, 1991.

Redford, Kent H. and Allyn MacLean Stearman. 'Forest-dwelling native Amazonians and the conservation of biodiversity: Interests in common or in collision?' *Conservation Biology* 7(2): 248–255, 1993.

Said, E. *Orientalism*. New York: Pantheon Books, 1979.

Shaner, David Edward. 'The Japanese experience of nature.' In *Nature in Asian Traditions of Thought: Essays in Environmental Philosophy*, J.B. Callicott and R.T. Ames, eds. Albany: State University of New York Press, 1989, pp. 163–182.

Shiva, Vandana. 'Recovering the real meaning of sustainability.' In *The Environment in Question. Ethics and Global Issues*, D.E. Cooper and J.A. Palmer, eds. London: Routledge, 1992, pp. 187–193.

Smith, Eric Alden and Mark Wishnie. 'Conservation and subsistence in small-scale societies.' *Annual Review of Anthropology* 29: 493–524, 2000.

Sponsel, Leslie E. 'Revisionism in ecological anthropology.' *Current Anthropology* 38(4): 619–622, 1997.

Sponsel, Leslie E. and Poranee Natadecha-Sponsel. 'The potential contribution of Buddhism in developing an environmental ethic for the conservation of biodiversity.' In *Ethics, Religion and*

Biodiversity. Relations Between Conservation and Cultural Values, L.S. Hamilton, ed. Cambridge: The White Horse Press, 1993, pp. 75–97.

Stearman, Allyn MacLean. 'Revisting the myth of the ecologically noble savage in Amazonia: Implications for indigenous land rights.' *Culture and Agriculture, Bulletin of the Culture and Agriculture Society* 49 (Spring): 2–6, 1994.

Suzuki, Daisetz. *Zen and Japanese Culture*. Tokyo: Charles E. Tuttle, 1988.

Totman, Conrad. *The Green Archipelago. Forestry in Preindustrial Japan*. Honolulu: University of Hawaii Press, 1989.

Turnbull, Colin. *The Forest People: A Study of Pygmies of the Congo*. New York: Simon and Schuster, 1961.

Vayda, Andrew. 'Holism and individualism in ecological anthropology.' *Reviews in Anthropology*: 295–313, Fall 1986.

Ward, Elaine. *Indigenous Peoples Between Human Rights and Environmental Protection – Based on an Empirical Study of Greenland*. Copenhagen: The Danish Centre for Human Rights, 1993.

WCED (World Commission on Environment and Development). *Our Common Future*. Oxford: Oxford University Press, 1987.

Wenzel, George. *Animal Rights, Human Rights: Ecology, Economy and Ideology in the Canadian Arctic*. Toronto: University of Toronto Press, 1991.

White, Lynn, Jr. 'The historical roots of our ecological crisis.' *Science* 155(3767): 1203–1207, 1967.

WWF (World Wildlife Fund). *The Assisi Declarations. Messages on Man and Nature from Buddhism, Christianity, Hinduism, Islam and Judaism*. Gland, Switzerland: WWF (World Wildlife Fund), 1986.

Yoshida, Teigo. '"The stranger as god": The place of the outsider in Japanese folk religion.' *Ethnology* 2(2): 87–99, 1981.

MICHAEL R. DOVE, MARINA T. CAMPOS, ANDREW SALVADOR MATHEWS,
LAURA J. MEITZNER YODER, ANNE RADEMACHER, SUK BAE RHEE
AND DANIEL SOMERS SMITH

THE GLOBAL MOBILIZATION OF ENVIRONMENTAL CONCEPTS: RE-THINKING THE WESTERN/NON-WESTERN DIVIDE

It is increasingly evident that the process of globalization is a more complex and conflicted one than has been thought to be the case. Former iconographic images of "one world" have come to be suspect (Ingold, 1993; Sachs, 1992), and predictions of the coming "global village" have receded in the face of increasingly prominent divisions between developed and under-developed countries, North and South, Western and non-Western (Huntington, 1996).[1] The first challenge of global governance, as the debate over global warming has demonstrated, is not to coordinate solutions to global environmental problems, but to agree on a definition of the problem in the first place (Dove, 1994). An apparent irony of the globalization process is that at the same time as it erases some barriers and boundaries it constructs and crosses others. The simultaneous construction and destruction of boundaries is evident in the new and unorthodox alliances and oppositions that global mechanisms like the World Trade Organization have fomented.

One prominent fault line running through these new global alliances and oppositions is that between the developed industrialized nations and the less-developed industrializing ones. Western and non-Western stances in these debates are often easily differentiated. This differentiation maps onto a more general distinction that has developed over the past generation between Western and non-Western systems of resource use and environmental knowledge. Through the 1960s and 1970s, Western scientists privileged their own views of the environment, perceived few alternatives, and assumed that their views would eventually hold sway over the world as a "global science". Then, in part because of the increasingly obvious unsustainability of certain of the resource-use systems underpinned by Western science, non-Western systems of environmental knowledge and practice began to receive some recognition. Previously pervasive deprecation of non-Western systems of resource use has been replaced in many quarters by valorization of these same systems. This reappraisal is a

19

H. Selin (ed.), Nature Across Cultures: Views of Nature and the Environment in Non-Western Cultures, 19–46.
© 2003 Kluwer Academic Publishers. Printed in Great Britain.

useful correction to the previously uniform approbation and disapprobation of Western and non-Western systems, respectively, in international circles. However, the perceived underlying division between the two is actually quite problematic (Agrawal, 1995; Dove, 2000).[2]

There is a critical literature on the social construction of global environmental problems (e.g., Lohmann, 1993; Taylor and Buttel, 1992). Recent contributions to this literature focus on the sociology of knowledge involved, in particular the way that environmental knowledge is transported and transformed, and in particular between Western and non-Western societies (e.g., Brosius, 1997; Dove, 1998; Gupta, 1998; Rangan, 1992; Tsing, 2000; Zerner, 1996). Most relevant to critiquing the validity of the division between Western and non-Western systems is recent research on the global circulation of environmental concepts. "Circulation" is actually an inadequate and misleading term to describe this process, because of what it implies for the agency of the concepts themselves as opposed to the people who hold and mold them. Environmental concepts do not travel independently from one place to another and impose themselves on agency-less people. Rather, the concepts of one part of the global community get appropriated, transformed, and contested by specific local actors when they move to another part, for which reason the term "deployment" or perhaps "mobilization" of ideas might be preferred to circulation (Tsing, 1999b, 2000). Transformation of concepts is made both possible by this movement and also necessary: concepts become powerful in a new setting only if they can be integrated into it, at the same time as a part of their power derives from continued identification with their prior setting. It is the non-fixity of the transported concept that allows it to draw on (as well as dispute) sources of authority in both its place of origin and its new setting.[3] The non-fixity of the transported concept is also a key to forging global coalitions: Tsing (2000) maintains that the key to successful global coalitions is the *mis*-translation of ideas. The movement of ideas is powerful, in part, because of this very reinterpretation and hybridization.

The process of the global mobilization of environmental knowledge is the subject of this chapter. The chapter is built around a number of different case studies, focusing on three themes. The first involves the valorization and villainization of resources users; the second focuses on the complexity of Western environmental discourses and how this affects their deployment in both Western and non-Western countries. The third theme concentrates on the historic processes by which Western and non-Western environmental concepts become hybridized.

HEROES AND VILLAINS IN THE TROPICAL FORESTS

In recent years, a problematic linkage has emerged between western environmentalists and politically marginal forest dependent communities in the tropics. Since the 1980s, some Western environmentalists concerned with conserving or sustainably managing tropical ecosystems have championed their cause in part through romanticized representations of forest-dwellers as "ecologically noble savages" or primitive environmentalists (Ellen, 1986; Redford, 1990; Brosius, 1997; Conklin, 1997). These representations exoticize forest-dwellers as timeless, egalitarian, wise, and natural stewards of the environment (Poffenberger and McGean, 1993; Lynch and Talbott, 1995). Some anthropological work criticizes these representations of forest-dependent communities, revealing their dubious authenticity, the Western environmentalist agendas that motivate them, and the political-economic consequences when one group of forest-dwellers rather than another captures the interest of western environmentalists (Brosius, 1997; Conklin, 1997; Li, 1999). Other scholarship has taken a different tack, examining not only the process of fashioning these representations, but also the benefits they provide to local people (Tsing, 1999; Li, 2000). [Editor's note: see Dudgeon and Berkes' paper on Traditional Ecological Knowledge and Roy Ellen on the Construction of Biological Knowledge in this volume.]

Primitive conservationists in Borneo

The first example comes from the work of the Center for International Forestry Research (CIFOR), which is developing forest co-management programs with Dayak communities and other stakeholders in East Kalimantan.[4] By investigating the possibilities of forest co-management, CIFOR has opened up new rhetorical space for Dayak to contest dominant state resource discourses and to represent their own resource rights and uses better. As Li (1999a: 24) explains,

> Instead of a dialogue between the state and its critics, a mirror effect simply inverts the categories (wise swiddener/destructive swiddener, valuable traditions/backward traditions) leaving the categories themselves essentialized and fundamentally unchanged. In between these opposing camps, uplanders must invent especially creative strategies in order to defend their livelihoods and advance their own agendas, attempting to turn both state and 'green' discourse to their own ends.

One example of these "creative strategies" was demonstrated during a two-week research workshop, which was attended by (among others) three local Dayaks. Each of the individuals represented a different ethnic group – Kenyah, Merap and Punan. All three men were well-respected individuals in their communities and had extensive contact with researchers. During the workshop, all three emphasized both the cohesion among the ethnic groups and their efforts to protect forest resources. For example, they explained that when they find *gaharu* (*Aquilaria* spp.), they attempt to extract the part of the tree that is infected, leaving the rest of the tree standing in hopes that it will recover.[5] This account contradicts previous descriptions by other Dayak of the *gaharu* harvesting process, in which the entire tree is felled. Further, these three Dayak men consistently spoke of the harmonious cooperation among their respective ethnic groups, presenting a picture of village social dynamics which dramatically differs from both the stories told previously by other villagers and personal observations during fieldwork in their villages.[6] This fieldwork revealed not only a lack of cohesion among the different ethnic groups, but also an explicit inter-group discourse of inferiority and aggression.

The key to understanding these contradictions lies in the shift in socio-political context between village and research station. At the research station – which represented a shift to the formal and public – these Dayak men were engaged in an event of formal documentation, the results of which were to be used by the international research organization. It was a "field of attraction". As Tsing (1999a: 162) explains, for rural minority leaders to flourish today as tribal elders, a representational strategy with political force is required: "a field of attraction must be created to nurture and maintain the relationship between the rural community and its [environmentalist and green development] experts ... that keeps experts coming back."

Another example of Dayak appropriation of Western environmentalist discourse involves the fashioning of resource rights claims in romanticized terms. In 1991, a number of Punan in this same part of East Kalimantan formed a foundation to serve as the official voice for Punan living throughout the province.[7] A document entitled "The Mission and Vision of the Community of Traditional Punan Dayak", written by the Punan Chief Customary Leader, exemplifies the use of metaphors of nature and ancient customary law that appeal to Western environmentalist fantasies of forest dwellers:

> The words of our Mission and Vision ... are as strong as ironwood and as hard as iron stone. The wisdom of customary law for the Punan community ... is not merely a theory or concept like the products of laws, presidential decrees and government regulations that ... have caused losses in the rights of customary communities. Anybody ... who does not respect the existence of custom means that he or she is not the creation of God, who said humans must live and have children and grandchildren just like the grass and wood that is above the earth, guard and be responsible for the protection of nature in the whole world.[8]

Even in less formal contexts, Punan in this region now invoke images of themselves that fit well with environmentalists' notions of the indigenous naturalist. For example, Punan who have worked closely with researchers (including ethnobotanists) and environmental and social justice NGOs may say, "The forest is to us as milk is to a baby", which is a completely non-traditional image.[9] Also, they frequently mention – without prompting – the abundance of medicinal plants in the area and their knowledge of them. This echoes Brosius' (1997: 62) account of Penan in Sarawak who had worked with environmental NGOs: "One of the more interesting consequences of the environmentalist rhetoric of medicinal plants is that this rhetoric has itself suffused back to the Penan and been adopted by them as their own." In contrast, Punan in East Kalimantan who have had little contact with Western researchers or with self-representation for outside audiences rarely articulate livelihood practices or beliefs in these terms (cf. Brosius, 1997). Nor do the non-Punan Kenyah and Merap deploy rhetoric that so closely matches Western environmentalist discourse.[10]

This analysis illustrates the ways in which forest dependent communities deploy western environmentalists' representations of them as points of political leverage. Its purpose is *not* to assess the degree of authenticity of these representations; rather it is to contribute to the understanding of identity formation and change through the articulation between forest dependent communities and Western environmentalists (see Tsing, 1999a; Li, 2000).

Migrant farmers on the Amazonian frontier

The principal Western perceptions and images of the tropical forest have undergone a paradigmatic change over the past generation, with "Green Hell" being replaced by "threatened and fragile Eden" and a view of "primitive and backwards" forest dwellers being replaced by a view of "wise bearers of ancient cultures" (Slater, 2000). The new images are often in contradiction, however, and this contradiction dominates current debates about tropical forest conservation. The idea of human presence being compatible with forest conservation is still the subject of a fierce debate. On the one hand, many tropical ecologists assume that forest people, who are characteristically portrayed as poor and hungry and with little organizational capacity (Alcorn, 1995), are the despoilers of forest resources, and thus they argue for the conservation of "pristine" examples of tropical forests in the form of parks and biological reserves (Redford and Sanderson, 2000; Terborgh, 1999). On the other hand, other natural and social scientists argue that forest people not only contribute to forest conservation but also actually enhance the biological diversity of forests and, as "natural conservationists" (Alcorn, 1995; Colchester, 2000), are potentially powerful political allies for conservationists (Balée, 1989; Alcorn, 1993, 1995; Schwartzman, Moreira and Nepstad, 2000). This debate over "parks versus people" has dominated discussions of tropical forest conservation in the Amazon and, in so doing, it has obscured other, equally important debates, notably one about differences between indigenous and non-indigenous peoples, in particular small migrant farmers. [Editor's note: See Balée on Amazonia in this volume.]

Commencing in the 1970s, both the indigenous tribal peoples and the long-settled rubber tappers of the Amazon have gained recognition as "natural conservationists" (Conklin and Graham, 1995). The high-profile nature of the struggles for land rights by these "green" actors has tended, however, to obscure the fate of many of the other people in the region, including the far more numerous small migrant farmers. In sharp contrast to the "green" image of indigenous forest peoples, small migrant farmers in the Amazon have always been viewed as the "villains" of the forest, and they have been virtually ignored by researchers and officials as potential allies in forest conservation. There are two reasons that migrant farmers have rarely been thought of as potential allies in forest conservation. First, they are not "native" and so are not thought to have any knowledge about the appropriate use and management of the forest. Second, the logic behind their existing use of the forest has been neither adequately nor sympathetically examined.

The small migrant farmers (locally known as *colonos*) who populate the Amazonian frontier hail from southern and northeastern Brazil and are a heterogeneous group (Moran, 1981). Their migration to the frontier represents a response to a variety of political and economic forces, including land availability, financial incentives through government programs, massive road building, economic opportunities, and economic failure in their place of origin (Hecht and Cockburn, 1989; Schmink and Wood, 1992; Hall, 1997). They have played an important symbolic role at the frontier for the Brazilian government, which has been able to successfully characterize them as the villains of Amazonian deforestation. By blaming the migrants for the ecologically disastrous consequences of its own development programs, the government has been able to deflect most blame from itself and the equally culpable private sector.

In the Brazilian Amazon, given the limitations of infertile soils, the high availability of cheap land, and the scarcity of capital and labor, the most rational way to practice agriculture is typically by using the forest as a source of nutrients, released through fire in extensive systems of swidden cultivation (Boserup, 1965; Nepstad *et al.*, 1999). The hypothetical benefits from conserving forests are often outweighed by the need to open the land and so establish a tenurial claim upon it (Brondízio *et al.*, 2002). In the absence of incentives and infrastructure to encourage sustainable land use, poor *colonos* are under heavy pressure to continue to mine the natural resource base rather than make long-term investments in it (Pichón, 1996). Agriculture intensification and a shift toward fixed-field production systems begin to make economic sense only in older frontier areas, where there is greater access to both markets and technology, and the scarcity of land and rising land prices make swidden agriculture as well as cattle raising less viable (Toniolo, 1996). In these areas, colonists who arrived not even two decades earlier have begun to experiment with natural forest management and agroforestry, partially reconstructing the tropical forest in their own farm plots. These non-indigenous farmers have started to develop their own systems of environmental knowledge, drawing on their background, culture and society, as well as their own experiences at the frontier and on knowledge acquired from other groups there (Moran, 1981; Hall, 1997). At the same time, these older migrants have started to organize themselves to fight for political legitimacy and recognition for their resource-use systems.

Two recent grassroots initiatives show the increasingly proactive stance being taken by migrants. One, developed in association with the Catholic Church, is the *Movimento pelo Desenvolvimento da Transamazônica e do Xingu* or MPDTX (Movement for Development on the Transamazon and Xingu Regions), which is an umbrella institution for forty different local organizations including rural unions, farmers' cooperatives, teachers' and health workers' organizations, and movements of women, youths, and blacks (Hall, 1997). The MPDTX's first major initiative, taken in concert with a number of other grassroots institutions, was to propose a development program in which government and farmers organizations, acting as partners, would try to reconcile the twin objectives of environmental conservation and smallholder agricultural production. The proposed program consists of plans to: (1) reorganize land tenure throughout the region; (2) disseminate sustainable agro-ecological technologies; and (3) establish major new conservation areas in the region. The second notable grassroots initiative by migrants involves a proposal to reformulate the government agricultural credit line known as *FNO Pro-Ambiente* (FNO Pro-

Environment). This new environmental credit line would provide incentives for sustainable production systems and extractive activities by compensating farmers for the expenses that they incur to protect natural watercourses, shift to permanent forms of agricultural production, and re-establish forest on cleared land that is not suitable for agriculture (Pereira and Faleiro, 2000).

These grassroots initiatives reflect the political astuteness of the migrants in trying to assume the newly powerful role of environmental steward. More broadly, they reflect the migrants' concern to be represented as neither the villains nor victims of development in the region but to structure for themselves their own, more active role in development. These and similar actions by the *colonos* are forcing observers not just within Brazil but around the globe to reassess preconceptions of the resource-use practices and policies of migrants and frontier farmers.

WESTERN DISCOURSES OF RESTORATION AND CONSERVATION

The circulation of global environmental discourses, which often originate in Western ideological and institutional contexts, has been widely scrutinized for its impact on non-Western politics (Rangan, 1992; Baviskar, 1996; Dove, 1998; Gupta, 1998). Through attention to transformations, recontextualizations (Dove, 1998), or hybridizations (Gupta, 1998) at local levels, such studies demonstrate the complexity of local appropriations of global environmental rhetoric. Less studied to date is how such appropriations are affected by the diversity of rhetoric contained within the Western or global category.

Ecological restoration in Kathmandu

Western environmental discourses contain dramatically divergent formulations of "ecology" that can be called upon by a variety of differently positioned actors. Rhetoric traceable to Western institutional production can be found not only among those with power to frame ecological issues locally, but also among those who might contest this framing. In exploring the diversity of rhetorical tools circulating in non-Western nations, we are led to think less monolithically about Western environmental rhetoric as well: it, too, contains a diverse assemblage of tactical and strategic tools that might be employed on behalf of a variety of environmental ideas.

This can be seen in Kathmandu, Nepal, which over the last decade has become one of South Asia's fastest-growing cities.[11] One of the urban environmental issues featured in contemporary political discourse is the plight of the urban reaches of the Bagmati and Bishnumati Rivers, which converge in the city. The rivers are perceived as suffering severe ecological degradation, characterized by extremely poor water quality, serious morphological changes, and, some argue, loss of the cultural and religious values traditionally attributed to the rivers. Comprehensive policy and development studies identify the main causes of river pollution inside the urban area as the discharge of untreated sewage and widespread dumping of solid waste into the rivers and on their banks. Excessive sand mining in river beds and banks, which supplies mortar

and cement materials to the city's construction industry, is blamed for significant morphological change and severely channelized flow patterns (see
IUCN, 1994).

In addition, most discussions identify human encroachment on the banks,
floodplains, and riverbeds exposed by falling water levels as a significant factor
in the degradation process. Urban growth in Kathmandu has catalyzed the
rapid spread of development over a large area and increased population density
throughout the city. For new migrants from Nepal's countryside as well as for
poorer city residents, participation in the current land and housing markets is
impossible. As a result, many new migrants as well as long-term Kathmandu
residents have joined the swelling numbers of *sukumbaasi* (squatter) settlements
along the rivers.[12] In 1991, these settlements were estimated to be growing by
12 percent annually, a rate twice that of the city itself (HMG/ADB, 1991). By
2000, the growth rate had slowed, but a significant portion of the urban riparian
corridor is lined with semi-permanent structures and settlers asserting their
right to the land they occupy.[13] The actual ecological impact of these riparian
communities, particularly in contrast to an entire city whose effluents discharge
directly into the rivers, is widely contested. State development planners routinely incorporate human encroachment into their degradation models, however, on the assumption that restoration might necessitate the forced removal
of existing riparian sukumbaasi settlements.

Recognizing the serious threats posed by state representations of degradation
and restoration, advocates of housing rights for riparian sukumbaasis try to
counter claims that settlers are an obstacle to restoration through a counter-
narrative with clear ties to Western ecological rhetoric. By emphasizing international development concepts such as "healthy cities" and "sustainable human
settlements", phrases with origins in the United Nations Habitat Agenda,[14]
they offer a narrative that inserts socioeconomic concerns into ideas about the
ecology of the rivers and the city itself. By expanding the connotations of
habitat to include both non-human and human populations, housing advocates
frame environmental improvement in terms of housing rights, settlement quality, and improved public health, education, and sanitation services. Although
they may not directly contest the main features of the official state narrative
of river degradation, housing advocates use their notion of ecology to argue
for precisely the opposite of the fate for riparian settlements called for in the
official restoration scenario: *upgrading* squatter settlements, rather than *remov-
ing* them, is understood through a "sustainable human habitat" rubric as the
key to realizing an ecologically healthy riverscape. A particular conceptualization of ecology, then, expressed through references to globally-circulating rhetoric, aids advocates in claiming a "natural", ecological place for people otherwise
marginalized by a dominant urban environmental discourse.

The housing advocates' rhetorical strategies can be traced to the United
Nations-sponsored *Future Cities World Habitat Day Conference* (FCWHD),
held in Kathmandu in 1997. Throughout the session, environmental terms like
"habitat" and "greening" were invoked, but never did these invocations imply
a threat to sukumbaasis on the urban riverscape. Rather than blaming the

squatters for river pollution, for example, insufficient urban infrastructure was criticized. Rather than being seen as the disproportionate cause of river degradation, sukumbaasis were discussed as the disproportionate sufferers of its consequences. The effects of urban pollution and degradation were regarded as immediate and serious threats to the sukumbaasis (personal communication, Manandhar, 1997 and Pradhan, 1997), the city residents most proximate to the rivers, rather than the other way around. By drawing on the UN rhetoric of "sustainable human habitat", ecology was expanded to include contemporary human rights and urban infrastructural development objectives. By repeatedly invoking a "healthy cities" model, conference presenters emphasized an urban ecology in which a healthy environment is assessed through its capacity to provide food, clothing, and shelter to its human inhabitants.

The prevalence of a Habitat Agenda framing of urban ecology was reinforced in field interviews, in which housing advocates resisted any discussion of the negative effects riparian settlements may have on the ecological integrity of the rivers. Proximity to the degraded resource seemed to implicate settlers in the degradation process, they agreed, but they added that in ecological terms sukumbaasis could play only a minor role. This is most obvious in the case of sewage, for example, which key documents identify as the single most important element in the degradation of the river sysem. Since Kathmandu lacks a comprehensive, functional sewage treatment system, effluent inputs originate throughout the city, implicating legal and illegal urban inhabitants alike. Housing advocates further asserted that since sukumbaasis consume relatively fewer goods than more wealthy urban inhabitants, their per capita contribution of chemical inputs and other by-products of industrial production to the river system is probably also relatively small (Pradhan, personal communication, 1997).

At the local level, many sukumbaasi communities were in fact actively engaged with river monitoring, performing tasks that constitute efforts at river quality improvement. This directly contradicts representations of sukumbaasi knowledge, attitudes, and practices included in official characterizations of river pollution (Rademacher, 1998: 46–63). Planting vegetation on the riverbanks was a common practice, ironically, since riparian re-vegetation is a goal of the state development narrative. Settlers described patrolling their settlements to prevent illegal riverside dumping, practiced widely at the time by the municipalities, and suggested that they should have more authority to watch for, and halt, solid waste dumping on riverbanks. This framing of urban ecology not only downplayed any deleterious effects the settlements might have, it constructed sukumbaasis as in many ways more ecological than their legal neighbors.

The construction of the "Northern Forest" in New England and New York

It is instructive to examine not only how the diversity of rhetorics in Western environmental discourses is utilized in non-Western nations, but also how it is utilized in the West itself. Most of the discussion regarding environmental relations between East and West assumes a divide between the two, which is attributed to endogenous environmental and socio-cultural characteristics. Yet the rhetoric used to promote conservation and development interventions in many domestic United States contexts, drawing heavily as it does on essentialized images of ruralism and wilderness, greatly resembles that often employed in non-Western nations. To a considerable degree, both narratives draw on an overarching discourse that objectifies natural environments and associated peoples as fundamentally "other". Exploring the discursive similarities between domestic environmentalism and East-West debates can shed light on the common roots of both and help reveal ubiquitous but problematic conceptual frameworks.

An example can be found in the Northern Forest of New England and New York, a socially constructed "region" that became a focus of conservation in the late 1980s (Klyza and Trombulak, 1994; Dobbs and Ober, 1995; Northern Forest Lands Council, 1994). A poster distributed in 1998 by a coalition of environmental organizations suggests its essence. A photograph, looking down on the sunrise from a mountain top, shows clouds dappled with shadows and shades of pink, an expanse of unbroken forest, and a remote pond tucked among the folds of a massive, fog-shrouded ridge. The poster invites us to "Explore the Northern Forest," which is identified on a map as a green mantle draped across the top of Maine, New Hampshire, Vermont and New York. We are told to "experience the landscape, the culture and the heritage of ... the largest and last continuous wild forest east of the Mississippi River," with its high mountains, "pristine lakes and rivers", and "remote wetlands". There are people here, too, who are said to have "grown up hunting, fishing, trapping, and walking in the woods ..." Although we are told that some now work in business or manufacturing, we are also told that they are "proud of their heritage, and a way of life so different than in the urban areas around them." Finally, we learn that there are problems here, like development of lakeshores, reduced recreational access, and loss of jobs in the forest industry. It is said to be "up to us" to save the Northern Forest, to "leave, for our children and grandchildren, a healthy forest and strong communities that can continue to support a way of life that has existed for generations."

Although this is a simplistic image, it draws on central themes that appear regularly in environmentalist literature. Of particular importance is the fundamental otherness of the Northern Forest, which constructs a symbolic and experiential opposite of the everyday and the mundane as the basic consumable resource of tourism (Urry, 1990). The Northern Forest is depicted as immense, wild, natural, and strikingly beautiful. It is represented as a contiguous, cohesive region that stands in opposition to the surrounding urban and suburban landscape so often lamented as artificial, confining, predictable, and unattractive. It is a place we can retreat to, where we can escape the stresses of modern life and explore new places and possibilities. Although people are included in this vision, they, too, are very different from "us". One gets a sense that the people here are frozen in time, or at least drastically slowed down. They hearken back to an idealized, imagined past and fit seamlessly into the natural landscape. And all of this is rare and endangered; it is something that must be saved.

These images have great appeal because they connect to popular narratives of the frontier as a land of new possibilities, of sublime nature as a means of transcending modernity, and of the fusion of both in the contemporary emphasis on wilderness as both recreational and spiritual retreat (Cronon, 1995; cf. Slotkin, 1973; Callicot and Nelson, 1998). These ideas are, however, highly problematic. Far from being isolated in time and space, for example, this northern border country has been tightly bound to surrounding urban centers for more than two centuries through social, cultural, and economic ties, including a pervasive influence of absentee ownership of land and capital (Luloff and Nord, 1993). Nor do the Northern Forest's boundaries fit neatly with biophysical or social indicators. The Appalachian Mountains and associated forest types bend southward through the length of New Hampshire and Vermont, while Lake Champlain and extensive farmland in northwestern Vermont create a sharp break between its eastern and western sections. On large corporate holdings within the Northern Forest, intensive forest practices and a dense network of logging roads have caused dramatic ecological changes. And communities that lie on either side of the region's borders may have more in common with one another than communities that lie within the region but are separated by state lines (especially by the border between New York and New England).

Moreover, life in these communities has changed markedly in the past several decades. There has been mechanization and job loss in the forest industry and a consequent increasing dependence on tourism, government, and the service industry. With improved transportation have come the growth of regional commercial centers, decreased rural isolation, and a loss of community cohesion and local economic activity. There has been a pronounced shift of political authority from local to state and federal governments (Bryan, 1974, 1981; Hays, 1987). Demographic changes include a long history of out-migration of youth and more recent in-migration of people seeking the amenities of a rural lifestyle. Finally, the very concept of an interstate region called the Northern Forest did not exist until 1988, when the sale of large tracts of industrial forest land in each of the four states drew the attention of both government and environmental groups. Only then was the region named and institutionalized for the purposes of policy studies and political advocacy (Reidel, 1994).

In the short space of a dozen years, the Northern Forest has become incorporated into the environmentalist lexicon and is now seen by many as a very real entity. The successful circulation of images that are so easily questioned suggests that they are not so much "mistaken" as ordered and power-laden constructions of knowledge (Ferguson, 1990). While there are, indeed, significant commonalities across many parts of the Northern Forest – mountains, infertile soils, low population density, recreational tourism centers, large private landholdings and industrial forestry – its status as a "region" was hardly inevitable. Rather, its construction has been undertaken largely by non-local organizations utilizing deep-seated popular conceptions of rurality and wilderness – conceptions that infuse the operations and goals of those organizations and appeal to their constituencies.

The construction of the Northern Forest has been neither seamless nor static, however. The environmental community has been the principal driver behind the Northern Forest concept since 1994, when a government initiative to promote the concept ended. As a strategic response to resistance to environmental initiatives among local residents, the environmental community has elevated local people to a central place in this imagined landscape. But even as some local concerns such as jobs and property rights have been added to the environmentalist agenda, others – including questions of absentee land ownership and loss of local political control – have been sidestepped. These latter concerns are missing, thus, from recent proposals for creating "healthy communities", which focus on ecotourism and, increasingly, heritage tourism.

For urban constituencies, the Northern Forest is principally a place to visit, to explore interesting natures and cultures. In the end, extra-local desires to assist Northern Forest communities are inseparable from the perceived function of those same communities as places for non-residents to enjoy through tourism, a process that by its very nature moves rural communities away from more traditional forms and towards more commodifiable, consumable ones. There may be benefits for local residents, but the Northern Forest remains in many ways a place to be controlled, utilized, and conserved by non-residents.

This example illustrates how marginal societies and environments in developed countries may be defined and objectified in the same ways as their counterparts in less-developed countries. This comparison can help alert us to the nature of conceptual lenses that, by accentuating the differences between East and West, obscure fundamental similarities between the two having to do with the relationships between economic core and periphery (O'Connor, 1989; Wallerstein, 1983).

HYBRID SYSTEMS OF KNOWLEDGE

A prominent feature of global environmentalism since the 1970s has been the discourse of indigenous environmentalism, in which indigenous peoples are portrayed as protecting nature due to their cosmology. In this same discourse, Western science is often posed as a polar opposite to indigenous knowledge, objectifying nature in order to manipulate it. Whereas this represented a necessary corrective to a century and more of virtual denial by the West of the existence of indigenous knowledge in non-Western regions, it nonetheless represents a simplistic understanding of scientific knowledge, and of the relationship between practice and theory (see Pickering, 1992). Studies of Western science have shown that practices may have local, rather than over-arching justifications (Fujimura, 1992), with some practices appealing to one theory and some to another, so that the same scientist may draw on conflicting theories in a patchwork of knowledge. Seen from this point of view there is no necessary opposition between Western and non-Western science, which may be combined in similarly patchy and eclectic ways (Agrawal, 1995); practices and ideas from both Western and non-Western traditions may be used by people who do not appeal to a unified theory and who feel no tension between them. (Similarly, different people within a community may have different ideas depending on their experiences and interests.)

A hybrid forestry system in the Sierra Juarez

The hybridity of knowledge systems is well illustrated by the forestry practices of the Zapotec communities of the Sierra Juarez of Oaxaca, Mexico. These communities have been widely praised for their sustainable forest management (Bray, 1991). The Mexican Forest Service has promoted them as outstanding examples of good forest management, awarding prizes to several of the most successful (Ramos, 2000). In some cases the communities of the Sierra Juarez have been able to use management ideas from modern forest science to bolster community solidarity and to protect their forests. To what can these successes be ascribed? Is it, as advocates of indigenous knowledge might suggest, due to an ethic of forest protection, based on their traditional agro-ecological knowledge? Tyrtania (1992), among others, has documented the impressive complexity of traditional resource-use systems among the Sierra Zapotec. And indeed throughout Mexico, as elsewhere, scholars and politicians in recent years have credited indigenous peoples with extensive ecological knowledge and the use of sophisticated techniques of forest management (Gomez-Pompa, Salvador Flores, and Sosa, 1987). To credit the Zapotec forestry successes to strictly local knowledge would represent a denial of history, however.

In the community of Ixtlán in the Sierra Juarez, research revealed that forestry practices had been learned by the community members who previously worked with outside logging companies, which also brought them into contact with forest service regulations and policies. The practices and theories thus acquired largely contradict traditional agricultural practices, but community members do not see this as problematic. They have largely accepted the view of Mexican scientific forestry that fire is destructive, although in traditional Zapotec swidden agriculture (milpa) fire was an important tool and was also used to encourage pasture growth for cattle and sheep.[15]

In the 1930s the Mexican forest service initiated a policy of active fire suppression that was influenced by contemporary U.S. forest service policies (Anonymous, 1930a; Anonymous, 1930b; Gutierrez, 1930; Mares, 1932; Simonian, 1995). The forest service imposed upon communities the duty to form fire-fighting brigades and to suppress fires. The degree to which the communities complied is unclear, but they did learn to employ accusations of fire setting to involve the forest service in their boundary disputes with neighboring communities (Various, 1942, 1945). Commercial logging began in the forests of Ixtlán in 1948. An outside company employed comuneros – villagers with property rights – as loggers, providing them with cash income and an alternative to subsistence cultivation. Preliminary evidence from research in the forests of Ixtlán shows a dramatic decline in fire frequency after approximately 1945, which is probably due to Ixtlecos' realization that the forest could be a valuable source of livelihoods and to their gradual abandonment of swidden agriculture. From 1956 onwards, the view that fire was destructive was further strengthened by the actions of the forest concessionaire FAPATUX, which built fire towers and organized fire brigades. In 1982, the community took responsibility for managing its own forests, largely continuing the fire management practices it had inherited from FAPATUX. During research in Ixtlán in 2000, community members involved in logging repeatedly described fire fighting as being a shared obligation and said that their willingness to fight fires set them apart from neighboring communities whom they described as lazy or contentious. In Ixtlán, traditional uses of fire for agriculture have become increasingly restricted, and comuneros describe fire as a destructive agent, let loose by malicious, stupid, or careless people, principally from neighboring communities. Willingness to fight fires is also a pre-requisite for working for the community logging company.

The policy of fire suppression has dissenters within the community; many citizens in Ixtlán are not comuneros, do not benefit from logging, and would be interested in continuing to farm, using fire. One non-comunero critiqued the community's stand against clearing and burning new swiddens, pointing out that pine trees came up spontaneously on old swidden fields. In fact, almost everyone in the community is aware that pines naturally regenerate both on old forest fire sites and the sites of former swiddens. Older comuneros can point to old swiddens that are now covered in forest and will even acknowledge that pine trees often came back after fires, but they continue to affirm the necessity of fire suppression and reforestation.

Successful forest protection in Ixtlán is based not only on the impact of modern scientific ideas about the forest but also on the community's success in incorporating forestry science into community management structures and practices. Not all communities have been so successful. Key factors in determining the community's ability to incorporate forest science are their political and bureaucratic skills, their political organization, and their tradition of local autonomy (Fox, 1995). The key factor is the ability to both engage with state programs and to hold the state at arms' length. The present-day community structures that provide this ability are the result of a profound re-ordering of community life during the colonial and post-colonial periods (Chance, 1998; Wolf, 1957; Wolf, 1986) and so are not in any simple sense "non-Western", although they contain elements of pre-colonial political traditions.

Critical factors in contemporary forestry protection in Ixtlán de Juarez are its large area of forest (19,000 ha) and its relatively small population (2,100). Its large territory reflects the continuing power of the community of Ixtlán within the Sierra Juarez, building upon its successful maneuvering during the political struggles of nineteenth and twentieth century Mexico. Ixtlán was a military supporter of presidents Benito Juarez and Porfirio Díaz and later of the ultimate winners of the Mexican revolution (Garner, 1988). In the nineteenth century Ixtlán was one of the few communities to successfully petition to have its boundaries surveyed, and it has since been able to retain much of its large land holdings in spite of attempts by sub-communities to break away.

More recently, Ixtlán has been selected as the location for a new secondary school and government offices, thereby creating a large pool of comuneros who are trained in forestry and accounting and affording them continued opportunities to learn how to manipulate government bureaucracies. At the same time, the community has been able to hold unwanted government services at a distance; a recent government land-titling project was only allowed to survey the external boundaries of the community, with internal boundaries being regarded as community business. This was justified not by appeals to cosmology or to a sacred bond with the land but by firm statements that the land is "ours" and that outsiders have no business with it.

In Ixtlán, comuneros have been able to incorporate modern forest science into community forest management through a mixture of pragmatism and political guile. They do not appear to hold a unified non-Western cosmology or science, proffering instead local explanations of specific practices and blending modern scientific forestry with traditional agricultural practices. However, there is also considerable difference of opinion within the community, reflecting the different interests and experiences of community members. These differences notwithstanding, and in spite of negative impacts from logging in the past, the community of Ixtlán has been able to combine modern forest management with its own traditional political organization to protect the forests of the Sierra Juarez.

Non-Western uses of mapping technology

Another way that non-Western systems of environmental knowledge become hybridized is through the adoption of Western methodologies for representing that knowledge. Although adopted methods today include writing, film, and public relations/outreach among others, one of the earliest, and still most important, is mapping. Both descriptive and normative, maps summarize the priorities of a state or of a society. They reflect what is of immediate importance to the makers and users of the product, including political boundaries, natural features, local resources, social structure, or cosmology (Scott, 1998; Thongchai, 1994; Peluso, 1995; Brody, 1982; Mundy, 1996). This political dimension notwithstanding, part of the power of maps in the modern era has derived from their self-representation as neutral and objective tools, typically employed by the West. This has given rise over the past decade or so to a critique and a counter-mapping movement among critical scholars and activists in many nations.

Historically, authorities have used mapmaking as a tool for gathering information, establishing borders, and projecting the administrative or development plans for a given area (Anderson, 1983; Kain and Baigent, 1992). Before the colonial era, indigenous maps were prevalent in some regions but very sparse in others. In Asia, China, Vietnam, and Burma have rich cartographic traditions, whereas there is virtually no record of pre-colonial maps for Cambodia, Laos, and insular Southeast Asia (Schwartzberg, 1994). Where local mapping traditions exist, local conceptions of space and territory can be revealed by analyses of these maps (Lewis, 1998; Harley and Woodward, 1994). Thus, Mundy (1996) examines how indigenous American maps reflected social relations and local cosmology, while Spanish cartography reflected the colonizers' aims in urban administration and in gaining information about local topography. Similarly, Thongchai (1994) describes the evolution of Thai cartography from traditional, centrally-focused representations of religious significance to more modern topographical and political depictions that defined outer frontiers in relation to the neighboring colonized regions.

With the spread of colonialism and the concomitant European effort to extend control over land, population, and production in Southeast Asia, maps took on a central importance for new rulers, with increased attention to detail and frequent revisions (Henley, 1995). Colonial use of mapping shifted from the early exploration of territory to efforts increasingly focused on delineation of the boundaries of administrative units (Thongchai, 1994). There were notable differences among the colonial powers in their attitudes toward traditional land claims and uses, and these differences are reflected in both mapped representations and in colonial land regulations (Furnivall, 1956). Toward the end of the colonial era, mapping also played a crucial role in shaping national and regional identities (Anderson, 1996; Thongchai, 1994; Henley, 1995).

Mapping in colonial Southeast Asia was largely an administrative project to facilitate control over human and natural resources. Modern states, and some conservation organizations as well, still use maps to limit local residents' land claims by defining such use as "encroachment" in conservation areas or in areas designated for some other form of development (Eghenter, 2000). For example, representing forests used by swidden cultivators as "empty" on official maps is a strategic move which ignores people's presence and denies the legitimacy of their land use, thereby bolstering the case for alternative claims (Li, 1996). The incidence of mapping increased in Indonesia during the 1990s, supported by state, corporate, and conservation initiatives, with parks and sites of state interventions being the subjects of particular attention (Momberg *et al.*, 1996).

However, local people also have begun to employ the Western technology of boundary mapping to communicate their resource claims in a medium acceptable to modern national authorities (Tsing, 1999a). Long an instrument of state control used to overwrite local land customs, mapping is now used to represent community land claims and to document local uses of forests and other resources (Peluso, 1995). Increasingly, non-Western actors are challenging Western scientific discourses of mapping as a state-directed activity which produces authoritative documents, with many mapmakers now defining their goals as the promotion of community interests (Sirait *et al.*, 1994). Community-level mapping uses participatory methodologies to reverse the flow of information from externally produced to locally informed maps, thereby challenging the authoritative claims of state maps.[16]

The development and popularization of computerized, digitized mapping technologies has created new opportunities and challenges in mapping. Historical data and satellite images can now be combined to analyze questions such as the causes of forest fires in Indonesia (Harwell, 2000; Rabindran, 2000) and the origin of forest islands in the African savanna (Fairhead and Leach, 1996). A challenge to the new methods of mapping is to reflect the dynamic nature of traditional land claims and use alongside other land designations (Fox, 1998). Otherwise, modern mapping in non-Western nations may, like some colonial maps, only serve to freeze and simplify what are otherwise mutable and layered boundaries and types of land use. Other concerns include the implications of fixing ethnicity to defined spaces (Li, 2000) and the difficulty of accurately representing complex local land classifications (Tsing, 1999a).

A case study from an interior valley in the central Bird's Head region of Papua, Indonesia illustrates the shifting purposes and multiple actors involved in mapping today. This region is inhabited by tribal clans who practice swidden agriculture alongside hunting and gathering in the forest. In 1999 the most topographically accurate maps available were Dutch maps from 1957, which were based on 1944 aerial photographs taken by the United States military. Villagers report that the Dutch used these maps in their regional planning for forestry development and plywood production in the valley, although this endeavor stalled during the 1960s due to local resistance to the plans and the imminent incorporation of Western New Guinea into Indonesia. Subsequent maps drew on the information in the Dutch maps with varying degrees of precision, but to this day many official Indonesian maps of the region inaccurately reflect the human settlements in the area, placing (e.g.) eastern villages in the west and southern villages in the north. In the 1990s, the planned construction of a road from the coast through the area refocused state attention on economic and social development planning in the valley. A review of the maps in use in different government departments showed the region variously categorized as protected area, slated for forestry development, a potential site for transmigration, or destined for conversion to agro-industry. Villagers, in partnership with a local legal aid society and the agriculture faculty of the provincial university, Cenderawasih, undertook a valley-wide mapping effort in the late 1990s to convey their forest resource use practices and understanding of traditional clan boundaries to Indonesian government officials. They used the 1957 Dutch topographical maps as a reference, supplemented by participatory mapping techniques, with plans to combine all of the data gathered using geographic information systems.

The political importance of maps has shifted somewhat as East-West colonial struggles for resources have been superseded by more complex power relations and struggles. Non-Western communities have taken up the Western concept of mapping and transformed it into a tool with the potential to challenge official views by presenting alternative views of resources and territories. In the arena of conservation of tropical forests and marine reserves, maps continue to play an important role in shaping Western awareness of and involvement in non-Western territories. Mapping is transformed, and transformative, as it is conducted with different actors, using different technologies, in different political environments.

The exoticization of swidden agriculture

A powerful symbol of Western views of non-Western environments is swidden agriculture, also known less accurately as "shifting cultivation" or more deprecatingly as "slash-and-burn agriculture". During the final quarter of the twentieth century, the image of a poor farmer standing in a swidden full of charred tree trunks became ubiquitous in Western representations of non-Western environmental degradation, especially tropical deforestation. This image explicitly stands for the alleged pressure of poverty on the environment; more implicitly, it stands for the purportedly short-term, irrational and destructive use of natural resources by non-Western farmers. This is a singular exception to the earlier-noted, popular and positive re-evaluation of the lifestyles of tropical forest peoples. The profoundly negative loading of this image is reflected in the role that the term "slash-and-burn" has come to play in Western language. Use of the term is largely confined to semantic domains where the required connotation is of ruthless and merciless behavior, recent examples of which included the corporate downsizings and government budget cutting by the Republican party in the United States in the 1980s and 1990s.

Most scholars abjure the use of the term "slash-and-burn agriculture" (or "shifting cultivation"), preferring instead to use the term "swidden agriculture". This is based on an archaic variant of old-English "swithen", meaning to singe, which was resurrected because no contemporary term was sufficiently neutral to be used or even rehabilitated.[17] Swithen/swidden was so archaic and indeed unknown as to have no prior connotations.

Over one-half century of sympathetic, systematic ethnographic research has shown, and continues to show, a picture of swidden agriculture quite unlike its popular representation in the contemporary Western world. Whereas the systems of agriculture most familiar to the contemporary West mine the soil, swidden agriculture does not. In tropical forests, the nutrient stores that can be exploited by agriculture lie mostly not in the soil but in the biomass atop it, and it is this that swiddens exploit. This is reflected in the term for swidden cultivation among Ibanic-speaking Dayak in Borneo, *bumai hutan* ("farming the forest").[18] The nutrients in the biomass are extracted through burning, which breaks down the biomass into a nutrient-rich ash that cultigens can easily access. Dayak in Borneo say that the burn is the most important single determinant of a good harvest; and burning the forest in the wet tropics is, popular misconceptions notwithstanding, far from easy.[19] Moreover, the burn is but a single moment in a long cycle that is otherwise devoted to encouraging natural processes of afforestation and maintaining a semi-natural forest cover on the land.

The reliance in swidden agriculture on natural forest dynamics to restore fertility after each cropping cycle conserves the use of both human and non-human sources of energy (Kleinman, Pimentel, and Bryant, 1995: 247–248). In fact, reliance on the dynamics of the forest gives swidden agriculture one of the greatest returns to labor (as opposed to land) known in agriculture. Dove (1985: 6) calculated that the return to labor in swiddens is 1–3 times as great as that of irrigated rice terraces, whereas Ruthenberg (1976) calculated it as 3–4 times as great.[20] Proof of the economic and ecological sustainability of swidden agriculture lies in its sheer persistence. Recent estimates suggest that it is practiced on 30% of the world's arable soils (Bandy, Garrity and Sanchez, 1993: 2) and supports as many as one billion people – 22 percent of the population of the developing world in tropical and subtropical zones (Thrupp, Hecht, and Browder, 1997: 1–4). [Editor's note: See J.L. Kohen's discussion on burning in his article on Australian Aboriginal people in this volume.]

When Western scientists and government officials penned accounts of swidden agriculture in the tropics in the nineteenth and early twentieth centuries, they typically noted how alien it appeared to their eyes and how much of a mental contortion they had to go through in order to understand it. This perceived alienness rests, however, on a curious forgetting of Western agricultural history, in particular the role of swiddens in Western Europe and North America. Thus, in the mid-sixteenth century the founder of modern Sweden, King Gustav I, is recorded urging his subjects to put less of their energies into resistance to the state and more into making swiddens in the forest (Weimarck, 1968). Similarly, in the Ardennes in France, swidden cultivation not only persisted into the twentieth century but was actually a more profitable occupation than paid labor in industry until early in that century (Sigaut, 1979: 685). And a swidden system based on a melding of Scots-Irish and Native American practices developed in the uplands of the southern United States, where it dominated through the nineteenth century and has persisted to the present day (Otto and Anderson, 1982). This recency and prevalence of swidden agriculture in the West is not reflected in either the scholarly or policy literatures, however. The paucity of attention is so marked that Sigaut (1979: 679) has asked, "What has anthropology [among other fields] missed by ignoring the European case ...?"

The most obvious reason for this historical erasure is the critical attitude of the modern state and its representatives toward swidden cultivation. Carl Linnaeus encountered an early example of this during his Scanian Travels in Southern Sweden in 1749 (from Weimarck, 1968: 40).[21] Linnaeus observed and commented favorably on the then-ubiquitous practice of swidden cultivation in this region, the Swedish term for which translates as "burn beating". He wrote "If the inhabitants ... were not allowed to have burn-beating, they would want for bread and be left with an empty stomach looking at a sterile waste" (Weimarck, 1968: 56). Burn beating was used in a system of crop rotation and land fallowing that produced first turnips, then rye, then hay, and then finally pasture (Weimarck, 1968: 52). The sponsor of Linnaeus' expedition, High Commissioner Baron Carl Harleman, did not appreciate Linnaeus' findings, however. He wrote "[Linnaeus] not only had not condemned burn-beating, so pernicious for the country, but even contrary to his own better judgment justified and sanctioned the undertaking" (Weimarck, 1968: 40). Linnaeus had to knuckle under to this criticism and substituted in the final draft of his report "harmless notes on manure" in place of his discussion of burn-beating (Weimarck, 1968: 40). The tenacity of this official anti-swidden discourse is reflected two and one-half centuries later in Kleinman, Pimentel, and Bryant's (1995: 235) puzzled remark on the "[continued] inability of domestic and international development agencies to consider slash-and-burn agriculture as a sound food production system."

The antipathy of the modern state to swidden agriculture is based, in part, on its "illegibility". Intensive, fixed-field, infrastructure-heavy agriculture tends to be favored by states because its product is visible, concentrated, and susceptible to state extraction, and its people are tied by capital investment to their fields. States tend not to favor swidden agriculture because, in contrast, its product is far less extractable and, in the absence of capital investment, its people are far more capable of evasion and flight. As Scott (1998: 282) writes, swidden is an "illegible form of agriculture", comprising "fugitive" fields and cultivators, and constituting "potentially seditious space." The state antipathy toward swidden agriculture is underpinned by the historic shift in Western economic development to a valorization of capital investment and returns to it versus returns to labor. Whereas the logic of intensive agriculture focuses on conserving land, the logic of swidden systems focuses on conserving and valuing human labor. The modern state's criticism of swidden agriculture is thus, in reality, a question not of agro-ecological development but of political-economic self-interest.[22]

The current antipathy of Western states toward swidden agriculture must be interpreted in light of the West's own swidden history. The erasure of this history and the rise of a critical, anti-swidden discourse appear to have occurred in the developed Western nations precisely as the practice of swidden was waning there and becoming concentrated in and identified with the less-developed non-Western nations. This coincidence suggests that Western deprecation of swidden agriculture is not so much a function of its geographic, historic, and technological distance as it is part of a political effort to *make* it distant. This socially constructed alienness enables Western nations to adopt a critical, self-empowering view of natural resource use in developing, non-Western nations. As recent scholarship suggests, any discourse of "under-development" privileges the part of the world that wields it and de-privileges the part that is characterized by it (Escobar, 1995; Ferguson, 1990).

*   *   *

We have endeavored here to take a new look at the "mobilization" of environmental concepts between Western and non-Western nations. One of our principal conclusions is that the purported divide between Western environmental science and non-Western systems of environmental knowledge, although continually represented as an important boundary marker heavy with symbolic meaning, is problematic. The two systems are historically inter-mingled. And there is also inter-mingling within each system. Neither is monolithic; both encompass multiple, diverse, and sometimes conflicting paradigms. The division between East and West or South and North, when discussing environmental knowledge and practice, is thus an essentialist fallacy. The linkages are more compelling than the divide. The boundary between East and West is more meaningful as a metaphor than as a geographic fact.

The linkages between East and West, and the mingling of different systems of knowledge, have implications for our views of the static versus dynamic qualities of knowledge. In most of the cases that we have discussed, mobilization of knowledge involves its transformation. Indeed, such transformation seems to characterize the mobilization of knowledge between East and West. This, in turn, has implications for power relations. The transformation of environmental knowledge seems to be inherently a political act, which is in keeping with the fact that all deployments of knowledge or mobilizations of concepts that we have examined have had winners and losers. The data presented here do not support a view of environmental knowledge as politically neutral.

Finally, we must acknowledge an epistemological difficulty inherent to this (and any similar) critique: namely, we have not been able to critique the conceptual divide between Western and non-Western systems of environmental knowledge without using this concept ourselves, in seeming contravention of our own critique. Derrida sees this paradox as one that is inherent in language and criticism, which he characterizes as "the problem of the status of a discourse which borrows from a heritage the resources necessary for the deconstruction of the heritage itself" (Derrida, 1978: 282). This paradox (and the point of Derrida's work is that it *is* a paradox) raises a number of analogous and additional questions about the sociology of knowledge, which are relevant to the subject of our chapter. To what extent, for example, do policy-makers and practitioners continue to use concepts like the divide between Western and non-Western environmental knowledge, long after scholars have abandoned them? Or, and bearing more directly on our subject, to what extent do non-Western scholars continue to use concepts (like this division) after Western scholars have critiqued them?[23] These dimensions of the life cycle of conceptual constructs are little studied as yet. More broadly, why is the distinction between Western and non-Western being reified precisely at the point in history when it seems to be losing whatever empirical validity it may ever have had? Of more specific relevance to this chapter, why have environmental relations and knowledge emerged as key dimensions of this distinction? Why, thus, are both Western and non-Western actors using the environment as a focal point for reiterating the myth of themselves in contradistinction to one another?[24]

NOTES

[1] Of these sets of related terms, we will for convenience and clarity use "Western versus non-Western", although we could equally well have used any of the others.

[2] Said (1978) laid the groundwork for much of this critique with his analysis of the origins of Orientalism in the West.

[3] An analogous process takes place in the movement of concepts from one scientific discipline to another (Dove, 2001).

[4] Dayak – an umbrella term for the indigenous, upland people of Borneo – are often located within the rubric of primitive environmentalists by conservation and community-based natural resource practitioners (see for example Poffenberger and McGean, 1993). Following King (1993), the term

Dayak here is used to include both swidden argiculturalists and the Punan, the traditionally nomadic people of Borneo.

[5] *Gaharu* is the resinous heartwood that results from a fungal infection (*Cytosphaera mangifera*) in some species of *Aquilaria*. The resulting aromatic heartwood is exported and used in perfumes and incense.

[6] Compare with Baviskar's (1996, 1997) commentary on down-playing of ethnic violence by researchers.

[7] The Punan foundation is supported by an indigenous rights/human rights NGOs based in the provincial capital.

[8] "*Bagaimanapun harus penekanan yang datang dari luar untuk melumpuhkan komunitas adat dapat bendung melalui setrategi Misi dan Visi penguatan Adat yang selama ini berjalan dan menjadi keharifan pandangan hidup masyarakat Dayak Punan dengan ibarat kata Misi dan Visi kita bagaikan tegaknya tunggal Ulin dan sekerasnya Lugem. Keharifan hukum adat bagi kehidupan masyarakat Dayak Punan, memang terbentuk dan berada didalam diri masing-masing insan manusia sejak semula, bukan tiori atau konsep semata sepertinya produk undang-undang, kapres dan peraturan pemerintah yang selama ini merugikan hak-hak masyarakat adat. Siapapun didunia ini yang tidak mengakui keberadaan Adat berarti dia bukan ciptaan Tuhan yang berkata Hai manusia hidup dan beranak cuculah kamu seperti rumput dan kayu di atas bumi ini jaga dan peliharalah pelestarian alam semesta ini, dengan baik dan manusia yang menghujat keberadaan adat adalah manusia yang kehilangan keseimbangan Moral*" (Lembaga Adat Punan Besar Dayak Punan KalTim, 1999).

[9] Western bio-physical researchers, participating in what Brosius (1997: 66) terms the "hall of mirrors of representation", often explicitly seek out Punan as field assistants because of their reputed indigenous, ethnobotanical knowledge.

[10] If they employ any metaphor at all to describe the meaning of the forest to them, it is apt to be the non-green metaphor of a "bank".

[11] According to the World Resources Institute (1996), the urban growth rate in the cities of the Kathmandu Valley was 7.1% over the period 1990–95, a figure considerably higher than the UNFPA's (1995) estimate of 6.5%. The WRI study estimates that by 2025 the percent of Nepal's population residing in urban areas will increase to 34% from the present 14%.

[12] *Sukumbaasi* is most commonly translated as "landless squatter", but this is a controversial translation, especially in Kathmandu. The landholding status of many occupants of urban *sukumbaasi* communities is publicly disputed, with government officials and others doubting the authenticity of some *sukumbaasis*' claims of landlessness. There are many stories of people who live in so-called *sukumbaassi* areas while renting out their city homes or flats and capitalizing on the city's skyrocketing housing market.

[13] In the fall of 1997, the total number of settlements characterized as *sukumbaasi* in Kathmandu was 54. Half of these were riparian – situated on the banks of the Bishnumati, Bagmati, or one of their larger urban tributaries. Of the total population of *sukumbaasis* in the Kathmandu Valley in 1996 – close to 9000 – 69% lived in riparian zones and about two-thirds of those occupied settlements on the banks of the Bishnumati or Bagmati Rivers (Tanaka, 1997).

[14] The United Nations Habitat Agenda, established at the United Nations Conference on Human Settlements in 1996, can be viewed at http://www.hsd.ait.ac.th/agenda/habitat.htm.

[15] Ironically, agricultural techniques were profoundly affected by the introduction of steel tools and large livestock in the sixteenth century (Nigh, 1975), so even this tradition has been affected by Western agricultural technology.

[16] The practice of participatory mapping was developed and popularized primarily in South and Southeast Asia (Chambers, 1997: 113 ff.).

[17] The earliest recorded uses of swithen in England date from early in the thirteenth century. The *Oxford English Dictionary* (1989, XVII: 401) cites Izikowitz (1951: 7) as the first modern usage of "swidden". Izikowitz himself attributes the term to Ekwall (cf. Ekwall, 1955).

[18] Cf. the title to Condominas' (1977) famous ethnography of Montagnard swidden cultivators, "We Have Eaten the Forest" (in the original French, *Nous Avons Mangé la Forêt de la Pierre-Génie Gôo*).

[19] The difficulty of burning wet tropical forest is reflected in the cultural prescriptions and proscriptions that surround burning in parts of Borneo, including freedom to ignore unfavorable

bird omens, lack of responsibility under customary law if a fire escapes to an adjoining swidden, proscriptions against drinking or bathing on the morning of a burn, and so on.

[20]   The relatively high labor productivity of swidden agriculture allows swidden communities to be, counter-intuitively, much more involved in cultivation of commodity crops for global markets than more intensive cultivators like irrigated rice farmers (Dove, 1993; Pelzer, 1978).

[21]   We are indebted to Pyne (1995: 86) for this reference.

[22]   This places state antipathy toward swidden fires in a different light. Pyne (1993: 256) writes, "As soon as it was politically and technically feasible, [colonial] foresters instigated fire control measures. As often as not, fire suppression was one of the most powerful means of controlling indigenes."

[23]   Note Agrawal's (1995) observation that the post-modern critique of essentialist constructions like this divide is much more prevalent within academic communities in more-developed than less-developed countries.

[24]   We are grateful to Carol Carpenter for this insight.

BIBLIOGRAPHY

Agrawal, Arun. 'Dismantling the divide between indigenous and scientific knowledge.' *Development and Change* 26: 413–439, 1995.

Alcorn, J.B. 'Indigenous people and conservation.' *Conservation Biology* 7(2): 424–426, 1993.

Alcorn, J.B. 'Economic botany, conservation, and development: What's the connection?' *Annals of the Missouri Botanical Garden* 82(1): 34–46, 1995.

Anderson, Benedict. *Imagined Communities*. London: Verso, 1983.

Anderson, B. 'Census, map, museum.' In *Becoming National: A Reader*, G. Eley and R.G. Suny, eds. New York: Oxford University Press, 1996, pp. 243–259.

Anonymous. 'Instrucciones para la campaña contra incendios de montes.' Vol. 893 *Expediente* 11, p. 1. Archivo General del Estado de Oaxaca: Asuntos Agrarios, Serie V Problemas por Bosques, 1930a.

Anonymous. 'Suplica se sirva dictar las ordenes conducentes a evitar la introduccion deproductos forestales fraudulentos.' Vol. 893 *Expediente* 2, p. 1. AGEO: Asuntos Agrarios, Serie V. Problemas por Bosques, 1930b.

Bajracharya, Sama and Lajana Manandhar. *World Habitat Data Report on the Celebration of World Habitat Day*. Lalitpur: Lumanti Support Group for Shelter, 1997.

Balée, W. and A. Gélly. 'Managed forest succession in Amazonia: The Ka'apor case.' *Advances in Economic Botany* 7: 129–158, 1989.

Bandy, D.E., D.P. Garrity, and P.A. Sanchez. 'The worldwide problem of slash-and-burn agriculture.' *Agroforestry Today* 5(3): 1–6, 1993.

Baviskar, Amita. 'Reverence is not enough: Ecological Marxism and Indian *adivasis*.' In *Creating the Countryside: The Politics of Rural and Environmental Discourse*, E.M. Dupuis and P. Vandergeest, eds. Philadelphia: Temple University Press, 1996, pp. 204–224.

Baviskar, Amita. 'Who speaks for the victims?' *Seminar* No. 451, 452, 1997.

Boserup, Ester. *The Conditions of Agriculture Growth*. Chicago: Aldine, 1965.

Bray, David Barton. 'The struggle for the forest: Conservation and development in the Sierra Juarez.' *Grassroots Development* 1991, pp. 1513–1525.

Brody, H. *Maps and Dreams*. New York: Pantheon Books, 1982.

Brondízio, E.S., S.D.M. Cracken *et al.* 'The colonist footprint: Towards a conceptual framework of land use and deforestation. Trajectories among small farmers in frontier Amazonia.' In *Deforestation and Land Use in the Amazon*, C. Woods and R. Porro, eds. Gainesville: University Press of Florida, 2002, pp. 133–161.

Brosius, J. Peter. 'Endangered forest, endangered people: Environmentalist representations of indigenous knowledge.' *Human Ecology* 25(1): 47–69, 1997.

Bryan, Frank M. *Yankee Politics in Rural Vermont*. Hanover, New Hampshire: University Press of New England, 1974.

Bryan, Frank M. *Politics in the Rural States*. Boulder, Colorado: Westview Press, 1981.

Callicott, J. Baird, and Michael P. Nelson, eds. *The Great New Wilderness Debate*. Athens: University of Georgia Press, 1998.

Chambers, R. *Whose Reality Counts? Putting the Last First.* London: Intermediate Technology, 1997.

Chance, John K. *La Conquista de la Sierra: Españoles e Indigenas de Oaxaca en la Epoca de la Colonia.* Oaxaca: Instituto Oaxaqueño de las Culturas, CIESAS, Fondo Estatal Para la Cultura y las Artes, 1998.

Colchester, M. 'Self-determination or environmental determinism for indigenous people in tropical forest conservation.' *Conservation Biology* 14(5): 1365–1367, 2000.

Condominas, G. *We Have Eaten the Forest: The Story of a Montagnard Village in the Central Highlands of Vietnam,* Adrienne Foulke, trans. New York: Hill and Wang, 1977 [1957].

Conklin, Beth A. 'Body paint, feathers, and VCRs: Aesthetics and authenticity in Amazonian activism.' *American Ethnologist* 24(4): 711–737, 1997.

Conklin, Beth, and B.A. Graham. 'The shifting middle ground: Amazonian Indians and eco-politics.' *American Anthropologist* 97(4): 695–710, 1995.

Cronon, William. 'The trouble with wilderness; or, Getting back to the wrong nature.' In *Uncommon Ground,* William Cronon, ed. New York: W.W. Norton, 1995, pp. 69–90.

Dahal, Dilli R. *Demolition of Squatter Houses at Kohiti and other Environmental Problems related to the Bishnumati Corridor: A Sociological Analysis.* Kathmandu: Asian Development Bank, 1997.

Derrida, Jacques. *Writing and Difference,* A. Bass, trans. Chicago: University of Chicago Press, 1978.

Dobbs, David and Richard Ober. *The Northern Forest.* White River Junction, Vermont: Chelsea Green, 1995.

Dove, Michael R. 'The agroecological mythology of the Javanese, and the political-economy of Indonesia.' *Indonesia* 39: 1–36, 1985.

Dove, Michael R. 'North-South relations, global warming, and the global system.' Special issue on 'Human Impacts on the Pre-Industrial Environment.' *Chemosphere* 29(5): 1063–1077, 1994.

Dove, Michael R. 'Local dimensions of 'global' environmental debates.' In *Environmental Movements in Asia,* Arne Kalland and Gerard Persoon, eds. Surrey: Curzon Press, 1998, pp. 44–64.

Dove, Michael R. 'The life-cycle of indigenous knowledge, and the case of natural rubber production.' In *Indigenous Environmental Knowledge and its Transformations,* Roy F. Ellen, Alan Bicker, and Peter Parkes, eds. Amsterdam: Harwood, 2000, pp. 213–251.

Dove, Michael R. 'Inter-disciplinary borrowing in environmental anthropology and the critique of modern science.' In *New Directions in Anthropology and Environment: Intersections,* C.L. Crumley, ed. Walnut Creek, California: AltaMira Press, 2001, pp. 90–110.

Eghenter, Christine. *Mapping Peoples' Forests: The Role of Mapping in Planning Community-Based Management of Conservation Areas in Indonesia.* Washington, DC: Biodiversity Support Program, 2000.

Ellen, Roy F. "What Black Elk left unsaid." *Anthropology Today* 2(6): 8–12, 1986.

Escobar, Arturo. *Encountering Development: The Making and Unmaking of the Third World.* Princeton, New Jersey: Princeton University Press, 1995.

Fairhead, James, and Melissa Leach. *Misreading the African Landscape: Society and Ecology in a Forest-Savanna Mosaic.* Cambridge: Cambridge University Press, 1996.

Ferguson, James. *The Anti-Politics Machine: 'Development,' Depoliticization, and Bureaucratic Power in Lesotho.* Cambridge: Cambridge University Press, 1990.

Fernandez, K., ed. 'Evictions in Nepal 1996.' In *Eviction Watch Asia 1996.* Singapore: Asian Coalition for Housing Rights, 1997, pp. 47–50.

Flyen, Cecile and Clementine Munch. *Historic City Core of Kathmandu: Changes and Upgrading of Public Areas.* Trondheim: Division of Town and Regional Planning/Norwegian Institute of Technology, 1990.

Fox, Jefferson. 'Mapping the commons: The social context of spatial information technologies.' *Common Property Resource Digest* 45: 1–4, 1998.

Fox, Jonathan. 'Governance and rural development in Mexico: State intervention and public accountability.' *Journal of Development Studies* 3: 21–30, 1995.

Fujimura, Joan.' Crafting science: Standardized packages, boundary objects and 'translation'.' In *Science as Practice and Culture,* Andrew Pickering, ed. London: University of Chicago Press, 1992, pp. 168–211.

Furnivall, J.S. *Colonial Policy and Practice: A Comparative Study of Burma and Netherlands India.* New York: New York University Press, 1956.

Garner, Paul. *La Revolucion en Provinciasoberania Estatal y Caudillismo en las Montañas de Oaxaca (1910–1920).* Oaxaca: Fondo de Cultura Economica, 1988.

Gomez-Pompa, Arturo, Jose Salvador Flores, and Victoria Sosa. 'The Pet Kot: A man-made tropical forest of the Maya.' *Interciencia* 12: 10–15, 1987.

Gupta, Akhil. *Postcolonial Developments: Agriculture in the Making of Modern India.* Durham, North Carolina: Duke University Press, 1998.

Gutierrez, Jose L. 'Se remite ejemplar del reglamento para constituir corporacion de defensa contra incendios' Vol. 892 *Expediente* 22, p. 4. AGEO: Asuntos Agrarios, Serie V. Problemas por Bosques, 1930.

Hagenstein, Perry. *A Challenge for New England: Changes in Large Forest Land Holdings.* Boston: The Fund for New England, 1987.

Hall, A. *Sustaining Amazonia.* New York: Manchester University Press, 1997.

Harley, J.B., and D. Woodward, eds. *The History of Cartography.* Vol. II, Book II: *Cartography in the Traditional East and Southeast Asian Societies.* Chicago: University of Chicago Press, 1994.

Harwell, Emily E. 'Remote sensibilities: Discourses of technology and the making of Indonesia's natural disaster.' *Development and Change* 31: 307–340, 2000.

Hays, Samuel. P. *Beauty, Health, and Permanence: Environmental Politics in the United States, 1955–1985.* New York: Cambridge University Press, 1987.

Hecht, S. and A. Cockburn. *The Fate of the Forest: Developers, Destroyers, and Defenders of the Amazon.* New York: Verso, 1989.

Henley, D. 'Minahasa mapped: Illustrated notes on cartography and history in Minahasa, 1512–1942.' In *Minahasa Past and Present: Tradition and Transition in an Outer Island Region of Indonesia,* Reimar Schefold, ed. Leiden: Research School CNWS, 1995, pp. 32–57.

His Majesty's Government of Nepal (HMG) and Asian Development Bank. *Environmental Policy Assessment: Kathmandu Urban Development Plans and Programs: Concept Plan for the Bishnumati Corridor.* Kathmandu: Halcrow Fox and Associates, 1991.

HMG Nepal Ministry of Housing and Physical Planning and Asian Development Bank. *Kathmandu Urban Development Project Preparation Report.* Kathmandu: HMG, 1992.

HMG Nepal Ministry of Housing and Physical Planning and Kathmandu Urban Development Project. *Community Participation in Development Programs: Working Paper.* Kathmandu: COWI Consult/Padco/Nepal Consult, 1996.

HMG Nepal. *National Shelter Policy.* Kathmandu: HMG, 1996.

Huntington, Samuel P. *The Clash of Civilizations and the Remaking of World Order.* New York: Simon and Schuster, 1996.

Ingold, Tim. 'Globes and spheres: The topology of environmentalism.' In *Environmentalism: The View from Anthropology,* Kay Milton, ed. ASA Monograph 33. London: Routledge, 1993, pp. 31–42.

International Union for the Conservation of Nature (IUCN), Stanley International (Canada), Matt-McDonald Ltd. (UK), and East Consult (Nepal). *Bagmati Basin Water Management Strategy and Investment Program: Final Report.* Kathmandu: IUCN, 1994.

Kain, R.J.P., and E. Baigent. *The Cadastral Map in the Service of the State: A History of Property Mapping.* Chicago: University of Chicago Press, 1992.

King, Victor T. *The Peoples of Borneo.* Oxford: Blackwell Publishers, 1993.

Kleinman, P.J., A.D. Pimentel, and R.B. Bryant. 'The ecological sustainability of slash-and-burn agriculture.' *Agriculture, Ecosystems and Environment* 52: 235–49, 1995.

Klyza, Christopher McG. and Stephen C. Trombulak. *The Future of the Northern Forest.* Hanover, New Hampshire: University Press of New England, 1994.

Kozak, Dave. 'Maintaining large forest landholdings in northern New England.' *Appalachia,* June 1989, pp. 22–44.

Lembaga Adat Punan Besar Dayak Punan KalTim. *Misi Dan Visi Masyarakat Adat Punan Dayak,* 1999.

Lewis, M.G. 'Future encounters in new contexts.' In *Cartographic Encounters: Perspectives of Native*

American Mapmaking and Map Use, M.G. Lewis, ed. Chicago: University of Chicago Press, 1998, pp. 273–284.

Li, Tania M. 'Images of community: Discourse and strategy in property relations.' *Development and Change* 27: 501–527, 1996.

Li, Tania M. 'Marginality, power and production: Analyzing upland transformations.' In *Transforming the Indonesian Uplands*, Tania M. Li, ed. Amsterdam: Harwood, 1999, pp. 1–44.

Li, Tania M. 'Articulating indigenous identity in Indonesia: Resource politics and the tribal slot.' *Comparative Studies in Society and History* 42(1): 149–179, 2000.

Lohmann, Lawrence. 'Green Orientalism.' *The Ecologist* 23(6): 202–204, 1993.

Luloff, A.E. and Mark Nord. 'The Forgotten of Northern New England.' In *Uneven Development in Rural America*, T.A. Lyson and W.W. Falk, eds. Lawrence: University Press of Kansas, 1993, pp. 125–167.

Lynch, Owen and Kirk Talbott. *Balancing Acts: Community-Based Forest Management and National Law in Asia and the Pacific*. Washington: World Resources Institute, 1995.

Manandhar, Lanjana. Personal Communication. Kathmandu, October 5, 1997.

Mares, Jose. 'Se remite plan general para la campaña contra incendios de montes.' vol. Legajo 895 *Expediente* 21, p. 6. AGEO: Asuntos Agrarios, Serie V. Problemas por Bosques, 1932.

Momberg, Frank, K. Atok, and Mertua Sirait. *Drawing on Local Knowledge: A Community Mapping Training Manual. Case Studies from Indonesia*. Jakarta: Ford Foundation, 1996.

Moran, Emilio F. *Developing the Amazon*. Bloomington: Indiana University Press, 1981.

Mundy, B.E. *The Mapping of New Spain: Indigenous Cartography and the Maps of the Relaciones Geográphicas*. Chicago: University of Chicago Press, 1996.

Nepstad, D.C., A. Moreira *et al. A Floresta em Chamas: Origins, Impacts e Prevencao de Fogo na Amazonia*. Brasilia: World Bank, 1999.

Nigh, Ronald Byron. *Evolutionary Ecology of Maya Agriculture in Highland Chiapas, Mexico*. Ph.D. dissertation. Palo Alto, California: Stanford University, 1975.

Northern Forest Lands Council. *Finding Common Ground: Conserving the Northern Forest*. Concord, N.H.: Northern Forest Lands Council, 1994.

O'Connor, James. 'Uneven and combined development and ecological crisis: A theoretical introduction.' *Race & Class* 30(3): 1–11, 1989.

Otto, J.S. and N.E. Anderson. 'Slash-and-burn cultivation in the Highlands South: A problem in comparative agricultural history.' *Comparative Study of Society and History* 24: 131–47, 1982.

Oxford English Dictionary. 2nd ed. Oxford: Clarendon Press, 1989.

Peluso, N.L. 'Whose woods are these?: Counter-mapping forest territories in Kalimantan, Indonesia.' *Antipode* 27(4): 383–406, 1995.

Pereira, C. and A. Faleiro. *Proposta de Criacao de uma Linha de Crédito Ambiental para a Amazonia*. Manuscript. Belém: FETAGRI/IPAM, 2000.

Pichón, F.J. 'Settler agriculture and the dynamics of resource allocation in frontier environments.' *Human Ecology* 24(3): 341–371, 1996.

Pickering, Andrew. 'From science as knowledge to science as practice.' In *Science as Practice and Culture*, Andrew Pickering, ed. Chicago: University of Chicago Press, 1992, pp. 1–26.

Poffenberger, Mark and B. McGean, eds. *Communities and Forest Management in East Kalimantan: Pathways to Environmental Stability*. Berkeley: University of California Press, 1993.

Pradhan, Prafulla Man. Personal communication. Kathmandu: September 29 and October 5, 1997.

Pyne, Stephen J. 'Keeper of the flame: A survey of anthropogenic fire.' In *Fire in the Environment: Its Ecological, Climatic, and Atmospheric Chemical Importance*, P.J. Crutzen and J.G. Goldammer, eds. New York: John Wiley, 1993, pp. 245–266.

Pyne, Stephen J. *World Fire: The Culture of Fire on Earth*. New York: Henry Holt, 1995.

Rabindran, Shanti. 'The role of large and small landholders during Indonesia's land fires: A GIS-econometrics analysis of satellite and land use data.' *Council on Southeast Asia Studies Seminar, Yale University*, 6 December 2000.

Rademacher, Anne. *Restoration as Development: Urban Growth, River Restoration, and Riparian Settlements in the Upper Bagmati Basin, Kathmandu, Nepal*. Master's thesis. New Haven, Connecticut: Yale School of Forestry and Environmental Studies, 1998.

Ramos, Fernando. 'En Sierra Juarez, obtienen premios al merito ecologico 14 comunidades.' *El Imparcial*, p. 7B. Oaxaca, 2000.

Rangan, Haripriya. 'Romancing the environment: Popular environmental action in the Garhwal Himalayas.' In *In Defense of Livelihoods: Comparative Studies in Environmental Action*, J. Friedmann and H. Rangan, eds. West Hartford, Connecticut: Kumarian Press, 1992, pp. 155–181.

Redford, Kent H. 'The ecologically noble savage.' *Cultural Survival Quarterly* 15: 46–48, 1990.

Redford, Kent H. and S.E. Sanderson. 'Extracting humans from nature.' *Conservation Biology* 14(5): 1362–1362, 2000.

Reidel, C. 'The political process of the northern forest lands study.' In *The Future of the Northern Forest*, Christopher McG. Klyza and Stephen C. Trombulak, eds. Hanover, New Hampshire: University Press of New England, 1994, pp. 93–111.

Ruthenberg, H. *Farming Systems in the Tropics*, 2nd ed. Oxford: Oxford University Press, 1976.

Sachs, Wolfgang. 'One world.' In *The Development Dictionary: A Guide to Knowledge as Power*, Wolfgang Sachs, ed. London: Zed Books, 1992, pp. 102–115.

Said, Edward W. *Orientalism*. New York: Vintage/Random House, 1978.

Schmink, M. and C.H. Wood. *Contested Frontiers in Amazonia*. New York: Columbia University Press, 1992.

Schwartzberg, J.E. 'Introduction to Southeast Asia cartography.' In *The History of Cartography*. Vol. II, Book I: *Cartography in the Traditional East and Southeast Asian Societies*, J.B. Harley and D. Woodward, eds. Chicago: University of Chicago Press, 1994, pp. 689–700.

Schwartzman, S., A. Moreira, and D.C. Nepstad. 'Rethinking tropical forest conservation: Perils in parks.' *Conservation Biology* 14(5): 1351–1357, 2000.

Scott, James C. *Seeing Like a State: How Certain Schemes to Improve the Human Condition Have Failed*. New Haven, Connecticut: Yale University Press, 1998.

Sigaut, F. 'Swidden cultivation in Europe: A question for tropical anthropologists.' *Social Science Information* 18(4/5): 679–694, 1979.

Simonian, Lane. *Defending the Land of the Jaguar*. Austin: University of Texas Press, 1995.

Sirait, M., S. Prasodjo, N. Podger, A. Flavelle, and J. Fox. 'Mapping customary land in East Kalimantan, Indonesia: A tool for forest management.' *Ambio* 23(7): 411–417, 1994.

Slater, C. 'Justice for whom? Contemporary images of Amazonia.' In *People, Plants, and Justice*, Charles Zerner, ed. New York: Columbia University Press, 2000, pp. 67–82.

Slotkin, Richard. *Regeneration Through Violence: The Mythology of the American Frontier, 1600–1860*. New York: Harper Collins, 1973.

Strommen, Katherine, HMG Ministry of Housing and Physical Planning, and the University of Trondheim. *Planning for Urban Development and Environmental Upgrading: an Environmental Impact Assessment of the Bisnumati Link Road*. Kathmandu: HMG, 1991.

Tanaka, Masako. *Conditions of Low Income Settlements in Kathmandu: Action Research in Squatter Settlements*. Kathmandu: Lumanti, 1997.

Taylor, Peter J. and Frederick H. Buttel. 'How do we know we have environmental problems? Science and the globalization of environmental discourse.' *Geoforum* 23(3): 405–416, 1992.

Terborgh, J. *Requiem for Nature*. Washington, DC: Island Press, 1999.

Thongchai Winichakul. *Siam Mapped: A History of the Geo-body of Siam*. Honolulu: University of Hawaii Press, 1994.

Thrupp, L. A., S. Hecht, and J. Browder. *The Diversity and Dynamics of Shifting Cultivation: Myths, Realities, and Policy Implications*. Washington, D.C.: World Resources Institute, 1997.

Toniolo, A. and C. Uhl. 'Perspectivas Economicas e Ecológicas da Agricultura na Amazonia Oriental.' In *A Evolucao da Fronteira Amazonica: Oportunidades para um Desenvolvimento Sustentável*, O.T. Almeida, ed. Belém: IMAZON, 1996, pp. 67–100.

Tsing, Anna. 'Becoming a tribal elder and other green development fantasies.' In *Transforming the Indonesian Uplands*, Tania M. Li, ed. Amsterdam: Harwood, 1999a, pp. 159–202.

Tsing, Anna. *Notes on Culture and Natural Resource Management*. Environmental Politics Working Paper Series, University of California, Berkeley, 1999b.

Tsing, Anna. 'The global situation.' *Cultural Anthropology* 15(3): 327–360, 2000.

Tura, L.R. 'Notas Introdutórias sobre os Fundos Constitutcionais de Financiamento e sua

Configuração na Região Norte.' In *Campesinato e Estado na Amazônia: Impactos do FNO no Pará*, L.R. Tura and F.d.A. Costa, eds. Brasília: Brasília Jurídica, 2000, pp. 29–46.

Tyrtania, Leonardo. *Yagavila un Ensayo en Ecologia Cultural*, 1st ed. Mexico: D.F. Universidad Autonoma Metropolitana, Unidad Iztapalapa, Division de Ciencias Sociales y Humanidades, 1992.

United Nations Population Fund. *The State of World Population*. New York: UNFPA, 1995.

United Nations. *Habitat Agenda and Istanbul Declaration on Human Settlements Summary: Road Map to the Future*. United Nations Department of Public Information. DPI/1846/HAB/CON, 1996.

Urry, John. *The Tourist Gaze: Travel and Leisure in Contemporary Societies*. London: Sage, 1990.

Valentin, Karen. *Urban Squatters in Kathmandu: Perspectives on Hierarchy and Heterogeneity*. M.A. thesis, University of Copenhagen, 1995.

Various. 'Los vecinos de Ixtlan de Juarez denuncian que los rancheros de Rancho Chivo han provocado un incendio.' vol. 900 *Expediente* 18, p. 9. AGEO: Asuntos Agrarios, Serie V. Problemas por Bosques, 1942.

Various. 'En que los comuneros de Xiacui denuncian el incendio de los bosques por los habitantes de La Trinidad, Ixtlan.' vol. 900, Exp 19, p. 9. AGEO: Asuntos Agrarios, Serie V. Problemas por Bosques, 1945.

Wallerstein, Immanuel. *Historical Capitalism*. London: Verso, 1993.

Weimarck, G. *Ulfshult: Invesitgations Concerning the Use of Soil and Forest in Ulfshult, Parish of Örkened, During the Last 250 Years*. Lund, Sweden: C.W.K. Gleerup, 1968.

Wolf, Eric. 'Closed corporate peasant communities in Mesoamerica and Central Java.' *Southwestern Journal of Anthropology* 131–18, 1957.

Wolf, Eric. 'The vicissitudes of the closed corporate community.' *American Ethnologist* 13: 325–329, 1986.

World Resources Institute. *The Urban Environment*. New York: Oxford University Press, 1996.

Yami, Hisla and Stephen Mikesell. *The Issues of Squatter Settlements in Nepal: Proceedings of a National Seminar*. Kathmandu: Concerned Citizens Group of Nepal, 1990.

Zerner, Charles. 'Telling stories about biological diversity.' In *Valuing Local Knowledge: Indigenous People and Intellectual Property Rights*, S.B. Brush and D. Stabinsky, eds. Washington, DC: Island Press, 1996, pp. 68–101.

ROY ELLEN

VARIATION AND UNIFORMITY IN THE CONSTRUCTION OF BIOLOGICAL KNOWLEDGE ACROSS CULTURES

This chapter examines the extent to which knowledge of biological entities and processes varies according to different human life experiences and cultural traditions.[1] It attempts to relate this to global, transmodern, scientific biology, with its origins in Western cultural history. What connects the first with the second is the increasingly well-documented recognition that all peoples share a basic way of apprehending the natural world, grounded in a common evolutionary history, even though this cognitive underpinning is everywhere filtered through the local particularities of environmental and cultural experience. Such a shared infrastructure of perception and cognition has been termed "natural history intelligence" and is linked to modular theories of the mind. What this means usually includes (1) a shared concept of basic natural kind (a species-like concept) reflecting a view of the biological world as a series of discontinuous entities; (2) an ability to recognise and respond to things as living matter, and more specifically an "algorithm for animacy" (Bulmer, 1970; Reed, 1988; Atran, 1998; Ellen, 1996; Boster, 1996); (3) a capacity to intuit certain kinds of behaviour based on expectations derived in part from common experiences linked to phylogenetic similarities or observations of human behaviour, and (4) strategies for classifying biological diversity (Atran, 1990; Boster, 1996; Keil, 1994; Mithen, 1996: 52–54). Because none of this is accessible other than through its local cultural versions, distinguishing what are shared human universals from what are simply culturally widespread is problematic. This has given rise to some lively debates.

DIFFERENT WAYS IN WHICH KNOWLEDGE IS CULTURALLY EMBEDDED

Although underlying cognitive strategies influence how people construct what they know about the biological world, most knowledge is culturally transmitted and shaped by environmental and social forces which vary from place to place. In part, what people know is constrained by local ecology, although what is uniquely human is the capacity for acquired biological knowledge to diffuse independently of what can be experienced in local habitats. Thus, people may

H. Selin (ed.), Nature Across Cultures: Views of Nature and the Environment in Non-Western Cultures, 47–74.
© 2003 Kluwer Academic Publishers. Printed in Great Britain.

have concepts for snakes, even if they have never seen one. Scientific biology is, in one sense, an extreme development of such an intuitive biology, augmented by the possibilities offered by effective cultural transmission, since the capacity to generalise and hypothesise is grounded in the way science aggregates knowledge of species and ecologies beyond what a scientist might have local first-hand experience of as a non-scientist.

Before going further it is useful to reflect on the relationship of culture to knowledge, knowledge to intelligence and on different kinds of knowledge. This has become necessary because of recent developments in anthropology and cognitive science. The default understanding of knowledge, at least in anthropology, is usually of what we might call "conscious", "cognised" or "reflective" knowledge: something we are aware of acquiring and using, and often do so purposefully in order to solve various technical and social problems. However, people also acquire knowledge unobtrusively and unreflectively as part of the process of socialisation and growing up. This is no less knowledge than that which we consciously articulate or recognise. One example of this kind of knowledge is "bodily knowledge" – knowledge acquired and stored as part of doing and recognising in particular practical contexts. An example is learning how to harvest rice with a Javanese finger knife, which requires sensory and motor skills which are often readily transmitted across generations but which are not explicitly formulated into a set of rules. Such techniques are, rather, acquired through mimicry, experience and informal apprenticeship. Much knowledge of the first (cognitive) kind is clearly encoded in language; in other words it is "lexical knowledge" (such as in plant and animal nomenclatures), and where this yields regularities in how people relate different living kinds, it translates into "classificatory knowledge". However, much knowledge, particularly of natural processes, is only partially lexically expressed. Where classificatory knowledge generates categories with no lexical markers, these are termed "covert categories" (Taylor, 1990: 42–51), but where knowledge is manifestly evident although not necessarily systematically expressed in language, we might speak of "substantive knowledge" (Ellen, 1999). Most knowledge of the biological world is substantive in this sense and classifications can be understood as codes to access and manipulate it.

There is another way of looking at the knowledge people have of objects and processes in their environment: not in terms of how they engage with nature, or the degree to which that engagement is encoded in cultural representations, but in terms of its division into empirically organised areas of substantive knowledge (the so-called "ethnosciences": ethnobotanical (plant) knowledge, ethnozoological (animal) knowledge, ethnoanatomical knowledge, ethnoveterinary knowledge and so on). Although people themselves may seem to divide their knowledge of the natural world in this way, this approach – displaying the bias of encyclopaedic, literary-based theoretical knowledge – is best reflected in the conventional partitioning of Western science, which in turn has influenced the development of ethnobiology. One of the great problems in researching how other people understand their biological worlds is ensuring that these conventional etic[2] divisions are not imposed on the subjects of our research. It is true that this framework for looking at pragmatic knowledges of biological form and process is helpful when seeking to make inventories of what people know about individual species or varieties. However, from the point of view of comparative study it is probably more useful to distinguish several kinds of knowledge organisation, irrespective of the type of organism or uses involved. Such an approach distinguishes (a) classificatory knowledge from (b) knowledge of anatomy, autoecology and processes with respect to individual organisms, or groups of organisms; from (c) knowledge of ecological systems (synecology: plant interaction, dynamics of various kinds of landscape, seasonality, food chains, pest ecology); and from (d) knowledge of the general principles of plant and animal biology. In the past research on local "folk" knowledge tended to emphasise the first of these (predominantly, the classification of macro-organism diversity), although increasingly it has become apparent that the application of insights from the second three may more than compensate for detailed knowledge of the first. However, how all this ethnoecological knowledge connects up into some larger whole presents considerable analytical difficulties, since it is less easy to disaggregate in local emic terms, partly because it is characteristically intermeshed with symbolic and aesthetic representations.

Another problem in studying biological knowledge cross-culturally is knowing to what extent we can generalise about the knowledge of particular populations, or indeed of societies or cultures. Knowledge is distributed geographically between populations, and it is also important to distinguish levels within the same population. Not all persons are equally expert, and important knowledge is always disseminated through social networks. For example, there are now some excellent demonstrations of the mechanisms which transmit genetic variability in *Manihot esculenta* (manioc, cassava, tapioca) amongst Aguaruna (Boster, 1986) and Guyanase Makushi women (Elias, Rival and McKey, 2000). Commonly applied knowledge, shared by all the members of the community, needs to be distinguished from more specialised knowledge shared by only one category of users. An example of one extremity of such a distribution is that of individual healers, where knowledge is hidden, secret, and transmitted to very few people. Important practical questions arise as to which of these – the individualised or the shared – are the most significant, or indeed what we mean by "significance", since this can be measured along a number of different (indeed, contrasting) axes (say, ecological versus social) and especially when it is evident that knowledge is dynamic and changing. Many descriptions of ethnobiological knowledge tend to aggregate knowledge obtained from different individuals in an unweighted fashion, or present the knowledge of a few individuals as if it were that of the entire population. When this methodological relationship between aggregated data and inference is transparent and its limitations understood, it can be described as the "omniscient speaker-hearer convention", but when the relationship is obviously misunderstood and abused through the drawing of false inferences, then we might speak of the "omniscient speaker-hearer fallacy" (Berlin, Breedlove and Raven, 1974: 58–59; Gardner, 1976).

But however we divide up different kinds of knowledge, they must always be understood in a broader context, both in terms of other kinds of knowledge and in terms of the context of social relations. Local knowledge of environmental resources is socially embedded, and only under very special conditions can it become modular, free-floating and transferable. It has become conventional to distinguish symbolic from technical (mundane) knowledge, following Durkheim and Mauss (1901). This distinction overlaps, although is not entirely equivalent to, the distinction between knowledge (as an abstract body of principles) and know-how (applied practice), or Geertz's (1966) "models of" and "models for". This convergence of cognitive and symbolic anthropology (Colby, Fernandez and Kronenfeld, 1980; Ohnuki-Tierney, 1981) is easy to understand when one realises that all human populations apprehend the social in terms of the natural world and the natural in terms of metaphors drawn from the social world. The two are intrinsically complementary, although in certain neurological pathologies they may conflate in unusual ways (as in varieties of autism); other kinds of confusion between the two may be perceived as culturally deviant. The classificatory language we use for plants and animals is derived from the way we talk about genealogical relations, and we understand the functional dynamics of both organisms and ecological systems in terms of our experience of participating in social systems, where technology provides numerous productive analogies: say, the heart as a pump, the blood vascular system as a thermostat or the brain as a computer. More generally people attribute meaning to parts of the natural world around them by investing them with human and spiritually anthropic qualities (animism). Increasingly, historical and cultural studies of scientific practices and thought are revealing this tendency.

Anthropologists, however, have had much more to say of the natural world as a source of symbols (Bulmer, 1979; Fox, 1971; Rosaldo, 1972; Rosaldo and Atkinson, 1975; Rival, 1998). For example, they have discussed how natural species are used to signify group difference (totemism), or why certain species should be used as symbolic reflections of fixed moral orders, while others should be prohibited (Douglas, 1966). They ask why certain species should be selected as symbols because they have properties which make them "good to think" with rather than necessarily being "good to eat" (Bulmer, 1967; Leach, 1964; Tambiah, 1969). On the whole, animals provide more, and more salient, primary symbols than plants, perhaps for anthropomorphic reasons. In all human populations some species or group of species predominate as symbols. This can be because they are not only economically important, such as zebu cattle (*Bos indicus*) amongst the pastoralist Bodi of southern Ethiopia (Fukui, 1996) or the palm *Borassus flabellifer* amongst the Rotinese of eastern Indonesia (Fox, 1977), but also often because of the visual characteristics and metaphoric possibilities particular species present. Examples are bowerbirds or birds of paradise in the highlands of Papua New Guinea (Healey, 1993; Hirsch, 1987), the powerful social and sexual imagery of the mudyi tree (*Diplorrhyncus condylocarpon*) amongst the Ndembu of Zambia (Turner, 1967), or the contrasting imagery of grains versus roots or trees versus lianas. Occasionally, symbolically salient organisms are those which have medicinal, including psychoactive, properties, such as the betel palm, *Areca catechu*, in the case of the Nuaulu of Seram in the Moluccan islands (Ellen, 1991), or ayahuasca (*Banisteriopsis caapi*) amongst the Jívaro cluster of the Amazonian foothills of Ecuador and northern Peru (Metzner, 1999).

CLASSIFICATORY KNOWLEDGE

The study of ethnobiological classification, or folk classifications of plants and animals, has tended to dominate anthropological approaches to the understanding of biological knowledge across cultures. This is partly historical, but it is also because understanding local classifications provides an essential framework for talking about other kinds of knowledge. Understanding folk classification is important for numerous reasons: (1) fieldwork necessitates learning the terms and concepts through which local people deal with the biological world, (2) direct translation into scientific nomenclature is not always possible, (3) scientific and vernacular categories do not always match, (4) names and categories provide important ethnobiological information, and (5) local processes of decision-making and environmental management can only make sense with respect to the categories employed by decision-makers. Additionally, irrespective of the practical role of classificatory approaches, they provide important data for cognitive and linguistic studies. But knowledge of folk classification is of limited value without scientific determinations, which permit proper cross-cultural comparison, generalisation and identification, including the degree to which folk categories deviate from scientific taxa and from each other. Phylogenetic classification therefore serves as a baseline and framework for analysis. For example, Nuaulu attach the label **sinsinte** to all kinds of *Codiaeum variegatum* (croton), a polychromatic waxy-leaved shrub with important symbolic qualities. However, they distinguish several sub-types: **sinsin totu onate, sinsin totu nawe, sinsin amasen, sinsin totu pukune, sinsin msinae,** and **sinsin matapai**. Since these are represented as species-like groupings, we can say that they "over-differentiate" the category. By contrast, earthworms, which they label **tumanai**, regardless of family or genus, are by the standards of Linnean classification, seriously "under-differentiated".

Basic organisation of classifying behaviour: words and categories

Most studies of ethnobiological classification approach the subject linguistically, because most data acquired in fieldwork settings are generated through interviews and by hearing people talk about wildlife, because this is how most people themselves share classificatory knowledge, and because many classificatory strategies are revealed through language. However, it has long been recognised that words are not always a good guide to the existence of categories: there may be several words which label the same category (synonyms), and the same word can be used for quite different organisms. Moreover, some categories may exist without being labelled.

The nomenclature for labelling categories tells us something both about

classificatory knowledge and also about the attributes which people find important in distinguishing different plants and animals. Most languages label plants
and animals below the "basic level" with some variant of the binomial system:
that is two terms, the first indicating a more inclusive category and the second
a less inclusive category, the two being linked by a "kind of" relationship.
Thus, for Nuaulu **sinsin msinae**, "red **sinsinte**", is a binomial. In this case the
more inclusive category is identified not only by its priority, but because it has
been lexically reduced: thus **sinsinte** becomes **sinsin**. Local linguistic conventions
have to be carefully observed, and it is important to note, for example, that
tobako sinsinte and **kasipi sinsinte** are not kinds of **sinsinte**, but are, respectively,
kinds of tobacco, *Nicotiana tabacum*, and manioc, *Manihot esculenta*. In this
linguistic context **sinsinte** becomes, instead, an adjectival qualifier. The kinds
of adjectival qualifiers used vary, from descriptions of diagnostic visual attributes, uses and smells to sounds. Birds and frogs, for example, are dispropor-
tionately distinguished using onomatopoeic references to their call (Berlin,
1992: 232–259). There have been several attempts to develop a typology of
lexemes to allow accurate description of ethnobiological nomenclatures
(Conklin, 1962; Berlin, Breedlove and Raven, 1973).

The structure of categories

Early attempts to understand how ethnobiological categories are established
and used employed a distinctive feature model, in which category A was
thought to be distinguished from category B in terms of a number of key
distinctive features. For example, birds have wings, feathers, beaks and fly, in
contrast to fish, which swim and have fins. This model was largely drawn from
lexicography and logic (Conklin, 1962). However, it was noted that the condition of contrast required for this model to work was not always evident. Thus,
category A might be linked to category B by one common attribute, and
category B linked to category C through a different common attribute, thus
linking categories A and C even though they had nothing in common: this is
known as "polythetic classification" (Ellen, 1979: 11–12). As work on ethnobiological classification expanded it became obvious that the digital distinctive
feature model was inadequate, and that a better way of modelling the cognition
of basic and more inclusive categories might be in analog terms, as cognitive
prototypes. In this model the brain has an image of, say, "birdness" or "treeness"
to which incoming perceptual images are matched; the presence or absence of
particular features is not an overriding consideration, only closeness of match
(Rosch, 1977). In this core-periphery model an image could be a close match
or a marginal match. Thus, in British English classification of birds, a robin
would match closely the core prototype, but an ostrich would be marginal. Of
course, in practice, both the notion of contrasting features and cognitive prototypes are necessary to understand how classifications work.

The relations between categories

It has become conventional, following the analytical procedures of cognitive
anthropology, to begin any analysis of classificatory knowledge of natural

entities by establishing a cognitive or semantic domain or field (Frake, 1969). The domain in question can be established at varying degrees of classificatory inclusiveness: thus it might de determined as "all living things", "plants", "trees" or "rice", depending on the focus of the analysis. Although the domain may be isolated for analytical reasons, and is to this extent arbitrary, its boundaries are generally understood to reflect distinctions which are empirically important for the population who share them. Thus, if a population has no concept of "tree", then such a category cannot be established as a cognitive domain. On the other hand, as we have noted, categories do not need to be labelled in order to exist, even at the level of domain. Where a cognitive domain has been established, it is usually understood that most categories which sub-divide it will be labelled, and a domain or field identified in terms of its labels is usually known as a "lexical field". Of course, the lexical field for plants may not correspond to the cognitive domain, because of the existence of covert categories at various levels of inclusiveness.

The earliest work on cognitive domains modelled the internal sub-divisions of a domain largely in terms of the taxonomic model: that is, in terms of a hierarchical model of contrast and class inclusion. This is partly because this form of classifying is so dominant in the literary and scientific tradition of the West and particularly because of the precedent of Linnean taxonomy. The work of Brent Berlin (1972, 1992; and Berlin with Breedlove and Raven, 1973 and 1974) developed the taxonomic idea further, putting forward a strong claim for it to be considered the general way in which ethnobiological classification works cross-culturally, hypothesising that a series of taxonomic ranks could be established, broadly reflected in the main ranks of the Linnean scheme: unique beginners, life forms, intermediates, generics, specifics and varietals (Figure 1). These terms have been widely adopted and are a useful way of structuring a discussion of variation in classificatory knowledge of the natural world. Other writers have preferred different terms (in part to avoid confusion with Linnean nomenclature) and these are also indicated in the figure.

"Unique beginners" define a cognitive domain. Thus concepts such as plant and animal are such, even where no labels exist. Many languages have no word for either of these, even though there is linguistic and non-linguistic evidence to confirm that people have the category. "Life forms" are the second rank, usually few in number and including almost all other categories in the domain. Thus, bird, tree and fish are obvious life forms, and their cognates exist in most languages. However, a problem arises because in some languages many life forms can be identified, in the sense that they themselves are not members of any other more inclusive group which might be called a life form, and which are seen directly as sub-categories of a particular unique beginner. Thus, frogs, bats, and bamboo, in some folk classifications, may contrast with other more salient life forms. This is a difficulty for the Berlin scheme, and to ensure that the life forms remain "few in number" they are reassigned to the category "unaffiliated generic". Most categories in a folk classification system are either generics or specifics, to use Berlin's terms, and these terms suggest strong correspondence with the Linnean ranks of genus and species. The lowest rank

Unique Beginner   Kingdom

Life form   Phylum and Class

(Intermediate)   Order and Family

Primary category   Generic   Genus

Secondary category   Specific   Species

(Tertiary category)   Varietal   Variety

Terminal category

Bulmer 1970   Berlin 1972, 1992   Major ranks in
Ellen 1993   classical Linnean
                                               taxonomy

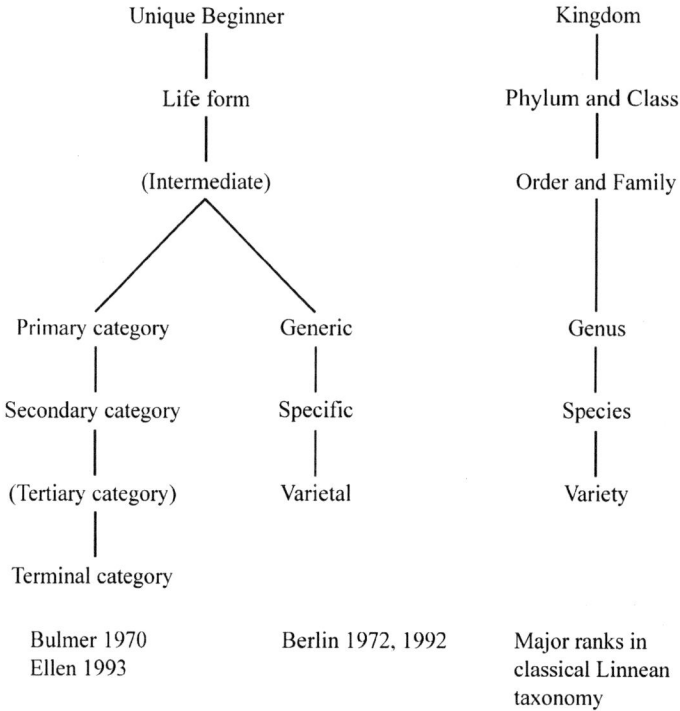

Figure 1   Some commonly used terms for classificatory ranks found in ethnobiological studies.

in the Berlin scheme is "varietal", which is mainly of significance when dealing with domesticates, and we shall return to it in that context. Finally, Berlin uses the term "intermediate" to refer to a rank between life form and generic. In the empirical examples with which he deals, this rank is weakly developed, although it is approximately equivalent to family in the Linnean scheme, where it is much more significant: e.g. rose-family versus ginger-family, felids versus canids, and so on. This may be because groups of this kind are not best perceived locally where biodiversity is limited but become apparent when many natural kinds are grouped together at a regional or global level.

There are difficulties with Berlin's universalist-evolutionist model: it assumes too rigid a notion of contrast in delineating categories, the levels are difficult to sustain given the way we know individuals use classificatory information, there is less hierarchical depth in practice than anthropological representations of aggregate folk knowledge imply, and although the mind appears to generate readily the notion of "basic category" (natural kind), its identification with the generic level is not always easy. Also problematic is the centrality given to general-purpose schemes and their purported cognitive independence of the specific functions to which classifications are put, as well as the underplaying of the role of variation, not only between people but between different occasions. These two latter issues will be returned to later. In response to these problems, critics have emphasised classificatory alternatives to taxonomy (e.g., paradigms, keys, fuzzy logics), and have distinguished between ways of representing the relations between organisms and storing knowledge (Ellen, 1993: 215–234; Hunn and French, 1984; Randall, 1976).

HOW ETHNOBIOLOGICAL CATEGORIES AND KNOWLEDGE VARY, CHANGE AND EVOLVE

Many early studies of ethnobiological classification paid relatively little attention to variation within a single population, sometimes providing a misleadingly uniform picture. Work over the last 30 years has identified the different ways in which knowledge varies within a population (as between, say, consistency, flexibility and sharing), how it is socially distributed according to age, gender, locality and division of labour and how it may be presented differently according to mundane versus symbolic schemes, or general-purpose versus special purpose schemes.

Berlin (Berlin, Breedlove and Raven, 1966) institutionalised the distinction between what we call "general-purpose" and "special-purpose" classificatory schemes. By the former Berlin indicated classifications based mainly on visual morphological and behavioural features, which he hypothesised were likely to have a high degree of constancy across cultures. By special-purpose functional classifications he indicated classifications which were related to the various uses to which fauna and flora might be put. Thus, classifications of animals as meat, or plants as medicinals, were special-purpose in this sense. The distinction is of major theoretical significance, because if it can be shown empirically to hold true then it provides the main grounds for supporting certain universalist and evolutionary characteristics of human cognition of the natural world (see above). Beyond this, however, the distinction enables us to examine how knowledge of the biological world is embedded in cultural knowledge, is stored and retrieved. Thus, even if it can be shown that there is an underlying universal general-purpose classificatory tendency, there can be little doubt that general-purpose schema are inadequate to understand all people know about individual species, much being embedded in particular special-purpose domains or discourses, such as medical knowledge.

Some progress has also been made in understanding how ethnobiological classification varies between different kinds of human population. It is now accepted, for example, that how much people know (especially as this is reflected in the names they supply for kinds of organisms) is strongly correlated to biodiversity. It has also been suggested that hunter-gatherers generally have less extensive nomenclatures for plants and animals, are less likely to use binomials and adopt more flexible classifications. Agricultural populations tend to encode classificatory information more systematically and lexically, perhaps because of larger group sizes and because of greater social sharing of knowledge in the course of production (Morris, 1976; Ellen, 1999). Where domesticates are clearly established and important it is usual to find extensive and complex varietal level classifications reflecting genetic diversity (e.g. Brush, 1992; Iskandar and Ellen, 1999). Farmers may possess a more extensive formal knowledge of forest products than foragers, which enables them to cope with the greater subsistence risks associated with agriculture.

Variation in people's classificatory knowledge is often the first stage of a process of change (Barrau, 1979; Nabhan and Rea, 1987). Short-term change and flexibility often arise through the semantic extension of categories to include new natural kinds and the assigning of low profile entities to residual categories. Where new categories are recognised these are indicated in language by marking. Thus, the introduction of *Lycopersicum esculentum* (tomato) into France in the seventeenth century was accompanied by the formation of the marked term **pomme d'amour** (love apple), contrasting it with the unmarked **pomme** (apple). Similarly, British English oak was qualified as **turkey oak**, to describe *Quercus cerris*, a non-native naturalised species from southern Europe and Asia Minor. In turn, the native *Quercus robur* was optionally then renamed the "common" **oak**, to effect the adjectival contrast. Sometimes a name moves with a plant. Thus, the Nuaulu for tomato is **tamati**, probably from the Dutch **tomaat**. Other new names are based on assumed place of origin, as in Ambonese Malay for *Manihot esculenta*: **ubi kastella** (Castilian/Spanish tuber/yam).

Work on long-term evolutionary changes in folk-biological classification has demonstrated how basic categories aggregate and segregate into folk biological ranks, the order in which they do so and, in particular, on how life forms evolve (Berlin, 1972). Brown (1984) has suggested that there is a regular sequence in which life form terms are added to language. This in part reflects basic shifts in subsistence behaviour and social organization, such as from gathering and hunting to agriculture, from minimal to elaborate divisions of labour, from non-centralised societies to states, from preindustrial to industrial economies, and from oral to literate traditions. What is contested is the extent to which there is a unilinear progression in life form encoding rather than a widespread evolutionary convergence in the way uses for plants and animals reflect this (Randall and Hunn, 1984). As has been noted, the numbers of plants named in agricultural societies is, for example, systematically larger than in non-agricultural societies, while more all-encompassing labels (life forms, intermediates, and so on) become more important, while basic level names become relatively less important in post-agricultural societies. The twin processes of domestication and cultivation give rise to observable differences between populations in their classificatory knowledge. Differences between wild and domesticated forms of the same natural kind are often lexically expressed (Nabhan and Rea, 1987), sometimes marking genetic differences, but sometimes simply location – e.g., that plants are grown in fields rather than in the forest. Indeed, in many societies the boundary between wild and domesticated, cultivated and non-cultivated is very fuzzy, especially in swiddening and agroforestry systems, where it seems more accurate to speak of degrees of management (the balance between simple extraction and purposeful or inadvertent regulation). This fuzziness is deliberately used in some indigenous management systems, which actively foster and exploit the interbreeding of wild and cultivated stock, both plant cultivars and animals. Recent exemplary discussions of the former include studies of *Manihot esculenta* amongst the Makushi in Guyana (Elias,

Rival and McKey, 2000) and *Ensete ventricosum* amongst the Ethiopian Ari (Shigeta, 1996).

KNOWLEDGE OF INDIVIDUAL ORGANISMS AND GROUPS OF ORGANISMS

There are now numerous studies of the substantive ethnoecological knowledge of different populations, and it would be impossible in this chapter to summarise them adequately. Any summary is likely to grossly underestimate what local people know. Here I refer to a few illustrative studies, but there are other useful collections (Johannes, 1989; Inglis, 1993; Nazarea, 1999; Williams and Baines, 1993) and monographical studies of the knowledge of particular peoples (e.g. Felgar and Moser, 1985; Friedberg, 1990; Heinz and Maguire, 1974; Hunn, 1990; Kocher-Schmid, 1991; Revel, 1990).

The classificatory knowledge employed to aggregate and segregate categories of different natural kinds is synthetic, meaning that the logic involved is inductive, moving outwards from the basic categories established for groups of discrete living creatures. This is in contrast to the analytic logic involved in understanding the relationship of different parts of organisms, where the subject starts with a single physical specimen and deductively classifies its parts. "Organism partonymy"[3] is at the basis of human understanding of how individual kinds of organisms appear, grow and reproduce. How much people know about the anatomy of a particular organism is closely related to how much people need to know to take advantage of its usefulness, although even the management of honey-producing bees and wasps amongst the Brazilian Kayapó hardly seems sufficient to explain the detail of their systematic nomenclature of parts of the head exoskeleton (Posey and Camargo, 1985). In some cases the use of particular partonyms (say words used for infloresence [flowering parts]) may provide clues as to how local people recognise different groups of organisms.

The densest knowledge of individual natural kinds is that which people have of domesticates and organisms which they husband: Peruvian Quechua potato knowledge, knowledge of rice amongst the Baduy in upland West Java, or Etoro knowledge of pigs in the New Guinea highland fringes (Kelly, 1988). This knowledge is often best reflected in local recognition of sub-specific genetic diversity, such as for the major starch staples (Boster, 1986; Brush, 1992; Iskandar and Ellen, 1999), in cultivation and management strategies, in folk genetics (Fukui, 1996), and what people know about feeding, growth, reproduction and behaviour, such as Baka perceptions of the growth cycles of Dioscorid yams (Dounias, 1993: 625). Similarly, Kayapó (Posey and Camargo, 1985) have an intimate knowledge of arthropod ontogenetic stages, the entrance tubes to Meloponinae (stingless bee) nests (Figure 2a), the internal structure of the nests (Figure 2b), and the relationship between nest structure and habitat niche. Such understandings of reproductive biology often feed into traditional strategies for conservation, as demonstrated by Johannes (1978) for marine resources in the Pacific, while detailed and perceptive knowledge of behaviour of large mammals is linked to the needs of human predation, such as amongst the !Kung of the Botswanan Kalahari (Blurton-Jones and Konner, 1976).

ABU

ME-Ê-KRÊ

NHIÊNH-DJÀ

NHUM-Ê-KRÊ

APYNHKRA-DJÀ

EIJKWA

KRA-KUNI

EIJKWA-KRÊ-KRÊ

KUPUDJÀ

KRA-KU-PU-DJÀ

PĪ-Ā-A-RI-A-DJÀ

KURORO

ABU-KRÊ-KRYRE

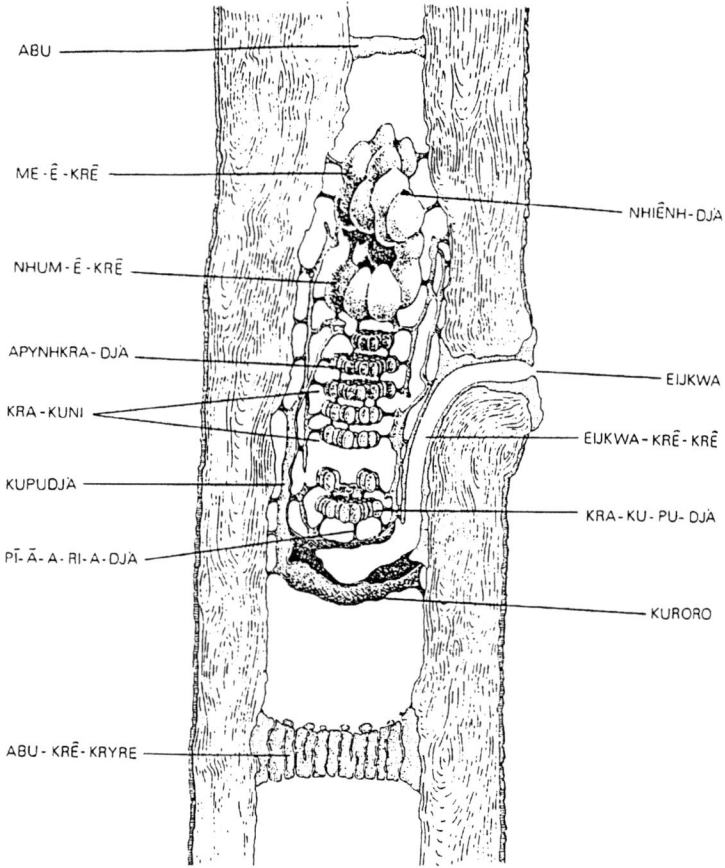

Figure 2a Schematic structures of Melipona nests with Kayapó nomenclature: *abu* (batumen), *me-ê-krê* (honey pot), *nhum-ê-krê* (pollen pot), *apynh kra-djà* (brood cell), *kra kuni* (brood comb), *kupu-djà* (involucrum), *pi-ā-ari-a-djà* (pillar), *abu-krê-kryre* (lower batumen with drainage channels), *nhiênh-djà* (pot opening), *eijkwa* (entrance structure), *eijkwa-krê-krê* (entrance gallery), *kra-ku-pu-djà* (cocoon), *kuroro* (shell of nest). From Posey and Camargo, 1985: 253.

Indeed, it is now well known that traditional peoples have many mechanisms for the protection, regulation and sustainable production of natural resources. Often these are reinforced by, or are part of, general ritual prohibitions. Some environmentalist literature has made improbable and untested claims for this knowledge, linking it to over-romanticised notions of traditional wisdom and edenic ecological harmony. While unsupported claims should be treated with caution, many practices do serve as effective and useful regulators, depending on knowledge of reproductive cycles or animal population dynamics of individual species in order to best determine closed seasons for harvesting or prohibitions on particular areas, populations, or species (Zerner, 1994).

Ethnobiological knowledge shows just how difficult it is to separate knowl-

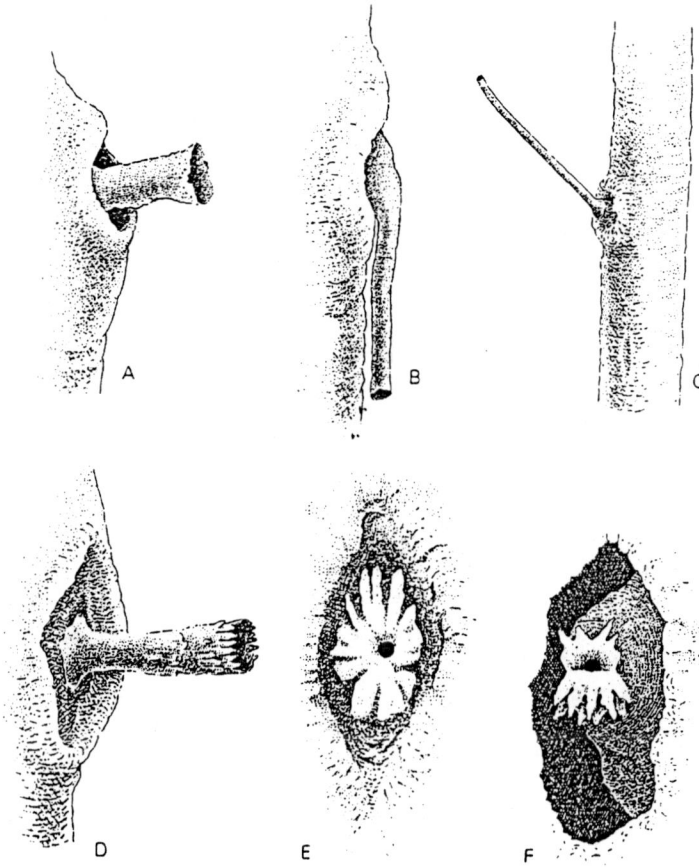

Figure 2b   Types of Meliponinae entrance tubes recognized by the Kayapó with their respective "focal species": (A) *imrê-ñy-kamrek* (*Scaptotrigona nigrohirta*), (B) *imrê-ti* (*S. polystica*), (C) * õ-i* (*Tetragona truncata*), (D) *udjy* (*Melipona seminigra pernigra*), (E) *menhire-udjà* (*M. melanoventer*), (F) *ngài-kumrenx* (*M. rufiventris flavolineata*). From Posey and Camargo, 1985: 254.

edge of one species from that of another, especially where relations of parasitism, symbiosis and mutualism are involved. Crop diseases are a case in point. Thus, in a classic study, Page and Richards (1977) have shown that in seeking solutions to *Zonocerus variagatus* infestation of manioc, Nigerian farmers accumulated a detailed knowledge of the life cycle of this pest. Similarly, using knowledge of the role of ants in the biology of semi-domesticated yams acquired by Baka Pygmies in southern Cameroon, McKey, *et al.* (1998) have discovered that several wild yam species of the forest understorey have complex biotic defences involving the production of nectar rich in amino-acids and sugar during its growth phase, which is highly attractive to ants. However, the presence of the ants also protects the apex of the new growing stem from attacks by herbivorous insects. These observations of mutualistic interrelation-

ships between yams and ants open up a new perspective on our understanding
of vine growth and the role of starch-rich reserves stored underground by tuber
plants, with concrete applications for pest control. Thus, to understand disease
and to diagnose it competently is to a large extent to understand the life cycle,
ecology and manifestations of pathogenic organisms (Whiteford, 1997), while
to understand the effects of medicinal plants involves understanding their
physiology, ecology and anatomy, where the best opportunities are usually
afforded by agricultural settings (Logan and Dixon, 1994).

KNOWLEDGE OF ECOLOGICAL SYSTEMS

A common feature of ethnobiological knowledge is the way in which knowledge
is structured in terms of networks of understanding, linking individual species
together in living contexts and entire landscapes, in contrast to formal science
in the West which historically reified the species and species-centred approach
to understanding early. This body of knowledge is sometimes called "ethnoecology", and may be systematically reflected in local classifications of vegetation
types and understandings of ecological relationships, knowledge of soil, topog-
raphies, environments which are often knowingly or inadvertently created and
maintained by humans, through the management of long fallows, and soil
restoration through use of additives and irrigation (Ellen, 1982: 211–226;
Johnson, 1974; Sillitoe, 1996; Lansing, 1991). More disconcerting is the way in
which these systemic understandings are closely interwoven with symbolic
constructions of the world (Hughes, 1983; Nelson, 1983), which has occasionally
led to their scientific credentials' being questioned (Diamond, 1987) by some
and elevated (Johannes, 1987) by others.
    In the same way as knowledge of individual organisms is closely linked to
experience through domestication and husbandry, so knowledge of ecological
systems arises through the requirement to manage resources. Increasingly, all
major environments which people inhabit are being reinterpreted as having
co-evolved with people and been managed consciously or inadvertently.
Kayapó (Posey, 1988: 89–90), for example, maintain buffer zones between
gardens and forests which contain plants with nectar-producing glands on their
foliage which inhibit aggressive ants and parasitic wasps from crops. They also
modify savannas by fire and by creating forest islands with concentrations of
useful plants. Indigenous forest-fallow cultivation and arboricultural practices
throughout the tropics have repeatedly been shown to maintain forest rather
than destroy it (Conklin, 1954; Balée, 1989; Dove, 1983; Fairhead and Leach,
1996), amplifying its diversity through the transmission of germplasm from
elsewhere, as well as the density of useful plants and animals. Stéphanie Carrière
(1999) has shown how Ntumu in southern Cameroun preferentially spare useful
trees and those which are characteristic of old secondary forests. Such practices
increase the number of these species over time and enhance the value of the
forest. Associations between trees and crops – now sometimes called "agroforestry systems" – have been shown to reduce the risks of declining soil fertility in
the face of increasing population pressure and to contribute to the regeneration

of fallow and mature forest. Carrière also shows that Ntumu understand the ecological principles upon which these strategies are based. Similarly, Laden (1993) and Ichikawa (1999) have illustrated how the density of species supplying non-timber forest products is higher along Congolese Mbuti trails in the Ituri than in unvisited forest. Local peoples have often had a long-term impact in creating distinctive patterns of biotopes.

This kind of systemic knowledge differs from biological science in emphasising long-term processes, including cyclical environmental change. In a few cases knowledge of irregular reproductive patterns has been exploited as a useful strategy in times of hardship. Dove and Kammen (1997) show how forest-dwelling peoples of Borneo understand the dynamics of mast fruiting of dipterocarps [a family of trees used for timber], triggered by slight climate fluctuation, in places attributable to the El Niño Southern Oscillation. These events are irregular and local, but result in the mass flowering and then fruiting of different dipterocarp species, which provides a windfall source of food for humans through direct consumption, the marketing of edible nuts, and indirectly through the additional food released for game animals upon which humans are dependent. [Editor's note: see the article, Central Andean Views of Nature and the Environment, by David Browman in this volume.] In the language of sustainability, the value of such long-term though irregular sources of food, which supplement normal subsistence practices, are greater than short-term timber extraction which destroys the possibility of the mast altogether. Comparing the extent to which knowledge is actually used may provide one measure of the danger of extinction of local knowledge. However, much (perhaps most) ethnoecological knowledge has only occasional and long-term adaptive advantages. Consequently, if knowledge and actual resources are allowed to erode because of perceptions of their short-term unimportance, this may be damaging for the long-term survival of populations (Dounias, 1996).

Ethnoecological knowledge systems also foster diversity (Nazarea, 1998), and for good reason. We now find that enclaves which have maintained a range of diverse traditional crop landraces have often been better at buffering instability. Much of the breadth of traditional knowledge of environmental resources, and the extent to which this knowledge is transferred between populations, arguably insures against long-term ecological oscillation, even if much of it seems irrelevant to survival at any one time. Diversification of crops in general and varied patterns of management tend to keep pest populations relatively low, even under conditions of intensive cultivation. Moreover, there is a strong correlation between biodiversity and cultural diversity, and where there has been cultural (including linguistic) erosion so local biological knowledge and associated management techniques have been depleted or replaced (Maffi, 2001), instituting a kind of poverty (of knowledge), diminishing control over local livelihoods and diminishing the options available for flexible response.

Just as knowledge of individual organisms is embedded in ecological knowledge of the relations between them, and the relationship of assemblages of living things is understood in wider landscape and functional contexts, so there

is a link between all culturally varied biological knowledges and local constructions of that aspect of the world we call "nature". We know enough now of
cross-cultural conceptions of nature to predict that it is everywhere defined in
relation to local social convention; its construction is everywhere diagnostic of
how people understand the world and their place within it. Nowhere is it
completely without ambiguity – sometimes positive, sometimes negative, sometimes reified and named, sometimes covert and implicit, by turns male and
female. It is influenced by the extent to which people consciously manipulate
and transform their surrounding environment (e.g. Reichel-Dolmatoff, 1976;
MacCormack and Strathern, 1980; Ellen, 1996). But at the same time, most
conceptions of nature are underpinned by conceptual universals. One is the
notion of what is "natural" (primordial, essence), second is the tendency to
contrast ourselves as humans and individuals with those biological others that
lie outside of and around us, and third is a compulsion to recognise and classify
natural kinds as things in ways which suggest that we are evolutionarily
adapted to cognise the natural world in broadly the same way. Thus, human
biological knowledge, in whatever cultural tradition it has developed, always
and simultaneously informs and reflects adaptive behaviour through flexible
cultural learning constrained by a common human cognitive framework and
is at the same time embedded in particular social worlds.

KNOWLEDGE OF THE GENERAL PRINCIPLES OF PLANT AND ANIMAL BIOLOGY

In terms of understanding human cultural adaptation to different environments,
knowledge of general principles of biology may be more important than breadth
of formal knowledge or depth of substantive knowledge of individual organisms.
What is central here is the ability to transfer general lessons learned with
respect to one organism to another. To some extent this may relate back to a
general module for natural history intelligence which predisposes us to recognise common aspects in the functioning of living things. But much substantive
knowledge of individual types of animals derives from analogical reasoning
with respect to human bodily functioning. Thus, knowledge of human anatomy
mutually reinforces knowledge of animal anatomy. Every time a Nuaulu hunter
dismembers a deer and removes its internal organs for food and augury, the
activity is serving as a proxy for human dissection. Knowledge of the human
body is, therefore, partly based on knowledge of animal bodies acquired in
hunting, food preparation and livestock keeping, while understanding of animal
physiology, pathology – and even psychology – derives from modified human
experience. However, it is necessary to distinguish the productive explanatory
use of analogy across species from the use of human anatomical nomenclature
to describe the parts of other organisms, as when, for example, Baka (Dounias,
1993: 624) describe yams in terms of human body parts.

Recent work has also demonstrated the capacity for culturally unrelated
people to innovate essentially similar understandings of ecological process. The
repeated discovery of the properties of nitrogen-fixing plants is one well-
reported example (e.g. Iskandar and Ellen, 2000). Sinclair and his associates

(in press; also Walker *et al.*, 1999) have shown resemblances in conceptualising the interaction between trees, agricultural crops and soils, in the contrasting agroecological and cultural conditions in Nepal, Sri Lanka and Thailand. They document similar understandings of how large water droplets falling from certain leaves cause splash erosion in both Nepal and Latin America. More generally, widely distributed "hot-cold" frameworks encode locally specific interactions amongst plants. For example, in central Sri Lanka, "cold" species such as banana (*Musa acuminata*) are said to have a positive effect on other species because the surrounding soil is moist, thereby providing a favourable microclimate for other plants. By contrast, "hot" species such as clove (*Myristica fragrans*) were perceived to have a negative effect on neighbouring plants, and the soil under clove trees was considered unsuitable for cultivation. Other patterns of convergence in general perception of biological properties are observable in relation to medicinal plants. The co-evolution of such plants and cultural systems must have been taking place for in excess of 10,000 years, and their pharmaco-logic is a fundamental species characteristic of humans. Regularities in the selection of taxonomically unrelated plants on the basis of chemical similarities, biases towards certain plant families displaying useful patterns of bioactivity, and a clear understanding that the properties of plants which make them toxic are the same as those which make them desirably bioactive, all provide evidence of this (Johns, 1990; Moerman, Pemberton, Keifer and Berlin, 1999).

The stereotyping of traditional biological knowledge as static is palpably false. What people know about plants and animals is constantly being tested and revised locally, and diffuses between populations. Some institutions of knowledge exchange connecting very different kinds of cultural groups have ancient roots, for example those between pygmy and Bantu in central Africa (Bahuchet, 1993). Plants whose bioactivity was discovered in other parts of the world have, since the sixteenth century and earlier, augmented European pharmacopoeias. For example, *Cinchona officinalis* (quinine) bark was introduced into Spain from the Andes in 1639. These are all examples of knowledge hybridisation, syncretisation or integration. When terms like this are used they often imply contact between science and folk science. Thus, the Baduy of upland west Java (Iskandar and Ellen, 2000) have introduced a previously tabooed leguminous tree, *Paraserianthes (Albizia) falcataria*. By alternating this commercially valuable perennial with rice, soil fertility is maintained, the socio-economic position of the Baduy improved, and swidden farming continues in a very nearly sustainable way. The mechanism for its successful introduction was its perceived similarity to existing nitrogen-fixing cultigens.

At a local level, much general biological knowledge is linked to the way organisms are grouped according to their usefulness. In all societies biological knowledge is innovated and embedded in applied contexts, and therefore all knowledge people have of organisms is, ultimately, because it is useful. But measurements of utility are tricky, and what is useful may include organisms which interact with those which are directly consumed or used in another way, or which are useful only because they are salient. Thus, when Nuaulu are hunting cassowaries (*Casuarius casuarius*) it is as important to have knowledge of plants on which cassowaries browse as much as of cassowaries themselves. Similarly, plants may be understood in terms of the technological uses to which they are put, emphasising qualities such as hardness in wood, the ductile strength of lianas, the engineering properties of bamboo internodes, or chemical properties in relation to dyeing and poisoning. Knowledge organised through functional modules may reveal intricate understandings of, say, the different chemical properties of the roots, stems and leaves of the same species, how bioactivity can only be achieved by combining different species or preparing the same organic ingredients in different ways. Knowledge of biological products used medicinally is evident from modes of preparation and treatment; for example, where detoxification processes are involved (Johns, 1990) or in the decisions to apply medicaments simply or as compounds (Berlin and Berlin, 1996). In the latter case, there may be knowledge of bioactivity which arises from the chemical changes which arise, although mixtures may also simply affect palatability or symbolic significance.

A very specific context of use is Western biological science, and it is to this that we must now turn.

CHARACTERISING BIOLOGICAL SCIENCE, FOLK KNOWLEDGE, AND SCHOLARLY KNOWLEDGE

Where the dividing line lies between scientific biology and other kinds of biological knowledge is by no means obvious. Simple, formal, definitions of what science is are always problematic because they end up excluding practices and kinds of knowledge which are, in common-sense terms, integral to how science works. Science is, sadly, not consistently "rational, objective and produced according to the canons of scientific method", but is rather "messy, contingent, unplanned and arational", a polythetic practice largely concerned with "trying to get the world to fit a particular kind of solution" (Turnbull, 2000: 6, 14). In the general sense of systematic knowledge, it was never uniquely Western, being dependent on the cross-fertilisation of different knowledge traditions (Turnbull, 2000: 227–228). In comparison with the kinds of knowledge systems which we have so far considered, science is undoubtedly in continuous rapid flux and in search of universal rather than local understandings (Hunn, 1993: 13–15), while, socially, "real" science is generated in laboratories, research stations and universities (Chambers and Richards, 1995: xiii). Of course, polythetic or essentialist, this is a model which scientists, decision-makers and administrators have now internalised throughout the world, and which often comes with a built-in assumption that other kinds of knowledge are less prestigious.

What is left, once we have defined "biological science", is ethnobiological knowledge, or "indigenous biological knowledge". But what this means is by no means clear, as terminologies, definitions and cognate concepts vary throughout their geographical, local-global and various historic and disciplinary refractions. There are many indigenous biological knowledges, each accessing the real world to various degrees of imperfection and subjectivity. These biological kinds of knowledge, which for well over 10,000 years have constituted the main body of our adaptive knowledge, are diverse, but in contrast to that scholarly and scientific knowledge self-consciously embodied in textual traditions, they have a number of broad common characteristics (Ellen and Harris, 2000: 4–5). They are rooted in particular places and sets of experiences, are generated by people living in those places, are mostly orally-transmitted or transmitted through imitation and demonstration, are a consequence of practical engagement in everyday life constantly reinforced by experience and error, are the product of generations of intelligent reasoning, and are often a good measure of Darwinian fitness. They are empirical rather than theoretical in character, orality to some extent constraining the kind of organisation necessary for the development of true theoretical knowledge. The redundancy which they embody aids retention and reinforces ideas; they are fluid and the outcome of continuous negotiation, constantly changing, being produced as well as reproduced, discovered as well as lost, although often represented as static. They are characteristically shared to a much greater degree than global biological science but are still socially clustered within a population, by gender and age, for example, and preserved through distribution in the memories of different individuals. Specialists may exist not only by virtue of experience but also by virtue of ritual or political authority. Although knowledge may focus on particular individuals and may achieve a degree of coherence in rituals and other symbolic constructs, it does not exist in its totality in any one place or individual, devolved not in individuals at all, but in the practices and interactions in which people themselves engage. As we have seen earlier, where local biological knowledge is at its densest, organisation is essentially functional. It is characteristically situated within broader cultural traditions, so that separating the technical from the non-technical, the rational from the non-rational, is problematic.

By comparison, the great scholarly ways of knowing come midway between these essentially local knowledges and biological science. They combine knowledge dependent on an agreed shared authority with that of the personal authority of a practitioner. They are often grounded in written texts and resemble the European scholarly traditions. Galenic, Chinese and Ayurvedic traditions of medicine differ from each other, but each have a notion of scholarship in common (Bates, 1995; Zimmerman, 1989). Where the scholarly and local folk traditions merge is unclear, and as in the European case there is historical evidence to suggest, for example, that the great Asian herbalist systems have been systematically absorbing and then replacing local folk knowledge. We see here something very reminiscent of the codifying and simplifying processes which accompanied the incorporation of European folk knowledge into the early modern scholarly traditions.

In Europe and the Mediterranean, codified pharmacopoeias such as the *De Materia Medica* of Dioscorides widely displaced local knowledge and oral tradition, but uncodified knowledge persisted and gradually filtered into organised texts as the number of modern remedies of European folk origin manifestly attest to. Western folk knowledge is just as important as it ever has been; it is just different, informed by science where appropriate and located in different contexts. We might contrast French rustic truffle collecting (Pujol, 1975) with high-tech Icelandic fishermen (Durrenburger and Palsson, 1986). These folk traditions have themselves become highly codified. During mediaeval and early modern Europe, proto-scientific knowledge of plants and animals superseded folk-knowledge by classification, analysis, comparison, dissemination (usually through books and formal learning) and thus generalisation. The process was not sudden; for a long time common experience, oral tradition, personal experience and learned authority contributed to the received wisdom upon which organised specialist knowledge, particularly medical knowledge, depended (Wear, 1995: 158–159). Delineating the boundaries between uncodified folk knowledge, professionally restricted organised knowledge, and proper scientific knowledge is not always easy. Neither are the ethnographic origins of incorporated elements of knowledge always straightforwardly evident. Sometimes ideas are of European folk origin (such as use of the foxglove, *Digitalis purpurea*, as a treatment for oedema [swelling]), but from the sixteenth century onwards European medicine increasingly incorporated herbal remedies of Asian and American origin. By the later middle ages and the beginnings of modern European global expansion, there emerged a self-consciousness about the desirability of obtaining new knowledge. The *Coloquias* of Garcia da Orta and the *Hortus* of Hendrik van Rheede tended to privilege strongly local medical and biological knowledge and to lead to effective discrimination against older Arabic, Brahmanical and European classical texts and systems of cognition in natural history (Grove, 1996). We can see a similar – although in terms of the epidemiology of ideas, less complex – process in the work and influence of George Rumphius. This resulted in the publication of scientific accounts of new species and revisions of taxonomies which, ironically, depended upon a set of diagnostic and classificatory practices which, although represented as "Western science", had been derived from earlier codifications of indigenous knowledge (Ellen and Harris, 2000: 8–10).

During the nineteenth and twentieth centuries local knowledge was increasingly tapped and codified, at home and abroad. Charles Darwin, for example, utilised the accumulated experience of pigeon fanciers in working out the details of natural selection, while colonial science systematically assimilated local knowledge of plants (e.g. Burkill, 1935). Such practices became so routinised that, once absorbed into scientific solutions, local biological knowledge disappeared from view, insufficiently real to merit any certain legal status or protection in the same ways which gave value and ownership to western scholarly knowledge and expertise. Even when the knowledge was clearly being utilised it was often redescribed in ways which eliminated any credit to those who had brought it to the attention of science in the first place. Thus, the boundaries between science, scholarly knowledge and folk knowledge, as these terms apply to biological phenomena, are constantly shifting, and the distinctions themselves are not always helpful. All knowledges are anchored in their own particular socioeconomic milieu; all are indigenous to a particular context, undermining what Agrawal (1995: 5) describes as the "sterile dichotomy between indigenous and Western".

Indeed, more generally we can see that modern natural history arose through a combination of such indigenous scholarship and field studies (Zimmermann, 1995: 312), field studies themselves drawing on the knowledge of local experts. Some have argued that the phylogenetic taxonomies of contemporary post-Linnean biology are based on a European folk template (Ellen, 1979; Atran, 1990) and, arguing a rather different tack, others have gone further by claiming that the European folk scheme and that of modern biology are no more than variants on a single cognitive arrangement to which all humans are predisposed through natural selection (Atran, 1998; Boster, 1996).

THE REDISCOVERY OF ETHNOBIOLOGY AND THE INVENTION OF "INDIGENOUS KNOWLEDGE"

From about the mid-1960s the tendency to marginalise local biological knowledges had begun to be put into reverse, prompted by a combination of romantic idealism and pragmatism (Conklin and Graham, 1995). This infectious combination has sometimes merged scholarly and local oral traditions, confusing ideal symbolic representations with hard-headed empirical practice, inevitably leading to a particular version of the "science wars" in which the contestants (put crudely) are those who see ethnobiology as a kind of science and biology as a kind of ethnobiology, against those for whom science represents a unique methodology for discovering the truth (see e.g. Diamond, 1987 versus Johannes, 1987; Anderson, 2000). Despite this, the demand for local biological knowledge from developers and industry at the present time shows no signs of ceasing (Sillitoe, 1998).

One of the main problems, though, with the development industry's appropriation of ethnobiology has been its transformation into a kind of context-independent knowledge mirroring the structure of Western science, parts of which can be conveniently modularised and transferred. Fairhead and Leach (1994: 75) argue that this risks overlooking broadly held understandings of agroecological knowledge and social relations. So, for example, research and extension agents examining tree management practices used by Kuranko farmers in the Republic of Guinea fail to take into account farmers' tree-related knowledge which involves a much broader range of knowledge: of crops, water, vegetation succession and the socioeconomic and ecological conditions which influence them. More radically, Richards (1993: 62) proposes that the range of skills and strategies employed by farmers often extends beyond simple applied knowledge into a fluid body of improvisations relevant to immediate needs, rather than the outcome of a prior stock of knowledge about inter-species ecological complementarity. By presenting agroecological knowledge as a decontextualised inventory of practices, all agency and creativity is drained, reducing it to a packageable commodity, secured and easily transferable from one place to another. Furthermore, as local knowledge is analysed and documented for use, it undergoes changes which necessarily result from the specific orientations, strategies and agendas of those using it, as well as from the transformations which inevitably occur through translation. Hobart (1993: 14) underlines some of the potential problems that can occur when knowledge is collected and codified into bite-sized chunks. And, as we have already seen, once ethnobiology is drawn within the boundaries of science it is difficult to know where to place the boundaries between the two. Indeed, changing the boundaries is often sufficient to redefine something as science, as what defines it is to a considerable extent determined by who practices it and in what institutional context the practices take place. The danger of turning local knowledge into global knowledge is that at the empirical level all local knowledge is precisely that, local, relative and parochial, no two societies perceiving or acting upon the environment in the same way; which is, of course, its applied strength. The corollary is that writing it down makes it more portable and permanent, but also changes some of its fundamental properties, all of which reinforce dislocation. Knowledge – as anthropologists repeatedly tell us, and as was demonstrated in the second section of this chapter – is grounded in multiple domains, logics and epistemologies.

Finally, there are important connections between local biological knowledge, identity and conceptions of property. Nowadays, many savvy local peoples see their knowledge as part of their patrimony. The disappearance of species, names for species and knowledge of their use and significance is increasingly a concern for local peoples themselves. This is not only a pragmatic matter but connects with people's sense of their own culture more generally. States and NGOs, as well as native people, have sought to protect rights to such knowledge, especially where there are threats of biopiracy; there are concerns about the expropriation of knowledge and intellectual property by pharmaceutical and other companies and agencies. This has given rise to a whole set of new issues, merging the philosophies, legal traditions and discourses of the West and of the rest. In some cases, cross-fertilisation of different local traditions and the reification of folk knowledge have occurred. Third World politicians, scientists and others have had to work out for themselves how indigenous or traditional knowledge is to be defined and whether its existence is altogether to be welcomed. When it becomes a means by which to flag problematic local minorities who seek to make political and cultural claims against a government, it is clearly threatening; if it can be defined in a more inclusive way and commoditised, it is a resource to be exploited. However it is constructed and represented, ethnobiological knowledge is self-evidently valuable, and understanding its range and intellectual foundations is no less important today than it ever has been, and in the context of the loss of so much biological and cultural diversity, much more so.

NOTES

[1]   The contemporary ethnobiological literature is huge, and I have here provided only selected bibliographical references, in some cases to flag historical benchmark studies, in others to illustrate some of the more interesting, accessible and influential work. For a general recent collection on ethnobiology, see Medin and Atran (1999). A useful bibliography covering the older literature,

particularly on folk classification, is Conklin (1971). On research methods in ethnobotany see Martin (1995), and in ethnozoology see Bulmer and Healey (1993).

[2] *Emic* is a perspective in ethnography that uses the concepts and categories that are relevant and meaningful to the culture under analysis, that is, a view from the inside. *Etic* is a perspective that uses the concepts and categories of the anthropologist's culture to describe another culture, that is, a view from the outside.

[3] A word or lexical item indicating that it is 'part-of' some whole. Thus, "leg" is a body partonym, and "root" a plant partonym. Thus, partonymy refers to the phenomenon of referring to parts of a whole through specialised terms. By extension, partonymy refers to the classification which underlies such lexical sets.

BIBLIOGRAPHY

Agrawal, A. 'Indigenous and scientific knowledge: some critical comments.' *Indigenous Knowledge and Development Monitor* 3(3): 5, 1995. Elaborated as 'Dismantling the divide between indigenous and scientific knowledge.' *Development and Change* 26: 413–439, 1995.

Anderson, E.N. 'Maya knowledge and "science wars".' *Journal of Ethnobiology* 20(2): 129–158, 2000.

Atran, S. *Cognitive Foundations of Natural History. Toward an Anthropology of Science.* Cambridge: Cambridge University Press, 1990.

Atran, S. 'Folk biology and the anthropology of science: cognitive universals and cultural particulars.' *Behavioural and Brain Sciences* 21: 547–609, 1998.

Bahuchet, S. *La rencontre des agriculteurs. Les Pygmées parmi les peuples d'Afrique centrale.* Paris: Peters-SELAF, 1993.

Balée, W. 'The culture of Amazonian forests.' In *Resource Management in Amazonia: Indigenous and Folk Strategies*, D.A. Posey and W. Balée, eds. Bronx, New York: New York Botanical Garden, 1989, pp. 1–21.

Barrau, J. 'Coping with exotic plants in folk taxonomies.' In *Classifications in Their Social Context*, R.F. Ellen and D. Reason, eds. London: Academic Press, 1979, pp. 139–144.

Bates, D., ed. *Knowledge and the Scholarly Medical Traditions.* Cambridge: Cambridge University Press, 1995.

Berlin, B. 'Speculations on the growth of ethnobotanical nomenclature.' *Language in Society* 1: 151–186, 1972.

Berlin, B. *Ethnobiological Classification: Principles of Categorization of Plants and Animals in Traditional Societies.* Princeton, New Jersey: Princeton University Press, 1992.

Berlin, E.A. and B. Berlin. *Medical Ethnobiology of the Highland Maya of Chiapas, Mexico.* Princeton, New Jersey: Princeton University Press, 1996.

Berlin, B., D. Breedlove and P.H. Raven. 'General principles of classification and nomenclature in folk biology.' *American Anthropologist* 75: 214–242, 1973.

Berlin, B., D. Breedlove and P. Raven. 'Folk taxonomies and biological classification.' *Science* 154: 273–275, 1966.

Berlin, B., D. Breedlove and P. Raven. *Principles of Tzeltal Plant Classification: An Introduction to the Botanical Ethnography of a Mayan-Speaking People of the Highland Chiapas.* New York: Academic Press, 1974.

Blurton-Jones, H. and M.J. Konner. '!Kung knowledge of animal behaviour.' In *Kalahari Hunter-Gatherers*, R. Lee and I. deVore, eds. Cambridge, Massachusetts: Harvard University Press, 1976, pp. 325–348.

Boster, J. 'Exchange of varieties and information between Aguaruna manioc cultivators.' *American Anthropologist* 88(2): 428–436, 1986.

Boster, J. 'Human cognition as a product and agent of evolution.' In *Redefining Nature: Ecology, Culture and Domestication*, Roy Ellen and Katsuyoshi Fukui, eds. Oxford: Berg, 1996, pp. 269–289.

Brown, C.H. *Language and Living Things: Uniformities in Folk Classification and Naming.* New Brunswick, New Jersey: Rutgers University Press, 1984.

Brush, S.B. 'Ethnoecology, biodiversity and modernization in Andean potato agriculture.' *Journal of Ethnobiology* 12(2): 161–185, 1992.

Bulmer, R. 'Why is the cassowary not a bird? A problem of zoological taxonomy among the Karam of the New Guinea Highlands.' *Man* 2: 5–25, 1967.

Bulmer, R.N.H. 'Which came first, the chicken or the egg-head?' In *Échanges et communications; mélanges offerts à Claude Lévi-Strauss*, J. Pouillon and P. Maranda, eds. The Hague and Paris: Mouton, 1970, pp. 1069–1089.

Bulmer, R. 'Mundane and mystical in Kalam classification of birds.' In *Classifications in Their Social Context*, R.F. Ellen and D. Reason, eds. London: Academic Press, 1979, pp. 57–79.

Bulmer, R. and C. Healey. 'Field methods in ethno-zoology.' In *Ecology for the 21st Century: The Relevance of Traditional Ecological Knowledge*, N. Williams and G. Baines, eds. Canberra: Centre for Resource and Environmental Studies, Australian National University, 1993, pp. 43–55.

Burkill, I.H. *A Dictionary of Economic Products of the Malay Peninsula*. London: Crown Agents for the Colonies, 2 volumes, 1935.

Carriere, St. *Les orphelins de la forêt. Influence de l'agriculture itinérante sur brûlis des Ntumu et des pratiques agricoles associées sur la dynamique du couvert forestier du sud Cameroun*. Thèse de doctorat, Université de Montpelier II, 1999.

Chambers, R. and P. Richards. 'Preface.' In *The Cultural Dimension of Development: Indigenous Knowledge Systems*, D. Michael Warren, L. Jan Slikkerveer and David Brokensha, eds. London: Intermediate Technology Publications, 1995, pp. xiii–xiv.

Colby, B., J. Fernandez and D. Kronenfeld. 'Toward a convergence of cognitive and symbolic anthropology.' *American Ethnologist* 8: 422–450, 1980.

Conklin, B. and L. Graham. 'The shifting middle ground: Amazonian Indians and eco-politics.' *American Anthropologist* 97(4): 695–710, 1995.

Conklin, H.C. 'An ethnoecological approach to shifting agriculture.' *Transactions of the New York Academy of Sciences* 17: 133–142, 1954.

Conklin, H.C. 'Lexicographical treatment of folk taxonomies.' *International Journal of American Linguistics* 28: 119–411, 1962.

Conklin, H.C. *Folk Classification: A Topically Arranged Bibliography of Contemporary and Background References through 1971*. New Haven, Connecticut: Department of Anthropology, Yale University, 1972.

Diamond, J. 'The environmentalist myth.' *Nature* 324: 9–20, 1987.

Douglas, M. *Purity and Danger*. London: Routledge and Kegan Paul, 1966.

Dounias, E. 'The perception and use of wild yams by the Baka hunter-gatherers in south Cameroon rainforest.' In *Tropical Forests, People and Food: Biocultural Interactions and Applications to Development*, C.M. Hladik, *et al.*, eds. Paris: UNESCO, 1993, pp. 621–632.

Dounias, E. 'Recrûs forestiers post-agricoles: perceptions et usages chez les Mvae du Sud-Cameroun.' *Journal d'Agriculture Tropicale et de Botanique Appliquée* 38: 153–178, 1996.

Dove, M.R. 'Theories of swidden agriculture and the political economy of ignorance.' *Agroforestry Systems* 1(3): 85–99, 1983.

Dove, M.R., and D.M. Kammen. 'The epistemology of sustainable resource use: managing forest products, swiddens, and high-yielding variety crops.' *Human Organization* 56(1): 91–101, 1997.

Durkheim, E. and M. Mauss. *Primitive Classification*, R. Needham, trans. Chicago: University of Chicago Press, 1963 (1900–1901).

Durrenberger, E. and G. Palsson. 'Finding fish: the tactics of Icelandic skippers.' *American Ethnologist* 13(2): 213–229, 1986.

Elias, M., L. Rival and D. McKey. 'Perception and management of cassava (*Manihot esculenta* Crantz) diversity among the Makushi Amerindians of Guyana (South America).' *Journal of Ethnobiology* 20(2): 239–265, 2000.

Ellen, R.F. 'Introductory essay.' In *Classifications in Their Social Context*, R.F. Ellen and D. Reason, eds. London: Academic Press, 1979, pp. 1–32.

Ellen, R.F. *Environment, Subsistence and System: The Ecology of Small-Scale Social Formations*. Cambridge: Cambridge University Press, 1983.

Ellen, R.F. 'Nuaulu betel chewing: ethnobotany, technique and cultural significance.' *Cakalele: Maluku Research Journal* 2(2): 97–122, 1991.

Ellen, R.F. *The Cultural Relations of Classification: An Analysis of Nuaulu Animal Categories from Central Seram*. Cambridge: Cambridge University Press, 1993.

Ellen, R.F. 'Introduction.' In *Redefining Nature: Ecology, Culture and Domestication*, Roy Ellen and Katsuyoshi Fukui, eds. Oxford: Berg, 1996, pp. 1–36.

Ellen, R.F. 'Modes of subsistence and ethnobiological knowledge: between extraction and cultivation in Southeast Asia.' In *Folkbiology*, D.L. Medin and S. Atran, eds. Cambridge, Massachusetts: MIT Press, 1999, pp. 91–117.

Ellen, R. and H. Harris. 'Introduction.' In *Indigenous Environmental Knowledge and its Transformations: Critical Anthropological Perspectives*, R.F. Ellen, P. Parkes and A. Bicker, eds. Amsterdam: Harwood, 2000, pp. 1–33.

Fairhead, J. and M. Leach. 'Declarations of difference.' In *Beyond Farmer First*, I. Scoones and J. Thompson, eds. London: Intermediate Technology Publications, 1994, pp. 75–79.

Fairhead, J. and M. Leach. *Misreading the African Landscape: Society and Ecology in a Forest-Savanna Mosaic*. Cambridge: Cambridge University Press, 1996.

Felger, R. and M.B. Moser. *People of the Desert and Sea: Ethnobotany of the Seri Indians*. Phoenix: University of Arizona Press, 1985.

Fox, J. 'Sister's child as plant, metaphors in an idiom of consanguinity.' In *Rethinking Kinship and Marriage*, R. Needham, ed. London: Tavistock, 1971, pp. 219–252.

Fox, J. *Harvest of the Palm: Ecological Change in Eastern Indonesia*. Cambridge, Massachusetts: Harvard University Press, 1977.

Frake, C.O. 'The ethnographic study of cognitive systems.' In *Cognitive Anthropology*, S.A. Tyler, ed. New York: Holt, Rinehart and Winston, 1969, pp. 28–41. First published in *Anthropology and Human Behavior*, T. Gladwin and W.C. Sturtevant, eds. Washington, DC: The Anthropological Society of Washington, 1962, pp. 72–85.

Friedberg, C. *Le savoir botanique des Bunaq: percevoir et classer dans le Haut Lamaknen (Timor, Indonésie)* (Mémoires du Muséum National d'Histoire Naturelle, Botanique 32). Paris: Muséum National d'Histoire Naturelle, 1990.

Fukui, K. 'Co-evolution between humans and domesticates: the cultural selection of animal coat colour diversity amongst the Bodi.' In *Redefining Nature: Ecology, Culture and Domestication*, Roy Ellen and Katsuyoshi Fukui, eds. Oxford: Berg, 1996, pp. 319–385.

Gardner, P. 'Birds, words and a requiem for the omniscient informant.' *American Ethnologist* 3: 446–468, 1976.

Geertz, C. 'Religion as a cultural system.' In *Anthropological Approaches to the Study of Religion*, M. Banton, ed. London: Tavistock, 1966, pp. 1–46.

Grove, R. 'Indigenous knowledge and the significance of south-west India for Portuguese and Dutch constructions of tropical nature.' *Modern Asian Studies* 30(1): 121–143, 1996.

Healey, C. 'The significance and application of TEK.' In *Traditional Ecological Knowledge: Wisdom for Sustainable Development*, N. Williams and G. Baines, eds. Canberra: Centre for Resource and Environmental Studies, Australian National University, 1993, pp. 21–26.

Healey, C. 'Folk taxonomy and mythology of birds of paradise in the New Guinea highlands.' *Ethnology* 32: 19–34, 1993.

Heinz, H.J. and B. Maguire. *The Ethno-Biology of the !Kõ Bushmen: Their Ethnobotanical Knowledge and Plant Lore*. Gaborone: The Botswana Society, 1974.

Hirsch, E. 'Dialectics of the bowerbird: an interpretative account of ritual and symbolism in the Udabe Valley, Papua New Guinea.' *Mankind* 17(1): 1–14, 1987.

Hobart, M. 'Introduction: the growth of ignorance?' In *An Anthropological Critique of Development*, M. Hobart, ed. London: Routledge, 1993, pp. 1–30.

Hughes, D.J. *American Indian Ecology*. El Paso: Texas Western University Press, 1983.

Hunn, E. *Nch'i-wána, 'The Big River': Mid-Columbia Indians and Their Land*. Seattle and London: University of Washington Press, 1990.

Hunn, E. 'What is traditional ecological knowledge?' In *Traditional Ecological Knowledge: Wisdom for Sustainable Development*, N. Williams and G. Baines, eds. Canberra: Centre for Resource and Environmental Studies, Australian National University, 1993, pp. 13–15.

Hunn, E.S. and D.H. French. 'Alternatives to taxonomic hierarchy: the Sahaptin case.' *Journal of Ethnobiology* 3: 73–92, 1984.

Ichikawa, M. 'Interactive process of man and nature in the Ituri forest of the Democratic Republic of Congo: an approach from historical ecology.' In *Central African Hunter-Gatherers in a Multidisciplinary Perspective: Challenging Elusiveness*, K. Biesbrouck, S. Elders, and G. Rossel, eds. Leiden: Research School for Asian, African and Amerindian Studies (CNWS), Universiteit Leiden, 1999, pp. 141–152.

Inglis, J.T. *Traditional Ecological Knowledge: Concepts and Cases*. Ottawa: International Development Research Centre, 1993.

Iskandar, J. and R. Ellen. 'In situ conservation of rice landraces among the Baduy of West Java.' *Journal of Ethnobiology* 19(1): 97–125, 1999.

Iskandar, J. and R. Ellen. 'The contribution of *Paraserianthes (Albizia) falcataria* to sustainable swidden management practices among the Baduy of West Java.' *Human Ecology* 28(1): 1–17, 2000.

Johannes, R.E. 'Traditional marine conservation methods in Oceania and their demise.' *Annual Review of Ecology and Systematics* 9: 349–366, 1978.

Johannes, R.E. 'Primitive myth.' *Nature* 325: 478, 1987.

Johannes, R.E., ed. *Traditional Ecological Knowledge*. Cambridge: IUCN, The World Conservation Union, 1989.

Johns, T. *With Bitter Herbs They Shall Eat*. Tucson: University of Arizona Press, 1990.

Johnson, A. 'Ethnoecology and planting practices in a swidden agricultural system.' *American Ethnologist* 1: 87–101, 1974.

Keil, F.C. 'The birth and nurturance of concepts of living things.' In *Mapping the Mind: Domain Specificity in Cognition and Culture*, L.A. Hirschfeld and S.A. Gelman, eds. Cambridge: Cambridge University Press, 1994, pp. 234–254.

Kelly, R.C. 'Etoro suidology: a reassessment of the pig's role in the prehistory and comparative ethnology of New Guinea.' In *Mountain Papuans: Historical and Comparative Perspectives from New Guinea Fringe Highland Societies*, J.F. Weiner, ed. Ann Arbor: University of Michigan Press, 1988, pp. 111–186.

Kocher Schmid, C. *Of People and Plants: A Botanical Ethnography of Nokopo Village, Madang and Morobe Provinces, Papua New Guinea*. Basel: Ethnologisches Seminar der Universität und Museum für Völkerkunde, 1991.

Laden, G.T. *Ethnoarchaeology and Land Use Ecology of the Efe (Pygmies) of the Ituri Rain Forest, Zaire: A Behavioural Ecological Study of Land Use Patterns and Foraging Behavior*. Ph.D. dissertation. Department of Anthropology, Harvard University, Cambridge, Massachusetts, 1992.

Lansing, J.S. *Priests and Programmers: Technologies of Power in the Engineered Landscape of Bali*. Princeton, New Jersey: Princeton University Press, 1991.

Leach, E. 'Anthropological aspects of language: animal categories and verbal abuses.' In *New Directions in the Study of Language*, E.H. Lenneberg, ed. Cambridge, Massachusetts: MIT Press, 1964, pp. 23–26.

Logan, M.H. and A.R. Dixon. 'Agriculture and the acquisition of medicinal plant knowledge.' In *Eating on the Wild Side: The Pharmacologic, Ecologic and Social Implictions of Using Noncultigens*, N.L. Etkin, ed. Tucson: University of Arizona Press, 1994, pp. 25–45.

MacCormack, C. and M. Strathern, eds. *Nature, Culture and Gender*. Cambridge: Cambridge University Press, 1980.

Mckey, D., B. Digiusto, L. Pascal, M. Elias, and E. Dounia. 'Stratégies de croissance et de défense anti-herbivore designames sauvages: leçons pour labronomie.' In *Ligname, plante séculaire et culture davenir*, J. Berthaud, N. Bricas and J.-L. Marchand, eds. Actes du séminaire international Cirad-Inra-Orstom-Coraf. 3–6 Juin 1997, Montpelier, France, 1998, pp. 181–188.

Maffi, L., ed. *On Biocultural Diversity: Linking Language, Knowledge and the Environment*. Washington, DC and London: Smithsonian Institution Press, 2001.

Martin, G.J. *Ethnobotany: A 'People And Plants' Conservation Manual*. London: Chapman and Hall, 1995.

Medin, D. and S. Atran, eds. *Folkbiology*. Cambridge, Massachusetts: MIT Press, 1999.

Metzner, R., ed. *Ayahuasca: Hallucinogens, Consciousness and the Spirit of Nature*. New York: Thunder's Mouth Press, 1999.

Mithen, S. *The Prehistory of the Mind: A Search for the Origins of Art, Religion and Science*. London: Thames and Hudson, 1996.

Moerman, D.E., R.W. Pemberton, D. Kiefer and B. Berlin. 'A comparative analysis of five medicinal floras.' *Journal of Ethnobiology* 19(1): 49–67, 1999.

Morris, B. 'Whither the savage mind? Notes on the natural taxonomies of a hunting and gathering people.' *Man* 11: 542–557, 1976.

Nabhan, G.P. and A. Rea. 'Plant domestication and folk-biological change: the upper Piman/Devil's Claw example.' *American Anthropologist* 89(1): 57–73, 1987.

Nazarea, V.D. *Cultural Memory and Biodiversity*. Tucson: University of Arizona Press, 1998.

Nazarea, V.D., ed. *Ethnoecology: Situated Knowledge/Located Lives*. Tucson: University of Arizona Press, 1999.

Nelson, R. *Make Prayers to the Raven*. Chicago and London: University of Chicago Press, 1983.

Ohnuki-Tierney, E. 'Phases in human perception/cognition/symbolization processes: cognitive anthropology and symbolic classification.' *American Ethnologist* 8(2): 451–467, 1981.

Page, W. and P. Richards. 'Agricultural pest control by community action: the case of the variegated grasshopper in southern Nigeria.' *African Environment* 2(3): 127–141, 1977.

Posey, D. 'Kayapo Indian natural-resource management.' In *People of the Tropical Rainforest*, J.S. Denslow and C. Padoch, eds. Berkeley: University of California Press, 1988, pp. 89–90.

Peeters, A. 'Nomenclature and classification in Rumphius's "Herbarium Amboinense".' In *Classifications in Their Social Context*, R.F. Ellen and D. Reason, eds. London: Academic Press, 1979, pp. 145–166.

Posey, D.A. and J.M.F. de Camargo. 'Additional notes on the classification and knowledge of stingless bees (Meliponinae, Apidae, Hymenoptera) by the Kayapó Indians of Gorotire, Pará, Brazil.' *Annals of the Carnegie Museum* 54(8): 247–274, 1985.

Pujol, R. 'Definition d'un ethnoecosystème avec deux exemples: étude ethnozoobotanique des cardères (*Dipsacus*) et interrelations homme-animal-truffe.' In *L'Homme et l'Animal: Premier Colloque d'Ethnozoologie*, Raymond Pujol, ed. Paris: Institut International d'Ethnosciences, 1975, pp. 91–114.

Randall, R.A. 'How tall is a taxonomic tree? Some evidence for dwarfism.' *American Ethnologist* 8: 229–242, 1976.

Randall, R.A. and E.S. Hunn. 'Do life-forms evolve or do uses for life? Some doubts about Brown's universals hypothesis.' *American Ethnologist* 11: 329–349, 1984.

Reed, E.S. 'The affordances of the animate environment: social science from the ecological point of view.' In *What is an Animal?*, T. Ingold, ed. London: Unwin Hyman, 1988, pp. 110–126.

Reichel-Dolmatoff, G. 'Cosmology as ecological analysis: a view from the rain forest.' *Man (N S)* 11(3): 307–318, 1976.

Revel, N. *Fleurs de Paroles: Histoire Naturalle Palawan*, vols. 1 and 2. Paris: Peeters/SELAF, 1990.

Richards, P. 'Cultivation: knowledge or performance?' In *An Anthropological Critique of Development*, M. Hobart, ed. London: Routledge, 1993, pp. 61–78.

Rival, L., ed. *The Social Life of Trees*. Oxford: Berg, 1998.

Rosaldo, M.Z. 'Metaphors and folk classification.' *Southwestern Journal of Anthropology* 28(1): 83–99, 1972.

Rosaldo, M.Z. and J.M. Atkinson. 'Man the hunter and woman: metaphors for the sexes in Ilongot magical spells.' In *The Interpretation of Symbolism*, R. Willis, ed. London: Malaby, 1975, pp. 43–75.

Rosch, E. 'Human categorisation.' In *Studies in Cross-Cultural Psychology*, N. Warren, ed. London: Academic Press, 1977, pp. 1–49.

Shigeta, M. 'Creating landrace diversity: the case of the Ari people and Ensete (*Ensete ventricosum*) in Ethiopia.' In *Redefining Nature: Ecology, Culture and Domestication*, R.F. Ellen and K. Fukui, eds. Oxford: Berg, 1996, pp. 233–268.

Sillitoe, P. *A Place Against Time: Land and Environment in the Papua New Guinea Highlands*. Amsterdam: Harwood, 1996.

Sillitoe, P. 'The development of indigenous knowledge: a new applied anthropology.' *Current Anthropology* 39(2): 223–252, 1998.

Sinclair, F.L., *et al.* 'General patterns in indigenous ecological knowledge.' In *Development and Local*

Knowledge: New Approaches to Issues in Natural Resources Management, Conservation and Agriculture, A. Bicker, ed. London: Routledge, in press.

Tambiah, S. 'Animals are good to think and good to prohibit.' *Ethnology* 8: 424–459, 1969.

Taylor, P.M. *The Folk Biology of the Tobelo People: A Study in Folk Classification*. Smithsonian Contributions to Anthropology No. 34. Washington, DC: Smithsonian Institution Press, 1990.

Turnbull, D. *Masons, Tricksters and Cartographers: Comparative Studies in the Sociology of Scientific and Indigenous Knowledge*. Amsterdam: Harwood Academic Publishers, 2000.

Turner, V.W. *The Forest of Symbols: Aspects of Ndembu Ritual*. Ithaca, New York: Cornell University Press, 1967.

Walker, D.H., P.J. Thorne, F.L. Sinclair, B. Thapa, C.D. Wood, and D.B. Subba. 'A systems approach to comparing indigenous and scientific knowledge: consistency and discriminatory power of indigenous and laboratory assessment of the nutritive value of tree fodder.' *Agricultural Systems* 62: 87–103, 1999.

Wear, A. 'Epistemology and learned medicine in early modern England.' In *Knowledge and the Scholarly Medical Traditions*, D. Bates, ed. Cambridge: Cambridge University Press, 1995, pp. 151–173.

Whiteford, L. 'The ethnoecology of Dengue fever.' *Medical Anthropology Quarterly* 11(2): 202–223, 1997.

Williams, N. and G. Baines, eds. *Traditional Ecological Knowledge: Wisdom for Sustainable Development*. Canberra: Centre for Resource and Environmental Studies, Australian National University, 1993.

Zerner, C. 'Transforming customary law and coastal management practices in the Maluku Islands, Indonesia, 1870–1992.' In *Natural Connections: Perspectives in Community-Based Conservation*, D. Western and R.M. Wright, eds. Washington, DC: Island Press, 1994, pp. 80–112.

Zimmermann, F. *Le Discours des Remèdes au Pays des Épices*. Paris: Payot, 1989.

Zimmermann, F. 'The scholar, the wise man, and universals: three aspects of Ayurvedic medicine.' In *Knowledge and the Scholarly Medical Traditions*, D. Bates, ed. Cambridge: Cambridge University Press, 1995, pp. 297–319.

ROY C. DUDGEON AND FIKRET BERKES

LOCAL UNDERSTANDINGS OF THE LAND: TRADITIONAL ECOLOGICAL KNOWLEDGE AND INDIGENOUS KNOWLEDGE

Much of the literature on traditional ecological knowledge (TEK) deals with similarities and differences between Western science and traditional knowledge (e.g., Johannes, 1989; Williams and Baines, 1993; Berkes, 1999). By contrast, little has been written about the relationship between TEK and indigenous knowledge (IK). These two areas constitute two closely related and broadly overlapping literatures. While each approach seeks to understand local knowledge of the land, there are both similarities and differences between the two. One of the primary differences is the insight provided by TEK regarding some new understandings in ecology and resource management.

This chapter deals with TEK and IK as two distinct approaches to the study of local knowledge of the environment. The chapter discusses the differences between these two approaches with respect to their understandings of the philosophy of science, especially in relation to the science of resource management. It concludes with a contrast between TEK and IK with regard to their implications for the policy and politics of development and their implications for the political autonomy of indigenous peoples.

The most obvious differences between TEK and IK are in the different names they choose to describe their research and in the definitions provided for these key terms. IK has been used to refer to the local knowledge of indigenous peoples or to the unique, local knowledge of particular cultural groups (Warren et al., 1995). As commonly used in the development literature, the term "indigenous" is meant to emphasize the culture of the original inhabitants of an area, as opposed to globalized culture. The term "knowledge" is meant to focus attention upon the contrast between local ways of knowing and interacting with one's environment versus the dominant understandings of economic development derived from modern understandings of development science. IK can be used as a synonym for "traditional knowledge", which recognizes that traditions are not static but continually changing and evolving over time, as cultural groups innovate, borrow and adapt their traditions to changing circumstances.

75

H. Selin (ed.), Nature Across Cultures: Views of Nature and the Environment in Non-Western Cultures, 75–96.
© 2003 *Kluwer Academic Publishers. Printed in Great Britain.*

Traditional *ecological* knowledge, on the other hand, has been defined as "a cumulative body of knowledge, practice and belief, evolving by adaptive processes and handed down through generations by cultural transmission, about the relationship of living beings (including humans) with one another and with their environment" (Berkes, 1999: 8). This definition, evolving from our earlier work (e.g., Berkes *et al.*, 1995), recognizes TEK as a knowledge-practice-belief complex and focuses on the ecological aspects of this knowledge. TEK may be viewed as a more specific focus within the larger IK literature. While sharing IK's emphasis on the area-specific and culture-specific nature of indigenous knowledge, the above definition of TEK also adds an explicitly ecological emphasis.

Consequently, TEK focuses explicitly not only upon the social patterns of relationship within the culture under study and those within the ecosystem, but also upon the patterns of relationship between the two. This includes local ways of knowing and interacting with the ecosystem. Such an explicitly holistic approach – informed by recent ecological philosophy, philosophy of science and political ecology – allows TEK to avoid some of the shortcomings of conventional development approaches and provides support for local political and economic autonomy for indigenous peoples.

IK AND DEVELOPMENT LITERATURE: SOME CRITIQUES

In providing a background for the discussion to follow, this section starts with a history of the IK approach and the substantive areas which have provided the central focus for this literature. It will also highlight several important criticisms which will form the subject matter of the rest of the discussion, including Brouwer's (1998) critique of Sillitoe (1998), who articulates one predominant view of the role of IK in development.

Sillitoe's review proposes that the central purpose of IK research is to "introduce a locally informed perspective into development" (1998: 224) and to make explicit connections between local understandings and practices and those of researchers and development workers. This understanding of local experiences and of the objectives of local people is undertaken in order to "link them to scientific technology" and to contribute to "positive change, promoting culturally appropriate and environmentally sustainable adaptations acceptable to the people as increasingly they exploit their resources *commercially*" (1998: 224, emphasis added). Consequently, Sillitoe argues for the necessity of anthropological participation in this research, since it requires that there be someone trained to mediate between the two cultures, people who can act as "knowledge brokers" (1998: 247).

The evolution of this trend away from top-down approaches to development, and towards those which include some attention to bottom-up approaches has, according to Sillitoe, been informed by two different strands of thought. The first is academic, including various studies in ethnoscience and human ecology which have been carried out over the last few decades. The second strand, which is development focused, has arisen in about the last decade and has largely applied itself to research concerning farming systems and participatory development (Sillitoe, 1998: 223). This view is reflected in the writings of other students of IK. For example, Agrawal (1995) observes that IK research has focused largely upon agricultural production systems and sustainable development, while Purcell (1998: 265) notes that much of the early IK work focused upon "agricultural and environmental practices, areas of immediate concern for survival".

Because of the turn towards bottom up research, Sillitoe (1998) suggests that there has been a "sea change" in the paradigms structuring concepts of development, in which approaches such as modernization theory and dependency theory have been replaced by either "market-liberal" or "neo-populist" approaches. Where the former "promotes market forces and decries state intervention", the latter "advocates [the] participation and empowerment" of local peoples (Sillitoe, 1998: 224).

The main point seems to be that anthropologists must participate in these various approaches to the development process, especially by making local knowledge relevant and understandable to development scientists. "The idea of harnessing anthropology to technical knowledge to facilitate development puts the discipline where it should be, at the centre of the development process" (1998: 231). However, to do so need not involve a critique of conventional development. Sillitoe seems to accept development science as it presently stands, for as he states,

> The implication of considering indigenous and scientific perspectives side by side is *not* ... that scientists need to revise their working suppositions regarding objectivity, positivism, reductionism, and so on, to accommodate other views (Sillitoe, 1998: 226, emphasis added).

Instead, the suggested partnership between ethnographers and development scientists "offers an opportunity to compare indigenous statements and explanations against scientifically measurable data" (1998: 227). Sillitoe does acknowledge that "the heretical idea is gaining currency that others may have something to teach us", as well as the possibility that their knowledge "may advance our scientific understanding" by challenging our received models of natural processes (1998: 227). It seems clear, however, that development science remains the final arbiter of the validity of knowledge (Brouwer, 1998: 351). After all, Sillitoe simply dismisses the arguments of "anti-positivistic social scientists" who seek to "undermine" natural scientists without further comment, suggesting instead that it is really a question of "seeking to make scientists work more effectively through partnership" with indigenous peoples (Sillitoe, 1998: 232).

The type of "development-oriented indigenous-knowledge work" which Sillitoe advocates makes no pretense of "understanding others as they understand themselves", that is, of attempting to understand their emic perspective on the world.[1] Rather, its goal is to understand and interpret "other cultures and their environments as the demands of development require" (Sillitoe, 1998: 229).

Similarly, Sillitoe's stance with regards to political and policy issues, which might involve questioning the aims rather than the methodology of development, is consistent with his views on development science. As he suggests, for example, while IK research "intimately and unavoidably involves political issues", anthropologists working in the area should merely "inform politicians and others about issues as they perceive them and leave the responsibility for policy decisions to them" (Sillitoe, 1998: 231). In Sillitoe's opinion, anthropologists are "not politicians, management consultants, or policy makers", and any pretense towards such ends would test "the limits of our disciplinary competence" as anthropologists (1998: 246–247).

Others have criticized the non-critical stance advocated by Sillitoe with regards to both development science and development policy. Both Stirrat (1998: 243) and Posey (1998: 241), for example, suggest that Sillitoe has ignored much of the "political context" in which IK research takes place. The point has been raised in more general critiques of the IK literature as well. For example, Agrawal (1995: 430) commented that the central objective of much IK research is politically inappropriate in that it involves the *ex situ* preservation of IK in centralized, bureaucratically organized databases. This, he claims, is "not just the preferred strategy ... it is almost always their only strategy" for preserving IK (1995: 430). IK researchers have tended to advocate the external preservation and exploitation of IK by development agencies, without insisting upon the concurrent preservation of the indigenous cultures which produce it. Nor have they supported indigenous peoples concerning their ownership of that knowledge.

As a way to approach political relationships between the global capitalist system and local systems of production, Agrawal (1995: 431) suggests that "it might be more helpful to frame the issue as one requiring modifications in political relationships". Thus, he suggests an alternate position, which advocates the *in situ* preservation of IK through changes to "state policies and market forces", and offers a detailed discussion of the arguments for and against such a course of action (Agrawal, 1995: 432). Stone echoes a similar view:

> Goodwill and inter-disciplinary open-mindedness will not be enough to change this highly bureaucratized system and world of development planning, implementation, and evaluation unless accompanied by institutional and policy change (1998: 243).

Brouwer raises yet another issue, which appears to be a response to Sillitoe's contention that IK research "is largely ethnographic reporting of others' production systems" (1998: 234). Such a view

> deals only with indigenous technical knowledge in the field of agriculture ... More important than this dimension, is an awareness of the epistemology on which the anthropologist's science is based and the one which informs the participant's knowledge (Brouwer, 1998: 351).

Such an emphasis would call for the development of an emic understanding of the culture under study; in contrast with the views of those in favour of studies which approach IK "as the demands of development require", and views which reject an "anti-positivist critique" (Sillitoe, 1998). These contrasting views indicate that there is more than one approach to IK research and to the issues which it raises. Sillitoe's contrast between market-liberal and neo-populist approaches is useful, but it does not adequately encompass the issues. Neither is a contrast between the IK and the TEK approaches adequate – not only because the latter is encompassed by the former, but also because none of the authors considered above explicitly identify themselves with TEK, although both Agrawal (1995) and Purcell (1998) make much more explicit reference to ecological issues.

Purcell provides a more useful suggestion with his contrast between the "indigenous knowledge" approach and the "ecoliberal" approach. The former approach is based upon an understanding of the fact that "capitalist transformation threaten[s] local communities and ecological systems and is therefore unsustainable" (Purcell, 1998: 265). The "ecoliberal" approach, by contrast, "aims at helping to integrate people into the capitalist market on their own terms" (1998: 267). While both might remain compatible with IK research, therefore, only the former explicitly questions the aims of development and the policies which guide it and, by implication, the epistemology and worldview of traditional development science. This is also a position which is suggested by the explicitly ecological focus provided by an emphasis upon TEK.

QUESTIONS OF IDEOLOGY AND EPISTEMOLOGY

While Sillitoe proposes that "the philosophy underlying indigenous-knowledge research is unexceptionable" (1998: 224), the same cannot be said of TEK, primarily because of its explicit emphasis upon the importance of ecological knowledge and issues. This allows TEK researchers to draw upon several decades of theoretical development within the broader area of ecological philosophy. The contrast drawn by Purcell between the indigenous knowledge approach and the ecoliberal approach to development, for example, is highly reminiscent of similar contrasts between two approaches to ecological problems in the earlier ecological literature.

Both Naess' (1973) contrast between shallow and deep ecology, and Bookchin's (1980: 58–59) contrast between environmentalism and ecology raise the same distinction. This is a contrast between an approach which seeks to solve ecological problems (or problems of "development") without questioning the premises of contemporary science and society and a more radical approach which sees the political, economic and scientific status quo as in large part the cause of global ecological problems. In fact, this contrast in approaches to ecological problems is traceable to the earlier works of Leopold and his discussion of the "A-B cleavage" amongst the conservationists of his day. Where one group saw the land merely as soil and its function as merely "commodity production", the other group saw it as something much broader, an ecosystem with a variety of functions (Leopold, 1949: 221).

This more radical approach to ecology has also developed an alternative understanding of what holism entails. As Ingold suggests, for example, while anthropologists have long espoused an holistic approach, "[t]hey have ... been inclined to take this holism as entailing an approach that focuses on 'wholes' – conceived as total societies or cultures – as opposed to their parts or members" (1992: 695).

By contrast, ecological philosophy has developed an explicitly relational understanding of holism, which arises from ecology's self-definition as the study of the interrelationships among living organisms, and between living organisms and the biophysical environment. For example, Bateson (1979: 100) has suggested that, "all serious holism, is premised upon the differentiation and interaction of parts" (1979: 100). This view proposes that holism consists of the study of patterns of relationship, rather than of abstract wholes of some type, including the relationships between theory and practice, between ideology and production systems, and between societies and their ecological circumstances.

The holistic perspective provided by TEK's explicit emphasis upon ecological knowledge allows it to draw upon a more sophisticated understanding of both epistemological and political issues. Informed by ecological philosophy and philosophy of science, some TEK research has emphasized not only the differences between the epistemologies and worldviews of modern science and indigenous philosophies, but also the similarities between indigenous worldviews and the alternative approaches to science.

This holistic perspective also leads to a more explicit focus upon political issues, because part of its mandate is to study the varying patterns of relationship between different social and ecological systems and their adaptive consequences. Once again, this line of inquiry suggests both a contrast between indigenous systems and those of global capitalism, as well as an explicit critique of the latter. Consequently, TEK has developed a more sophisticated critique of current patterns of development and has proposed policy alternatives as well.

All of these issues – including the critique of conventional resource management practices, its greater emphasis upon proposing political and policy alternatives, and upon comparing and contrasting diverse systems of knowledge – are well illustrated by a recent debate over the inclusion of TEK in an environmental impact assessment process in the Northwest Territories of Canada. This debate responded to the position of the Government of Canada's Environmental Assessment Panel, which proposed that TEK must be given "equal consideration with scientific research in assessing the environmental and socio-economic impacts" of a proposed diamond mine in the Northwest Territories (Howard and Widdowson, 1996: 34).

On one side of the debate are its instigators (Howard and Widdowson, 1996, 1997), who have argued that TEK is unscientific and should not be made a mandatory part of environmental impact assessments because it is spiritually based. Their argument is twofold. First, they object to the mandatory inclusion of TEK in impact assessments because they consider that it implies "the imposition of religion" upon Canadian citizens (Howard and Widdowson, 1996: 34). Second, and more importantly with regards to the present argument, they suggest that TEK is unscientific, since "spiritualism is obviously inconsistent with scientific methodology". They conclude, therefore, that TEK "hinders rather than enhances the ability of governments to more fully understand ecological processes", since there is no way in which "spiritually based knowledge claims can be challenged or verified" (Howard and Widdowson, 1996: 34).

On the other side of the debate are those who argue in various ways for the importance and value of TEK and, most importantly to the present question, suggest that the scientific community and environmental managers may have something to learn from a study of TEK (Berkes and Henley, 1997; Fenge, 1997). Berkes and Henley (1997: 30), for example, argue for pluralism and "the recognition of indigenous knowledge as a legitimate source of information and values", as opposed to the "the simplistic view that there is such a thing as objective, value-free science".

Although they explicitly deny that their arguments are intended to promote the idea of a "universal truth", or a "totally objective science", however, Howard and Widdowson's (1997: 46–47) reply to their critics only serves to reinforce the fact that they do. They contend, "There are not different ways of knowing. There are different beliefs about the same phenomena" (Howard and Widdowson, 1997: 46). Clearly, the only legitimate way of knowing which they recognize is that of conventional science, in the same way that some IK proponents, such as Sillitoe (1998), consider development science to be the final arbiter of the validity of all knowledge, as seen in the debate analyzed in the previous section.

There are remarkable parallels between the two debates in terms of the epistemological questions raised. While Sillitoe clearly recognizes the value of indigenous or traditional knowledge, at least insofar as it contributes to the development process, his understanding of science, and of its role in development, differs little from that of Howard and Widdowson. As Sillitoe suggests, for example, "the perspective of natural science has proved successful in promoting the kinds of interventions that development demands" (1998: 226). Yet whether the types of interventions which development demands have proven to be successful adaptations, or whether they are ecologically appropriate, is a question which is never raised. Sillitoe's reasons for not questioning the scientific orthodoxy concerning development issues, however, are quite clear, for as he states, "It is unrealistic to think that the scientific community could be persuaded to abandon its successful orthodoxy; it would be unable to make sense of the natural world without it" (1998: 232).

Such a position assumes two things: first, that scientific orthodoxy is successful, and second, that there is no alternative to the dominant approach to development science. An ongoing critique of the scientific orthodoxy from the perspective of ecological philosophy and TEK research, however, challenges both of these assumptions. The two sides in this debate within the philosophy of science (concerning the nature of science, its aims, methods and worldview) also closely parallel the contrast drawn by Purcell. The following discussion will particularly highlight the way in which this debate applies to ecological and resource management issues, its implications for policy, and the manner in which it supports the importance of traditional or indigenous knowledge in developing a better understanding of contemporary ecological problems.

On one side of the debate are conventional views of science, which arose with the advent of modern philosophy in the sixteenth century, and which remain predominant today. Based upon a combination of Newton's physics, Descartes' philosophy, and Bacon's experimental method (Holling et al., 1998:

344), this view has several characteristics. "The general conception of reality from the seventeenth century onward saw the natural world as a multitude of separate material objects assembled into a huge machine". Its methodology is essentially analytic or reductive, in that it believes that natural phenomena can best be understood by breaking them down into parts for separate study, in isolation from their larger context (Holling *et al.*, 1998: 344). Through experimental manipulation of the parts, it then seeks to learn to predict and control phenomena through the discovery of causal laws governing the motion of a mechanical universe. This reductionistic philosophy has also been institutionalized with the historical fragmentation of "natural philosophy", as science was originally called, into many separate and specialized disciplines.

On the other side of the debate is an alternative view of science arising from a variety of sources, including biological ecology (Holling, 1973; Holling *et al.*, 1998; Gunderson and Holling, 2001), ecological philosophy (Berman, 1984; Capra, 1982; Griffin, 1988; Merchant, 1980), and anthropology (Bateson, 1972, 1977; Bateson and Bateson, 1988; Ingold, 1990, 1992, 2000; Harries-Jones, 1992, 1995; Rappaport, 1979, 1994). This alternative view can be seen in the context of the paradigm shifts suggested by Kuhn (1970) who argued that the history of science has been characterized by a series of scientific revolutions through which its understandings of the nature of the world, and of science itself, have been fundamentally altered. Essentially, this alternative view calls for a paradigm shift away from mechanistic understandings of science and towards more organic models (Berman, 1984).

The most pragmatic reason for this is that mechanical models and methods appear to be substantially challenged by the types of questions which contemporary ecological problems have brought to the forefront. Many of these problems have no simple solutions; there are scale issues, time lags in response, and the scientific evidence is never clear-cut and always incomplete, making it very difficult to apply cause-effect science of the conventional kind (Kates *et al.*, 2001). As Holling *et al.* (1998: 352) suggest, "there is a worldwide crisis in resource management because the existing science that deals with the issue seems unable to prescribe sustainable outcomes".

This proposed paradigm shift, therefore, needs to involve a variety of changes in the aims and methodology of science. Unlike conventional science, this approach is fundamentally "integrative" and "interdisciplinary", focusing upon the study of complex systems (Holling *et al.*, 1998: 345). It focuses largely upon the study of patterns of relationship, including "the interaction of social systems with natural systems" (1998: 346). In this way it challenges the dualism of conventional science, which is "disciplinary, reductionist, mechanistic, and detached from people, policies and politics" (1998: 345–346).

The new science responds to the ecological dilemma posed by Ingold (2000: 3) – "human beings must simultaneously be constituted both as organisms within systems of ecological relations, and as persons within systems of social relations". Rather than dealing with culture and nature as essentially separate phenomena, it proposes that social systems must be viewed as a part of larger living systems, upon which they depend for their survival. Such a participatory view also flies in the face of traditions which consider science to be value-free and objective, because when society is considered to be a part of a larger living system, human values and their variant effects upon that system become an important component of the system to be studied. Rather than the analytic and reductive methodology of traditional science, this stream proposes a more synthetic and holistic approach, in the sense that it focuses upon studying the patterns of relationship between the parts, rather than the parts themselves (Bateson, 1979: 100; Holling et al., 1998: 346).

Bateson provides perhaps the most articulate statement of this new approach in his discussion of the concept of "abduction" (Bateson, 1979: 149, 153–155; Bateson and Bateson, 1988: 37, 174–175). Bateson used this term in order to distinguish his own methodology from that of induction and deduction. The methodology he proposed was essentially qualitative and metaphorical, and consistent with a relational understanding of holism, in that it involved a comparison of patterns of relationship with one another and an evaluation of their symmetry or asymmetry.

This is the same methodology that is applied in the system of scientific classification in biology, for example, through which individual animals and plants are classified into species, genera and the like on the basis of just such similarities of patterning in their phenotypes. Not surprisingly, this methodology is also perfectly tailored to the central question with which this view of science is concerned – the adaptation of human patterns of social and economic relationship to the larger patterns of relationship exhibited by ecological systems.

When applied to resource management issues, each view has very different consequences for scientific practice. The reductionist model is based upon utilitarian premises, which view nature merely as a collection of commodities which have no value until humans make use of them (Berkes and Folke, 1998). The conservation practices suggested by this tradition have, since the 1930s, tended to rely upon a calculation of the maximum sustainable yield for any particular resource. This approach not only attempts to predict and control the abundance of the resources harvested, but is also reductive in two different senses. First, it concentrates upon single species, in isolation from the ecosystem within which they are embedded. Second, it reduces the question of sustainable management to an equation which seeks to calculate the maximum possible harvest. This approach appears, therefore, to have adapted itself quite effectively to the monetary calculus of economics, but rather less effectively to the eco-systems which it exploits, as the collapse of Canada's east coast cod fishery, among many other examples, seems to illustrate (Rogers, 1995).

The synthetic and relational approach, on the other hand, proposes the development of a type of "adaptive management" (Holling et al., 1998: 358), which seeks to integrate not only the science of resource management, but also human institutions, techniques and values into the larger system of which they are a part. One of the premises of this view "is that knowledge of the system we deal with is always incomplete. Surprise is inevitable" (1998: 346).

There are two reasons for this unpredictability in complex organic systems.

First, the complexity of the social-ecological system is such that we can never have complete knowledge of all of the relevant variables. Second, an important characteristic of complex organic systems is their "non-linearity" (Holling *et al.*, 1998: 352–354), in the sense that the same stimulus, applied to the same system at a different time, may produce strikingly different results. Consequently, "adaptive management ... treats policies as hypotheses, and management as experiments from which managers can learn" (1998: 358).

Ecological and resource management issues thus necessitate an approach that does not fit well with the conventional mechanistic, linear science of the Age of Enlightenment. As Kates *et al.* (2001) argue, a new science of sustainability must differ fundamentally from most science as we know it. The common sequential analytical phases of scientific inquiry such as conceptualizing the problem, collecting data, developing theories and applying the results do not work. We must also consider the consequences of our actions within a complex organic system and the non-linear effects which these actions have upon ourselves (Bright, 1999). In the emerging sustainability science, the parallel functions of social learning, adaptive management and policy as experiment become key.

This new kind of science recognizes the need to act before scientific uncertainties can be resolved. This is not only because it is difficult to get agreement among experts, but because some uncertainties are not resolvable by science. Hence, it becomes important to design institutions and processes that can facilitate cooperation between scientists and resource users. For example, the participation of fishers in management decisions not only increases the likelihood that they agree to these decisions, but it also ensures that the parties share the risk in decision-making in a fundamentally uncertain world (Berkes *et al.*, 2001).

Such a new paradigm of resource and environmental management promises a much humbler role for the manager. It also changes the role of science in dealing with ecological crises. Scientific orthodoxy comes under question, and expert-knows-best science is abandoned in favour of a kind of science in which "research must be created through processes of co-production in which scholars and stakeholders interact to define important questions, relevant evidence, and convincing forms of argument" (Kates *et al.*, 2001). The implications of such a paradigm are far reaching. Not only are the roles of science and experts altered, but the processes of knowledge making and decision making are also changed. This, in turn, has implications for the empowerment of resource users and other stakeholders in areas such as the management of small-scale fisheries (e.g., Berkes *et al.*, 2001) and for the recognition of the value of the knowledge they hold.

TRADITIONAL ECOLOGICAL KNOWLEDGE AND NEW UNDERSTANDINGS OF RESOURCE MANAGEMENT

The possibility of learning from the values, epistemologies and practices of non-Western cultures has not been lost on scholars dealing with complex organic systems and adaptive management (Berkes and Folke, 1998; Folke and

Colding, 2001; Gunderson and Holling, 2001). Conventional science and management has a questionable record with regard to long-term sustainability, whereas some indigenous or traditional peoples have developed systems that seem more sustainable than our own (Bodley, 2001). Traditional resource management systems may thus be viewed as experiments in successful living, and drawing upon knowledge of these alternatives may provide insights and "speed up the process of adaptive management" (Holling *et al.*, 1998: 359).

There are several similarities between traditional or indigenous management systems and adaptive management. If the orderly and rational science of the Age of Enlightenment is replaced by a new paradigm along the lines of adaptive management, the chasm between indigenous knowledge and Western science essentially evaporates. In support of this view, at least one Native American scholar (Cajete, 2000) has recently gone so far as to describe TEK as a "Native science" which focuses upon the study of "natural laws of interdependence".

Many of the prescriptions of traditional knowledge and practice are consistent with adaptive management as an integrated method for resource and ecosystem management (Berkes *et al.*, 2000). Adaptive management, like many TEK systems, emphasizes processes that are part of ecological cycles of renewability and regards human use of the environment in terms of how well it fits these cycles. Like many TEK systems, adaptive management considers change as inevitable and assumes that nature cannot be controlled and yields cannot be predicted. In both adaptive management and TEK, uncertainty and unpredictability are considered to be characteristics of all ecosystems, including managed ones; in both, social learning appears to be the way in which societies respond to uncertainty. Often this involves social learning at the level of society or institutions.

TEK, based on detailed observation of the dynamics of the natural environment, feedback learning, social system/ecological system linkages, and resilience-enhancing practices, bears a strong resemblance to adaptive management (Berkes *et al.*, 2000). In a sense, adaptive management may be seen as a "rediscovery of traditional systems of knowledge". Even though there are no doubt major differences between the two, adaptive management may be viewed as the scientific analogue of TEK. Drawing on management practices based on TEK and understanding the social processes behind them may speed up the process of designing alternative resource management systems.

Table 1 provides a list of management practices documented from TEK systems from around the world (Berkes *et al.*, 2000). The first heading of the list itemizes five practices that are found in both TEK systems and in Western science, including species conservation and habitat conservation, for example, through taboos and sacred areas (Ramakrishnan *et al.*, 1998; Folke and Colding, 2001). The second identifies three practices (multiple species management, rotation, succession management) that have been abandoned by those Western management methods which emphasize yield maximization but are demonstrably alive and well in many TEK systems, for example, including those that use fire for forest succession management (Boyd, 1999).

The third heading in Table 1 is most interesting because it identifies five

Table 1 Social-ecological practices in traditional knowledge and practice (Berkes *et al.*, 2000, adapted from Folke *et al.*, 1998)

Management practices based on ecological knowledge

(a) Practices found both in conventional resource management and in some local and traditional societies
- Monitoring resource abundance and change in ecosystems
- Total protection of certain species
- Protection of vulnerable life history stages
- Protection of specific habitats
- Temporal restrictions of harvest

(b) Practices largely abandoned by conventional resource management but still found in some local and traditional societies
- Multiple species management; maintaining ecosystem structure and function
- Resource rotation
- Succession management

(c) Practices related to the dynamics of complex systems, seldom found in conventional resource management but found in some traditional societies
- Management of landscape patchiness
- Managing ecological processes at multiple scales
- Responding to and managing pulses and surprises
- Nurturing sources of ecosystem renewal
- Watershed-based management

---

resource management practices documented from TEK systems but seldom found in Western resource management. Ecologists have discussed one of these, landscape patchiness, but its management has not to any extent become part of conservation practice. The next three (managing ecological processes at multiple scales, responding to pulses, nurturing renewal sources) deal with kinds of ecosystem dynamics that have been discussed only with the development of the notion of adaptive management and adaptive renewal cycles (Holling *et al.*, 1998; Gunderson and Holling, 2001). The existence of these practices in some TEK systems is further evidence of the similarity between TEK and adaptive management.

Take for example the management of ecological processes at multiple scales. Based on ethnohistorical information and current practice, Cree hunters of James Bay in subarctic Canada seem to be simultaneously managing beaver populations on a 4–6 year time scale, lake fish on a 5–10 year scale, and caribou on a 80–100 year scale, with a well established code of practice appropriate for each resource type. The rules and practices of resource use, for both temporal and spatial scales, and the kinds of environmental information that provide feedback, for example, to relocate to a new fishing area, have been documented extensively through participatory research with the Cree (Berkes, 1999).

The fifth item in the third heading, watershed based management, has been known to ecologists at least since the watershed experiments of the 1970s (Bormann and Likens, 1979), even though it is not yet being used extensively in conservation and management practice. A watershed unit provides the biogeographic boundary for a terrestrial ecosystem; as such, it is the starting point for ecosystem management. The use of watershed units in TEK systems provides evidence for the existence of ecosystem-like concepts among several Amerindian, Asia-Pacific, European and African cultures. The rediscovery of such ecosystem concepts among traditional peoples has been important in the appreciation of TEK among scientists.

Table 2 provides examples of the application of ecosystem-like views in TEK systems. Southeast Asia and Oceania had, and to some extent still have, a wealth of these traditional ecosystem management practices. Examples in Table 2 include TEK systems from the Pacific Northwest, Southeast Asia, Japan, and West Africa. Watershed units are commonly used in TEK systems in North America to mark out tribal boundaries and hunting territories. For example, tribal chiefs of the Gitksan (Gitxsan) and Wet'sewet'en of the Pacific Northwest describe their land boundaries as "from mountain top to mountain top". They orient themselves by two directional axes within this watershed, vertically up and down from valley bottom to mountaintop, and horizontally, upstream and downstream (Tyler, 1993). Similarly, family hunting territories among the James Bay Cree are based on watersheds. The height of land between adjacent river systems provides a convenient and enforceable way of delimiting territorial boundaries.

Some of the most highly developed ecosystem applications were found in the Asia-Pacific region. The ancient Hawaiian *ahupua'a* (Costa-Pierce, 1987), the Yap *tabinau*, the Fijian *vanua*, and the Solomon Islands *puava* (Ruddle et al., 1992) all refer to generically similar watershed-based management systems. In ancient Hawaii, valleys within watersheds were used for integrated

Table 2 Examples of traditional applications of the ecosystem view (Berkes et al., 1998)

System	Country/region	Reference
1. Watershed management of salmon rivers and associated hunting and gathering areas by tribal groups	Amerindians of the Pacific Northwest	Williams and Hunn (1982)
2. Delta and lagoon management for fish culture (*tambak* in Java) and the integrated cultivation of rice and fish	South and Southeast Asia	Lasserre et al. (1983)
3. *Vanua* (in Fiji), a named area of land and sea, seen as an integrated whole with its human occupants	Oceania, including Fiji, Solomon Islands, ancient Hawaii	Ruddle and Akimichi (1984); Baines (1989)
4. Family groups claiming individual watersheds (*iworu*), as their domain for hunting, fishing, gathering	The Ainu of northern Japan	Watanabe (1973); Ludwig (1994)
5. Integrated floodplain management (*dina*) in which resource areas are shared by social groups through reciprocal access arrangements	Mali, Africa	Moorehead (1989)

farming. The ecosystem unit extended from upland forests protected by taboo, through several agricultural zones, downstream to the coral reef and lagoon. Similarly in the Solomon Islands, a *puava* in the widest sense includes all resources and land in a watershed, from the top of the mainland mountains to the open sea outside the barrier reef (Hviding, 1996). [Editor's note: See Hviding's article in this volume.] In each of these cases, the social group inhabiting the ecosystem unit was considered to be part of the system, and affiliation with a particular area was considered to be part of a person's identity.

These characteristics highlight the fact that ecosystem-like concepts in TEK systems are fundamentally different from those of scientific ecology. The scientific concept of ecosystem that emerged in the postwar period was very much in the positivistic tradition, "a machine theory applied to nature"; the ecosystem was often conceived as a mechanistic device and represented as a computer model (Golley, 1993: 2). By contrast, many TEK systems depict ecosystems not as lifeless, mechanical and distinct from people, but as fully alive and encompassing humans. In all cases, TEK makes sense only in the context of active engagement with the land, which Ingold (2000) calls the "dwelling perspective".

Ecosystems are in part socially constructed, and resource management and conservation practices in TEK systems are based on a variety of social processes. Table 3 summarizes four clusters of social processes that inform the ecological practices used in TEK systems. One set deals with the generation, accumulation and transmission of TEK. A second set concerns the structure

Table 3   Social-ecological processes in traditional knowledge and practice (Berkes *et al.*, 2000, adapted from Folke *et al.*, 1998)

Social processes behind management practices
(a) Generation, accumulation and transmission of local ecological knowledge
• Reinterpreting signals for learning
• Revival of local knowledge
• Folklore and knowledge carriers
• Integration of knowledge
• Intergenerational transmission of knowledge
• Geographical diffusion of knowledge
(b) Structure and dynamics of institutions
• Role of stewards/wise people
• Cross-scale institutions
• Community assessments
• Taboos and regulations
• Social and religious sanctions
(c) Processes for cultural internalization
• Rituals, ceremonies and other traditions
• Cultural frameworks for resource management
(d) Worldview and cultural values
• A worldview that provides appropriate environmental ethics
• Cultural values of respect, sharing, reciprocity, and humility

and dynamics of institutions, including leadership and rule making. A third set is about rituals and ceremonies that provide cultural processes for the internalization of TEK practices. A fourth set deals with the worldview and cultural values of the group in question. Each of the processes outlined in Table 1 has in fact been described in the TEK literature from various places in the world.

Much of this literature recognizes TEK not only as "knowledge" or "technique", but also as a knowledge/practice/belief complex in which the context is provided by culture and history (Berkes, 1999). Thus, in the study of the importance of TEK to the conservation of biodiversity, for example, one cannot merely learn from traditional techniques of biodiversity conservation outside of their cultural context (e.g., Berkes *et al.*, 1995). Nor can one discuss, in a decontextualized way, the possible contribution of TEK to sustainable land use (Preston *et al.*, 1995), to environmental assessment (Stevenson, 1996), or to ecological restoration (Kimmerer, 2000). Indeed, there is an increasing recognition of the potential contributions of indigenous knowledge and indigenous cultures in a number of areas towards the creation of a sustainable world.

In this, TEK, as Western academics understand it, again betrays the legacy which it owes to earlier works in the ecological literature, which have long demonstrated a willingness to learn about ecological patterns of living and thinking from non-Western cultures. In this vein, for example, Capra (1975) studied in detail the similarities between the emerging scientific paradigm discussed above and Daoism. Similarly, deep ecologists often draw explicit links between their own position and Buddhism (Devall and Sessions, 1985: 100–101; Fox, 1990: 11–12; Zimmerman, 1994: 313–317). Many other ecological thinkers explore the possibility of learning about environmental ethics and ecological understanding from an examination of the traditional views of Native Americans and other Aboriginal peoples (Bird-David, 1993; Booth and Jacobs, 1990; Callicott, 1982; Hughes, 1983, 1991; Reed, 1991; Salmon, 2000). Perhaps the most comprehensive survey to date is provided by Callicott (1994), with his consideration of ecological ethics from around the world.

What TEK research adds to this tradition, which has largely remained focused upon ideological, ethical and epistemological issues, is its clear emphasis upon practical matters such as resource management and biodiversity conservation. Perhaps one of the more important lessons provided by a study of indigenous or traditional approaches to the management of resources, however, is as much theoretical as it is practical. This is that indigenous resource management systems tend to be non-dualistic, in the sense that values are explicitly incorporated into the system. As Berkes *et al.* (1998) suggest, for example, in traditional management systems, morality and ethics are explicitly a part of the management system; in Western scientific systems they are merely implicit. The newer approach to science is attempting to make these values explicit as well, for two reasons. First, following Kuhn, any scientific paradigm needs to be based upon premises about the nature of the world. Second, as social systems are embedded within ecological systems, the ecological implications of various value systems are an important research question. Given these premises, the system of values enacted within the global socioeconomic system, and within processes of development, cannot remain immune from scrutiny.

Therefore, unlike those who consider that there is but a single way of knowing or understanding "truth", the proponents of this new scientific paradigm have long suggested that there is much to be learned not only from a study of other cultural practices, but from other cultural ways of *knowing* as well. As Bateson once suggested with regard to contemporary ecological problems, "it is possible that some of the most disparate epistemologies which human culture has generated may give us clues as to how we should proceed" (Bateson and Bateson, 1988: 136). Further,

> other attitudes and premises – other systems of human 'values' – have governed man's relation to his environment and his fellow man in other civilizations and at other times ... In other words, our way is not the only possible human way. It is conceivably changeable (Bateson, 1972: 493).

IMPLICATIONS FOR POLITICS, POLICY AND PRACTICE

The broader approach to indigenous knowledge, including understandings of ecology, resource management and environmental values, would seem to suggest that there is much more to be learned from traditional peoples than simply techniques for improving the success of ongoing development initiatives. Rather than promoting research which uncritically serves the development process, the alternative approach seeks, as Purcell suggests, "the application of anthropological knowledge to ecological problems – both indigenous and non-indigenous" (1998: 267).

There is a rich global heritage of local environmental knowledge developed over millennia, and a diversity of worldviews (Callicott, 1994). Just as biodiversity provides the raw material for ecological evolution, cultural diversity provides the raw material for the evolution of sustainable relations between humans and their biophysical environment (Gadgil, 1987). "The alternative world views of traditional peoples could provide insights for redirecting the behavior of the industrial world towards a more sustainable path" (Berkes *et al.*, 1995: 299).

This emphasis upon identifying both the diversity of opinion which characterizes the scientific and development communities, and the similarities between its own understandings and those of indigenous peoples, allows TEK to transcend the opposition between Western and indigenous knowledge which Agrawal (1995: 414) claims has "seduce[d] modernization and indigenous theorists alike". For "it may be more sensible to accept differences within these categories and perhaps find similarities across them" (1995: 427).

Yet this type of cross-cultural study, which examines both the worldviews and knowledge, as well as the production systems and techniques, of indigenous peoples, is not without its political obstacles. As Purcell observes, for example, "interest in the study and application of IK – or indigenous praxis as a transformative process – logically implies indigenous people assuming relative autonomy" (1998: 260). Such autonomy would imply not only that indigenous people achieve greater political control, or self-government, but also that they assume much more control over resource management decisions in their traditional territories.

Such a community-based and decentralized pattern of resource management, however, is almost the polar opposite of the highly centralized patterns of development which are currently practiced and is not without its detractors. Howard and Widdowson, for example, suggest that allowing aboriginal peoples to control their own resource management strategies leads to a conflict of interest. "Aboriginal groups are the main harvesters of renewable resources in the north; clearly they should not be placed in a position of regulating these resources" (1996: 35).

This quote helps to bring into focus the politics of TEK. Giving legitimacy to TEK may be seen to be in support of self-determination and local political autonomy, including decision-making in the area of land and resource management. However, it does not fit the world of top-down decision making. The question of ownership or authority over resources is a fundamental issue in the debate over the management of common property resources (e.g., Ostrom, 1990). The use and control of TEK by the indigenous peoples themselves helps them manage their own resources (community-based management) or become partners in decision-making (co-management). The sharing of power and responsibility between aboriginal groups, governments and public development agencies in such a co-management situation "implies a partnership of equals. One of the ways for creating an equitable relationship lies with the recognition of indigenous knowledge as a legitimate source of information and values" (Berkes and Henley, 1997).

A major problem in this regard is the pressure to use TEK outside of its cultural context. The self-proclaimed goal of much of TEK research in the North, Nadasdy points out, is to integrate it with scientific knowledge. Nadasdy (1999: 15) argues that

> framing the problem as one of 'integration' [of TEK with science] automatically imposes a culturally specific set of ideas about 'knowledge' on the life experiences of aboriginal people. The goal of knowledge-integration forces TEK researchers to compartmentalize and distill aboriginal people's beliefs, values and experiences according to external criteria of relevance, seriously distorting them in the process.

Nadasdy could just as easily have made this statement on the issue of collecting IK for development. The very idea of IK as technique to be incorporated into Western-style development, or the very idea of aboriginal knowledge as data that can be integrated into resource management science "implicitly assumes that knowledge is an intellectual product which can be isolated from its social context" (Nadasdy, 1999: 11). Such an approach "also takes for granted existing power relations between aboriginal people and the state by assuming that traditional knowledge is simply a new form of 'data' to be incorporated into already existing management bureaucracies" (Nadasdy, 1999: 15).

Challenging this power relation, rather than taking the status quo for granted, requires either empowering the indigenous group to carry out their own research or developing participatory research methodologies that serve the indigenous group as well as the researcher. A rich literature base is developing in both of these approaches (Taiepa *et al.*, 1997; Beaucage, 1997; Alcorn and Royo, 2000; Nabhan, 2001; Riedlinger and Berkes, 2001). A review of such participatory TEK research and the forces that have led to its development is beyond the scope of this chapter. Suffice it to say, however, that recent developments in international law and policy regarding the rights of indigenous peoples and local communities to their knowledge have changed the nature of indigenous knowledge research, especially regarding biodiversity. This, in turn, radically changes older concepts of IK as the "common heritage of mankind" and increasingly requires that policy and management be made with the full participation of indigenous and local communities (Mauro and Hardison, 2000).

*   *   *

An explicitly ecological approach to the study of traditional knowledge appears to answer many of the criticisms which have been made concerning IK research. This is largely due to the holistic or relational approach of TEK and to the theoretically sophisticated ecological literature upon which it is able to draw. Such a relational approach also forces this type of ecologically oriented approach to examine patterns of relationship as diverse as the relationships between varying ideologies and production systems; allows it to compare the symmetry and asymmetry of various worldviews; and, most importantly, allows it both to critique the dominant patterns of development promoted by global capitalism and to suggest alternatives. For "it is economic development based on the logic of unbridled growth that destroys indigenous territories" (Purcell, 1998: 265). This is also the case with the cultures and systems of knowledge – as praxis – of local cultures and communities.

As indigenous knowledge researchers, our choice of methodology and focus is as much ethical and political as it is practical. Shall we study indigenous knowledge "as the demands of development require", or shall we promote its importance for the benefit of the people who possess it? Purcell (1998) and Nadasdy (1999), for example, suggest the importance of either fully collaborative research or the promotion of IK research conducted by indigenous peoples themselves, for their own use and benefit, rather than that of agents of development. Therefore, researchers of indigenous knowledge "must chose between being facilitators of local autonomy ... as the indigenous perspective demands, or be agents of hegemonic 'progress'" (Purcell, 1998: 267).

From an ecologically informed perspective, such as that which TEK brings to indigenous knowledge research, the choice between these two options is clear – to support local autonomy and local participation in resource management. Diversity, after all, is a universal characteristic of complex organic systems; *cultural* diversity is required for successful adaptation to our diverse ecological circumstances. We must support local cultures in their efforts to continue to live sustainably within a wide variety of ecological contexts, so that they may continue to adapt to change.

NOTE

[1] *Emic* is a perspective in ethnography that uses the concepts and categories that are relevant and meaningful to the culture under analysis, that is, a view from the inside. *Etic* is a perspective that uses the concepts and categories of the anthropologist's culture to describe another culture, that is, a view from the outside.

BIBLIOGRAPHY

Agrawal, Arun. 'Dismantling the divide between indigenous and scientific knowledge.' *Development and Change* 26: 413–439, 1995.

Alcorn, Janis B. and Antoinette G. Royo. *Indigenous Social Movements and Ecological Resilience: Lessons from the Dayak of Indonesia.* Washington, DC: Biodiversity Support Program, 2000.

Baines, Graham B.K. 'Traditional resource management in the Melanesian South Pacific: A development dilemma.' In *Common Property Resources,* F. Berkes, ed. London: Belhaven, 1989, pp. 273–295.

Bateson, Gregory. *Steps to an Ecology of Mind.* New York: Ballantine Books, 1972.

Bateson, Gregory. *Mind and Nature: A Necessary Unity.* New York: Bantam Books, 1979.

Bateson, Gregory and Mary Catherine Bateson. *Angels Fear: Towards an Epistemology of the Sacred.* New York: Bantam Books, 1988.

Beaucage, P. and Taller de Tradición Oral del Cepec. 'Integrating innovation: The traditional Nahua coffee-orchard (Sierra Norte de Puebla, Mexico).' *Journal of Ethnobiology* 17: 45–67, 1997.

Berkes, Fikret. *Sacred Ecology. Traditional Ecological Knowledge and Resource Management.* Philadelphia and London: Taylor and Francis, 1999.

Berkes, Fikret, Johan Colding and Carl Folke. 'Rediscovery of traditional ecological knowledge as adaptive management.' *Ecological Applications* 10: 1251–1262, 2000.

Berkes, Fikret and Carl Folke. 'Linking social and ecological systems for resilience and sustainability.' In *Linking Social and Ecological Systems: Management Practices and Social Mechanisms for Building Resilience,* F. Berkes and C. Folke, eds. Cambridge: Cambridge University Press, 1998, pp. 1–25.

Berkes, Fikret, Carl Folke and Madhav Gadgil. 'Traditional ecological knowledge, biodiversity, resilience and sustainability.' In *Biodiversity Conservation,* Charles Perrings, Karl-Göran Mäler, Carl Folke, C.S. Holling, and Beng-Owe Jansson, eds. Dordrecht, The Netherlands: Kluwer Academic Publishers, 1995, pp. 281–299.

Berkes, Fikret and Thomas Henley. 'Co-management and traditional knowledge: threat or opportunity?' *Policy Options,* March 1997, pp. 29–31.

Berkes, Fikret, Mina Kislalioglu, Carl Folke and Madhav Gadgil. 'Exploring the basic ecological unit: ecosystem-like concepts in traditional societies.' *Ecosystems* 1: 409–415, 1998.

Berkes, Fikret, R. Mahon, P. McConney, R.C. Pollnac and R.S. Pomeroy. *Managing Small-Scale Fisheries: Alternative Directions and Methods.* Ottawa: International Development Research Centre, 2001.

Berman, Morris. *The Reenchantment of the World.* New York: Bantam Books, 1984.

Bird-David, Nurit. 'Tribal metaphorization of human-nature relatedness: a comparative analysis.' In *Environmentalism: The View from Anthropology,* Kay Milton, ed. London: Routledge, 1993, pp. 112–125.

Bodley, John H. *Anthropology and Contemporary Human Problems.* Toronto: Mayfield Publishing Company, 2001.

Bookchin, Murray. *Toward an Ecological Society.* Montréal: Black Rose Books, 1980.

Booth, Annie L. and Harvey M. Jacobs. 'Ties that bind: native American beliefs as a foundation for environmental consciousness.' *Environmental Ethics* 12: 27–43, 1990.

Bormann, F.H. and G.E. Likens. *Pattern and Process in a Forested Ecosystem.* New York: Springer-Verlag, 1979.

Boyd, Robert, ed. *Indians, Fire and the Land in the Pacific Northwest.* Corvallis: Oregon State University Press, 1999.

Bright, Chris, 'The nemesis effect.' *World Watch*, May/June 1999, pp. 12–23.

Brokensha, David. Comment on 'The development of indigenous knowledge: A new applied anthropology' by Paul Sillitoe. *Current Anthropology* 39(2): 236–237, 1998.

Brouwer, Jan. 'On indigenous knowledge and development.' *Current Anthropology* 39: 351, 1998.

Cajete, Gregory. *Native Science: Natural Laws of Interdependence*. Santa Fe, New Mexico: Clear Light Publishers, 2000.

Callicott, J. Baird. 'Traditional American Indian and western European attitudes towards nature.' *Environmental Ethics* 4: 293–318, 1982.

Callicott, J. Baird. *Earth's Insights: A Survey of Ecological Ethics from the Mediterranean Basin to the Australian Outback*. Berkeley: University of California Press, 1994.

Capra, Fritjof. *The Tao of Physics: An Exploration of the Parallels Between Modern Physics and Eastern Mysticism*. London: Fontana, 1975.

Capra, Fritjof. *The Turning Point: Science, Society, and the Rising Culture*. New York: Bantam Books, 1982.

Costa-Pierce, B.A. 'Aquaculture in ancient Hawaii.' *BioScience* 37: 320–330, 1987.

Devall, Bill and George Sessions. *Deep Ecology: Living as if Nature Mattered*. Salt Lake City, Utah: Peregrine Smith Books, 1985.

Fenge, Terry. 'Ecological change in the Hudson Bay bioregion: a traditional ecological knowledge perspective.' *Northern Perspectives* 25(1): 2–3, 1997.

Folke, Carl, Fikret Berkes, and Johan Colding. 'Ecological practices and social mechanisms for building resilience and sustainability.' In *Linking Social and Ecological Systems*, F. Berkes and C. Folke, eds. Cambridge: Cambridge University Press, 1998, pp. 414–436.

Folke, Carl and Johan Colding. 'Traditional conservation practices.' In *Encyclopedia of Biodiversity*, Simon Levin, ed. Vol. 5. San Diego, California: Academic Press, 2001, pp. 681–693.

Fox, Warwick. *Towards a Transpersonal Ecology: Developing New Foundations for Environmentalism*. Boston: Shambhala, 1990.

Gadgil, Madhav. 'Diversity: cultural and biological.' *Trends in Ecology and Evolution* 2(12): 369–373, 1987.

Golley, Frank B. *A History of the Ecosystem Concept in Ecology*. New Haven and London: Yale University Press, 1993.

Griffin, David Ray, ed. *The Reenchantment of Science: Postmodern Proposals*. Albany: State University of New York Press, 1988.

Gunderson, Lance H. and Holling, C.S., eds. *Panarchy: Understanding Transformations in Human and Natural Systems*. Washington, DC: Island Press, 2002.

Harries-Jones, Peter. 'Sustainable anthropology: ecology and anthropology in the future.' In *Contemporary Futures: Perspectives from Social Anthropology*, Sandra Wallman, ed. London: Routledge, 1992, pp. 157–171.

Harries-Jones, Peter. *A Recursive Vision: Ecological Understanding and Gregory Bateson*. Toronto: University of Toronto Press, 1995.

Holling, C.S. 'Resilience and stability of ecological systems.' *Annual Review of Ecology and Systematics* 4: 1–23, 1973.

Holling, C.S., Fikret Berkes and Carl Folke. 'Science, sustainability and resource management.' In *Linking Social and Ecological Systems: Management Practices and Social Mechanisms for Building Resilience*, F. Berkes and C. Folke, eds. Cambridge: Cambridge University Press, 1998, pp. 342–362.

Howard, Albert and Frances Widdowson. 'Traditional knowledge threatens environmental assessment.' *Policy Options*, November 1996, pp. 34–36.

Howard, Albert and Frances Widdowson. 'Revisiting traditional knowledge.' *Policy Options*, April 1997, pp. 46–48.

Hughes, J. Donald. *American Indian Ecology*. El Paso: Texas Western Press, 1983.

Hughes, J. Donald. 'Metakuyase.' *The Trumpeter* 8(4): 184–185, 1991.

Hviding, Edvard. *Guardians of Marovo Lagoon: Practice, Place and Politics in Maritime Melanesia*. Honolulu: University of Hawaii, 1996.

Ingold, Tim. 'An anthropologist looks at biology.' *Man* (N.S.) 25: 208–209, 1990.

Ingold, Tim. 'Editorial.' *Man* (N.S.) 27: 693–696, 1992.

Ingold, Tim. *The Perception of the Environment: Essays on Livelihood, Dwelling and Skill*. London and New York: Routledge, 2000.

Johannes, Robert E., ed. *Traditional Ecological Knowledge: A Collection of Essays*. Gland: International Conservation Union (IUCN), 1989.

Kates, Robert W., William C. Clark, R. Corell *et al*. 'Sustainability science.' *Science* 292: 641–642, 2001.

Kimmerer, Robin W. 'Native knowledge for native ecosystems.' *Journal of Forestry* 98(8): 4–9, 2000.

Kuhn, Thomas S. *The Structure of Scientific Revolutions*. 2nd ed. Chicago: University of Chicago Press, 1970.

Lasserre, Pierre, *et al*. *Traditional Knowledge and Management of Marine Coastal Systems*. Biology International, Special Issue 4, 1983.

Leopold, Aldo. *A Sand County Almanac: And Sketches Here and There*. Oxford: Oxford University Press, 1949.

Ludwig, N.A. 'An Ainu homeland: an alternative solution for the Northern Territories/Southern Kuriles imbroglio.' *Ocean and Coastal Management* 25: 1–29, 1994.

Mauro, Francesco and Preston D. Hardison. 'Traditional knowledge of indigenous and local communities: international debate and policy initiatives.' *Ecological Applications* 10: 1263–1269, 2000.

Merchant, Carolyn. *The Death of Nature: Women, Ecology and the Scientific Revolution*. San Francisco: HarperCollins, 1980.

Moorehead, Richard. 'Changes taking place in common-property resource management in the Inland Niger Delta of Mali.' In *Common Property Resources*, F. Berkes, ed. London: Belhaven, 1989, pp. 256–272.

Nabhan, Gary P. *Sonoran Desert Sense of Place Project*, 2001. http://www.desertmuseum.org.

Nadasdy, Paul. 'The politics of TEK: power and the 'integration' of knowledge.' *Arctic Anthropology* 36: 1–18, 1999.

Naess, Arne. 'The shallow and the deep, long-range ecology movement.' *Inquiry* 16: 95–100, 1973.

Ostrom, Elinor. *Governing the Commons: The Evolution of Institutions for Collective Action*. Cambridge: Cambridge University Press, 1990.

Posey, Darrel A. Comment on 'The development of indigenous knowledge: A new applied anthropology' by Paul Sillitoe. *Current Anthropology* 39(2): 241–242, 1998.

Preston, Richard J., Fikret Berkes and Peter J. George. 'Perspectives on sustainable development in the Moose River Basin.' In *Papers of the Twenty-Sixth Algonquian Conference*, David Pentland, ed. Winnipeg: University of Manitoba Press, 1995, pp. 379–394.

Purcell, Trevor W. 'Indigenous knowledge and applied anthropology: questions of definition and development.' *Human Organization* 57: 258–272, 1998.

Ramakrishnan, P.S., K.G. Saxena and U.M. Chandrashekara, eds. *Conserving the Sacred for Biodiversity Management*. New Delhi: Oxford and IBH Publishing, 1998.

Rappaport, Roy A. *Ecology, Meaning and Religion*. Berkeley, California: North Atlantic Books, 1979.

Rappaport, Roy A. 'Humanity's evolution and anthropology's future.' In *Assessing Cultural Anthropology*, R.F. Borofsky, ed. New York: McGraw-Hill, 1994, pp. 153–166.

Reed, Gerard. 'A Sioux view of the land: the environmental perspectives of Charles A. Eastman.' *The Trumpeter* 8(4): 170–173, 1991.

Riedlinger, Dyanna and Fikret Berkes. 'Contributions of traditional knowledge to understanding climate change in the Canadian Arctic.' *Polar Record* 37: 315–328, 2001.

Rogers, Raymond A. *The Oceans are Emptying: Fish Wars and Sustainability*. Montréal: Black Rose Books, 1995.

Ruddle, Kenneth, Edvard Hviding and Robert E. Johannes. 'Marine resources management in the context of customary tenure.' *Marine Resource Economics* 7: 249–273, 1992.

Ruddle, Kenneth and Tomoya Akimichi, eds. *Maritime Institutions in the Western Pacific*. Osaka: National Museum of Ethnology, Senri Ethnological Studies No. 17, 1984.

Salmon, Enrique. 'Kincentric ecology: indigenous perceptions of the human-nature relationship.' *Ecological Applications* 10: 1327–1332, 2000.

Sillitoe, Paul. 'The development of indigenous knowledge: a new applied anthropology.' *Current Anthropology* 39: 223–252, 1998.

Stevenson, Marc G. 'Indigenous knowledge in environmental assessment.' *Arctic* 49: 278–291, 1996.

Stirrat, R. L. Comment on 'The development of indigenous knowledge: a new applied anthropology' by Paul Sillitoe. *Current Anthropology* 39: 242–243, 1998.

Stone, M. Priscilla. Comment on 'The development of indigenous knowledge: a new applied anthropology' by Paul Sillitoe. *Current Anthropology* 39: 243, 1998.

Taiepa, Todd, Phil Lyver, P. Horsley, J. Davis, M. Bragg and H. Moller. 'Co-management of New Zealand's conservation estate by Maori and Pakeha: a review.' *Environmental Conservation* 24: 236–250, 1997.

Tyler, Mary Ellen. 'Spiritual stewardship in aboriginal resource management systems.' *Environments* 22(1): 1–8, 1993.

Warren, D. Michael, L.J. Slikkerveer and D. Brokensha, eds. *The Cultural Dimension of Development: Indigenous Knowledge Systems*. London: Intermediate Technology Publications, 1995.

Watanabe, H. *The Ainu Ecosystem, Environment and Group Structure*. Seattle: University of Washington Press, 1973.

Williams, Nancy M. and Graham Baines, eds. *Traditional Ecological Knowledge: Wisdom for Sustainable Development*. Canberra: Centre for Resource and Environmental Studies, Australian National University, 1993.

Williams, Nancy M. and Eugene S. Hunn, eds. *Resource Managers: North American and Australian Hunter-Gatherers*. Washington, DC: American Association for the Advancement of Science, 1982.

Zimmerman, Michael. *Contesting Earth's Future: Radical Ecology and Postmodernity*. Berkeley: University of California Press, 1994.

RICHARD W. STOFFLE, REBECCA TOUPAL, AND NIEVES ZEDEÑO

LANDSCAPE, NATURE, AND CULTURE: A DIACHRONIC MODEL OF HUMAN-NATURE ADAPTATIONS

The relationship between culture and nature has been a topic of intense discussion for thousands of years. Since who we are depends in part on where we are, it is no wonder that people have argued about how much influence nature has had in shaping our lives, and, in turn, about how much freedom we have had in creating nature to fit our lifeways. Basic to this discussion are the concepts of human adaptation, local knowledge, environmental values, place attachments, and cultural landscapes. Critical too is a discussion of what nature is and what role it has in placing limits or even directing human adaptation. We discuss each of these issues in an effort both to explain the intellectual course of this debate and to point to directions where it may lead. The essay will focus on American Indians and their history and culture.

ROLE OF HUMANS IN NATURE

The relationship between humans and nature has been the foundation of a Western philosophical debate that extends back in time to the early Greeks. This debate is beyond the scope of this essay, which is designed to shed light on non-Western views of this issue. Nonetheless, in Western philosophical thought humans are sometimes viewed as being a part of and sometimes separate from nature. Casey (1993, 1996, 1997) has recast this debate with a focus on the role that the concept of place has as a critical aspect of nature.

Biologists and ecologists also have wrestled with this concern. They have attempted to put humans back into the study of nature, especially the biological components of ecosystems (McDonnell and Pickett, 1993). At a seminal conference in 1991 they resolved (to their satisfaction) this issue by showing that humans always have both dramatic and subtle effects on their environments and thus can never be viewed as being separate from nature. In fact most believe that humans are keystone species (Castilla, 1993). Recently a noted evolutionary biologist published an article in *Science* (Palumbi, 2001) which seems to have capped the debate.

At the same time, human culture seems always to be defined in part by

97

H. Selin (ed.), Nature Across Cultures: Views of Nature and the Environment in Non-Western Cultures, 97–114.
© 2003 *Kluwer Academic Publishers. Printed in Great Britain.*

where people live. Early debates attempted to explain differences in social structure in terms of the limits placed on them by various environmental factors. They argued that tropical versus temperate climate affected socioeconomic factors. Some even thought that the complexity or simplicity of the topography would influence people's mental development (Mills, 1887).

Julian Steward (1938, 1955), an early founder of modern human ecology, pointed out that only certain aspects of each culture are connected with the environment and then only with specific features. He argued that each culture has a core composed of beliefs, values, and norms which is where these nature-culture connections primarily lie. The culture core should become the focus of human ecological studies. Steward's theory strengthened previous environmental deterministic theories by insisting studies specify which aspects of nature were influencing which aspects of culture. While his views were less deterministic than previous environmental theorists, he still concentrated on what nature did to people rather than the reverse. In addition he believed that subsistence strategies would be most likely to constitute the culture core that is connected to nature and thus failed to allow for cultures' ceremonial and spiritual activities centrally connected to nature. These more abstract activities, he maintained, were not critical to and were generally unrelated to human adaptation. This is a point we wish to rectify in this essay.

Contemporary human ecology studies demonstrate the importance of ceremonial and spiritual activities in relationship to the environment. These activities can serve as critical vehicles for conservation and can symbolize a society's respect for its environment (Rappaport, 1984). Today, many studies of human-nature relationships look at ceremonial relationships, especially as these occur at identified places and among complexes of these places called cultural landscapes.

Places

While debates about the mutual relationships between cultures and ecosystems persist, many scholars have taken the simple idea of a place and a people and tried to understand how the two are intertwined and how these relationships came into being. Tuan's (1977) book, *Space and Place*, may be the most cited on the subject. He combines complex philosophical issues with clear illustrations to demonstrate that it is through experience with places that nature is made real and becomes part of human culture. Basso's (1996) *Wisdom Sits in Places* is an analysis of American Apache Indian place naming as a process of documenting where and how Apaches learned about the environment and how they incorporated these names into social and environmental ethics. Basso's work is a further grounding of Tuan's theory of place-making and a study of how people adapt.

Cultural landscapes

The idea of a cultural landscape as a meaningful way to organize cultural data about places and their relationships with each other has emerged over the past few decades. Today, we refer to larger, integrated, and more abstract phenomena about places and their connections as cultural landscapes. Recent scholarship (Dewey-Hefley et al., 1998; Stoffle, Halmo, and Austin, 1997; Stoffle et al., 2000a,b; Zedeño, 1997, 2000; Zedeño, Austin, and Stoffle, 1997) sheds light on methodological, analytical, and theoretical issues remaining to be resolved before cultural landscapes are understood as networks of connected places.

The concept of cultural landscape derives from the notion that the land exists in the mind of a people and that their imagery or knowledge of the land is both shared among them and transferred over generations. All human groups develop and come to share cultural landscapes. The concept implies that many cultural groups (ethnic groups) can hold different, even conflicting, images of the same land. The imagery of the land that is held by a people is a result of their past experiences with the land and other cultural perspectives of the people themselves.

A cultural landscape expands the idea that a special place can have dozens of cultural meanings. Central to the concept is the notion that not all places within it have the same culture value even for a single ethnic group. The places may derive their value from interactions between people and natural phenomena (Zedeño, 2000). Tilley (1994: 34) distinguishes between the concepts of place and landscape, with the former emphasizing difference and singularity and the latter encompassing commonalties or relationships among singular locales and events. A cultural landscape should make sense as a kind of culturally defined single area, defined by a common logic and composed of unique and connected places.

Cultural landscapes are nested (Stoffle, Halmo, and Austin, 1997; Tilley, 1994: 20). They exist at different scales but are integrated into a whole. For many American Indians, for example, these levels include, from broadest to narrowest scale, an Eventscape, a Holy Land, songscapes, regional landscapes, ecoscapes, and landmarks. The topographic criteria for defining these categories range from their fit with the natural terrain (i.e., an ecoscape) to a spiritual landscape that exists in terms of its own criteria with minimal reference to the topography of the land (i.e., a songscape).

People may attach more than one cultural landscape to a place. We call this "cultural landscape layering". Layered cultural landscapes may have very different meanings. One landscape layer may be composed of places visited by a spiritual being. Another may involve an event such as a forced march following military conquest, as in the trail of tears for the Cherokees, the march to Bosque Redondo for the Navajos, or the march to Fort Independence for the Owens Valley Paiutes and Shoshones, all American Indian cultures.

HUMAN ADAPTATION: A DEVELOPMENT CYCLE MODEL

One way to understand human adaptation is by using a diachronic[1] model of human adaptation that begins with the hypothetical arrival of a group of people in a new place and observes their initial adaptation to this ecosystem. Eventually, this group begins to make certain modifications in the environment.

The model then moves forward in time over generations to observe how these people become fully adapted and how the place becomes the center of their lives both physically and spiritually. The model is informed by the direct analysis of actual human adaptation and resulting environmental changes (Figure 1).

Three cautionary points need to be made before proceeding. In this discussion we are scaling up; that is we are increasing the variables of space, time, and complexity of typical human ecological analysis. We are moving beyond what is confidently known in order to build a diachronic model with heuristic value for situating and perhaps guiding how we think about these issues. Most human-nature studies have a narrow time frame and focus on only a few species interactions. "Very little ecology deals with any processes that last more than a few years, involve more than a handful of species, and cover an area of more than a few hectares" (Pimm, 1991: xi). Nonetheless Pimm builds food web and temporal variability ecological models from narrow time and focus studies.

Social and natural scientists have had few opportunities for long-term joint research projects; thus we tend not to understand each other's theories, methods, variables, and analytical techniques. As a consequence, most studies of how humans adapt to and change their environments are conducted by social scientists who know the human side of the equation well but study a very limited aspect of the natural system. Similarly, studies by ecologists that look at the impacts of humans on nature tend to view people as simple consumers of the environment who are without complexity and lack the capacity to modify

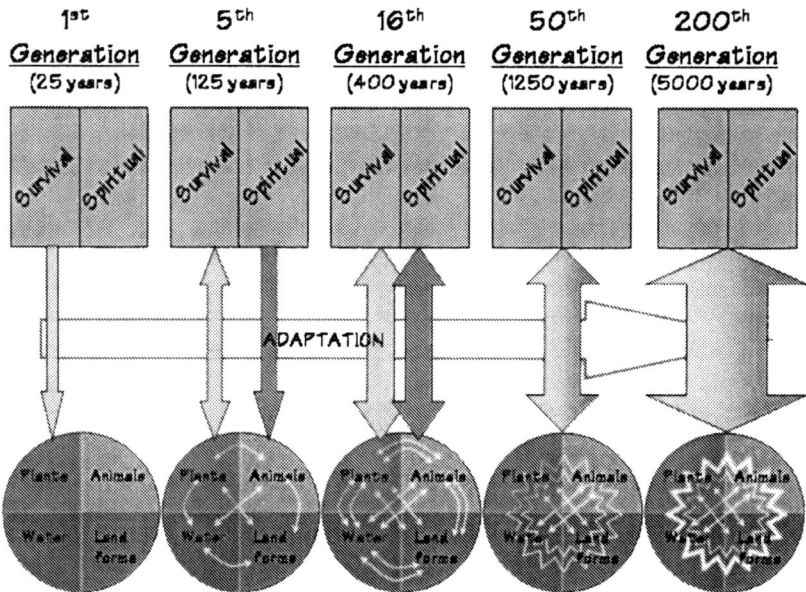

Figure 1   Development cycle model of human adaptation.

their behavior. This is something like the basic predator-prey models in natural science. For this reason, the current analysis describes and explains the human dimensions of ecology.

Finally, the model is intended to sequence types of adaptations so we can begin to imagine the cultural implications of adaptive behavior. So, for example, when the Southern Paiute Indians say the Creator made them in Las Vegas Wash below their origin mountain, it can be viewed as the end result of thousands of years of living in this area rather than some inherent aspect of their culture. There are many peoples around the world who have embedded portions of the landscape into their culture and their definition of self. Our model suggests it takes great periods of time to produce Creation embeddedness, but obviously there is no time rule. For example, the Navajo Indians of the American Southwest maintain they emerged from an underworld in the southwestern corner of the state of Colorado, which is similar to a Pueblo Indian theory of origin (McPherson, 1992). Yet archaeologists and linguists maintain these people arrived from what is now Canada no more than 600 years ago (Towner, 1996). If this is true, then the Navajo people have developed a topographically embedded origin knowledge in record time. The example actually fits sequences of the diachronic model, because the creation knowledge was to follow hundreds of years of occupancy and adaptation to this new land. They only adapted faster than has been recorded for other peoples.

First generation – arriving in a new land

The first generation arrives in a new land.[2] It has high mountain ranges flanking low narrow valleys. Small streams come from the peaks and produce both intermittent and small permanent streams. The area is highly diverse in terms of microniches, which mostly vary by elevation. It is a land that is generally semi-arid with riverine oases in the arid valleys and rain shadows on the eastern sides of the mountains. Because of past geologic events, many soil types exist, including special ones on the flanks of the inactive volcanoes. Animals and plants vary according to econiche.

Approximately 200 people come into this area, which is about 500 square miles in size. This is a population density of about 2.5 persons per square mile, which is defined as the natural carrying capacity of this unaltered land. The people have all the social organization, cultural elaboration, and technology we would expect in a human group after the upper Paleolithic. This base culture/society must contain the following:

• A social structure including hierarchy;
• A knowledge of the supernatural that has been translated into a religion, including a complex ceremonial cycle and religious specialists;
• An understanding of key life cycles – birth, maturity, aging, death – and ceremonial responses to each stage;
• Knowledge of their former local ecosystem including both its biotic and abiotic dimensions;

- Experience studying and learning from the abiotic and biotic environment and distributing this knowledge, and
- Technology to build homes, make clothes, and acquire and process food.

Although many aspects of this base culture/society will not be useful in the new lands, they remain as background experience carried into and influencing future adaptations. These immigrants are pre-agricultural but have extensive knowledge of plants, animals, and climate.[3] They do not bring domesticated plants, but they have the dog.

How does environmental learning occur? The first generation actually learns much from native plants and animals. This foreshadows supernatural explanations of environmental knowledge as having been taught by the land that will become a central feature of their culture many generations in the future. How can animals and plants teach? Plants tell the observant viewer about rainfall and the subsurface distribution of water. Longer-term climatic stories are told by where plants do or do not grow near or in the intermittent streams, which are subject to both unpredictable and cyclical El Niño-type catastrophic flood events. Animals move between econiches according to weather shifts and availability of food resources.

Why can a culture learn more from mistakes than from successes? It may be that when they fail they have reached a critical limit. On the other hand, they may succeed and never know why because they are working within the limits of the resource being used. People learn by hurting nature and seeing the undesired consequences. When they kill too many large mammals, they learn not to repeat those behaviors because they rely on that resource.

The first generation makes mistakes, learns how not to cause unwanted damage, and then tells family, friends, and perhaps the community. Some first generation lessons remain at the family level.

Second generation – do what parents tell you to do

The second generation can build upon the lessons learned by the first generation. The study of biotic resources continues. The first generation has scrambled to survive; the second generation begins to develop theories of how to learn about the environment. Lessons come from some event: perhaps a person follows a bee to water, or perhaps a person builds his home near a dry wash only to be flooded. Perhaps a person observes an animal healing itself with special white mud. The second generation will develop theories of how data moves between people and the world and becomes useful for short-term survival and longer-term adaptation.

The second generation receives these lessons from living people who can be questioned about the event. Knowledge comes from deep understandings of non-intuitive aspects of the world. Few of these are expected in the first two generations.

We see at this stage what is called "adaptive behavior". Bennett (1969: 11) defines adaptive behavior as coping mechanisms or ways of dealing with people and resources in order to attain goals and solve problems. Our emphasis here is on patterns of behavior: problem solving, decision making, consuming or not consuming, inventing, innovating, migrating, and staying. Bennett documented how different people at different times constructed unique cultures in a single ecosystem in central Canada.

The fifth generation – do what great grandparents tell you to do

The fifth generation (approximately 125 years) has a firm information base about the environment and knows how to use the land without hurting it. Families have experienced birth, life, and death in the new homeland, and these events emotionally connect the people with specific places. They have pushed a number of environmental boundaries, made mistakes, and changed their behavior. By the fifth generation various aspects of successful and sustainable behavior have been put into place.

Environmental learning seems to be directly related to the amount of natural resource scarcity, especially when this occurs in annual and decennial cycles. People mostly learn to hunt and gather when basic resources are regularly available. It is not until the limits of the resource are reached that they have an opportunity to learn what parameters drive the system. It is at exactly these moments that people decide to modify their environmental resource use patterns in order not to drive other resources into extinction. By this time, they have reduced the possibility of Hardin's (1968) tragedy of the commons (McCay and Atcheson, 1987).[4] If they fail to adopt resource conservation procedures and to build these into the rules by which they govern themselves, then the carrying capacity of the environment will be reduced.

Learning from plants and animals continues during this generation, but the lessons are less intuitive and probably about natural processes that are less accessible to human view. Plants often serve as calendars, like the rabbitbrush (*Chrysothamnus nauseosus*) that blooms when it is time to go into the mountains to harvest pine nuts (*Pinus monophylla*). Smaller animals and insects provide information about pollination, food webs, and landscapes (Nabhan, 1997). Beaver dams keep the ecosystem from eroding and retain water during arid seasons, so the people minimize their consumptive use of beaver. People learn to stay away from water sources because it disturbs the animals.

Proactive adaptive behavior will be normal by this time. This might include selectively increasing the yields of certain species by burning or pruning. Also proactive is the planting of certain species along or diagonally across intermittent and small streams to assist the beavers' efforts to reduce erosion while at the same time retaining the water and moving it to places along the streams where it is needed. By creating a specialized patch ecology that does not occur naturally, the carrying capacity of the land will increase (Lewis, 1982; Lewis and Ferguson, 1999).

The people begin to have adaptive strategies. Bennett (1969: 14) defines these as the patterns formed by the many separate adjustments that people devise in order to obtain and use resources and to solve the immediate problems confronting them. The rules of adaptation become culturally embedded. Bennett

(1969: 16) notes that as time passes, the many separate adjustments that have become patterned as strategies can also "enter into culture". As repetitive patterns of actions the people view them as traditions – behavior defined as "right" or "good". These embedded traditions form part of a group's cultural style. The extent to which adaptive behavior becomes culturally sanctioned varies according to the demands placed on the society by various external factors. If the major changes in the natural system are recognized and occur in predictable cycles there will be a tendency for a given strategic regime to become "sacred".

The population grows as the fit between technology and resources increases the carrying capacity of the land. Basic issues of village life, including the need for seasonal movements of the community or community members, exist.

The sixteenth generation – do it because "We do it like this"

There is a point in the development of a society when the origins of ancestral lessons are vaguely remembered but firmly established as the correct way to treat the environment. This is a time (perhaps by the sixteenth generation or approximately 400 years) when lessons are taught and maintained as general principles. The name of the person who originally learned the lesson probably is not attached to it. People teach their children to behave in certain ways because "it is how we (the members of this community) do it." If you follow "our ways", you will always have food, shelter, and health, and the environment will be in balance.

By now the people have amassed sufficient data from the natural environment so that they begin to develop knowledge about what is happening around them. They recognize pollination of plants by certain species. They have learned many lessons about medicine. They have seen sufficient climate variability that they know with some confidence what the cycles provide in terms of opportunities and threats. There are still other forces that are unclear to them but appear to be variables within the system. The rules for engaging the environment are now well tested and those rules, which have repeatedly proven useful for maintaining and improving the productivity of the environment, are increasingly defined as sacred and not subject to debate.

The population grows, so there are more people to organize, to teach about "how we live here", and to make regulations. The natural resources of the environment are further stressed. New boundaries are reached and perhaps broken. Certain long-term climatic events have occurred a few times, and there is the recognition that one has to prepare for events that may not come within the lifetime of a generation.

The 50th generation – environmental ethics defined and sanctioned by supernatural forces

The model now skips to the 50th generation, about 1,250 years since arrival. People have lived so long in this ecosystem that they have clear adaptive strategies that have survived a wide variety of temporal and biotic shifts in the ecosystem. Over the period of 1,250 years many natural resource changes also will occur. New plants will come into the ecosystem, and old ones will become extinct. Streams will flow more or less depending on climate change and use.

By this time they have embedded these adaptive strategies in their culture. They have moved key values into the realm of the supernatural, with both the lessons and the sanctions being supernaturally defined. There is a confluence between science and religion, as the scientific findings of past generations have been recognized as essential to life and moved to the realm of the sacred and thus beyond human control.

200 generations – we were created here

After 200 generations, approximately 5000 years, in the same ecosystem many kinds of complex human adaptations can be expected. Only a few biotic features will remain unchanged. Climate changes reduce or even eliminate most fauna and flora and certainly radically alter the food webs (Grayson, 1993). Even the topography can change. Sea levels will rise and fall. Volcanoes will build up the land, and erosion will tear down and transform it. Long wet periods will alternate with long dry periods. What does it mean to have adaptive strategies, which define, organize, and maintain a human group's adaptations to an environment, if this environment itself changes?

Those things that persist in structure and function through a five thousand year period will not only remain but will become increasingly central in the culture. For example, mountain peaks and ranges should remain the same and draw rain from the sky. This is what is called the "sky-island function" (see Crowley and Link, 1989). Mountain ranges are central in the lives of people dependent on related resources. For example, Baboquivari Peak and Mountains are central in the lives of the Tohono O'odham people (Toupal, 2001) and are primary rain makers in southern Arizona.

Biotic and biologic evolutionary forces cause some changes in the environment, while growth in human population size and technological innovations create others. The people know how to listen to the environment and change their behavior.

CASES OF ADAPTATION

The following cases illustrate humans adapting to specific environments. These cases both inform and stretch the diachronic model of cultural adaptation as they follow the course of adaptation in approximately the same time periods. In these cases we see that people usually arrive in areas already occupied by other people, causing a complexity in adaptation which is not considered by the model. This complexity has two major influences on adaptation. The people who already live there can be useful to the immigrants by teaching them and building an informed foundation upon which to adapt (Atran et al., 1999). They can also compete for space and resources, causing a threat to the newcomers that itself may become the focus of adaptation. Our cases also demonstrate that the newly arrived human groups bring important cultural knowledge from previous places. Such knowledge is impossible to model, because it may or may not yield positive adaptations. Newcomers may initially believe they have ways to improve their new land, but these behaviors often cause dramatic natural resource mistakes. Examples of such false adaptations in North America include the suppression of American Indian land and forest burning (Boyd, 1999), the draining of wetlands by eliminating beavers (Cronon, 1983), and the channelization of Western rivers by destroying associated riparian habitats (Dobyns, 1981).

100 years of adaptation – Scandinavian-American folk fishers

In the upper reaches of Lake Superior, approximately thirteen miles from the Canada-United States border, a jagged sliver of volcanic uplift forms what is known as Isle Royale. Now a National Park, its wetlands and tree-covered ridges have provided resources for humans and habitat for caribou, coyotes, moose, wolves, several fur-bearing species, birds, and waterfowl for thousands of years. Its terrestrial-lacustrine interface is comprised of rocky shorelines, small islands, and many harbors, which accommodated fishermen for hundreds of years. While the island is part of the same geologic formation as the Upper Peninsula of Michigan, its proximity and similarity to the North Shore of Minnesota contributed to its becoming a social, cultural, and economic extension of that area.

When a wave of Scandinavian immigrants came to Minnesota in the 1880s, many settled along the North Shore. Many of them were fishermen (Toupal, Stoffle, and Zedeño, 2002). Finding the area similar to their homelands, they settled in the protected coves and inlets from which they established fishing areas and subsistence farms. They also developed communities within which they could communicate in their native languages and continue their traditions. They preferred the company of those with whom they could enjoy traditional foods and who had the same style of thinking and work ethic.

Isle Royale attracted many of these fishermen. When they first came to fish the waters, they found other fishermen from diverse ethnic backgrounds including German, Irish, English, French-Canadian, Chippewa, and American (Karamanski, Zeitlin, and Derose, 1988). These people were displaced within five years of the arrival of the Scandinavians who afterwards and for the next one hundred and twenty years maintained exclusive usufruct rights. These Scandinavian-Americans preferred the solitude and independent lifestyle of fishing on the island (Jentoft and Mikalsen, 1994). They and their families established their fish camps and began adaptations to their new environment that would become the foundations of a new folk culture.

The Scandinavian-American fishermen developed a herring industry. Instrumental in its success was the establishment of an exclusive relationship with the Booth Company, which instituted credit relationships and provided them with provisions and equipment in the spring in exchange for their fish throughout the season (Kaups, 1975).

Their Scandinavian fishing heritage had some benefits in the new environment in spite of differences between ocean and lake fishing. Saltwater seine
nets had been used to bring catches into the shore, and hooklines had been let
down to the ocean bottom. The rocky shorelines of Isle Royale ruled out use
of the seine nets, so they began using gill nets of different sizes placed at
different depths to catch trout, whitefish, and herring. The lake waters were
too deep to fish the bottom with hooklines, so they suspended them with floats
and weights to depths up to 200 feet (Kaups, 1975).

Another adaptation occurred with boats. Many of the Swedish immigrants
had been boat builders, and they quickly adapted boats like the Mackinaw to
handle the conditions of Lake Superior (Toupal, Stoffle, and Zedeño, 2002).
Some fishermen built their own herring skiffs, which resembled the *sjekte* used
along the inner coast of eastern and southern Norway. These boats ranged
from fifteen to seventeen feet long, four to five feet wide, and about two feet
deep. The fishermen seldom went more than two miles from shore, the approximate extent of the herring fishery (Kaups, 1975). In some instances, a fisherman
might have two boats, a nineteen-footer for fishing closer to shore, and a
twenty-four-footer for going further out (Toupal, Stoffle, and Zedeño, 2002).

From these boats, they learned to manage their nets and hooklines in the
temperamental lake waters, to read the currents and lake bottom for rock reefs
and passages, and to read changes in the weather that might indicate the onset
of severe storms. Living on the island from April to November, the fishermen
and their families soon learned to read wind, cloud formations, and fog conditions that promised a difficult if not dangerous lake and to gauge how long
they might be able to work their nets safely.

Several of the Scandinavian-American fishermen worked with Minnesota
fish hatcheries, providing milt and spawn in the fall and planting young fish
in the spring prior to the first fishing season. While most of the fishermen used
hooklines and gill nets, a few who could afford to do so experimented with
pound nets. They found the pound nets so effective that, particularly when
used in spawning areas, they could decimate a fish population within a few
seasons. They soon abandoned the pound nets for the traditional gill nets, in
order to keep their supply of fish.

The fishermen continued to learn about the underwater environment. Small
islands, for example, were good net areas because of shallow water, reefs, shelter
from the wind, and proximity to deeper waters (Toupal, Stoffle, and Zedeño,
2002). The reefs had bottom structures beneficial for lake trout spawning. They
tracked fish populations by reef, noting increases and declines from one season
to the next. They could detect changes in fish populations and behavior caused
by seasonal and climatic conditions such as equinox disturbances, storms,
squalls, and full moons (Toupal, Stoffle, and Zedeño, 2002). This allowed them
to determine when a reduced catch was due to natural conditions and when it
was due to fishing pressure, the latter resulting in changes in set times or mesh
size. They made changes that included mesh size and/or the length of time they
would leave nets on the reefs so that fewer and larger fish were taken and the
population could recover.

As a five-generation example of adaptation, the Scandinavian-American fishermen of Isle Royale exhibit the characteristics of adapting knowledge of one environment to another in order to obtain and use resources, and of developing new strategies, which ultimately became folk traditions. Succeeding generations increasingly embedded these strategies in their culture. Many families had natural resource connections through "pet" birds, foxes, mink, and moose. Places became special because of their topographic uniqueness, viewscapes, and community and family histories. Collectively the places and natural resources became their homeland, which can be understood as a cultural landscape.

400 years of adaptation – Afro-Caribbean peoples

African people have been in the Caribbean for 16 generations (approximately 400 years). The geography of the Caribbean ranges from volcanoes rising from the ocean floor throughout the Lesser Antilles, to large, low, flat expanses of limestone in the Bahamas and Barbados, to the large complex mountains on islands like Jamaica and Hispañiola.

African people were brought into the Caribbean as slaves. They immigrated with neither tools, animals, nor domestic plants, yet they were to make many adaptations, which are central features of their contemporary culture. Today there is a debate as to whether these derived from memories of Africa or were gleaned from lessons learned in the Caribbean. The Rastafari of Jamaica are a people who farm the rocky hillsides of mountains. In their gardens, they use swidden horticulture, which turns the limestone rocks into soil. By using terraced walls and rotating fields, they are able to improve the soil. They are known to protect many species and varieties of certain crops like yams, which are planted in different niches as a hedge against fluctuations in rainfall, and to take advantage of different types of soils.

In the Bahamas, African peoples developed an elaborate system of gardens in which a variety of drought resistant and irrigable crops were planted (Eldridge, 1975). Their understanding of the environment is reflected in how they remove land crabs from their gardens. One local plant, Nicker Bean (*Caesalpinia bonduc*), has a hard round marble-like seed. When the land crab makes a burrow in a garden, instead of digging him out or killing him, the seed of this plant is dropped into the burrow. Being hard and round, the seed cannot be moved out the burrow by the crab, who is so finicky that he cleans all foreign objects out of his home. Eventually frustration causes the land crab to move away from the garden and burrow elsewhere.

1,250 years of adaptation – people of Bali

In the middle of the volcanic archipelago of Indonesia is the island of Bali, with habitats ranging from montane forests in the highlands to terraced rice fields on the slopes, to wetlands and coral reefs along the coast. Rice, developed over one thousand years ago, has become a staple food to the Balinese. Fifty generations (approximately 1,250 years) of religious and social practices that sustain rice cultivation represent a series of rational adaptations that have survived centuries of social instabilities such as shifting kingship boundaries.

The production of rice is embedded with rituals for planting, maintenance, irrigation, and harvesting (Lansing, 1991). The season commences with temple festivals and pilgrimages to the lakes to obtain holy water to sprinkle on the fields. Planting begins by walking the fields with water buffalo to prepare the soil. Ceremonies for planting the fields involve carrying young stems of rice cultivated for this purpose to each field. Water temple priests work with the local *subaks*, or farmer associations, that determine local cropping patterns and arrange the irrigation and planting schedule. This relationship of religion and ecology depends upon the farmers' acceptance of and reliance on the water temple priests who must coordinate irrigation of hundreds of terraced rice fields among hundreds of farming communities.

Beginning at the highest volcanic lake on the island, the water is metered down through many *subaks* in a staggered planting and irrigation cycle. The farmers who receive the water first voluntarily stop irrigating when directed by the water temple priests, and the farmers at the next level, who have just prepared their fields, begin irrigating. This pattern is repeated until the last farmers have received their water. The system results in optimal water sharing, minimal pest problems, and some of the highest rice yields in the world.

As an adaptive system, water temple irrigation combines natural and cultural cycles. Seasonal rains, long growing seasons, rice paddy ecosystems, and pests are integrated with religion to control timing and amounts of irrigation water, which influences planting, care, and harvesting of the rice fields.

5,000 years of adaptation – Numic people of the Great Basin

The Numic-speaking people of the Western United States[5] have a system of beliefs for the desert ecosystem where they live. They believe that all they know about, use from, and owe to the environment was defined at the moment they were created. In a sense, they and their environment could not survive without each other.

The Numic people believe they were created in this region when all the plants, animals, wind, minerals, and water also came into existence. The Creator made everything and gave each element specific rights and responsibilities, which if exercised and fulfilled will maintain ecosystem-wide balance.

The concept of a "living universe" is an epistemological foundation of Numic culture, or what Rappaport (1999: 263–271; 446) calls an "ultimate sacred postulate". These terms mean that the concept of a living universe is so basic in Numic culture that you cannot understand many other aspects of culture without first fully recognizing it. A living universe is alive in the same way that humans are alive. The universe has physically discrete components that we will call "elements" and something like energy that these people call *puha*. These are a few general statements that we can make about *puha*:

- *Puha* exists throughout the universe, but like differences in human strength, power will vary in intensity from element to element.
- *Puha* varies in what it can be used for, and so it determines what different elements can do.
- *Puha* is networked, so that different elements are connected, disconnected, and reconnected in different ways, and this occurs largely at the will of the elements that have the power.
- *Puha* originally derives from Creation and permeates the universe in a thin scattering and in definite concentrations with currents, generally where life is also clustered.
- *Puha* exists and can move between the three levels of the universe: upper (where powerful anthropomorphic beings live), middle (where people now live), and lower (where super-ordinary beings with reptilian or distorted humanoid appearance live).

The fundamental meaning of a Numic place derives from the *puha* it exhibits. This *puha* is dispersed in a network of relationships among the elements of the universe – relationships that most resemble spider webs. At various points in this web, power is concentrated, producing powerful places. Zedeño (2000) observes that places are "made" because they are the loci of human interactions or nature experiences. Therefore, power accumulates at a place as people live or re-live those experiences. Power is cumulative.

Powerful places tend to attract other powerful elements. So, for example, during studies of storied rock sites (rock peckings and paintings), American Indian people tend to look first at the rock on which the painting and peckings occur and then look around for medicinal plants. The basic assumption of interpretation is that the place had to be powerful before the rock paintings or peckings were made there. Indian tobacco often grows out of the cliff face where the peckings have been placed. The Indian interpretation is that the powerful plant went to live on the powerful cliff. Humans derive power from places, causing them to be even more central in Indian culture (Stoffle *et al.*, 2000a).

Numic cultural landscapes are composed of sets of different places, which are connected by webs of *puha* and recognized by humans as collectively having certain ceremonial functions (Stoffle, Halmo, and Austin, 1997). Such landscapes, like the one near Hoover Dam on the Colorado River at the mouth of Black Canyon, may be a combination of a song cave, red paint mineral deposit, hot spring, and a medicine mountain, which when used in sequence permit healing ceremonies that keep individuals, groups, and the world in balance (Stoffle *et al.*, 2000b).

Environmental lessons were defined at Creation and are thus not subject to debate by present generations. A most basic rule is never to touch anything without talking with it about your intentions and asking its permission to move or pick it. Once the rock, plant, animal, mineral, spring, or cave grants permission, then human activities governed by specific rules are permissible. For example, when a woman reaches menses she is taken to an isolated place where older women teach her about female relationships with the world. One of these is how to "whip the trees". Women are specifically charged with talking to, praying over, and increasing the productivity of pine nut trees. They accomplish the latter by gently whipping the ends of all branches on the tree with a 12-foot long thin pole. This action breaks the ends of all the branches, much like pruning, which causes the branches to grow many more shoots and have many more pinecones. Given that Numic women did this for thousands of years in every pine nut forest in the region, it is probably the case that they increased the carrying capacity of the land and contributed to its biodiversity. The genetic structure of some plants was modified because of its being regularly selected for certain characteristics and moved to places more useful to humans (Nabham *et al.*, 1981).

* * *

When people arrive in a new place, they probably will not be good for the land. According to Steadman (1997), as many as 2000 bird species became extinct following human colonization of the Pacific islands. Pimm *et al.* (1995) estimate that approximately half of the native avifauna was exterminated after the arrival of humans. Whittaker's (1998: 255) review of these issues caused him to conclude that in the Pacific and on islands around the world, so many species died soon after colonization that often the people died and the islands were abandoned. Whittaker further concludes that the notion that the "noble savage" lived in balance with nature is illusory, that where something like it was achieved, it was only accomplished by trial and plenty of error along the way. He also concludes that people do try to resolve the nature-human problems their presence causes. It seems likely, according to Whittaker, that the Polynesian taboo system, which functions to prevent the over exploitation of natural resources, was started as a result of the extinctions and misuse of other resources (Whittaker, 1998: 233).

Despite the damage to the environment that humans can cause, these well-documented events of early arrival impacts also support the model presented here. Each involves the needed components for adaptation and conservation practices.

This model and these cases document a process of cultural adaptation to the local environment, which seems to be sequential over generations. The process of learning from the environment continues, because the environment is always changing. Certain lessons, however, do not have to be relearned by each generation; some aspects of cultural adaptation are removed from consideration by the living generations and placed under the protection of the supernatural. This analysis suggests that local knowledge, environmental values, place attachments, and cultural landscapes are all functionally interdependent.

There is always a difference in the adaptation of a people to a new ecosystem. In general though, when people come into a new environment they are like the first generation in our hypothetical model. If they have power and succeed in overwhelming the native population, they may come to believe that the locals do not know about the environment. Or perhaps the newcomers just do not care to listen to the people they dispossessed.

In recent centuries many environmental mistakes have been made, and these are often enormous in scale compared to any that occurred before. Rivers have been dammed, old growth forests have been cut, air and water systems have been polluted, the earth has been mined, and the great fresh water lakes have been connected to the ocean. Now the challenge is to stop or even reverse ecosystem-damaging activities. Perhaps this can be achieved more quickly by having the lessons of traditional peoples become available to and considered by members of dominant societies.

NOTES

[1]   A good definition is the study of the development of something, especially a language, through time.

[2]   The hypothetical location of this case is drawn from the ecology of the basin and range topography and arid ecosystem of southern Arizona. Long-term climate changes with resulting fauna and flora changes have actually occurred in this region over the past 10,000 years.

[3]   Dillehay's (2000) *The Settlement of the Americans* documents that all of these cultural characteristics were present when early Native Americans arrived in the New World more than 14,000 years ago.

[4]   Ecologist Garrett Hardin's "tragedy of the commons" (1968) is a useful concept for understanding how we have come to be at the brink of numerous environmental catastrophes. Hardin asks us to imagine the grazing of animals on a common pasture. Individuals are motivated to add to their flocks to increase personal wealth. Yet, every animal added to the total degrades the commons a small amount. Although the degradation for each additional animal is small, if all owners follow this pattern the commons will ultimately be destroyed. Therein is the tragedy. Each man is locked into a system that compels him to increase his herd without limit in a world that is limited.

[5]   Numic-speaking people include Owens Valley Paiutes, Northern Paiutes, Western Shoshone, Southern Paiutes, and Utes. They live in four major ecosystems: the western Colorado Plateau, the Great Basin, the Mohave Desert, and the Owens Valley of California.

BIBLIOGRAPHY

Atran, Scott, *et al.* 'Folkecology and commons management in the Maya Lowlands.' *Proceedings of the National Academy of Sciences* 96(13): 7598–7603, 1999.

Basso, Keith H. *Wisdom Sits in Places: Landscape and Language Among the Western Apache.* Albuquerque: University of New Mexico Press, 1996.

Bennett, John. *Northern Plainsmen: Adaptive Strategy and Agrarian Life.* Chicago: Aldine Publishing Company, 1969.

Boyd, Robert, ed. *Indians, Fire and the Land in the Pacific Northwest.* Corvallis: Oregon State University Press, 1999.

Casey, Edward. *Getting Back into Place: Toward a Renewed Understanding of the Place-World.* Bloomington: Indiana University Press, 1993.

Casey, Edward. 'How to get from space to place in a fairly short stretch of time: phenomenological prolegomena.' In *Senses of Place*, Steven Feld and Keith Basso, eds. Santa Fe, New Mexico: School of American Research Press, 1996, pp. 13–52.

Casey, Edward. *The Fate of Place: A Philosophical History.* Berkeley: University of California Press, 1997.

Castilla, Juan. 'Humans: capstones strong actors in the past, present, coastal ecological play.' In *Humans as Components of Ecosystems: The Ecology of Subtle Human Effects and Populated Areas*, Mark J. McDonnell and Steward T. Pickett, eds. New York: Springer-Verlag, 1993, pp. 158–162.

Cronon, William. *Changes in the Land: Indians, Colonists, and the Ecology of New England*. New York: Hill and Wang, 1983.

Crowley, Kate and Mike Link. *The Sky Islands of Southeast Arizona*. Stillwater, Minnesota: Voyageur Press, 1989.

Dewey-Hefley, Genevieve, M. Nieves Zedeño, Richard W. Stoffle, and Fabio Pittaluga. 'Piecing the puzzle: perfecting a method for profiling cultural places.' *Common Ground*, Winter 1998/Spring 1999, p. 15.

Dobyns, Henry. *From Fire to Fire*. Socorro, New Mexico: Ballena Press, 1981.

Eldridge, Joan. 'Bush medicine in the Exumas and Long Island, Bahamas: A field study.' *Economic Botany* 29(4): 307–332, 1975.

Grayson, Donald K. *Desert's Past: A Natural Prehistory of the Great Basin*. Washington, DC: Smithsonian Institution, 1993.

Hardin, Garrett. 'The tragedy of the commons.' *Science* 162: 1243–1248, 1968.

Jentoft, Svein and Knut H. Mikalsen. 'Regulating fjord fisheries: Folk management or interest group politics?' In *Folk Management in the World's Fisheries: Lessons for Modern Fisheries Management*, Christopher L. Dyer and James R. McGoodwin, eds. Niwot: University Press of Colorado, 1994, pp. 287–316.

Karamanski, Theodore J. and Richard Zeitlin with Joseph Derose. *Narrative History of Isle Royale National Park*. Chicago: Mid-American Research Center, Loyola University of Chicago, 1988.

Kaups, Matti. 'Norwegian immigrants and the development of commercial fisheries along the North Shore of Lake Superior: 1870–1895.' In *Norwegian Influence on the Upper Midwest: Proceedings of an International Conference*. Duluth: University of Minnesota Continuing Education and Extension, 1975, pp. 21–34.

Lansing, J. Stephen. *Priests and Programmers: Technologies of Power in the Engineered Landscape of Bali*. Princeton, New Jersey: Princeton University Press, 1991.

Lewis, Henry T. *A Time for Burning*. Edmonton: The University of Alberta, Boreal Institute for Northern Studies, 1982.

Lewis, Henry T. and Theresa A. Ferguson. 'Yards, corridors, and mosaics: how to burn a boreal forest.' In *Indians, Fire, and the Land in the Pacific Northwest*, Robert Boyd, ed. Corvallis: Oregon State University Press, 1999, pp. 164–184.

McCay, Bonnie J. and James M. Atcheson. *The Question of the Commons: The Culture and Ecology of Communal Resources*. Tucson: University of Arizona Press, 1987.

McDonnell, Mark J. and Steward T. Pickett, eds. *Humans as Components of Ecosystems: The Ecology of Subtle Human Effects and Populated Areas*. New York: Springer-Verlag, 1993.

McPherson, Robert S. *Sacred Land, Sacred View*. Salt Lake City, Utah: Brigham Young University, Charles Reed Center for Western Studies, 1992.

Mills, T. Wesley. 'The study of a small and isolated community in the Bahamas Islands.' *American Naturalist* 21(10): 875–885, 1887.

Nabhan, Gary. *Cultures of Habitat: On Nature, Culture, and Story*. Washington, DC: Counterpoint, 1997.

Nabhan, Gary, Alfred Whiting, Henry Dobyns, Richard Hevly, and Robert Euler. 'Devil's Claw domestication: Evidence from Southwestern Indian fields.' *Journal of Ethnobiology* 1: 135–164, 1981.

Palumbi, Steven. 'Humans as the world's greatest evolutionary force.' *Science* 293 (5536): 1786–1790, 2001.

Pimm, Stuart. *The Balance of Nature? Ecological Issues in the Conservation of Species and Communities*. Chicago: University of Chicago Press, 1991.

Pimm, Stuart, M.P. Moulton, and L.J. Justice. 'Bird extinctions in the Central Pacific.' In *Extinction Rates*, J.H. Lawton and R.M. May, eds. Oxford: Oxford University Press, 1995, pp. 75–87.

Rappaport, Roy A. *Pigs for the Ancestors: Ritual in the Ecology of a New Guinea People*. New Haven, Connecticut: Yale University Press, 1984.

Rappaport, Roy A. *Ritual and Religion in the Making of Humanity*. Cambridge: Cambridge University Press, 1999.

Steadman, D.W. 'Human-caused extinctions of birds.' In *Biodiversity II: Understanding and*

Protecting our Biological Resources, Marjorie L. Reaka-Kudla, Don E. Wilson and Edward O. Wilson, eds. Washington, DC: Joseph Henry Press, 1997, pp. 139–161.

Steward, Julian H. *Basin-Plateau Sociopolitical Groups.* Salt Lake City: University of Utah Press, 1938.

Steward, Julian H. *Theory of Culture Change: The Methodology of Multilinear Evolution.* Urbana: University of Illinois Press, 1955.

Stoffle, Richard W., David Halmo, and Diane Austin. 'Cultural landscapes and traditional cultural properties: A Southern Paiute view of the Grand Canyon and Colorado River.' *American Indian Quarterly* 21(2): 229–249, 1997.

Stoffle, Richard W., Larry Loendorf, Diane Austin, David Halmo, and Angie Bulletts. 'Ghost Dancing the Grand Canyon: Southern Paiute rock art, ceremony, and cultural landscapes.' *Current Anthropology* 41(1): 11–38, 2000a.

Stoffle, Richard W., *et al. Ha'tata (The Backbone of the River): American Indian Ethnographic Studies Regarding the Hoover Dam Bypass Project.* Report prepared for the Federal Highway Administration. Tucson: Bureau of Applied Research in Anthropology, University of Arizona, 2000b.

Tilley, Christopher. *A Phenomenology of Landscape: Places, Paths, and Monuments.* Oxford: Berg, 1994.

Toupal, Rebecca S. *Landscape Perceptions and Natural Resources Management: Finding the 'Social' in the 'Sciences'.* Unpublished Ph.D. Dissertation. Tucson: School of Renewable Natural Resources, University of Arizona, 2001.

Toupal, Rebecca S., Richard W. Stoffle, and M. Nieves Zedeño. *The Isle Royale Folkefiskerisamfunn: Familier som levde av fiske: An Ethnohistory of the Scandinavian Folk Fishermen of Isle Royale National Park.* Report for the National Park Service, Midwest Regional Office. Tucson: Bureau of Applied Research in Anthropology, University of Arizona, 2002.

Towner, Ronald H., ed. *The Archaeology of Navajo Origins.* Salt Lake City: University of Utah Press, 1996.

Tuan, Yi-Fu. *Space and Place: The Perspectives of Experience.* Minneapolis: University of Minnesota Press, 1977.

Whittaker, Robert J. *Island Biogeography: Ecology, Evolution, and Conservation.* Oxford: Oxford University Press, 1998.

Zedeño, M. Nieves. 'Landscape, land use, and the history of territory formation: An example from the Puebloan Southwest.' *Journal of Archaeological Method and Theory* 4(1): 67–103, 1997.

Zedeño, M. Nieves. 'On what people make of places: a behavioral cartography.' In *Social Theory in Archaeology,* M.B. Schiffer, ed. Salt Lake City: University of Utah Press, 2000, pp. 97–111.

Zedeño, M. Nieves, Diane Austin, and Richard Stoffle. 'Landmark and landscape: A contextual approach to the management of American Indian resources.' *Culture and Agriculture* 19(3): 123–129, 1997.

MARY EVELYN TUCKER

WORLDVIEWS AND ECOLOGY: THE INTERACTION OF COSMOLOGY AND CULTIVATION[1]

The growing alliance of religion and ecology within the academic world and within religious communities is bringing together for the first time diverse perspectives from the world's religious traditions regarding attitudes and ethics toward nature with implications for policy. Scholars of religion from various parts of the world have begun to identify symbolic, scriptural, and ethical dimensions within particular religions in their relations with the natural world. They are examining these dimensions both historically and in response to contemporary environmental problems. Religious practitioners and environmentalists are utilizing these resources as a source of inspiration and activism to motivate long-term changes regarding the environment in many parts of the world. The *State of the World 2000* report notes that in solving environmental problems, "all of society's institutions – from organized religion to corporations – have a role to play" (Brown, 2000: 20). That religions have a role to play along with other institutions and academic disciplines is also the premise of this emerging alliance of religion and ecology.

Several qualifications regarding the various roles of religion should be mentioned. First, we do not wish to suggest that any one religious tradition has a privileged ecological perspective. Rather, multiple perspectives may be the most helpful in identifying the contributions of the world's religions to the flourishing of life for future generations. Second, while we assume that religions are necessary partners in the current ecological movement, they are not sufficient without the indispensable contributions of science, economics, education, and policy to the varied challenges of current environmental problems. Third, we acknowledge that there is frequently a disjunction between principles and practices: ecologically sensitive ideas in religions are not always evident in environmental practices in particular civilizations. Many civilizations have overused their environments, with or without religious sanction. Finally, we are keenly aware that religions have all too frequently contributed to tensions and conflict among ethnic groups, both historically and at present. Dogmatic rigidity, inflexible claims of truth, and misuse of institutional and communal power by religions have led to tragic consequences in various parts of the globe.

115

H. Selin (ed.), *Nature Across Cultures: Views of Nature and the Environment in Non-Western Cultures*, 115–127.
© 2003 *Kluwer Academic Publishers. Printed in Great Britain.*

Nonetheless, while religions have sometimes preserved traditional ways, they have also provoked social change. They can be limiting but also liberating in their outlooks. In the twentieth century, for example, some religious leaders and theologians participated in progressive movements such as civil rights for minorities, social justice for the poor, and liberation for women. More recently, religious groups were instrumental in launching a movement called Jubilee 2000 for debt reduction for poor nations.[2] Although the world's religions have been slow to respond to our current environmental crises, their moral authority and their institutional power may help effect a change in attitudes, practices, and public policies. As key repositories of enduring civilizational values and as indispensable motivators in moral transformation, religions have an important role to play in projecting persuasive visions of a more sustainable future. This is especially true because our attitudes toward nature have been consciously and unconsciously conditioned by our religious worldviews. Over thirty years ago the historian Lynn White observed this when he noted: "What people do about their ecology depends on what they think about themselves in relation to things around. Human ecology is deeply conditioned by beliefs about our nature and destiny – that is, by religion" (White, 1967: 1204).

NATURE AS METAPHOR AND MATRIX

We are constantly interacting with our natural environment and drawing on its vast power. It is in the midst of the relational field of the natural and human worlds that a larger numinous reality is experienced and named in the world's religions. We may discuss this in terms of transcendence and immanence, but I am seeking a language that gets us beyond this kind of dualism, useful as it may be. All religions have developed rich and distinctive expressions for the matrix of relationality we see in the natural world. Indeed, many of the most powerful symbols and rituals of our religious traditions are dependent on our primal encounter with nature. This is true in the Christian understanding of the birth of Christ coordinated with the solstice and the return of the light as well as with Easter associated with the spring equinox and rebirth of life. We can multiply these kinds of examples in other religions as well. Because the cosmologies are different the ethics that arise will also be distinctive. The question remains, where do they converge? I would suggest that while the approaches are different they converge in two areas. One is in seeing nature as metaphor – a steppingstone to the divine; the other is seeing nature as matrix – a meeting place for the divine. In both of these perspectives nature is valued and cherished.

What we will explore here then is the nature of religion as containing cosmological symbols, myths, and rituals that orient humans to a natural world of meaning and mystery, pointing even beyond itself. This is what we would call the numinous experience of nature both as metaphor and as matrix. Our thesis is that in our contemporary world we have manipulated our environment so extensively that it is difficult for us to experience this numinous character of the natural world. Bill McKibben (1989) has written of the "end of nature" because of our extensive tampering with it. Indeed, Thomas Berry (1988) has said we have become autistic in encountering the natural world and thus have become caught in a technological trance. Our fascination with technology has many aspects, from television, movies, and video games to the internet and virtual reality. Somehow we have shifted the focus of our imaginative creativity from the encounter with nature and other humans to an insatiable obsession with technology and with the entertainment it provides as a new kind of magical realm continually feeding the senses.

How then to reexperience in fresh ways the mystical dimensions of nature is a primary challenge for all of us and for our educational and religious institutions as well. Without this the destruction of the environment will no doubt continue unabated. With it we may reclaim the voices of religions in understanding the sacred dimensions of nature, appreciating the rich creativity of nature's cosmological processes, and identifying our special role in fostering that continuing creativity for a sustainable future. Thus through a fuller exploration of the cosmology of the world's religions we may reinvigorate our understanding of how natural processes and religious symbol systems are deeply and subtly intertwined. In so doing we would hope to suggest a context for rethinking human-earth relations in a more comprehensive manner with implications for both social and environmental ethics. Indeed, this is part of activating our imaginative entry into the relational field of the human, the earth, and the divine – what has been called an anthropocosmic perspective as contrasted with an anthropocentric perspective.

What I am suggesting here is that the interconnection of cosmology and cultivation may be a fruitful perspective for understanding the role of religions in fostering a relationship to the natural world that may be helpful for our current environmental crisis. If we put this in more personal terms what I am asking is for us to consider how we reflect on questions in our own lives such as: where do we come from, why are we here, where are we going. By the same token, we can think not just of these large cosmological questions, but also of how we respond to moments in nature that move us – how are we affected by the changing of the seasons, how does a beautiful sunset touch us, why do we seek places of rest in nature such as the mountains or oceans, what is it we experience when we see a line of geese flying overhead or a great blue heron landing at a pond? Why are children so fascinated with nature and how and why do we lose that fascination? How does nature draw us into its mysteries as well as its overwhelming power that can be both destructive and creative? These may be described as experiences of the numinous dimensions of the natural world – nature as both alluring and inspiring as well as awesome and fearful.

DEFINING COSMOLOGY AND CULTIVATION

We may speak, then, of religion as a means of orientation in the midst of the powers of the universe and a means of relationship in the midst of human affairs.[3] Yet it is always contained in the mystery of the cosmological patterning of life itself; that patterning appears to be embedded in both order and chaos, as we now know from chaos theory.

We seek to connect to the deep inner cosmological patterning of things in nature and in human life. This drive to see and understand pattern, coherence, and chaos in the universe is in part what motivates many scientists. It moves the astronomers and the microbiologists to seek the mysteries of matter in its far reaches and in its inner depth. Religions promise something of that connecting link through myths and symbols, rituals and prayers. They seek to weave a web from our inner structures to those complex structures that hold life together. This patterning is called by many names in various religions. In Hinduism and Buddhism it is *dharma* or law; in Confucianism it is *li* or principle; in Daoism it is the *Dao (Tao)* or the Way; in Judaism it is *seder bereishit* or order of creation; in Christianity it is *logos* or word; in Islam it is *shariah* or law. For Native American Algonguins it is *manitou* or spirit presence. Among the Warlpiri people of Australia it is *tjukurpa*, sometimes translated as "dreamtime" but meaning in fact law or pattern in the landscape.

To apprehend and support this patterning we balance ourselves between the outward pull of cosmological processes and the inward pull of the wellsprings of personal authenticity and collective communion. Many religious traditions organize themselves around the patterns they perceive in nature. For example, indigenous peoples seek to embody the cosmos in their own person as well as in the structures they create in bioregions such as subsistence activities of agriculture and organization of habitat (Anderson, 1996). Hindu society has organized itself into sacrificial ritual patterns analogous to the great sacrifice at the origin of the world. Chinese religious thought has developed complex rules of correspondences based on how to live in relation to the patterns of nature. This is evident in such practices as geomancy or orientation to local landscape (*fengshui*) and exercises for health and circulation of the life force or *qi* (*taji, qigong*). [Editor's note: See the article on Fengshui in this volume.] Religions mediate between the patterns of nature and the individual by creating stories of our origins, rituals and practices of cultivation to insure continuity through the various stages of life from birth to death, and codes of behavior which aim to maximize harmonious relations and thus survival itself. In its simplest form, then, religion consists of worldviews embracing cosmology and cultivation. These are linked by patterns (or rituals) connecting self, society, nature and the larger field of being in which they exist.

In further defining terms, it may be helpful to distinguish here between "worldview" (*weltanschauung*) and "cosmology". Although these are sometimes used interchangeably, we take worldview to be a more general and less precisely defined perspective, while cosmology is a more specific focused description of reality often associated with story or narrative. Thus, worldview refers to a broad set of ideas and values, which helps to formulate basic perspectives of societies and individuals.[4] Cosmology is more specifically linked to an explanation of the universe (mythical or scientific) and the role of humans in it. Cosmology may or may not include cosmogony (a story of origins). Cosmologies of particular world religions, however, usually include explanations for the way things are in the universe or the way things ought to be. Science includes the former but not the latter. Cosmologies of world religions, however, imply a metaphysics and/or an ethics which give both orientation and meaning to human life. In this sense, cosmologies of religions contain "principles of order that support integrated forms of being" (Sahlins, 1992: x) and thus give moral direction to a person's life.[5] This is true because of both the orientation and openness that religious cosmologies provide for self-cultivation. This is a term from Chinese religions, which I am defining as a means of personal moral and spiritual development which unifies inner and outer lives.

In attempting to formulate the theoretical grounds for describing cosmology and its functions, Gregory Schrempp writes:

> What do we mean by 'cosmology'? In part we seem to point toward formulations that involve a quest for *ultimate* principles and/or grounds of the phenomenal world and the human place in it. But cosmology often – and this aspect stems perhaps from the Greek notion of kosmos – seems to carry for us a concern with wholeness and integratedness, as if cosmological principles are not only ultimate principles, but also principles of *order* in the broadest sense, that is, principles engendering and supporting a way of being that is cognitively and emotionally integrated and whole. In these two kinds of concerns – the impetus to seek the 'ground' of the present order, and the impetus toward integratedness and wholeness – there is already a potential tension, since the quest for a ground is implicitly a resting of one thing on another, and thus involves a regression from any given state, whereas the impetus toward wholeness may engender the task of finding closure, as a condition for wholeness (Schrempp, 1992: 4).

Something of this tension between grounding and growing is what interests me in proposing a dialectic of cosmology and cultivation. If cosmology in Schrempp's sense has within it an is/ought tension, then cultivation is the working toward resolving that tension or living that tension through an on-going deepening and broadening of one's personhood and ethical life. The deepening is the inner grounding while the broadening is the growing outward.

More than ever before our contemporary challenge is to reorient ourselves to the universe, to know its vastness and its limits and to attune our rich inner space to its rhythms. That is, in essence, a religious act of boldness, of imagination, and of courage in the midst of staggering odds and enormous obstacles. This is the challenge of the evolution of religion to respond to the complex story of the universe and the environmental crisis of our time.

COSMOLOGY AND CULTIVATION IN THE WORLD'S SCRIPTURES

When we examine the early scriptures of many of the world's religions we can see that the great cosmological movements of the universe inspired them – that nature was both teacher and guide. Even as historical traditions arose in distinctive contexts, such as Judaism and Christianity in the Middle East, they were always seen in relation to the larger dynamics of nature and seasonal cycles.

As we read these early scriptures we see into the world of nature not simply as backdrop to human action but as inspiration and animator, as the vehicle for recognizing deeper truths, for exploring greater mysteries of the field of Being. The natural world is not only that which has given humans birth and sustenance, it is that which sustains humans psychically and spiritually in very tangible ways. It is, indeed, the container of mystery that is, both here and beyond, always captivating us with the promise of transformation.

So we return to these two fundamental directions of the religious experience – toward resonance with the universe which lures us forward and toward resilience within ourselves and in our ethical choices in the midst of constant change. We link ourselves, sometimes unwittingly, to an emerging evolving universe, and yet we also draw back into the pulsations of the personal and the demands of the communal for creating sustainable societies. As we look at the scriptures of the world's religions, we sense this inner and outer dynamic joined by patterns. The *Psalms* in Israel, the *Vedas* in India, and the *Yijing* (*Book of Changes* or *I Ching*) in China are some of the oldest written scriptures known to the human community, dating back to the first and second millennium BCE. They reflect a sense of longing for identification with a comprehensive cosmology, and at the same time they signal the needed component of personal cultivation and communal responsibility. They suggest that early river civilizations which were engaged in agriculture were concerned not just with dominion, as has often been suggested, but with cooperation and harmony with nature as well.

At the heart of all of these early scriptures is a profound sense of the dynamic flow of life in the midst of both change and continuity. It is harmonizing with this life pattern which is both within things and yet beyond that characterizes these scriptures. How to affect reciprocal relations with the transformations of life is the challenge they present. This is underscored by rituals which mitigate the unseen forces and which call forth sustaining energy. To open up the transformative powers of the universe in the midst of change is part of the challenge of these early scriptures and the concurrent ritual practices for social organization. In this context, human history fits into this great sweep of cosmological powers, not the other way around.

Cosmology and justice: Israel

The *Book of Psalms* contains 150 prayer-poems probably intended to be sung or accompanied by music. In Hebrew the *Book of Psalms* means "Praises" (*tehillin*) reflecting affirmation or trust in God even in the midst of sorrow. Although most likely complied in the post-exilic period (550 BCE) for temple rituals, many of the themes stretch back much earlier.

As we look at the Psalms we see these songs of nature worshipping, praising, and invoking the Creator and his creation. These are profoundly linked concepts that go beyond constructed dualisms of Western monotheism that divide God and humans. Instead, there is in the psalms a lively dynamic exchange of Creator, Creation, and Creatures. There is a sense of dependence of the Creatures on both Creation and Creator. This is more than simply a static monotheism within a historical trajectory. It is a worldview showing us a God of care and compassion as well as omnipotence and justice. But it is a God deeply engaged in Creation, not simply directing it from afar. This is a God involved in both cosmos and history. It is a God who offers justice to his chosen people who in turn yearn for affirmation, mercy and forgiveness. Thus, there is a sense of Israel's history as woven into a co-existence with God, as Abraham Heschel puts it. For the Israelites, history was seen as the epiphany of God, and Israel was chosen for "converse" with Yahweh (Anderson, 1986: 541). All of this is set against the background of Yahweh as cosmic King and Creator, as enthroned over all of creation, yet intimately connected with it. Indeed, this is celebrated in the enthronement psalms that were part of the cult establishing a throne-ascension festival. This was held each New Year when Yahweh's rule over Israel, over other nations, and over the cosmos was celebrated with hymns and rituals.

The Psalms are divided into hymns of praise, of lament, and of thanksgiving. In terms of Psalms of praise, in Psalm 136: 4–9, we have a striking depiction of the Creator and Creation as deserving utmost respect, wonder, and awe.

> To the One who alone does great wonders,
> who by understanding made the heavens,
> who spread out the earth upon the waters,
> who made the great lights,
> the sun to rule over the day,
> the moon and stars to rule over the night (Anderson, 1986: 550–551).

The order of creation is celebrated and the power and majesty of the Creator is underscored. Yet God's continual creativity in history is noted, as in Psalm 104: 27–30 (Anderson, 1986: 551).

> All of them [animals and humans] look to you
>   to give them their food in its season.
> When you give to them, they gather up,
> when you open your hand, they are satisfied to the full.
> When you hide your face, they are disturbed,
> when you take away their breath, they expire
> and return to their dust.
> When you send forth your spirit, they are [re]created,
> and you renew the surface of the soil.

The Psalms that reflect lamentation and thanksgiving might be seen as part of the cultivation side of the dyad. As injustice occurs there is a call for deliverance, and as blessings are received thanksgiving pours forth. Justice and mercy are reasserted against the forces of oppression and sorrow.

In short, the cosmological world is the container of the history of the chosen people. As the people cultivate their relationship with the Creator of the Cosmos they will experience justice and peace. While in this model history becomes a key element, nonetheless, maintaining a proper relationship to Creation and the cosmic order is the container for all the history of the Chosen People.

Cosmology and sacrifice: India

This great sweep of cosmological powers is also present in the Vedas in India, which are hymns to celebrate the gods of nature. Here we have a richly textured universe presided over by a variety of gods. This is not a model of monotheism as we see in the Psalms, but rather henotheism as described by Max Mueller. This term describes a worldview in which there is a pantheon of gods with no strict hierarchy (Zysk, 1989: 10–11). While there is more emphasis on cosmos than on history, still there is a sense that the gods are involved in human concerns and need sacrifices to maintain order in the universe and to support human action. The sense of the awesome powers of nature as depicted in these Vedic hymns resonates down to the present period in India,[6] where the hymns are still recited and on occasion the Vedic sacrifices are still offered. In fact, sacrifice is the structured pattern of the universe itself.

The Vedic hymns were composed between 1600 and 600 BCE and thus constitute the oldest written literature in India. They were transmitted orally for almost 3000 years until some Brahmins in Calcutta were reluctantly persuaded to write them down in the 1780s. There are four principal texts, the oldest of which is the *Rigveda*, a collection of over 1000 hymns. Although there are current historical debates about early Indian history, the Vedas are generally attributed to the Aryans, a nomadic and horse riding people who moved into central Europe and India during the second millennium. As they settled in the Indus Valley region they gradually took up agriculture. Thus, these hymns are a fascinating collection from a people moving from hunting and gathering to farming.

In relation to our overarching theme of the interaction of cosmology and cultivation, we might say that there are two types of cosmological hymns in the Vedas, hymns of creation and hymns celebrating natural phenomena. There is one type which, broadly speaking, deals with cultivation, in this case sacrificial rituals.

The first kinds of cosmological hymns are those which deal with creation and origins. In other words, they are cosmogonic hymns. One of enormous importance is the *Mahapurusha*, which describes the birth of the universe from the sacrifice of a Great Person. This becomes a key link to modes of communal cultivation in sacrifice. The correspondence of the person to the universe provides the pattern for the relationship of all humans to the cosmos. (*Rigveda*, Book 10)

> When they divided the Man,
> into how many parts did they divide him?
> What was his mouth, what were his arms,
> what were his thighs and feet?
>
> The brahman was his mouth,
> of his arms was made the warrior,
> his thighs became the vaisya [peasant]
> of his feet the sudra [serf] was born.
>
> The moon arose from his mind,
> from his eye was born the sun,
> from his mouth Indra and Agni,
> from his breath the wind was born.
>
> From his navel came the air,
> from his head there came the sky,

from his feet the earth, the four quarters from his ear,
thus they fashioned the worlds.

With Sacrifice the gods sacrificed to Sacrifice –
these were the first of the sacred laws.
These mighty beings reached the sky,
where are the eternal spirits, the gods (Zysk, 1989: 25).

Second major types of cosmological hymns are those celebrating the power of natural phenomena. In these Vedic hymns there are three kinds of deities – those of the heavenly, atmospheric, and earthly realms. These include Indra, the god of war and rain, who overcomes the evil serpent to release the waters for the benefit of humans. Varuna is the sky deity who restores and guards the order of the universe (*rta*). Also significant are the earthly deities of Agni, the god of fire, and Usas, the goddess of dawn, who brings refreshing hope and renewal to each day.

This light, most radiant of lights, has come; this gracious one who illumines all things, is born. As night is removed by the rising sun, so is this the birthplace of the dawn.

We behold her, daughter of the sky, youthful, robed in white, driving forth the darkness. Princess of limitless treasure, shine down upon us throughout the day.

Rigveda I, 113 (Berry, 1971: 22)

Finally, in terms of cultivation there are hymns that define and celebrate ritual sacrifice as key to the maintenance of cosmological order in the natural and human realms. The power of ritual action is evident throughout these hymns. The early Creation hymn of sacrifice previously mentioned sets the stage for the importance of ritual in this worldview. Prayers and offerings are key to placating the powers of the universe such as wind and rain, thunder and lighting. Moreover, there is a need to maintain balance and order through correct speech, through the sacrifices of the present, through the consecration of the kings, through the great horse sacrifice, and through the offerings of ghee, of soma, and of fire itself.

To placate the powers of nature, to maintain order, to obtain material benefits, and to establish moral coherence – all of these are reasons for the importance of ritual sacrifice. While this is true in many early societies, it takes on a particular importance in India in relation to the cosmos itself.

Cosmology and harmony: China

In turning to China we see in the *Yijing* (*I Ching, The Book of Changes*) the dynamics of change and continuity in the universe celebrated not as gods of nature but as patterns, namely hexagrams, which can be read as symbols guiding human affairs. To discern correct action humans must be in relationship to the movements of the universe. The first hexagram, the Creative, *qian*, illustrates this well.

COMMENTARY ON THE TEXT: Vast indeed is the sublime Creative Principle, the Source of All, co-extensive with the heavens! It causes the clouds to come forth, the rain to bestow its bounty and all objects to flow into their respective forms. Its dazzling brilliance permeates all things from first to last; its activities symbolized by the component lines, reach full completion, each at the proper time. (The Superior Man), mounting them when the time is ripe, is carried heavenwards as though six dragons were his steeds! The Creative Principle functions through Change; accordingly, when we rectify our way of life by conjoining it with the universal harmony, our firm persistence is richly rewarded. The ruler, towering above the multitudes, brings peace to all countries of the world (Blofeld, 1965: 85).

The creativity of the universe is manifest throughout the natural order. For a society deeply engaged with agriculture such as China was, the changes of the seasons were seen as key cyclical patterns mirroring transformation in human life. They were revelatory of the birth and death processes of nature and of humans. The constancy of the seasons gave guidance to human affairs. Indeed, it was often said that the seasons did not err and that therefore the great person took them as a model for action and behavior (Wilhelm, 1977: 18). Thus, the emperor, for example, was considered a pole star, the exemplar for the entire society. He offered sacrifices at the great temples of Heaven and Earth in the capital city, Beijing. He ritually planted the rice in the fall and harvested it in the late summer. (This still takes place for the Japanese emperor in Tokyo.) Moreover, throughout Asia there are elaborate systems of geomancy (*fengshui*) which orient persons, houses, public buildings, and graves to the most auspicious direction and balance with nature.

In terms of the broad dialectic of cosmology and cultivation, Chinese religious thought concentrates on connecting biological processes of growth and transformation with particular virtues to be cultivated. There are four attributes of creativity: sublimity, success, furtherance, and perseverance. These are metaphors for the life process of beginning, duration, advantage, and flourishing. They are linked to the virtues of humaneness, faithfulness, righteousness, and wisdom. Thus, to be receptive to these cosmological processes of the life cycle one must cultivate virtue. The result of this dynamic process of cosmological creativity finds its counterpoints in the receptive cultivation of the individual. The person is linked to the cosmos through life generating patterns reflecting both order and change in the universe. A person can thus penetrate the Dao of Heaven and Earth.

> The Book of Changes contains the measure of heaven and earth; therefore it enables us to comprehend the tao of heaven and earth and its order.

> Looking upward, we contemplate with its help the signs in the heavens; looking down, we examine the lines of the earth. Thus we come to know the circumstances of the dark and the light. Going back to the beginnings of things and pursuing them to the end, we come to know the lessons of birth and death. The union of seed and power produces all things; ...

> The result is not only the growth of knowledge but also the growth of virtue:

> Since in this way man comes to resemble heaven and earth, he is not in conflict with them. His wisdom embraces all things, and his tao brings order into the whole world; therefore he does not err. He is active everywhere but does not let himself be carried away. He rejoices in heaven and has knowledge of fate, therefore he is free of care. He is content with his circumstances and genuine in his kindness, therefore he can practice love (Wilhelm, 1960: 69).

The result of this penetration of the changes in the cosmos and the cultivation of virtue in the self is that humans are able to both cooperate and collaborate with heaven and earth. In this way they form a triad with heaven and earth, completing their transforming and nourishing powers.

Indeed, the sense of completing and harmonizing with the fecundity of life that underlies the *Yijing* is at the heart of this dynamic system of cosmology and cultivation. For what the *Yijing* aims at is how to release the transformative energies in nature so as to be resonant with the creativity of human energies. Here the overflowing power of material force or *qi* comes into play.

Qi is that which unites all life from atoms, plants, animals, and humans, to the cosmos itself. This vitalistic principle of life holds within it the great transformative potential of life. These are the patterns that connect, buried deep within the storehouse of human knowing. Our genes contain these patterns of knowing that link us to all other atoms in the universe.

*   *   *

To reignite that link between our inner patterned storehouse and that of the natural world is what is needed in all the major religious traditions. It is what is called for now from out of the fire of groping toward patterns of order and meaning and purpose. These ancient scriptures are examples of how that linkage was fostered in earlier times and how the numinous encounter with nature was reaffirmed. Now we need to weave new linkages to both time and space within the context of the epic of evolution. If the epic is the warp, religions may be seen as the woof, as the theologian Philip Hefner (1998) suggests. The patterns and design are still emerging. In terms of developmental time, we are seeking our place in this vast sweep of evolution. In particular, then, we need to examine the cosmological dimensions of the world's religions so that our efforts at moral cultivation will include reciprocity with the natural world as the relational field on which our life is completely dependent.

In terms of space, we are seeking appropriate modes of ecological design – how to live with the river patterns, how to tap into solar energies, how to flow with the rhythms of water, how to move with the currents of air. All of this means harmonizing with the deep cosmological rhythms in nature, not controlling them or harnessing them in manipulative ways, but learning once again nature's inner ordering principles. We have to understand the patterning imprinted in the cosmos and cultivated in ourselves. This is the way of the religious imagination, from its earliest pulsations to the present.

There are many hopeful signs that such changes are occurring as we begin to learn that progress needs to be redefined, that growth for the sake of growth is not necessarily desirable, that ecological economics is a growing field, that green businesses are emerging, that alternative energies are feasible and most of all that many people care deeply about preservation of the environment. Indeed, we find in this time of overload with bad news there are many insightful and even heroic voices calling for change. One of the foremost of these is Thomas Berry, whose book *The Great Work* (1999) evokes a vision of mutually enhancing human-earth relations within the context of the unfolding process of our evolving universe. He sees this broadened context of the universe story with its rich cosmological dimensions as a primary means of reorienting humans to living in a universe of meaning and mystery. Our response, he suggests, should be one of reciprocity, reverence, and care for the earth and for future generations.

The questions we need to ask are can we listen, can we see, can we touch anew with a feeling for the organism – with a deep resonance without and abiding reverence within? If so, not only will we survive but the planet itself will flourish. And it will do so if we trust the transforming and nourishing powers of the cosmos.

As Thomas Berry writes at the conclusion of his essay, "The New Story":

> ... The basic mood of the future might well be one of confidence in the continuing revelation that takes place in and through the earth. If the dynamics of the universe from the beginning shaped the course of the heavens, lighted the sun, and formed the earth, if this same dynamism brought forth the continents and seas and atmosphere, if it awakened life in the primordial cell and then brought into being the unnumbered variety of living beings, and finally brought us into being and guided us safely through the turbulent centuries, there is reason to believe that this same guiding process is precisely what has awakened in us our present understanding of ourselves and our relation to this stupendous process. Sensitized to such guidance from the very structure and functioning of the universe, we can have confidence in the future that awaits the human venture (Berry, 1988: 137).

NOTES

[1] This chapter arises in part from a conference series that I organized with John Grim from 1996–1998 at Harvard's Center for the Study of World Religions, the Harvard Center for the Environment, and the Harvard-Yenching Institute. The conferences brought together over 800 scholars of the world's religions along with environmental activists to rethink views of nature and the potential for environmental ethics in the world's religions.

We have now begun a Forum on Religion and Ecology that is designed to continue this research and to publish the papers from the conferences in a ten volume series distributed by Harvard University Press. In addition, we are encouraging educational initiatives in religion and ecology on the college and high school level and to this end have conducted teacher workshops. We have also established a comprehensive web site under the Harvard Center for the Environment (http://environment.harvard.edu/religion) that is designed to promote religion and ecology as a field of study and to foster outreach to other disciplines such as science, education, economics, and public policy. With the scientist Brian Swimme we have been conducting a yearly summer seminar exploring the intersection of the cosmology of science and the cosmology of world religions. All of this is part of the background context for this paper that is exploring some of the differences among cosmologies of religions.

The Harvard conferences and the Forum arose out of a concern for the burgeoning environmental crisis and the need to identify the possible responses of the world's religions. Clearly the moral force of religion may be instrumental in alerting people to the ethical implications of what we are doing to the planet by unbridled industrialization, wanton use of resources, and spreading pollution of air, water, and soil.

[2] The movement, which began in Britain, has had demonstrable influence on the decisions of the World Bank and other lending organizations to reduce or forgive debts in more than twenty countries. See http://www.jubilee2000uk.org.

[3] This section of the paper has been published in a slightly different form in Stephen Kellert and Timothy Farnham, eds., *The Good in Nature and Humanity: Connecting Science, Religion, and Spirituality with the Natural Environment* (Washington DC: Island Press, 2002).

[4] In using "worldview" I am indebted to Clifford Geertz's articulation of worldview and ethos as well as Ninian Smart's use of the idea of worldview to describe religious traditions.

[5] Robin Lovin and Frank Reynolds have edited a volume of provocative essays from cross-cultural perspectives on this topic titled *Cosmogony and Ethical Order* (Chicago: University of Chicago Press, 1985).

[6] We must be careful not to conflate what is known about the beliefs and behaviors of ancient Indians with modern India. What worked for a few thousand people has obviously not worked for India's billion. "While they inherit their past, they must live in the present" (see Booth, this volume).

BIBLIOGRAPHY

Anderson, Bernhard W. *Understanding the Old Testament.* 4th ed. Englewood Cliffs, New Jersey: Prentice Hall, 1986.

Anderson, Eugene N. *Ecologies of the Heart.* New York: Oxford University Press, 1996.

Berry, Thomas. *The Dream of the Earth.* San Francisco: Sierra Club Books, 1988.

Berry, Thomas. *The Great Work: Our Way into the Future.* New York: Bell Tower, 1999.

Berry, Thomas. *Religions of India: Hinduism, Yoga, Buddhism.* Beverly Hills, California: Bruce Publishing Co., 1971.

Blofeld, John, trans. and ed. *I Ching: The Book of Changes.* New York: E.P. Dutton, 1965.

Brown, Lester R. 'Challenges of the new century.' In *State of the World 2000.* New York: Norton, 2000.

Chapple, Christopher and Tucker, Mary Evelyn, eds. *Hinduism and Ecology: The Intersection of Earth, Sky, and Water.* Cambridge, Massachusetts: Center for the Study of World Religions and Harvard University Press, 2000.

Hefner, Philip. 'The spiritual task of religion in culture.' *Zygon* 33(4): 535–545, 1998.

McKibben, Bill. *The End of Nature.* 2nd ed. New York: Anchor Books, 1999.

Sahlins, Marshall. 'Foreword.' In *Magical Arrows: The Maori, the Greeks, and the Folklore of the Universe,* Gregory Schrempp, ed. Madison: University of Wisconsin Press, 1992, pp. ix–xii.

Schrempp, Gregory. *Magical Arrows: The Maori, the Greeks, and the Folklore of the Universe.* Madison: University of Wisconsin Press, 1992.

Tucker, Mary Evelyn and John Berthrong, eds. *Confucianism and Ecology: The Interrelation of Heaven, Earth, and Human.* Cambridge, Massachusetts: Center for the Study of World Religions and Harvard University Press, 1998.

Tucker, Mary Evelyn and Duncan Williams, eds. *Buddhism and Ecology: The Interaction of Dharma and Deeds.* Cambridge, Massachusetts: Center for the Study of World Religions and Harvard University Press, 1997.

Tucker, Mary Evelyn and John Grim, eds. *Worldviews and Ecology.* Lewisburg, Pennsylvania: Bucknell University Press, 1993. Paperback edition, Orbis Books, 1994. (Sixth edition, 2000.)

White, Jr., Lynn. 'The historical roots of our ecologic crisis.' *Science* 155(10 March): 1204, 1967.

Wilhelm, Helmut. *Change: Eight Lectures on the I Ching,* Cary F. Baynes, trans. New York: Harper Torchbooks, 1960.

Wilhelm, Helmut. *Heaven, Earth, and Man in The Book of Changes.* Seattle: University of Washington Press, 1977.

Zysk, Kenneth G., ed. *A.L. Basham: The Origins and Development of Classical Hinduism.* Boston: Beacon Press, 1989.

SUSAN M. DARLINGTON

THE SPIRIT(S) OF CONSERVATION IN BUDDHIST THAILAND

Two contradictory images strike the traveler in northern Thailand. First is the lush, forested mountains rising beyond expanses of rice paddy land and small farming villages. Second is the spotty appearance of the mountains, denuded of primary growth in large areas and filled instead with economic crops such as cabbages or corn. Both images are set against the backdrop of congested cities, particularly Bangkok and Chiang Mai (the largest city in the north), through which all travelers pass before seeing rural areas. The contrasts inherent in these scenes point to a major tension in Thailand between the push to develop economically and efforts to conserve and protect the nation's natural resources.

The struggle to find a balance between development and conservation occurs in many arenas, including culture and religion. Various actors for both pro-development schemes and conservation projects use cultural and religious beliefs, practices and attitudes toward nature to promote their positions. Religious beliefs themselves are not intrinsically either ecologically or developmentally oriented, but they can be interpreted in ways to support either conservation or development. The Thai government, for example, has used Buddhism to promote rapid national economic development; on the other hand, a handful of Buddhist monks and non-government activists incorporate both indigenous spirit beliefs and Buddhist practices to foster an environmental ethic on a local level. Both approaches claim to be based on intrinsically "Thai" understandings of the world as well as basic Buddhist teachings. Yet, as will be shown below, both are reinterpretations of religious beliefs and practices and, particularly in the case of activist monks, examples of cultural creativity designed to promote specific political agendas.

In the search for "Thai" approaches to environmental conservation, a key cultural source is Buddhism. The Thai population is approximately 95 percent Buddhist, the remainder being Christian, Muslim or animist. It is important to note, however, that for Thais, Buddhism incorporates spirit and Brahmanic beliefs and practices (Kirsch, 1977; Tambiah, 1970). Yet scholars and activists often downplay spirit and Brahmanic beliefs in their environmentalist efforts,

H. Selin (ed.), Nature Across Cultures: Views of Nature and the Environment in Non-Western Cultures, 129–145.

opting primarily to look at the ecological knowledge of rural peoples and at how Buddhism can be used to promote an environmental ethic (see Chatsumarn, 1990 and 1998; Darlington, 1997a, 1997b, 1998; Davies, 1987; Kaza and Kraft, 2000; Pipop, 1993; Sponsel and Natadecha, 1988; Sponsel and Sponsel-Natadecha, 1995, 1997, and this volume).

Buddhism alone, however, does not encompass the full range of concepts and attitudes of rural Thais towards nature and the forest. A complementary set of cultural beliefs that influence attitudes of and behavior towards the natural environment in Thailand is spirit beliefs. In particular, the interplay between Buddhism and spirit beliefs affects rural people's concepts of the forest and the ways in which they approach or value it. Similar beliefs are found throughout the Tai cultural region (which includes northern Thailand, Laos, southern Yunnan Province, China, and parts of Burma and northern Vietnam; see Pei, 1985).

This syncretic religious system can be clearly seen in northern Thailand where, as across the nation, it contributes to an elaborate cosmology that includes the natural, human and spiritual environments (see Kirsch, 1977). This sacred geography provides the framework for efforts to use cultural and religious concepts for both developmental and environmental – both inherently political – ends.

SPIRITUAL GEOGRAPHY

There is no traditional word in the various Thai language dialects for "nature" (Davis, 1984: 85). The word used today, *thammachaat*, is a combination word borrowed from Sanskrit. Its root words are *thamma*, or "dhamma", meaning truth (or the teachings of the Buddha, in a more specific context), and *chaat*, meaning life or rebirth (from the concept of reincarnation). Davis argues that Thais adopted the concept of "nature" underlying this term because of European influence (1984: 85). This concept did not correspond with any indigenous ideas. It is, according to Stott (1991: 144), "too refined and wide in its meaning, embracing as it does all natural phenomena, such as rain, wind and sun, and even natural human behaviour." Applying the concept of *thammachaat* to the forest sanitizes the barbaric qualities of the wild, bringing it closer within the control of people (Stott, 1991: 150). The effects of this process, as we will see, can be either pro-conservation or pro-development. It can enable people to take advantage of the forest resources without fear or concern over maintaining a balance between the civilized and wild worlds, by emphasizing domination and control of the former over the latter. It also has the potential to influence people's attitudes in the direction of responsibility toward, rather than domination over, the natural environment.

Before the introduction of the concept of *thammachaat*, there was a distinction between the human-built civilized world (*muang*) and the wild, uncivilized forest (*paa thu'an*) (Stott, 1991). The civilized world centered on human settlements, particularly the walled cities of northern Thailand in which the kings and princes lived. In a careful linking of religion and geography, civilization radiated outward through the ruler's religious merit from the muang (here meaning the city) to nearby towns and outlying farming villages (Stott, 1991: 145). The forest, on the edge of the social and religious domain of the muang and the ruler's influence, was the world of wild, dangerous and unpredictable beings, including tigers, bears, gods, spirits and non-Tai hill peoples.

Even today, lowland Thais (in the north called *khon muang* or "people of the muang") often consider the hill peoples backwards or uncivilized because their ways of life are different from those of Tai peoples. The hill peoples follow their own cultures, speak their own languages, eat different foods, and, especially, are not Buddhist. They are predominantly animist, each group having its own elaborate set of beliefs and practices that the lowland, Buddhist Thais tend to see as primitive. Buddhism for Thais is an element of their cultural identity and a key component of the sacred geography that has defined their civilization for centuries.

At the same time, most Thai Buddhists also believe the universe is inhabited by spirits and gods. Buddhism as a religion has never denied their existence. The classic treatise on Thai cosmology, *The Three Worlds According to King Ruang* (from the 13th century, translated by Reynolds and Reynolds, 1982), describes the interplay between the heavens (the realms of *theewadaa* or celestial beings), earth (where the humans live), and the hells (the place of *phii* [spirits] or lower level beings living out the consequences of negative behavior). All sentient beings, according to this Buddhist cosmology, can be reborn at any level of these three worlds depending on the merit of their actions in each life. Even the gods, who lead lives of pleasure and comfort, will eventually pass away and be reborn. Only humans can achieve enlightenment, release from the cycle of rebirth and suffering.

While Thais view the three worlds as distinct, they are not isolated from each other. Usually the theewadaa and phii of the upper and lower realms are benign and do not harm humans. If provoked, however, the phii in particular may harass or even possess some humans. Thais believe some phii are malevolent and go out of their way to avoid or appease all spirits. Other phii may simply appear to people as their paths cross. Guardian spirits (considered both theewadaa and phii, illustrating the vague nature of these concepts) monitor and protect particular places. Some are considered "lords of the land", associated with the muang or principalities of the past (see Shalardchai, 1984). People fear the spirits, theewadaa and phii alike, because of their supernatural qualities and powers.

Many different kinds of spirits inhabit the northern Thai cosmos. There are ghosts, or *preta*. This ambiguous term covers the spirits that all people become at death before they are reborn into another form of existence. It also denotes the hungry ghosts, phii that are constantly hungry and thirsty because their tiny mouths cannot open fully. People believe these phii were greedy in a former life and are now paying the consequences of their actions through insatiable hunger. (This case is a good example of *kamma*, or karma, at work, the belief that all intentional actions, good or bad, have consequences that usually come to bear on how someone is reborn in future lives.)

Other kinds of phii include tutelary spirits that help people who ritually respect them and follow the ethics that they enforce. Some of these are guardian spirits of particular places, such as a village or a household. Spirits live in forested areas, occupying trees, fields and streams. People make offerings to these spirits before they cut down a tree or clear a field for their crops. Every year villagers hold a series of agricultural rites, including offerings to the spirits of the fields, continually thanking them for allowing the forest to be cleared and the crops to grow. Spirits also look after the cremation grounds and sacred groves that surround some villages. In some areas these are the only patches of forest that have not been cut down. They have been left standing due to people's fear of and respect for the spirits believed to live there.

A number of deities or celestial beings also share the northern Thai world. These theewadaa also receive ritual offerings and propitiations for their powers to influence the human world. There are four lords who rule over the four cardinal directions, overseen by Indra. (Indra is a god borrowed from Indian cosmology and was probably introduced into the region before Buddhism's arrival.) Two female spirits affect people's interactions with the natural world. These are Mae Phosop, the rice goddess, and Mae Thoranee, or Mother Earth. These theewadaa can either help farmers with their rice and other crops or punish those who offend them.

The annual rituals performed in honor of the various theewadaa are examples of how northern Thais interact with supernatural beings. Humans do not have control over these beings, but exist in a balanced and dynamic relationship with them. People can influence the phii and theewadaa for temporary periods but must continually renew their relationship through rituals.

An example of these rituals is an annual rite performed in April during the Thai New Year in honor of Lord Indra, the Four Lords and Mae Thoranee. A layman trained in Brahmanic practices, often a former monk, usually conducts the rite. Even though technically there are no Buddhist elements to the ritual, I have seen the rite done by a monk. Northern Thais do not make such distinctions within their religious system, and none of the villagers, including the monk himself, saw any contradictions in his performing the rite.

The ritual involves propitiating the six deities with offerings of food, rice, betel nut, cigarettes, and clay models of buffalo, elephants and other useful and powerful animals. After the formal ritual, conducted in the center of the village, trays made of banana trunk and bamboo, loaded with the offerings, are carried to the four cardinal points of the village boundary for the Four Lords. Two trays are left in the center of the village, one on a pole for Lord Indra, the other sitting on the ground at the pole's base for Mae Thoranee. In addition to honoring the deities, the ritual exorcises the village of evil or harmful forces. These negative forces are carried to the edge of the village with the trays for the Four Lords and are exiled to the wild space beyond the settlement. The Four Lords help guard the village boundaries, as well as looking after the region of each direction. Besides the annual performance for the entire village, ritual specialists will perform this rite when serious illnesses or calamities occur within the village or individual households.

Probably the most feared of the spirit world are the phii that inhabit the forests. As the paa thu'an (wild forest) represents the antithesis of the civilized world, people see the spirits that live there as especially dangerous. The divide between muang and paa thu'an, and between Buddhism and animism, however, is not absolute. Even the forest can be used by people if they take the proper precautions, including conducting rituals to propitiate and appease the gods and spirits that dwell there. The influence of such rituals is temporary, however. The rites need to be done on a regular basis or every time someone wants to use forest resources such as trees for building houses or animals for food.

BUDDHISM AND SPIRIT BELIEFS

Buddhism does not teach that spirits do not exist. Instead, its teachings emphasize developing one's ability to control fear, such as the fear of spirits, through recognizing the impermanence of all things, including one's self. This is a learned skill, however, often developed through years of practice and confronting fear (see Kamala, 1997).

When I first began to study Thai Buddhism in the mid-1980s, I did not expect to find many monks who believed in spirits. Living in northern Thailand and studying with monks, however, I soon realized the extent of the interplay between Buddhism and animism. One monk's story in particular highlights the dynamic relationship between the two.

As a novice in his teens, living in a small, rice-farming village in northern Thailand, this monk walked several kilometers daily to attend classes on Buddhism. His journey took him along a long stretch of road that ran through the rice fields and forest between his village and the next. Often he traveled this road at night when his lessons ran late. He walked with trepidation as the villagers told tales of a hungry ghost that haunted the fields and forest at night.

One dark, moonless night, he related to me years later, as he walked the road, he saw a light across the field. No houses sat in that area, and it was not the season for villagers to be hunting frogs at night. The novice froze in his steps as the light glided across the field towards him. "The hungry ghost," he thought, hardly able to breathe.

As the light came closer, the novice tried to still his panic. All the warnings people had given him about walking this stretch of road at night flooded his mind and he regretted not heeding them. Then he remembered why he walked this way – to study Buddhist scripture in the temple in the district town. "I'm a monk," he thought, "I'm not supposed to fear ghosts. This ghost can't harm me."

With that thought, the light stood still a few meters into the field. The novice smiled at his own silliness and continued on his way home. He never told anyone of his experience until years later, after the forest had been mostly cut down and the road was more traveled, even at night. No point in fostering these fears, he told me, as his story would have only confirmed people's belief in the hungry ghost.

The tale of the hungry ghost involves only one kind of the many phii and theewadaa in northern Thai cosmology. Nevertheless, it shows how belief in a spirit causes fear and can potentially keep people out of certain places. It also illustrates the belief in the superiority of Buddhism over spirits in the cosmological hierarchy, a belief that has been used to change the spiritual geography, civilize untamed areas and develop the forest.

BUDDHISM AND FOREST DEVELOPMENT

Buddhism has contributed to civilizing the forest in the modern era through subduing the wild forest and its spirits, enabling human settlements to expand beyond city boundaries. In the late 19th and early 20th centuries, forest-dwelling monks bridged the space between muang and paa thu'an (Stott, 1991: 149; see also Kamala, 1997; Tambiah, 1984; J. Taylor, 1993a). These Buddhist monks chose to follow thirteen ancient ascetic practices, including forest dwelling, called *thudong*, that aided meditation and purification. Their spirituality enabled them to face the fears of the wild that most lay people held and enter and even live within the forest for extended periods (Kamala, 1997). Forest monks brought elements of the civilized world to the forest, blurring the distinctions between the two. Their presence helped quell villagers' fears of the forest spirits, leading them to move into the forest, settling previously wild, dangerous and unlivable areas. If the monks could live there, people believed, then, under the protection of their spiritual powers, so could devout lay people.

Kamala (1997) points to the presence of forest monks as one of the factors that enabled the Thai government to "invade" and then "close" the forest in the 20th century. She refers to two relevant eras. First was the Forest-Invasion Period (1957–1988), in which the military-backed government manipulated the Buddhist Sangha (or monkhood) in order to promote economic development that led to significant deforestation across Thailand (Kamala, 1997: 229–243). The second was the Forest-Closure Period (1980-present), in which the government, in an effort to retain control of dwindling forest resources, banned monks – and most lay people – from living in the forests (Kamala, 1997: 243–249).

The governmental efforts to develop and control forest resources, and the use of the Sangha in this process, was epitomized by Field Marshall Sarit Thanarat, the autocratic ruler who, after coming to power through a coup in 1958, pushed Thailand into an intensive development policy. Aiming to bring Thailand into the global economy, Sarit promoted agricultural intensification and expansion based on export and industry. He emphasized a shift toward cash cropping and bringing more forest land under cultivation, continuing the concept of "civilizing" the wild forest and making it useful for humans. He also drew on traditional cultural values to promote his development agenda. Yoneo Ishii comments,

> Sarit thought that national integration must be strengthened to realise national development. To attain this goal he planned to start with fostering the people's sentiment for national integration through the enhancement of traditional values as represented by the monarchy and Buddhism (Ishii, 1968: 869).

Sarit incorporated Buddhism into his development campaign through community development and missionary programs involving monks (Keyes, 1971; Tambiah, 1976: 434–471). These programs included *thammathud*, which sent monks as missionaries to politically sensitive and economically poor border provinces; *thammacarik*, through which monks worked among minority hill peoples to convert them from animism and civilize them (bringing them into the national economy); and community development programs sponsored by the two national Buddhist universities.

The involvement of monks in development programs provided legitimacy and encouraged Buddhist lay people to participate. At the same time as they contributed to national economic growth, however, these monks inadvertently were a factor in the accompanying environmental degradation. The balance of the spiritual geography shifted. The elements of fear of and respect for spirits that limited encroachment into forested land in the past were negated by monks' engagement in and support of national economic development (see Somboon, 1977 and 1982).

The contribution of individual Buddhist monks to the destruction of the forest can be seen on the local level as well, through the act of subduing forest spirits. The following quote from an ethnography done in a northeastern Thai village in the 1960s illustrates villagers' beliefs in forest spirits and the potential role Buddhism plays in overcoming them so that people can use forest resources. It concerns a particular kind of tree, the *takien* tree, in which spirits live.

> *Takien* trees are most well-known for female malevolent spirits. Nevertheless, the villagers have to overlook their harm whenever they wish to cut them down. A certain person with an especially strong *mantra* will have to conduct a ritual to subjugate the spirits. After a rite is performed, the tree may be cut down.
>
> If a person is uncertain as to whether or not it would be right to cut down the tree, he would cut the trunk with an axe and leave the axe stuck in the tree overnight. If the axe is found fallen from the tree in the morning, the man would be able to cut the tree unharmed. Sometimes if the government wants a particular tree, the cutter would put an official emblem on it. This would supposedly bring no danger to the cutter.
>
> When the *takien* tree is cut, one would usually hear the cry or groan of a woman. Sanguan Srisuwarn [a villager] recounts that once he went to carry a *takien* tree which had already been cut. He could not even get close to the tree. Big flies were swarming the tree so thickly that it was impossible to do anything with the tree. This is supposed to have been the doing of the *phi*. So they had to ask a monk from Wat Nong Yang to chase the *phi* away. The monk made holy water and sprinkled it all around with a *takien* branch. After a while the flies were gone (Kingkeo Attagara, 1967: 44–5; quoted in Stott, 1978: 16).

A professor at Chiang Mai University told me of stories of villagers asking monks to exorcise spirits from forested areas so that they could cut down the trees and establish farms in the northeast. I never directly met anyone who had witnessed such actions, but the logic of such actions makes sense. The accounts of the government using forest monks – directly and indirectly – to facilitate economic development of forested areas (Kamala, 1997: 229–249) and involving other monks in community development and Buddhist missionizing activities since the 1950s (Keyes, 1971; Tambiah, 1976: 434–471), together with descriptions such as the quote above, suggest that the use of Buddhism to civilize the paa thu'an was a common and accepted event.

SPIRITUALITY AND CONSERVATION

In more recent times, however, a small number of monks are working to use Buddhism to protect the forest and the natural environment. These self-proclaimed "environmentalist monks" (Thai: *phra nak anuraksa*) promote environmental conservation based on Buddhist principles such as compassion and the interdependence of all beings. They are concerned about the suffering resulting from unmonitored economic development and environmental damage. They cite the Buddha's close connection with the forest, including stories that his birth, enlightenment and *parinibbana* (physical passing) all occurred in the forest, to support their calls for people to protect forested areas. Monks have undertaken projects to create protected community forests, clean up polluted rivers, and challenge illegal logging in national parks.

While environmentalist monks emphasize Buddhist principles in their work, they are aware of the importance of the cultural contexts in which they work. Some incorporate local beliefs in their environmental work. For example, Phrakhru Pitak Nanthakhun of Nan Province in northern Thailand recognizes the potential of spirit beliefs to support conservation projects and gain villagers' commitment to them. In 1990, he sponsored a ritual that symbolically ordained a tree in his home village in order to establish and sanctify a protected community forest (Darlington, 1998). Before the tree ordination, the villagers held a ceremony asking the village guardian spirit, one of the region's "lords of the land", for permission to create the community forest and his help in protecting it. They established a shrine at the base of the tree to be ordained and made offerings to the guardian spirit. Phrakhru Pitak did not attend the ceremony, but neither did he oppose it. He commented later on the combined effect of the two rituals:

> Holding a tree ordination, establishing a shrine for the guardian spirit and placing a Buddha image as the "president" of the forest to forbid cutting trees are all really clever schemes. It's not true Buddhism to conduct such rituals. But in the villagers' beliefs, they respect the Buddha and fear some of his power. Thus we can see that there is nothing so sacred or that the villagers respect as much as a Buddha image. Therefore we brought a Buddha image and installed it under the tree which we believe is the king of the forest and ordained the tree. In general, villagers also still believe in spirits. Therefore we set up a shrine for the guardian spirit together with the Buddha image. This led to the saying that "the good Buddha and the fierce spirits help each other take care of the forest." This means that the Buddha earns the villagers' respect. But they fear the spirits. If you have both, respect and fear, the villagers won't dare cut the trees (Quoted in Arawan, 1993: 11; my translation).

Most villagers would agree with the assessment that since the tree ordination there has been greater cooperation in protecting the community forest and less encroachment within it. "Ordaining the tree and asking the spirits to help have equal success. The spirits and the Buddha work together to protect the forest," one village elder commented to me.

While many of the villagers developed an understanding of the principles of ecological conservation through the education sessions held by Phrakhru Pitak prior to the ceremony, the impact of both the Buddhist principles applied to forest conservation and the power of the spirit's charge to guard the forest cannot be denied. The local people's indigenous knowledge of forest ecology, which is fostered and encoded in their spirit beliefs, their awe for the sacred aspects of the project, both Buddhist and animist, their respect for Phrakhru Pitak and the newly introduced concepts of conservation all work together to heighten the villagers' cooperation and responsibility to preserving the remaining forest.

Nevertheless, the element of fear may have had more of an impact than Phrakhru Pitak intended. Over the two years following the ordination ceremony, four deaths and several illnesses occurred in the forest that villagers attributed to retribution from the spirits for violation of the terms protecting the community forest. The people who died or became ill were all believed to be cutting wood or hunting within the protected areas. Their misfortunes were determined by spirit ritual specialists to have been caused by forest spirits (whether the guardian spirit himself or unnamed phii inhabiting the woods was never stated) who were offended by these people's actions.

Although some other monks have criticized Phrakhru Pitak for using the fear of spirit beliefs to achieve the ends of environmental conservation (due to both the use of fear and because many monks deny spirit beliefs as being counter to the teachings of Buddhism), he sees it differently. The villagers believe in and respect the spirits of the forest. Their relationship with these spirits, along with the various theewadaa and other phii that share their world, defines and reaffirms their understanding of how the world works – in its natural, supernatural and human aspects. The monk also recognizes that northern Thai belief systems are not static. They have evolved over time, adapting to incorporate various elements of Indian culture, including Buddhism, that have entered the region over several centuries. He recognizes and uses the traditional sacred geography rather than trying to alter it to achieve his aims.

Just as Buddhism was used throughout the 20th century by the government to encourage a shift in villagers' attitudes toward the forest in the service of development, Phrakhru Pitak sees the potential for using religious and cultural beliefs to change people's attitudes again – only this time in the service of conservation. Recognizing the continual evolution of people's beliefs and practices, Phrakhru Pitak integrates traditional beliefs with Buddhist practices to promote an ecological ethic. While the government and pro-development forces emphasize what they see as "pure" Buddhism (thereby invalidating non-Buddhist beliefs), the monk seeks to incorporate all attitudes toward the natural world. Rather than civilizing the forest through imposing Buddhist concepts in order to conquer and use it, he hopes to instill values that recognize the importance of the forest as an integral component of the environment, which includes humans (Buddhist and non-Buddhist alike), animals and spirits. Through the Buddhist concept of interdependence (Pali, *paticca-samuppada*, or dependent origination), he teaches that humans need the forest for a well-balanced life.

Accompanying such rituals as tree ordinations, Phrakhru Pitak offers villagers concrete methods for achieving this balance. He acknowledges that he cannot merely forbid people from using forest resources as these provide their livelihood. He works with farmers to develop appropriate, sustainable and organic agricultural techniques, including integrated agriculture. This method is based on using complementary plants and animals that sustain and support each other, rather than growing a single cash crop that requires chemical fertilizers or pesticides. Phrakhru Pitak encourages villagers to farm for their own subsistence first, only selling any surpluses that may remain. In this way, he hopes they can avoid getting into debt through the high costs and risks of growing a single cash crop.

Phrakhru Pitak does not only work with Buddhists. He involves the non-Buddhist hill peoples in Nan Province in his environmental education programs, recognizing their relationship with the forest and the potential demise of their traditional lifestyles through rapid economic development and deforestation. As with the lowland Buddhist villagers, Phrakhru Pitak emphasizes the importance of integrated agriculture and sustainability among hill peoples and respects and uses their spirit beliefs in engendering their understanding of their relationship with the natural environment and their cooperation in conservation efforts. (It should be noted, however, that ultimately Phrakhru Pitak hopes to convert hill peoples to Buddhism, even as he respects their cultural traditions. He trained in both the government-sponsored thammacarik missionary program among hill peoples and a non-government Buddhist missionary program.)

Phrakhru Pitak's tree ordination program is only one example of the kind of work he and other environmentalist monks undertake. In conjunction with the tree ordinations, for example, he also incorporates modified *thaut phaa paa* ceremonies, which are traditionally performed by villagers to offer "forest robes" (*phaa paa*) to monks in merit-making rituals. Phrakhru Pitak expanded the rite to include people's giving tree seedlings along with the robes. The seedlings, after being accepted and sanctified by the monks, are then given back again to the villagers to plant in deforested areas. The trees selected are kinds such as fruit trees that are productive without having to be cut down (Darlington, 1998).

In 1993, Phrakhru Pitak expanded his work to increase awareness of the importance of water for all life through a project designed to show the polluted state of the Nan River and garner support for cleaning up and maintaining the quality of its water and wildlife. Again incorporating local beliefs, Phrakhru Pitak performed a traditional "long-life" ceremony for the river in a two-day event in Nan City attended by villagers, government officials, military personnel, non-government organizations, and over 200 monks from across northern Thailand. At the same time, he established a fish sanctuary, marking the boundaries of a section of the river within which fishing was not permitted and where the fish would be fed. Here he followed the model of a village upriver from the city; in this village, within a year of creating a fish sanctuary, villagers claimed the numbers of fish increased rapidly, and they were better able to balance between protecting the fish and providing food for their families.

The "long-life" ceremony itself, called *syyp chaa taa* in the Northern Thai dialect, is an exorcism ritual usually performed to rid a person of negative forces and promote a long and healthy life. While Phrakhru Pitak performs several such rites a month for his followers, he says he does not actually believe in the physical effectiveness of the ritual. Rather, he views it as an opportunity to teach people about ways to live their lives following the teachings of the Buddha, thereby decreasing the problems and suffering they face in their lives and ultimately living in greater harmony with the natural world. Adapting the ceremony to the river afforded a similar opportunity to teach the importance of caring for the river and its resources, both water and fish.

Phrakhru Pitak skillfully included elements of Thai society that often oppose the work of the environmentalist monks – government officials, military leaders and businessmen. All of these groups often have stakes in promoting economic development along the lines laid out through government policy, as discussed in the previous section. Phrakhru Pitak believes, however, that these people are not inherently bad nor do they aim purposely to degrade the natural environment and the quality of life of the people who depend upon it. They have merely been led astray through the traditional Buddhist evils of greed, ignorance and anger, on which most environmentalist monks blame the problems of rapid economic development. Through involving the government, military and businessmen in their projects, activist monks such as Phrakhru Pitak hope to teach them as well as the villagers the value of living an ecologically balanced life.

Activists in southern Thailand used a similar approach in organizing several Dhamma Walks (*Dhammayatra*) around Songkhla Lake (Santikaro, 2000). The purpose of these walks was to bring attention to the environmental degradation of Thailand's largest lake, a unique and complex ecosystem, to local people (who include both Buddhists and Muslims) and local and regional government officials. Participants aimed to strengthen the voice of local, usually poor and marginalized, people. The issues identified by local people included a lack of fish to eat, bad water and reduced water levels, theft of water taken before it drains into the lake, loss of land through nearby urban population increases, and the breakdown of community and loss of traditional livelihoods (Santikaro, 2000: 208–209). Buddhist monks, members of a national, small but growing network of activist monks called Phra Sekhiyadhamma (of which Phrakhru Pitak is also a member), led the walks, and included local people, village leaders, non-governmental organization workers, foreign environmental activists, and some local government officials. The participants drew on local culture, both Buddhist and Muslim, in order to foster sympathy and commitment to protecting the lake's ecosystem (including the people whose lives are integrally entwined with it). Unfortunately, few Muslims actually participated in the walks, pointing to perhaps too great an emphasis on Buddhism rather than truly integrating local cultures in the project. Nevertheless, the group succeeded in building confidence among the lake's people, involving local monks who previously did not engage in environmental conservation and gaining awareness nationwide for the problems unique ecosystems such as Songkhla Lake face due to unmonitored development. Anthropologist Ted Meyer, who chronicled the walks, observed the potentials of the activities for creating change:

> I believe the lake walks have created a unique kind of public space in which a very broad range
> of issues can be explored by participants and observers. These issues include not only the
> current state of human relationships with nature at the local and global level, but also the
> meaning and significance of Buddhist practice, the possibilities and problems of cultivating
> relations of trust between people of very diverse backgrounds, and the challenges of
> designing effective strategies for social change. The breadth and openness of this space makes
> it possible, I believe, for a unique kind of spiritual and social creativity to take place. That same
> breadth and openness, however, also increases the range of difficulties that may be encountered
> (quoted in Santikaro, 2000: 213–214).

Demonstrating further cultural creativity and the willingness to face the potential difficulties, environmentalist monks undertake numerous other activities
across the country. These programs include a monk who teaches environmentalism to young people through a bird watching club in northeastern Thailand.
Another leads children on Dhamma Walks in the forest surrounding their
village in northern Thailand while taking the opportunity to talk with them
about both the Buddha's teachings and the ecology of the forest. This same
monk (the hero of the hungry ghost story) also runs a model integrated
agriculture farm designed to teach villagers sustainable methods and encourage
a shift away from cash cropping.

Even monks identified as "forest monks" (a category that indicates ascetic
monks' retreat from the social world to meditate in the wilderness rather than
any inherent activist stance toward the forest) protect and maintain the forest
within their temple compounds even while everything surrounding them is
being cut down (see J. Taylor, 1993a: 246–252). As Jim Taylor noted,

> Presently the only remaining primeval forests in the northeast with a semblance to that described
> by elderly informants and pre- and post-war bibliographical texts are small pockets dotted here
> and there throughout the relatively less accessible parts of the countryside. The forest monastery,
> the ancestor/spirit forest (*paadorn puutaa*) and cremation grounds (*paa chaa*) are normally the
> only forested areas in close proximity to villages (J. Taylor, 1993a: 250).

The presence of monks, ancestors, and spirits, whether those believed to inhabit
the forest naturally or ghosts of the recently deceased, all contribute to protecting forested land. Thai villagers respect and even fear both monks and spirits;
thus they avoid destroying forested areas connected with both. This level of
protection is not enough, however, as Taylor pointed out. These areas tend to
be small, often surrounded by denuded land. The quality of the environment
within these patches of forest often deteriorates as well due to the loss of the
larger ecosystem that used to enclose them. And ironically, as noted earlier,
even the presence of forest monks can often empower people to overcome their
fears and cut the forest. A more proactive approach is needed to maintain a
balance in the spiritual geography and protect the natural environment.

To these ends, and through a desire to spread, strengthen and support
Buddhism as well as encourage monks actively to take responsibility for the
physical as well as spiritual condition of their nation, a couple of non-governmental organizations and activist monks founded Phra Sekhiyadhamma, the
network of activist monks. This informal group supports and educates monks
across the nation working on a range of social issues, with particular focus on
environmental problems. The group, backed by non-governmental organiza-

tions such as the Thai Inter-religious Commission for Development, runs seminars for activist monks several times a year. At these conferences, anywhere from 20 to 200 monks come together to discuss their environmental projects, how they became involved in conservationist activities, and the challenges and obstacles they have faced in their efforts. Phrakhru Pitak, among a handful of other well-known environmentalist monks, is a frequent speaker at these conferences, telling his story with the hope of inspiring other monks to take the risks involved in going beyond expectations of Buddhist monks to engage in social issues. While prioritizing Buddhism, most of these monks also emphasize incorporating local beliefs and practices, acknowledging the value of local culture for initiating and enacting social change that is meaningful and not harmful for the intended recipients, the local people.

* * *

The two approaches described here reflect a (self-)conscious shift in the application of religious beliefs and practices toward influencing people's attitudes toward the natural world. The case of Thailand illustrates the various ways religion, tradition and culture have been used to influence people's relationship with nature, in favor of both economic development and nature conservation. While neither is intrinsically superior to the other – arguments are made on each side of the need for either development or conservation for the good of the country – the potential for affecting the future is clear. Given the severity of environmental degradation and deforestation in Thailand, largely attributed to rapid economic development (see Hirsch, 1993, 1996; Hirsch and Warren, 1998; Lohmann, 1991, 1993, 1995; Pinkaew and Rajesh, 1991; Rigg, 1995; Rigg and Stott, 1998), I find it encouraging that some monks and activists, such as Phrakhru Pitak, are engaging in the creative use of all Thai traditions and beliefs – Buddhism and spirit beliefs – for the long-term benefit of both people and nature. The spiritual geography of Thailand has been altered for political ends, but hopefully not to the point of irreversible damage to either the natural environment or the people who live within it.

ACKNOWLEDGEMENTS

The research for this paper was undertaken with the generous support of the Joint Committee on Southeast Asia of the Social Science Research Council and the American Council of Learned Societies with funds provided by the National Endowment for the Humanities and the Ford Foundation; the Association of Asian Studies Southeast Asian Council, with funds from the Luce Foundation; and a travel grant from the Ford Foundation Comparative Scientific Traditions Program of Hampshire College. Thanks to Barbara Ito, Jeffrey Hagen and the editors of this volume for critical feedback on drafts of this paper. All responsibility for accuracy remains with the author.

BIBLIOGRAPHY

Arawan Karitbunyarit, ed. *Rak Nam Naan: Chiiwit lae Ngaan khaung Phrakhru Pitak Nanthakhun*
  (Sanguan Jaaruwannoo) [Love the Nan River: The Life and Work of Phrakhru Pitak Nanthakhun

(Sanguan Jaaruwannoo)]. In Thai. Nan: Sekiayatham, The Committee for Religion in Society, Communities Love the Forest Program, and The Committee to Work for Community Forests, Northern Region, 1993.

Chai Podhisitra. 'Buddhism and Thai world view.' In Traditional and Changing Thai World View, Chulalongkorn University Social Research Institute/Southeast Asian Studies Program, ed. Singapore: Institute of Southeast Asian Studies, 1985, pp. 25–53.

Chatsumarn Kabilsingh. 'Buddhist monks and forest conservation.' In Radical Conservatism: Buddhism in the Contemporary World, Thai Inter-Religious Commission for Development and the International Network of Engaged Buddhists, ed. Bangkok: The Sathirakoses-Nagapradipa Foundation, 1990, pp. 301–309.

Chatsumarn Kabilsingh. Buddhism and Nature Conservation. Bangkok: Thammasat University Press, 1998.

Chulalongkorn University Social Research Institute/Southeast Asian Studies Program, ed. Traditional and Changing Thai World View. Singapore: Institute of Southeast Asian Studies, 1985.

Darlington, Susan M. Buddhism, Morality and Change: The Local Response to Development in Thailand. Ph.D. dissertation, University of Michigan, 1990.

Darlington, Susan M. 'The earth charter and grassroots ecology monks in Thailand.' In Buddhist Perspectives on the Earth Charter, Amy Morgante, ed. Cambridge, Massachusetts: Boston Research Center for the 21st Century, 1997a, pp. 47–52.

Darlington, Susan M. 'Not only preaching – the work of the ecology monk Phrakhru Pitak Nanthakhun of Thailand.' Forest, Trees and People Newsletter 34: 17–20, 1997b.

Darlington, Susan M. 'The ordination of a tree: the Buddhist ecology movement in Thailand.' Ethnology 37(1): 1–15, 1998.

Darlington, Susan M. 'Rethinking Buddhism and development: the emergence of environmentalist monks in Thailand.' Journal of Buddhist Ethics 7, 2000. [http: //jbe.gold.ac.uk/7/darlington001.html]

Darlington, Susan M. 'Practical spirituality and community forests: monks, ritual and radical conservatism in Thailand.' In Nature in the Global South, Anna L. Tsing and Paul Greenough, eds. Durham, North Carolina: Duke University Press, forthcoming-a.

Darlington, Susan M. 'Thai Buddhist monks.' In Encyclopedia of Religion and Nature, Bron Taylor and Jeffrey Kaplan, eds. New York: Continuum International Pubs., forthcoming-b.

Davies, Shann, ed. Tree of Life: Buddhism and Protection of Nature. Hong Kong: Buddhist Perception of Nature, 1987.

Davis, Richard B. Muang Metaphysics: A Study of Northern Thai Myth and Ritual. Bangkok: Pandera Press, 1984.

Demaine, Harvey. 'Kanpatthana: Thai views of development.' In Context, Meaning and Power in Southeast Asia. M. Hobart and R. Taylor, eds. Ithaca, New York: Southeast Asia Program, Cornell University, 1986, pp. 93–114.

Harris, Ian. 'How environmentalist is Buddhism?' Religion 21: 101–114, 1991.

Harris, Ian. 'Getting to grips with Buddhist environmentalism: a provisional typology.' Journal of Buddhist Ethics 2: 173–190, 1995.

Hirsch, Philip and Carol Warren. 'Introduction: through the environmental looking glass: the politics of resources and resistance in Southeast Asia.' In The Politics of Environment in Southeast Asia: Resources and Resistance, Philip Hirsch and Carol Warren, eds. London: Routledge, 1998, pp. 1–25.

Hirsch, Philip, ed. Seeing Forests for Trees: Environment and Environmentalism in Thailand. Chiang Mai: Silkworm Books, 1996.

Hirsch, Philip. Political Economy of Environment in Thailand. Manila: Journal of Contemporary Asia Publishers, 1993.

Hirsch, Philip. 'Environment and environmentalism in Thailand: material and ideological bases.' In Seeing Forests for Trees: Environment and Environmentalism in Thailand, Philip Hirsch, ed. Chiang Mai: Silkworm Books, 1996, pp. 15–36.

Ishii, Yoneo. 'Church and state in Thailand.' Asian Survey VIII(10): 864–871, 1968.

Kamala Tiyavanich. *Forest Recollections: Wandering Monks in Twentieth-Century Thailand.* Honolulu: University of Hawai'i Press, 1997.

Kaza, Stephanie and Kenneth Kraft, eds. *Dharma Rain: Sources of Buddhist Environmentalism.* Boston: Shambala Publications, Inc., 2000.

Keyes, Charles F. 'Buddhism and national integration in Thailand.' *Journal of Asian Studies* 30(3): 551–567, 1971.

Keyes, Charles F. *Thailand: Buddhist Kingdom as Modern Nation-State.* Bangkok: Editions Duang Kamol, 1989.

Kirsch, A. Thomas. 'Complexity in the Thai religious system: an interpretation.' *Journal of Asian Studies* 36(2): 241–266, 1977.

Lohmann, Larry. 'Peasants, plantations and pulp: the politics of eucalyptus in Thailand.' *Bulletin of Concerned Asian Scholars* 23(4): 3–17, 1991.

Lohmann, Larry. 'Thailand: land, power and forest colonization.' In *The Struggle for Land and the Fate of the Forests*, Marcus Colchester and Larry Lohmann, eds. London: Zed Books, 1993, pp. 198–227.

Lohmann, Larry. 'Visitors to the commons: approaching Thailand's 'environmental' struggles from a western starting point.' In *Ecological Resistance Movements: The Global Emergence of Radical and Popular Environmentalism*, Bron Raymond Taylor, ed. Albany: State University of New York Press, 1995, pp. 109–126.

Mulder, J. A. Niels. *Monks, Merit, and Motivation: Buddhism and National Development in Thailand.* Special Report No. 1, 2nd rev. ed. DeKalb: Center for Southeast Asian Studies, Northern Illinois University, 1973 [1969].

Pei Sheng-ji. 'Some effects of the Dai people's cultural beliefs and practices on the plant environment of Xishuangbanna, Yunnan Province, Southwest China.' In *Cultural Values and Human Ecology in Southeast Asia*, Karl L. Hutterer, A. Terry Rambo, and George Lovelace, eds. Michigan Papers on South and Southeast Asia, No. 27. Ann Arbor: Center for South and Southeast Asian Studies, University of Michigan, 1985, pp. 321–339.

Pinkaew Leungaramsri and Noel Rajesh, eds. *The Future of People and Forests in Thailand After the Logging Ban.* Bangkok: Project for Ecological Recovery, 1991.

Pipop Udomittipong. 'Why should Buddhist monks be involved in conservation?' *Seeds of Peace* 9(1): 6–7, 1993.

Pipop Udomittipong. 'Thailand's ecology monks.' In *Dharma Rain: Sources of Buddhist Environmentalism*, Stephanie Kaza and Kenneth Kraft, eds. Boston: Shambhala Press, 2000 [1995], pp. 191–197.

Queen, Christopher S. and Sallie B. King, eds. *Engaged Buddhism: Buddhist Liberation Movements in Asia.* Albany: State University of New York Press, 1996.

Reynolds, Frank E. 'Dhamma in dispute: the interactions of religion and law in Thailand.' *Law and Society Review* 28(3): 433–451, 1994.

Reynolds, Frank E. and Mani B. Reynolds. *Three Worlds According to King Ruang: A Thai Buddhist Cosmology.* Berkeley Buddhist Studies Series 4. Berkeley, California: Asian Humanities Press/Motilal Banarsidass, 1982.

Rigg, Jonathan and Philip Stott. 'Forest tales: politics, policy making, and the environment in Thailand.' In *Ecological Policy and Politics in Developing Countries: Economic Growth, Democracy, and Environment*, Uday Desai, ed. Albany: State University of New York Press, 1998, pp. 87–120.

Rigg, Jonathan, ed. *Counting the Costs: Economic Grow and Environmental Change in Thailand.* Singapore: Institute of Southeast Asian Studies, 1995.

Rigg, Jonathan. 'Counting the costs: economic growth and environmental change in Thailand.' In *Counting the Costs: Economic Grow and Environmental Change in Thailand.* Jonathan Rigg, ed. Singapore: Institute of Southeast Asian Studies, 1995, pp. 3–24.

Santikaro Bhikkhu. 'Dhamma walk around Songkhla Lake.' In *Dharma Rain: Sources of Buddhist Environmentalism*, Stephanie Kaza and Kenneth Kraft, eds. Boston: Shambhala Press, 2000, pp. 206–215.

Santita Ganjanapan. 'A comparative study of indigenous and scientific concepts in land and forest classification in Northern Thailand.' In *Seeing Forests for Trees: Environment and*

Environmentalism in Thailand, Philip Hirsch, ed. Chiang Mai: Silkworm Books, 1996, pp. 247–267.

Seri Phongphit. *Religion in a Changing Society: Buddhism, Reform and the Role of Monks in Community Development in Thailand.* Hong Kong: Arena Press, 1988.

Shalardchai Ramitanondh. *Phii Cawnaaj [The Spirit Lords].* In Thai. University Monograph Project, Monograph Series No. 18. Chiang Mai: The Library of Chiang Mai University, 1984.

Siam Society. *Culture and Environment in Thailand: A Symposium of the Siam Society.* Bangkok: Siam Society, 1989.

Somboon Suksamran. *Political Buddhism in Southeast Asia: The Role of the Sangha in the Modernization of Thailand.* London: C. Hurst and Co., 1977.

Somboon Suksamran. *Buddhism and Politics in Thailand.* Singapore: Institute of Southeast Asian Studies, 1982.

Somboon Suksamran. 'A Buddhist approach to development: the case of "development monks" in Thailand.' In *Reflections on Development in Southeast Asia*, Lim Teck Ghee, ed. Singapore: ASEAN Economic Research Unit, Institute of Southeast Asian Studies, 1988, pp. 26–48.

Sponsel, Leslie E. and Poranee Natadecha. 'Buddhism, ecology, and forests in Thailand: past, present, and future.' In *Changing Tropical Forests*, John Dargavel, Kay Dixon, and Noel Semple, eds. Canberra: Centre for Resource and Environmental Studies, Australian National University, 1988, pp. 305–325.

Sponsel, Leslie E. and Poranee Natadecha-Sponsel. 'The role of Buddhism in creating a more sustainable society in Thailand.' In *Counting the Costs: Economic Grow and Environmental Change in Thailand*, Jonathan Rigg, ed. Singapore: Institute of Southeast Asian Studies, 1995, pp. 27–46.

Sponsel, Leslie E. and Poranee Natadecha-Sponsel. 'A theorectical analysis of the potential contribution of the monastic community in promoting a green society in Thailand.' In *Buddhism and Ecology: The Interconnection of Dharma and Deeds*, Mary Evelyn Tucker and Duncan Ryuken Williams, eds. Cambridge, Massachusetts: Harvard University Center for the Study of World Religions, 1997, pp. 45–70.

Stott, Philip. 'Nous avons mangé la forêt: environmental perception and conservation in mainland South East Asia.' In *Nature and Man in South East Asia*, P. A. Stott, ed. London: School of Oriental and African Studies, University of London, 1978, pp. 7–22.

Stott, Philip. '*Mu'ang* and *pa*: elite views of nature in a changing Thailand.' In *Thai Constructions of Knowledge*, Manas Chitakasem and Andrew Turton, eds. London: School of Oriental and African Studies, University of London, 1991, pp. 142–154.

Sulak Sivaraksa. *A Buddhist Vision for Renewing Society.* Bangkok: Tienwan Publishing House, 1986.

Sulak Sivaraksa. *Religion and Development.* Bangkok: Thai Inter-Religious Commission for Development, 1987 [1976].

Sulak Sivaraksa. *Seeds of Peace.* Berkeley, California: Parallax Press, 1992.

Sulak Sivaraska. 'Development as if people mattered.' In *Dharma Rain: Sources of Buddhist Environmentalism*, Stephanie Kaza and Kenneth Kraft, eds. Boston: Shambhala Press, 2000a [1992], pp. 183–190.

Swearer, Donald K. *The Buddhist World of Southeast Asia.* Albany: State University of New York Press, 1995.

Swearer, Donald K. 'The hermeneutics of Buddhist ecology in contemporary Thailand: Buddhadasa and Dhammapitaka.' In *Buddhism and Ecology: The Interconnection of Dharma and Deeds*, Mary Evelyn Tucker and Duncan Ryuken Williams, eds. Cambridge, Massachusetts: Harvard University Center for the Study of World Religions Publications, 1997, pp. 21–44.

Tambiah, Stanley J. *Buddhism and the Spirit Cults in North-East Thailand.* Cambridge: Cambridge University Press, 1970.

Tambiah, Stanley J. *World Conqueror and World Renouncer: A Study of Buddhism and Polity in Thailand against a Historical Background.* Cambridge: Cambridge University Press, 1976.

Tambiah, Stanley J. *The Buddhist Saints of the Forest and the Cult of Amulets.* Cambridge: Cambridge University Press, 1984.

Taylor, Bron Raymond, ed. *Ecological Resistance Movements: The Global Emergence of Radical and Popular Environmentalism*. Albany: State University of New York Press, 1995.

Taylor, Jim. *Forest Monks and the Nation-State: An Anthropological and Historical Study in Northeastern Thailand*. Singapore: Institute for Southeast Asian Studies, 1993a.

Taylor, Jim. 'Social activism and resistance on the Thai frontier: the case of Phra Prajak Khuttajitto.' *Bulletin of Concerned Asian Scholars* 25(2): 3–16, 1993b.

Taylor, Jim. ' 'Thamma-chaat': activist monks and competing discourses of nature and nation in northeastern Thailand.' In *Seeing Forests for Trees: Environment and Environmentalism in Thailand*, Philip Hirsch, ed. Chiang Mai: Silkworm Books, 1996, pp. 37–52.

Taylor, Jim. 'Community forests, local perspectives and the environmental politics of land use in northeastern Thailand.' In *Land Conflicts in Southeast Asia: Indigenous Peoples, Environment and International Law*, Catherine J. Iorns Magallanes and Malcolm Hollick, eds. Bangkok: White Lotus Press, 1998, pp. 21–55.

Tegbaru, Amare. 'Local environmentalism in northeast Thailand.' In *Environmental Movements in Asia*, Arne Kalland and Gerard Persoon, eds. Surrey: Curzon Press, 1998, pp. 151–178.

Thai Inter-religious Commission for Development and the International Network of Engaged Buddhists. *Radical Conservatism: Buddhism in the Contemporary World*. Bangkok: The Sathirakoses-Nagapradipa Foundation, 1990.

Trébuil, Guy. 'Pioneer agriculture, green revolution and environmental degradation in Thailand.' In *Counting the Costs: Economic Grow and Environmental Change in Thailand*, Jonathan Rigg, ed. Singapore: Institute of Southeast Asian Studies, 1995, pp. 67–89.

Tucker, Mary Evelyn and Duncan Ryuken Williams, eds. *Buddhism and Ecology: The Inter-connection of Dharma and Deeds*. Cambridge, Massachusetts: Harvard University Center for the Study of World Religions Publications, 1997.

Williams, Duncan Ryuken. 'Bibliography on Buddhism and ecology.' In *Buddhism and Ecology: The Interconnection of Dharma and Deeds*. Mary Evelyn Tucker and Duncan Ryuken Williams, eds. Cambridge, Massachusetts: Harvard University Center for the Study of World Religions Publications, 1997, pp. 403–426.

D.P. CHATTOPADHYAYA

INDIAN PERSPECTIVES ON NATURALISM

Naturalism, like spiritualism, is an ambiguous term. In some cases, in India and elsewhere, naturalism has been contrasted with spiritualism or idealism, and naturalism at times has been used as a synonym for materialism. Equally widespread is the use of the word shamanism, derived either from Sanskrit *śraman* (priest) or Tunguso-Manchurian *saman* (knowledgeable person) or both. Shamanism is a concept in which the elements of both naturalism and spiritualism are present. Interestingly enough, both in their European and Indian contexts, one finds some thinkers who do not recognize any dualism between naturalism and spiritualism. Nature is spiritually informed or permeated, and spirit has its influential presence in nature. Spirit and nature are also said to be two different hemispheres or aspects of one and the same reality, God or Brahman. Naturalism and spiritualism developed in India at times side by side and often in and through mutual interaction. In India the Sāṁkhya system, (number-or *saṁkhyā*-based system), is recognized as paradigmatically naturalist. In fact one of its first principles, *prakṛti*, literally means nature. Those who are pro-materialist are inclined also to designate *lokāyata/cārvāka* as a naturalist system. Literally *lokāyata* means that which is pervasive among the people (*loka*), and the expression *cārvāka* is a combination of *cāru* (sweet) and *vāk* (words).

That there is no sharp demarcation between naturalism and spiritualism is evident even in contemporary India. There are many people who are Jain or Buddhist and do not believe in any personal God and are not theist in the accepted sense. Certainly many tribal religious cults which are not committed to the idea of any omnipotent, omnipresent and omniscient God should also be taken to be naturalist in their basic tenor. However they believe in the existence of spirits, ghosts or *geist*. In the Himalayan foothills and Northeast India there are many peoples who subscribe to some form of shamanism. But naturalism and spiritualism are not necessarily antithetical.

NATURALISM IN THE VEDAS AND UPANIṢADS

The earliest extant sacred literature of India consists of four groups: (1) the *Vedas* (sacred hymns), (2) the *Brahmaṇas* (ritual treatises), (3) the *Upaniṣads*

147

H. Selin (ed.), Nature Across Cultures: Views of Nature and the Environment in Non-Western Cultures, 147–159.
© 2003 Kluwer Academic Publishers. Printed in Great Britain.

(philosophical works) and (4) *tantras* (psychosomatic principles of attaining superhuman powers). Initially, this vast literature was composed and transmitted orally. That is why it is said to be *śruti* (that which is heard): knowing by being told. Later on it was written down.

Tantra, to start with, had a distinct cultural lineage, but later on it got integrated with other components of sacred literature. The two main forms of Indian sacred literature, *nigama* (mainly Vedic) and *āgama* (mainly non-Vedic), are compendia or collections (*saṁhitā*), not authored by any single person or group.

The interaction between the Vedic and non-Vedic schools of beliefs and practices gave rise to two types of naturalism, theist (*āstika*) and atheist (*nāstika*). Atheism, in the Indian context, does not necessarily imply not believing in the spirit. Buddhism and Jainism, for example, do not believe in the existence of a single God but are both spiritual. Even the non-spiritualistic varieties of naturalism, like *ājīvikas* and *cārvāka*, had well-developed views on reality, knowledge and ethics.

The dating of the earliest form of Vedic and non-Vedic forms of literature is not free from controversy. Most authorities are inclined to agree that these were created in the first quarter of the second millennium BC, if not earlier. They developed before the advent of Jainism and Buddhism around 500 BC.

In Vedic naturalism the forces of nature – *ādity*, *agni*, or *tejas* (fire, sun and light), *vāyu* or *marut* (wind and air) *pṛthivi* or *kṣiti* (earth), *ap* or *salila* (water) and *vyom* (space) – are among the most frequently mentioned elements. Very frequently these forces were personified, deified and worshipped. Elaborate hymns were composed and sung to worship and propitiate them.

These natural forces also shaped human life. Adjustment to the environment is a prime necessity for human survival. For example, production of food grains is impossible if there is no rainfall. Therefore, people prayed for the approval of the gods and goddesses of the sky, water and clouds. There is also a distinct symbolic aspect in all the descriptions of natural forces. Most of the words designating such forces of nature as *agni* and *fire* have striking phonetic and semantic counterparts in all the Indo-European family of languages. For example, Sanskrit words like Dev, Divā or Devatā have Greek synonyms in Zeus, Theos and Jupiter; in English we find words like *day* and *deity*. Sanskrit *Agni*, Latin *Ignis*, English *Ignition*, Lithunian *Ugnos* and Russian *Ogon* stand for the same natural force, fire.

During the Vedic period the dominant tenor of the hymns was naturalist in form but spiritualist in content. However, there are also passages which allude to a kind of skepticism about the existence of gods and spirit as supreme reality. [*Rgveda Saṁhita* Mandala I, Hymn 185, Verse 1; Mandala III, Hymn 54, Verses 5–6]. Questions were raised about the justifiability of sacrifices and gifts to *brāhmaṇas* (priests). These acts of faith (*śraddhā*) have been questioned; reference was made to those who were faithless (*aśraddhā*) and to the repudiators of *Indra* (*anindra*). Authorities like F. Max Müller and Maurice Bloomfield are of the view that these skeptical streaks found in the Vedic literature are remnants of the previous age in which, it appears, anti-Vedic ideas had found relatively more widespread popular acceptance. This explains the long and sustained interplay between the forces of pronaturalism and those of antinaturalism.

This clash of ideas is also evident in the *Upaniṣads*. The Upaniṣads are distilled forms of the Vedic hymns. While in the Vedic hymns one finds naturalist symbolism with clear spiritual implications, Upanisadic literature is relatively more abstract and conceptual in character. Its language, though poetic, is not marked by symbolic or metaphorical profusion. A conscious effort was made to secularize and philosophize the seminal ideas of the Vedic corpus.

As in the Vedas, one finds distinct traces of naturalism in the Upanisadic literature. The earliest Upaniṣads appeared in the first millennium BC. Upaniṣads like *Maitrī*, *Śvetāśvatara* and *Chāndogya* presented views disputing the existence of God, Heaven, Hell and several other beliefs in the supernatural. They expressed doubt about the efficacy of time, nature, necessity, chance, or the elements as the cause of the emergence of human beings in the process of evolution. Since most of the writers of the Upaniṣads were themselves pronouncedly spiritualist, their presentations of antinaturalism and skepticism are mostly unfavourable.

The first well-known naturalist in the Indian tradition is said to be Uddālaka (7th century BC). This was the time when the Ionian naturalism of Thales and others came to India. We do not know whether Uddālaka is a real or fictitious personality. To him *vāyu* (breath-wind) was the ultimate stuff of reality. Both in the Ṛgvedic literature and in Hellenic thought *vāyu* or *aer* is considered ultimate reality.

Uddālaka's main contributions are:

* rejection of Vedic spiritualism,
* refutation of the teleological view of nature,
* defense of the mechanistic view of nature, and
* drawing a clear line of distinction between what is material and what pertains to life.

In his cosmology the components which receive distinct recognition are breath, water, fire or heat, earth, and thought. All these elements are also recognized in the Vedic literature. It is in this context that one has to understand the debate between the spiritualist Yājñavalkya and the naturalist Uddālaka. The main difference between the two is that while the former affirms a spiritual teleological view of the universe and its evolution, his naturalist critic opts for the mechanistic hypothesis.

FROM THE VEDIC (*ĀSTIKA*) TO THE NON-VEDIC AND ANTI-VEDIC (*ÑĀSTIKA*) SCHOOLS

Vedic doctrines are mixed in character. Elements of both naturalism and spiritualism coexist not only in the Vedas but also in the Upaniṣads. At least two different interpretations of these mixed views gained ground and gradually, depending on the changing social settings, established themselves as two broad approaches to life. These two approaches are generally referred to as orthodox (*āstika*) and heterodox (*nāstika*). Acceptance or rejection of the Vedic authority was the main point of difference between these two competing approaches, and an auxiliary consideration was whether or not to go back to the pre-Vedic ideas. Neither the orthodox nor the heterodox views are homogenous at all in character, but we can still refer to some general characteristics of the orthodox schools:

1. belief in the transmigration of the soul or immortality;
2. the doctrine of *karma*, suggesting that humans are affected by their merits and demerits and acts of commission and omission;
3. belief in the possibility of attaining disembodied liberation or salvation, i.e., going beyond the realm of thought and action, enjoyment and suffering; and
4. belief that life is full of suffering.

Naturally those who wanted a different way of life criticized and rejected these fundamental tenets of the orthodox approach. Among the main heterodox groups were the Ājīvikas and Cārvākas. They seem to have flourished between 600–300 BC. They accorded primacy to the non-living material world, and apparently their approach and language of presentation was very persuasive and popular. The emergence of these popular thinkers presupposes a relatively dismal social situation which may be attributed to the decline of Vedic spirituality consequent upon the Aryans' settlement in the Ganganetic Valley, their easy ways of life and their practice of excessive ritualism and strong commitment to conventionalism. Mahāvīra (Jains) and Buddhists formed anti-metaphysical and pro-practical movements of protest against the Vedic orthodoxy. Both the Ājīvikas and Cārvākas were critical not only of the Vedic legacy but also of the many tenets of Jainism and Buddhism.

However, they all adhered to some basic principles.

1. They thought the infallibility claim of the sacred Vedas must be questioned and rejected.
2. The supernatural and the metaphysical are enemies of a sound understanding of life and its goals.
3. The belief in the immortality of soul is a myth; after death nothing like a soul survives.
4. The doctrine of causation in the context of karma is not only illusory but also impossible to establish.
5. The forces and actions of material elements (*mahābhūtas*) cause all things and beings of the world.
6. For their operation and interaction these elements need no external impetus; motion is inherent (*svabhāva*) in them.
7. What we call human intelligence is derived from the increasingly complex interaction of all these material forces.
8. All sources of knowledge, inference, analogy etc. are questionable and fallible; sense perception (*pratyakṣa*) is the only reliable basis of knowledge.
9. Ritual, spiritual and ethical injunctions which are derived from the so-called sacred scriptures and are not in accord with the practical needs of life

should be discarded; the mind should be free from superstitions and prejudices.

10. The life of the senses is the ideal life to pursue.

Though Mahāvīra is regarded as the chief preacher (*tīrthaṅkar*) of the Jaina view of the world, it is traditionally believed that 23 other preachers preceded him. He was a contemporary of Gautam Buddha, who lived and preached in and around the same area of northern India, from the present Uttar Pradesh to Bihar, if not a little beyond.

The universe, in the Jaina view, consists of two uncreated entities, consciousness (*jīva*) and extended matter (*ajīva*). *Jīva* is the eternal consciousness; also it is used in the sense of soul or self. There are two types of *ajīva*: stuff with form or matter (*rūpa*) and stuff devoid of form or matter (*arūpa*). The extended material world has in it motion (*dharma*); rest (*adharma*); two kinds of space: *lokākāśa*, space filled up by material things, and (*alokākāśa*), space which is absolutely void; and time (*kāla*).

The Jaina thinkers also spoke of three levels of *loka* or realm: the upper realm (*ūrdha loka*) populated by celestial beings, the middle realm (*madhya loka*) inhabited by humans and other creatures, and hell (*adho loka*). Time is supposed to be quasi-substantive in character, persisting through successive movements of the world. Time may be eternal (*kāla*) or it may be relative, marked by a beginning and end and with temporal divisions. In the latter case it is called *samaya*. Matter or material substance is called *pudgala*. *Pudgala* is divided into sound, grossness, fineness, shape, union, divisibility, darkness, image, luminosity and heat. *Pudgala* is the conductor of the two forms of kinetic energy – simple motion (*parispanda*) and development (*pariṇāma*). Matter may be gross or subtle. It may be so subtle that at times it is beyond the reach of human senses. All physical objects, consisting of gross or subtle atoms, are called *paramāṇūs* (sub-atomic entities). Atoms or *aṇu* possess weight, are infinitesimal, eternal, ultimate, uncreated, indestructible and formless. Atoms have primary and secondary qualities that may appear and disappear and therefore are not permanent. Aggregation or combination of atoms is called *skanda*. There are different kinds of combination – binary, tertiary or of higher orders. Atoms of subtle matter create a subtle body within the human soul. According to Jainas, karma is material in nature. There are two types – *jīva-karma* (karma-in-soul) and *ajīva-karma* (no-soul-karma). This karmic or motion-like character of particles is responsible for the soul's bondage. Unless and until the soul becomes absolutely perfect and free from the karmic effects, bondage remains.

The Jaina ethic is remarkably free from transcendental trappings or supernatural presuppositions. Its emphasis is on non-violence or love (*ahiṁsā*); chastity in thought, word and deed; and renunciation of worldly attachments and interests. Jaina thinkers sought to reconcile naturalism and humanism. God does not figure in the system as the presiding power of the world of karma. The karmic law is self-sustaining and self-fulfilling. The Jain epistemology is also naturalistic in the sense that it believes that external objects are given to our senses. It is a very unified system. Only in its doctrine of perfection or liberation of the soul are elements of supernaturalism discernible.

NATURALIST ELEMENTS IN BUDDHISM

There are different types of naturalism found in Buddhism. Buddhism has been interpreted differently at different stages of its development and in different countries. It emerged as a protest movement against the Vedic culture, but it carries in it distinct traces of both Brahmanism and Tantra. Buddhism is understood mainly under two broad interpretations, Hīnayāna (literally, small vehicle) and Mahāyāna (large vehicle). Historically speaking, Hīnayāna, or the orthodox school, developed earlier and spread from India to Sri Lanka and different countries of Southeast Asia. Mahāyāna, a later development, was more common in Central Asia, Tibet, China, Mongolia, Korea and Japan. Both the schools claim to have developed the basic teachings of Buddha. Also notable is the difference between the sub-schools within each one of the schools. The main tenets of Buddhism were written and rewritten in languages such as Pali, Sanskrit, Tibetan and Chinese. The three aspects of the Buddhist canons which deserve special mention are found in the *Tripiṭaka* (Three Baskets): rules of discipline, doctrine and the thoughts attributed to the Buddha soon after he attained enlightenment (*Buddhatva*).

The Hīnayāna group had a very strongly pro-realist school in it, called Sautrāntika. This school recognized the reality of external objects without mental modification. The Hīnayānists are also known as Theravādins. Both Sautrāntikas (followers of formulae, – *sūtra*) and Sarvāstivādins (those who believe *sarva* exists in everything) are kindred in their realistic inclinations. According to the Sautrāntikas, all things, not only sentient beings, may be viewed under two categories, primary existents (*dravyasat*) and secondary existents or conceptual constructs (*prajñaptisat*). Though these realist Buddhists did not completely disown the momentariness of the object-ward experience, it is in this school that realism receives strongest support. They did not accept the schism between realism and idealism, regarded as metaphysical in the Buddhist orthodoxy. The Hīnayānist refers to Buddha in support of his assertion of the reality of the sense-given world. The other, apparently opposite, aspect of Buddhism is its recognition of intuition as a mode of knowledge, which is accepted by all of the schools. Buddha warned his disciples against the dangers of overly admiring the importance of sense experience. For proper understanding of the world of objects we are required to recognize the importance both of form (*rūpa*) and name (*nāma*). It is in terms of them that the specific peculiarities of objects are identified, discerned and grasped. Objects are the furniture of the external world of which humans themselves are a part.

External reality of the objective world is proved by how our process of perception is resisted (*patigha*) by the world outside. The external world, though in a sense independent of the human body, is not concrete or continuous; it is marked by discrete events. It is impermanent. The world of sense consists of four elements – earth, air, fire, water and ether (*ākāśa*). Every material thing consists of different elements but ultimately disintegrates and passes away. The materiality of the world is recognized simultaneously with its impermanence. Later on this doctrine of impermanence yielded to that of momentariness (*kṣaṇabhanguravād*).

Aligned to the doctrine of momentariness is the doctrine of dependent origin (*pratītya-samutpādavād*). The crux of the view is that there is no thing-in-itself. Every entity is dependent or conditional in its origin, transformation and destruction. Cause and effect are neither identical nor different. The most authoritative exposition is Nāgārjuna's *Mādhyamika Kārikā*.

The naturalism of Buddhism is evident also in what is called no-selfism (*nairātmyavād*). The doctrine of the non-existence of self rests on several arguments. It has no form (*rūpa*) like the material body. Its appearance is due to the aggregation of mental states like feelings, perceptions, dispositions and intellect. But none of these constituents of mental nature can be taken to be self. Even these perish and are not real. If this no-selfism is pushed to its logical end, then the notions of action, previous births and fruits of action become untenable. In view of this fact, Buddhists, both the Hīnayānist and the Mahāyānist, developed a doctrine of surrogate self. Human individuality is explained in terms of a collection of changing aggregates. Additionally, it is claimed, these collections or their near-analogues can be captured by imagination (*kalpanā*).

Further, the Buddhist concept of body confirms its naturalism. The body includes sense organs and an awareness and reception of the sense information and the resulting feelings of pleasure and pain. The body is credited with the power to have ideation and to form mental images about the external world. It is also endowed with the power of volition or desiring the objects of sense contacts and what arises out of them.

Buddhist naturalism is ambivalent. Its early form (Hīnayāna) was totally against the supernatural. That came from its firm commitment to the reality of the sense-given external world. Later on, however, one observes a gradual but clear emergence of a kind of supernaturalism. The Buddha, who was treated earlier as a human person, was, with the passage of time, increasingly mythologized and deified. An all-pervasive cosmic body (*dharmakāya*) was also ascribed to him. This shift of emphasis, from the concrete to the abstract, from the sensible to the spiritual, reminds one of the lingering Vedic influence on Buddhism.

Also notable in this connection is the Buddhist's original recognition of the traditional Vedic gods such as Indra, Agni and Varuna. Simultaneously, their existence is downplayed by saying that this is to be understood only as rooted in traditional reality and not strictly existential in nature. Denial of the ontological existence of Vedic gods and affirmation only of their traditional reality may be cited as evidence in support of the naturalism of the later Buddhists.

This point is further buttressed by an important double-sided commitment of the later Śūnyvādins. They recognize the reality of the empirical world (*saṁsāra*) available to sense and science. At the same time, they recognize the reality of *nirvāṇa*, a state of liberation, which is absolutely indeterminate by anything worldly. What is more, they assert the essential unity of the empirical *saṁsāra* and the transcendental *nirvāṇa*. Some modern commentators on Buddhism, like T.R.V. Murti, have likened this position to the Kantian maxim that the world of science and sense are empirically real but transcendentally ideal (Murti, 1980). [Editor's note: See Sponsel and Natadecha-Sponsel's article on Buddhist Views of Nature and the Environment for a detailed exploration of this issue. Also see Darlington on Buddhist Thailand, for a study of contemporary Buddhists and their ecological activities.]

THE SĀMKHYA VIEW OF NATURALISM

One very promising way of approaching and understanding the main forms of Indian naturalism is to look into three paired concepts: *puruṣa* (self) and *prakṛti* (nature), *brahman* (absolute) and *māyā* (materializing principle) and *īśvara* (god) and *śakti* (power). The first pair represents the Sāmkhya view. The second two are associated with the Advaita Vedānta. The third is ascribed to varieties of pro-theistic and theistic schools.

The Sāmkhya account of naturalism is very ancient, in certain respects maybe even pre-Vedic, comprehensive and systematic. The word "sāṁkhya" means number in Sanskrit. The Sāmkhya recognizes 25 truths or *tattvas*. These are:

1. pure consciousness or self (*puruṣa*),
2. primordial materiality or nature (*mūlaprakṛti*),
3. intellect (*buddhi*),
4. ego (*ahaṁkāra*), and
5. mind (*manas*), both as perceiver and actor.

The five sense capacities are:

6. hearing (*śrotra*),
7. touching (*tvac*),
8. seeing (*cakṣus*),
9. tasting (*rasana*), and
10. smelling (*ghrāṇa*).

The five actions are:

11. speaking (*vāc*),
12. grasping (*pāṇi*),
13. walking (*pāda*),
14. excreting (*pāyu*), and
15. procreating (*upastha*).

The five subtle elements (*tanmātras*) are:

16. sound (*śabda*),
17. contact (*sparśa*),
18. form (*rūpa*)
19. palatable essence (*rasa*), and
20. smell (*gandha*).

And the five gross elements (*mahābhūtas*) are:

21. space (*ether or ākāśa*),
22. wind or air (*vāyu*),
23. fire (*tejas*),
24. water (*ap*), and
25. earth (*pṛthivī*).

Of these 25 real entities only two, self and nature, are said to be independent. Nature is intended to be the object of enjoyment (*bhogya*) for the self (*pūruṣa*). Nature consists of three elements (*guṇas*): *sattva*, exhibiting the traits of luminosity, expressiveness and ordering; *rajas*, marked by energy and spontaneous activity; and *tamas* characterized by gravity, density and inertia. In everything real all three *guṇas* or elements are present. Depending upon the dominance of one over the other two qualities, the concerned *tattva* appears to be bright and expressive, dynamic and active, or slow, sluggish and static. All these characterizations are relative, since in everything real energy (*rājasikguṇa*) is present, and nothing can remain strictly static. This is the reason that the Sāṁkhya view has often been described as "Process Naturalism" or "Process Materialism". The process is believed to be real, not illusory. That explains why the Sāṁkhya theory on transformation (*pariṇāmavād*) is distinguished from the Vedāntic theory of "evolution" or self-alienation (*vivarttavād*). The former view asserts that, although the process of transformation is continuous, the transformed real objects are genuinely new. In contrast, the Vedānta theory maintains that throughout the empirical and visible process of change, the essential reality (*brahman*) which sustains the process does not itself undergo any change. The Sāṁkhya form of naturalism is indeed radical. Contrary to popular belief, it views intellect, ego, mind and even our sense and action capacities as aspects of nature. In other words, these real things (*tattvas*) are all natural at bottom. They differ only in their degree of manifestness or non-manifestness. Nature is both potential and actual. Many of these tattvas gradually, through the process of real transformation, assume actual forms.

Some interpreters, like K.C. Bhattacharyya, are of the view that three guṇas or constituents of nature are three modes of feeling, which should be understood in terms of freedom. Only in a subsidiary way should these guṇas be taken as qualities of the object. In view of the fact that all tattvas, according to the Sāṁkhya, come from nature, the first alternative interpretation seems to be strained. But the view that these guṇas have three grades of feeling vis-à-vis the objective world appears to be plausible.

VAIŚEṢIKA NATURALISM

In the Indian tradition, the Sāṁkhya and Yoga systems are generally studied together. This is also the case with Nyāya and Vaiśeṣika ideas. But there are considerable differences between these groups of thoughts. In the context of naturalism perhaps it is best to present the Vaiśeṣika views independently, but this does not deny that on many issues their views are similar. There are numerous views about the antiquity and origins of Vaiśeṣika naturalism. Although Kaṇāda is generally acknowledged as the founder of this system, scholars point out its similarity, in some respects, with Jaina atomism and Sāṁkhya naturalism (see Larson, 1987). Some thinkers have pointed out its affinity with Lokāyata (Cārvāka) views and the ideas found in the *Carakasaṁhitā*, the treatise on the Indian system of medicine known as *Āyurveda*. None of the recognized Indian systems of thought has developed in a clear-cut, autonomous and systematic manner. The order that is ascribed to this or that particular cluster of philosophic-scientific views is mainly based on subsequent compilers and interpreters of different and interacting schools (Sampradāya). There is even doubt about the historical authenticity of the name Kaṇāda. It is generally assumed that Kaṇāda, whoever he was, did not live before *ca.* 300 BC. But A.L. Thakur (2002) speaks of a much older origin. Besides Kaṇāda, the scholar who is extensively referred to in the context of Vaiśeṣika naturalism is Praśatapāda (4th century AD). Kaṇāda's basic ideas are found in *Vaiśeṣkasūtras* and Praśatapāda's *Padārthadharmasaṁgraha*.

In Vaiśeṣika naturalism, physical and metaphysical ideas received close attention. The six basic categories enumerated by Kaṇāda are substance (*dravya*), quality (*guṇa*), motion (*karma*), genus (*sāmānya*), species (*viśeṣa*), and inherence (*samavāya*). Substances are characterized by their motion and qualities and are said to be inherent causes of motion and quality. The nine substances are earth, water, fire, air, ether, time, place, self and internal organs (*manas*). It is interesting to note that mind, which is ordinarily believed to be conscious, has been combined with natural entities like earth, water and fire. Similarly, among the 17 qualities enumerated by the Vaiśeṣika thinkers, we find not only psychological and physiological qualities like colour, taste, smell and touch but also abstract entities like number and size. In addition, we find psychological entities like pleasure, pain, desire, aversion and effort. When one recalls that the characteristic feature of quality consists in its residing in substance, one realizes that the enumerated qualities are entwined with, adjuncts to, or modifications of the recognized substances. To understand Vaiśeṣika realism, it is important to recall the five forms or modifications of motion (*karma*): upward movement, downward movement, contraction, expansion and going. There is an important distinction between motion and action. In Sanskrit the word *karma* is or, at least may be, used in both senses. Commonly, motion is regarded as a physical notion that need not be attributed to any one person. In contrast, action presupposes a psychological state such as pleasure or pain, desire or aversion. It is difficult to think of action without effort (*prayatna*), unless, of course, we refer to reflex, habitual or instinctive actions. This line of argument illustrates that a strong streak of naturalism runs throughout the Vaiśeṣika system of thought.

At the same time, naturalism does make use of some abstract notions like being and inherence. The simple truth that without some abstract or theoretical concepts a rich naturalist system could not be developed was clear to the Vaiśeṣika thinkers. Therefore, we find in their thought a judicious blend of observational and theoretical entities. In this respect their departure from the

Cārvāka School of naturalism is clear. For example, the Vaiśeṣika writers assert that self is known to exist by means of inference. Self is inferred from such bodily signs as breathing, ocular movement, the movement of *manas* as the internal organs, desire, aversion and effort.

This inferred view of self enlarges and enriches the scope of Vaiśeṣika naturalism of the realistic kind. But there is a problem when they claim that self is eternal and independent of the body. Selves are said to be numerous and eternal. To establish the eternal character of self, the authority of scripture was invoked. This also marks the departure of Vaiśeṣika naturalism from its classical counterparts. Within the Vaiśeṣika system, even an atom is an abstract entity. It is abstract in the sense that it is uncaused, unperceived and eternal. Yet, the fact that it has to be admitted inferentially from its effect shows that the Vaiśeṣika fold of naturalism is liberal. Authorities over the centuries have debated how atoms are combined, and they have generally agreed that atoms are combined to form compound substances and eventually common objects.

NATURALISM IN VEDANTIC THOUGHT

Naturalism and anti-naturalism are not necessarily antithetical. Their relation can be understood in a culture-specific or thinker-specific sense or, preferably, situating both within a given socio-historical perspective. Free thinkers in every age and culture articulated at least two broad interactive trends – pro- and anti-naturalist. The pro-naturalist cannot ignore the anti-naturalist and vice versa. The Vedantic worldview, although it generally highlights the anti-naturalist and transcendental aspects of Vedic thought, does take note not only of the pro-naturalist elements of the Vedas but also those of the anti-Vedic schools. Both in the Vedic corpus and the Vedantic fold there are different views on naturalism and anti-naturalism and their epistemic correlates. Also within the broad framework of Buddhism one can see the radical realism of Sautrāntika and the radical idealism of the Mādhyamik schools. However, elements of idealism are present in the former view and those of realism or naturalism are in the latter.

In this respect the Buddhist school, Mādhyamik, and mainstream Vedic school, the Vedānta, are instructively analogous. That explains why Śaṁkara, the chief proponent of the Vedantic view, has often been described as crypto-Buddhist. Both the Mādhyamik and the Vedantic thinkers take pains to affirm their belief both in the empirical and in the transcendental. Also they are similar in asserting the reality of the world of sense and science (*saṁsāra*), however illusory it may be, and also in that of Brahman or *śūnya*, i.e., *guṇa-śūnya*, the indeterminate or Absolute. Contrary to widespread belief, the Buddhist concept of *śūnya* is neither ṅihil" nor "void". It is to be understood only in the sense of indeterminateness (*nirguṇatā*); i.e., it cannot be expressed in terms of any predicable or general attribute (*guṇaśunyatā*). The ultimate reality in its fullness is available only in intuition. The Vedantin, like the Buddhist, recognizes the natural world which is said to be caused by *māyā*. Māyā is neither mysterious nor illusory. It is the principle of materialization, making ultimate reality natural or material and sensible. The root of māyā is "mā", and it carries the sense of measurement and quantity. The natural world is measurable and quantifiable and subject to the determinations of space, time and causality. It consists of five elements. Māyā is the material cause (upādāna-kāraṇa) of the natural world. The Vedantins recognize five subtle matters: ākāśa (ether or space), vāyū (air or wind), tejas (energy or fire), ap (water) and kṣiti or pṛthivi (earth). The subtle matters (sūkṣma bhūtas) through the process of quintuplication (pañcskaraṇa) take the perceptible forms of gross matter (mahābhūtas). If ākāsa is symbolized by AK, vāyū by V, tejas by T, ap by AP and earth by E, then the process of forming mahābhūtas, gross matter, out of subtle matter may be formulated in the following manner:

$$AK = ak_4(v_1 \cdot t_1 \cdot ap_1 \cdot e_1), \quad ak_1 \text{ being the radical}$$

$$K = v_4(ak_1 \cdot t_1 \cdot ap_1 \cdot e_1), \quad v_1 \text{ being the radical}$$

$$T = t_4(ak_1 \cdot v_1 \cdot ap_1 \cdot e_1), \quad t_4 \text{ being the radical}$$

$$AP = ap_4(ak_1 \cdot v_1 \cdot t_1 \cdot e_1), \quad ap_4 \text{ being the radical}$$

$$E = e_4(ak_1 \cdot v_1 \cdot t_1 \cdot ap_1), \quad e_4 \text{ being the radical}$$

From the process of quintuplication spelt out by the Vedantins it is clear that they took immense pains to clarify their philosophical account of naturalism in distinctly empirical terms.

In the different forms of Indian naturalism, both orthodox and heterodox, there is an intimate relation between the living world, the mental world and the natural world. All forms of life and consciousness are shaped by nature. The close relation between forms of life and the environment was specially highlighted in medical treatises such as the Caraka Saṁhitā, Suśruta Saṁhitā and Aṣṭāṅga Āyurveda. What was philosophically asserted by different systems of thought was explicated in detail in Indian indigenous medical and pharmacological ideas.

BIBLIOGRAPHY

Bhattacharyya, K.C. Studies in Philosophy, G. Bhattacharyya, ed. 2nd ed. Delhi: Motilal Banarsidass, 1983.
Chattopadhyaya, Debiprasad, ed. Cārvāka/Lokāyata: An Anthology of Source Materials and Some Recent Studies. New Delhi: Indian Council of Philosophical Research, 1990.
Dasgupta, S.N. A History of Indian Philosophy, 5 vols. Cambridge: Cambridge University Press, 1922–1955.
Larson, G.J. and Bhattacharya, R.S., eds. Encyclopedia of Indian Philosophies, vol. IV. Delhi: Motilal Banarsidass, 1987.
Murti, T.R.V. The Central Philosophy of Buddhism. London: Unwin Paperbacks, 1980, pp. 293–301.
Nakamura, Hajime. Indian Buddhism. Delhi: Motilal Banarsidass, 1996.
Potter, Karl H., ed., Encyclopedia of Indian Philosophy, Vol. II: Nyāya-Vaiśeṣika. Delhi: Motilal Banarsidass, 1977.
Potter, Karl H., ed. Encyclopedia of Indian Philosophies, Vol. VIII: Abhidharma Buddhism to 150 A.D. Delhi: Motilal Banarsidass, 1998.
Radhakrishnan, S. Indian Philosophy. 2 vols. 2nd ed., London: Allen & Unwin, 1927.

Sinha, J.N. *Indian Epistemology of Perceptions.* Calcutta: Sinha Publishing, 1969.
Thakur, A.L. 'Philosophy of Vaiśeṣika.' In *Project of History of Science, Philosophy and Culture in Indian Civilization* (PHISPC), D.P. Chattopadhyaya, ed. Vol. II, Part 3. Delhi: Munshiram Manoharlal, forthcoming 2002.

JOHN A. TUCKER

JAPANESE VIEWS OF NATURE AND THE ENVIRONMENT

ETYMOLOGICAL INTRODUCTION

In traditional Japan, the word *shizen*, also pronounced *jinen*, meant naturalness, or the mode of being which is natural. It did not, however, signify "nature", at least not as that word has been commonly understood in modern times, i.e., as the objective, material world existing alongside of – often outside of, or in opposition to – the more subjective realms of humanity, art, and culture. Most literally, the meaning of *shizen* is "from itself (*shi/ji*) thus it does (*zen/nen*)". The first component of *shizen* (*shi/ji*) is often read *onozukara* and refers to what is "spontaneously or naturally so". Etymologically, *shizen* is traceable to the ancient Chinese word *ziran*, which was used in various religio-philosophical texts, especially the Daoist classics, to refer to the spontaneous way of birth, growth, and transformation. The *Zhuangzi*, for example, promotes an ethic of "following natural spontaneity (*ziran*)" (ch. 5), and of "complying with the natural spontaneity of things" (ch. 7). More formulaically, the *Daodejing* situates *ziran* at the most fundamental level of the Chinese "great chain of being", stating, "Humanity is based on earth; earth is based on heaven; heaven is based on the Way (*dao*); and the Way is based on what emerges naturally (*ziran*)" (ch. 25). If nature is understood simply in terms of naturalness and spontaneity, the word *ziran* captures those nuances well.

The Chinese term *ziran* was introduced to Japan, along with the Chinese system of writing and Chinese literature of all sorts, in the mid-sixth century, via Buddhist missionaries sent by the ancient Korean state, Paekche. According to the twentieth-century philosopher, Saigusa Hiroto, *shizen*, as understood by Japanese, came to denote "a spontaneous process (*onozukara shikaru mono*)" wherein humanity is absorbed in "the becoming of things" (Piovesana, 1997: 236–237). In explaining why Asian thought never developed a tradition of materialist theorizing (*yuibutsuron*) as appeared in the West, Saigusa cites the closeness between nature and humanity in Japan, while in the West, humanity was supposedly viewed as the conqueror of nature, and nature itself was deemed an objective material reality, existing in concrete opposition to the human subject. The implication of Saigusa's analysis is that Japanese were too

H. Selin (ed.), Nature Across Cultures: Views of Nature and the Environment in Non-Western Cultures, 161–183.
© 2003 *Kluwer Academic Publishers. Printed in Great Britain.*

immersed in the natural world of spontaneity to conceptualize a materialistic nature existing apart from that of humanity. No doubt, the term *shizen* did circulate throughout Japanese literary history, but it never became "the" conceptual category exclusively signifying nature, materialistic or otherwise, in literature, poetry, philosophy, or religion. The well-known Japanese veneration of nature, which was undoubtedly real, flourished more within a richly pluralistic, nuanced set of discourses wherein any number of terms such as *michi* (the natural Way), *tendō* (the Way of heaven), *tenka* (the world-below-heaven), *tenchi* (heaven and earth), *kawa* (rivers), *yama* (mountains), etc., served to signify the manifold diversity and bounty of the natural world and its spontaneous processes.

In Meiji Japan (1868–1912), as Japanese translated Western ideas into their language and increasingly formulated a lexicon which more or less corresponded with that of the leading Western nations, *shizen* came to designate "nature," a word found in virtually all European languages. The Western "nature" was derived from the Latin *natura* and the Greek term *gnascor*, meaning to be born, grow, emerge, and originate. *Gnascor*, related to the notion of genesis, is close in meaning to the traditional Sino-Japanese notion of *ziran/shizen*, especially in its transformative, process-related nuances. As reinterpreted by Romans, *natura* was assigned the semantic value that had been associated with the Greek notion *physis*, signifying everything that exists, i.e., "being as such in its entirety". While humanity was, strictly speaking, included in this realm, it was not the primary element, and was often implicitly bracketed from it. The modern Japanese word *shizen* similarly conveys the primary sense of nature as the environment encompassing all between heaven and earth, especially the sky, earth, seas, mountains, rocks, rivers, flora and fauna.

Ironically enough, although the Japanese had borrowed the word *shizen* from the Chinese over a millennium earlier and used it traditionally to signify "naturalness" and "natural spontaneity", after Japanese lexicographers of the Meiji period decided that *shizen* would denote the Western conception of "nature," that relatively modern nuance subsequently entered the Chinese lexicon in the 1920s, having been imported from Japan. It is even more ironic that as the modern global lexicon has been recently challenged by postmodern reinterpretations, the word "nature" increasingly has come to be understood not in terms of an objective reality existing outside of the human cultural realm, but as the inseparable, environmental crucible of humanity, art, and culture. In this sense, postmodern conceptions of nature return to nuances conveyed by the ancient Sino-Japanese term, *ziran/shizen*, before it was tagged, first by Meiji lexicographers redefining traditional concepts along Western lines, as the East Asian analog of the Western notion of nature.

The modern Japanese word for environment, *kankyō*, has similarly ancient origins in traditional Chinese literature. There the term is read *huanjing* and traces back to the *Yuan shi* (*History of the Yuan Dynasty*) where it was used to denote, in a more localized fashion, one's surroundings. Significantly, the word *huan* (Japanese: *kan*) appeared in a variety of ancient Confucian texts, including the *Shijing* (*Book of Poetry*), the *Liji* (*Book of Rites*), *Zuozhuan* (*Zuo's*

Commentary on the Spring and Autumn Annals), and the *Xunzi*, referring to one's surroundings, even to the extent that those surroundings signified the world at large. Still, *kankyō* as a very modern notion signifies a more objective, scientifically defined entity than these ancient texts intended.

Only a Eurocentric commentator would suggest that pre-modern Japanese lacked something simply because they had no single word signifying nature or the environment as they have come to be understood in modern times. After all, it is doubtful that any one word could have spawned the myriad nuances that were indeed communicated in traditional and modern Japanese literature celebrating the natural world in religious, poetic, philosophical, and literary terms, not to mention quasi-scientific ones which prefigured modern movements in environmentalism. Like Saigusa, the philosopher Sakamaki Shunzō has suggested that Japanese did not coin any single word for nature signifying something apart and distinct from humanity precisely because they considered themselves to be integral parts of the whole, closely identified with the elements and forces of the world around them. While that may be, an understanding of traditional conceptions of nature as the natural world can be easily garnered by examining Japanese thoughts about particular aspects of nature such as heaven, earth, mountains, rivers, the Japanese archipelago, its trees, flowers, fields, and fauna. In a somewhat similar way, Edwin O. Reischauer has linked the supposed Japanese love of nature to the geography of the island nation: no area, he notes, is more than seventy miles from the sea, and mountains are always within view. Because there is ample rainfall, the archipelago is "luxuriantly green and wooded." Given these somewhat idyllic ecological circumstances, Reischauer observed that the "strong awareness of the beauties of nature" among Japanese should be no mystery.

ANCIENT VIEWS OF THE NATURAL WORLD

Ancient Japanese evinced an unabashed intimacy with the natural world in their earliest poetry as compiled in the eighth-century anthology, the *Man'yōshū* (*Collection of Myriad Leaves*). Its poems speak of the world of mountains, rivers, flora and fauna in anthropomorphic, animistic terms, investing each entity with a living personality infused with *kami*, or mysterious spiritual energy and presence. Mountains were deemed most divine. Indeed, in their verses the ancient poets immortalized Mt. Fuji as a peerless deity. Rivers were considered living forces, manifesting immense spirituality in their rushing flows of clean, life-giving water. The sea was viewed as a more awe-inspiring, fearsome spiritual force, one supplying not only sustenance but also destruction. References to birds and beasts fill the poems, evoking a sense of the four seasons and the human feelings linked to them. Moreover, the formulaic eroticism distinguishing many of the *Man'yōshū* poems is often cast in terms of allusions to the natural world and its fecundity. Finally, the anthology conveys a political agenda, that of legitimizing the ruling Yamato line, and partially fulfills it by propitious references to aspects of the natural world, the sun, moon, mountains, and rivers, and the rightness and justice of their physical bounty. Even the poems themselves were likened directly to the natural world: the *Man'yōshū* included, as the title suggests, myriad *yō*, or "leaves" of poetry.

The *Man'yōshū* is even more significant because it pioneered an alliance between poetry and the natural world that has since dominated the Japanese poetic tradition. The poems included in it are generally referred to as *waka* (Japanese poems), as opposed to *kanshi* (Chinese-style poems) written by Japanese. *Waka* are typically divided into two categories, the *tanka* (short poem) and the *chōka* (long poem). *Tanka*, which predominate (4,200 of the 4,516 poems) in the *Man'yōshū*, are structured syllabically with lines typically running 5-7-5-7-7 in syllable count, so as to total thirty-one syllables. Later, as this structure was simplified in the seventeenth century, *haiku*, or the 5-7-5 syllable poem, emerged from the *tanka* to become the most popular poetic form among Japanese. Apart from the structural rule regarding syllable count, the *haiku* is, almost without exception, directed towards some aspect of the natural world, capturing in its very brevity a fascinating slice of it. In first celebrating aspects of this world through its poetry, and then spawning a powerful poetic genre, that of *haiku*, the *Man'yōshū* both codified and molded the Japanese infatuation with the natural world in literary ways with which few other ancient classics could compare.

In certain respects, the *Man'yōshū* celebrated the natural world not as a politically amorphous, transcendent entity, existing above and beyond the realms of rulers, but instead as the world of heaven and earth comprising the land of Yamato and ruled by the nascent imperial line establishing its authority over that realm, in part by poeticizing its natural beauties. For example the first poem, attributed to Emperor Yūryaku (r. 456–479), presents the sovereign asking the name of a maiden picking herbs on a hillside. In the final lines, the emperor announces that "this land of Yamato, seen by the gods on high – it is all my realm, in all of it I am supreme." The second poem, supposedly composed by Emperor Jomei (r. 629–642), opens stating, "Countless are the mountains in Yamato, but perfect is the heavenly hill of Kagu." The poem concludes, "A splendid land is the dragonfly island, the land of Yamato." In both poems, emperors celebrate the natural beauty of their political realms, ensconcing their rule in poetic grace as if to legitimize themselves as sovereigns through verse. Explicit in the first poem, and implied in the second, is the divine status of the land so lavishly praised throughout the *Man'yōshū*, the Yamato realm. In this respect, the imperial poems were not simply exercises in verse extolling nature, but a means by which emperors reinforced, through poetic reiterations, their claims to rule, by sanction of the *kami*, or gods, a purportedly divine land, the Yamato plain and all territories surrounding it. It might be added that the name most commonly used to refer to this polity, Yamato, referred not to the surname of the imperial line, nor to any deity who supposedly sanctioned its rule, but instead, apparently, to the natural terrain, and perhaps the natural beauty of the place itself. Yamato, as Ian Hideo Levy explains, "seems to have the etymological implication of 'the place between the mountains'" (*Man'yōshū*, 1981: 7).

The sacred quality of the natural, often linked timelessly (and anachronistically) with the Japanese perspective on nature, is more correctly associated with the natural world as it existed within the Japanese archipelago and more specifically within the politico-cultural unit first known as Yamato. While there are hints of this throughout the *Man'yōshū*, the *loci classicus* relating the divinity of the Japanese politico-natural world appear in the *Kojiki* (*Records of Antiquity*, 712) and *Nihon shoki* (*Chronicles of Japan*, 720), both of which open with narratives describing a cosmogonic vision, not unlike ones produced in Malayo-Polynesian cultures, that came to be associated with Shinto (the Way of the Kami, or Spiritual Forces), the most indigenous form of Japanese religiosity. After heaven and earth congealed out of an undifferentiated, egg-like mass, successive generations of personified kami begat one another. Finally, the kami pair, Izanagi and Izanami, created the first island, Onogorojima, by letting the brine drip off a spear they had plunged into the ocean's depths. There, Izanami gave birth to the elements of the natural world including rivers, mountains, birds, beasts, and flowers and trees, as well as forces such as fire. Because of the divinity of the progenitors, Japanese traditionally considered their archipelago and all within it sacred. Again, however, it seems that the *Kojiki* and *Nihon shoki* accounts, if interpreted along strictly literal lines, allow less for an overall view of nature as a sacred realm than they do for the sacredness of the specific geopolitical unit whose existence was being delineated via cosmogony, the nascent Yamato state.

According to one account, Izanagi and Izanami also procreated Amaterasu no mikoto, the Sun Goddess, and her brother, Susanoo no mikoto, the impetuous god of storms. Because Amaterasu, a benevolent kami, and Susanoo, a mischievous if not malevolent one, are deemed divine beings, Japanese have not typically viewed destructive natural forces as being radically evil. Nor have they branded acts defiling the natural world as necessarily wrong. Indeed, Shinto myths relate that Susanoo wreaked havoc in his sister's rice fields, implying that similar actions contrary to the general good of the natural world might still have some divine sanction via Susanoo's example. Perhaps this partly explains why many Japanese, despite their close religious, poetic, and mythic ties to nature as defined by their politico-cultural realm, have at times tolerated egregious abuses of it.

Pollution, however, was considered anathema, and was dealt with via purification rites that came to characterize much of Shinto practice. Pollution did not necessarily mean physical dirtiness or noxious environmental conditions, yet analogous forms were recognized among the more spiritual nuances associated with emerging Shinto notions. Traditional abhorrence of physical and spiritual pollution perhaps explains, in part, the energetic opposition of many Japanese, although certainly not all, to environmental pollution.

Amaterasu and Susanoo parented the imperial line and were the ancestral stock of all its human subjects. Belief in the divinity of their islands, their imperial family, and everything within their natural environment thus results from the supposed ancestry of Japanese in Izanagi and Izanami. This became a basic tenet of Shinto, one regularly repeated throughout Japanese history. In the *Engi shiki* (*Religious Regulations of the Engi Era*, 927), ritual prayers record a similar vision of the natural world as infused with divine spirituality. Kitabatake Chikafusa's (1293–1354) political treatise, *Jinnō shōtōki* (*The Legitimate Succession of Divine Emperors*, ca. 1340), opens with the declaration, "Japan is a sacred land (*shinkoku*)." Japanese Buddhists and Confucians have accepted the same, essentially Shinto doctrine, almost unanimously. Belief in the divine status of Japan was also increased, at least among samurai, in the late-thirteenth century, when the Kamakura shogunal regime was able to survive two massive maritime attempts, launched by Mongol forces, to invade Japan. Incredibly enough, on both occasions the armadas were destroyed by typhoons, subsequently dubbed *kamikaze* (divine winds), sent by the gods to protect Japan from barbarian invasion.

Belief in the divinity of the geo-political archipelago encouraged some Japanese to xenophobia: foreigners, typically considered barbarians, were feared as potential agents of pollution. Japanese soil, they felt, would be violated if the barbarians were allowed on it. Although Western missionaries and traders were allowed somewhat free access inland during the second half of the sixteenth century, as the country emerged from a prolonged period of chronic civil war an attempt was made to localize Western presence in the archipelago by restricting it to the artificial island of Dejima, created just off the shore of Nagasaki bay. Dejima was thus devised as a compromise, one preventing direct, prolonged contact between polluted foreigners and the divine land, even while allowing their presence nearby, on soil dredged up from the harbor. During the Meiji period, and especially since 1945, Japanese have become more accustomed to foreigners in their country. Still their concern for nature often seems Japan-specific: some Southeast Asian forests, for example, have been depleted to accommodate the preference, among many Japanese, for disposable, wooden chopsticks.

Even in the ancient texts, especially the *Nihon shoki*, hints appear suggesting that where there might be a contest between the aggressive forces of human development and those of natural spirituality, the former must prevail. For example, during the reign of Emperor Keiko, the legendary hero Yamato-takeru was supposedly commissioned to "subdue" the Emishi tribes living in the northeast portion of the archipelago (*Nihon shoki*, 1972: 203). The Emishi were among the indigenous stock of Japan, but were considered by the emerging Yamato state and its ruling elite to be non-Japanese peoples. As the Yamato state grew in territory, its imperial policy was to incorporate, enslave, or eliminate these peoples. In commissioning Yamato-takeru to lead the mission against the Emishi, Emperor Keiko vilified the latter, stating that in their mountains are "malignant deities, on the moors, there are malicious demons, who beset the highways and bar the roads, causing men much annoyance." In this case, when the spiritual forces of the natural world blocked the progress of the Yamato state, they became "malignant" and "malicious" and had to be conquered by "the kingly civilizing influences". Thus ancient Japanese views of the natural world as divine clearly did not apply far and wide, even within the archipelago, but were more part of a worldview promoted from within the natural realm of the Yamato state extending to the borders of its territories but not far beyond.

Traditional prose literature, especially from the Heian period (794–1185), also provides evidence at every turn of the Japanese infatuation with what would later be called "nature". Perhaps the classic example is Lady Murasaki Shikibu's (fl. ca. 1000) *Tale of Genji* (*Genji monogatari*, early 11th century), which portrays an aristocratic society obsessed with and continually alluding to natural phenomena such as the moon, the noise of crickets, babbling brooks, rushing streams, flowers in bloom, the break of dawn, the passing of the seasons, etc. Although never the main focus of the novel, Lady Murasaki's frequent references to natural entities reveal that one crucial aspect of aristocratic imperial culture was associated with a heightened sensibility vis-à-vis the natural world, its transformations, and its analogs in the world of human romance and emotion. Although with far less grace and subtlety, essentially the same, very refined religio-aesthetic appreciation of the natural world is evident in the opinionated writing of another Heian authoress, Sei Shōnagon's (fl. late 10th century), the *Pillow Book* (*Makura no sōshi*).

BUDDHISM AND THE NATURAL WORLD

Introduced to Japan in the mid-sixth century, Buddhism advanced various attitudes towards the natural world. The Four Noble Truths, the original teachings of the historical Buddha, Siddhartha Gautama (563–483 BCE), characterize existence as suffering, implying that *samsara*, or the environment of reincarnation, was similar. Their religious solution, *nirvana*, which literally means "putting out the flame" (of existence within this world), seems to offer an escapist otherworldliness which might have permitted relative disengagement from nature.

The Mahāyāna Buddhist ideas of the Indian thinker Nāgārjuna (ca. 150–250), which were accepted by most Japanese Buddhists, affirmed the natural order. Nāgārjuna equated *samsara* with *nirvana*, disallowing otherworldliness of any sort. Nāgārjuna's view was based on the doctrine that everything is empty (*sunya*). Empty here means empty of self-sustaining substance, i.e., substance existing in and of itself. Nāgārjuna insisted that everything that exists does so through spatial, temporal, and causal relations with the remainder of the universe. Nothing exists independently of anything else. His ideas could be construed as anticipating the ecologist's belief that all life is interrelated, and that destruction of even small niches will endanger the entire ecosystem. Nāgārjuna's thinking surely facilitated a more positive appraisal of the natural world by Japanese Buddhists.

Traditional Buddhist cosmology, however, claims that the world is subject to creation and disintegration just as humans experience cycles of death and rebirth. Arguably this view could allow a cavalier attitude towards the natural environment, since regardless of one's efforts the world will inevitably disintegrate and then begin anew. The Buddhist belief that attachment to things leads to suffering might also have vitiated wholehearted involvement in an environmental ethic geared toward conservation. The Buddhist two-level theory of truth, assigning ultimate status to *sunya*, and relegating common sense to a secondary level of validity, allows for a concern for the natural world, but not in any primary way.

Otherworldly tendencies appeared in the popular Jōdo (Pure Land) School. Based on the Indian writing *Sukhāvatīvyūha-sūtra* (*Discourse on Paradise*), Jōdo posits both a heaven called the Pure Land, presided over by Amida Buddha, and a multileveled hell where sinners suffer eternally. Yet some theorists claim that the Pure Land and hell are merely "expedients" meant to motivate non-believers to meditate on Amida Buddha as the way to salvation. If so, then the Pure Land becomes a symbol of *nirvana*, or existential extinction, while hell becomes a hyperbole of the Buddha's claim that life in this world, the natural world, entails suffering.

Yet the ideas of many Japanese Buddhists evinced a religiously based concern for nature. The Kegon, or Flower Garland, School asserted that every particle of existence was infused with Buddha-nature, making the natural universe a spiritual one as well, one to be saved from suffering. The *Konkōkyō* (*Sutra of the Golden Light*), an important text in early Japanese Buddhism, claimed that rulers who promoted the Buddha's teachings would be protected, as would be their domains, by the Four Deva Kings, tutelary divinities who protected Buddhism throughout the universe.

Compassion for all sentient beings, the core ethic inculcated by Mahāyāna Buddhists, instilled in many a concern for the natural world. Buddhist poets like Saigyō (1118–1190) even extolled the natural realm as the primary arena of Buddhist values. Earlier Buddhists, in China and then Japan, had debated whether plants and trees, although nonsentient beings, could actually attain Buddhahood. Notably, Saichō (766–822), the founder of the Tendai School, first broached the topic in an affirmative way, declaring, "trees and rocks have Buddha-nature" (*mokuseki busshō*) (La Fleur, 1989: 186). Kūkai (774–835), founder of the Shingon School of esoteric Buddhism, similarly declared that "trees and plants" could attain enlightenment. Ryōgen (912–985), a Tendai abbot, even authored a treatise, "Account of How Plants and Trees Desire Enlightenment, Discipline Themselves, and Attain Buddhahood" (*Sōmoku hosshin shugyō jōbutsu ki*). The claims of Saichō, Kūkai, and Ryōgen were all based on the *Lotus Sutra* (*Saddharma-Pundarīka*), one of the most doctrinally authoritative Mahāyāna texts, known especially for its declaration that all will ultimately attain Buddhahood.

Other Japanese Buddhists also claimed that the natural world possessed a healing, and even soteriological capability. Probably influenced by Shinto beliefs, the Shugendō (Order of Mountain Ascetics) movement had its practitioners make pilgrimages to sacred mountains to glimpse scenery foreshadowing the Pure Land. Buddhist temples aesthetically enhanced the environment by their communion with the natural, often being built along the slopes of mountains. With their rock gardens, moss gardens, bamboo gardens, cherry and plum orchards, and vegetable gardens, temples were practically involved in local environmental improvement as a way of meditation and as a means of approximating, through natural beauty, the spiritual beauty of nirvana.

Zen Buddhists in particular saw enlightenment as an experience to be had in this world and in this body. Disregard of the natural world, therefore, could not be allowed. Dōgen (1200–1253), founder of the Sōtō School of Zen Buddhism, even declared that "the ocean speaks and mountains have tongues – that is the everyday speech of Buddha ... If you can speak and hear such words you will be one who truly comprehends the entire universe" (Shaner, 1989: 172). Despite this, Buddhists were not generally known for authoring agricultural or environmental tracts that would promote a more harmonious symbiosis between humanity and the natural sphere. It was the Confucian and Neo-Confucian scholars who, in addition to admiring natural beauty and revering it spiritually, made the natural world in which they lived the focus of proto-scientific research designed to conserve the environment for future generations.

NEO-CONFUCIANISM AND EARLY MODERN VIEWS OF THE NATURAL WORLD

Neo-Confucianism, the last major philosophical force to emerge in traditional Japan, consistently encouraged scientific interest in and ethical concern for the natural world. Prompted by the sophisticated Buddhist metaphysics with its (to Neo-Confucians) repulsive doctrine of emptiness, Neo-Confucians reformulated their originally socio-political thought along novel metaphysical lines so as to refute the Buddhist challenge. Originally a Chinese movement of the Song dynasty (960–1279), Neo-Confucianism ultimately became a more pan-Asian movement, decisively affecting China, Korea, Japan, and Vietnam and other East Asian areas through the modern period. In Japan, it was a dominant force during the Tokugawa period (1600–1867), often referred to as the *kinsei* (early modern) period of Japanese history.

Neo-Confucians endorsed common sense, declaring that the natural environment was both substantial and fully real. In their view, there was no other world, nor would there ever be one. Rejecting emptiness, they asserted that everything consisted of a quasi-material, psycho-physical energy called *ki*. Giving this energy its rationale was another ontological element, *ri* (principle). The fusion of *ki* and *ri* accounted for both diversity and unity within the natural world. The latter was the ongoing creation of heaven and earth, which engaged in constant production and reproduction as its Way. Most Neo-Confucians recognized the complementary forces of *yin* and *yang* as the cardinal modes of material being. They also admitted the five elements of earth, wood, fire, water, and metal, as essential processes defining all developments within the natural world. Because nature and humanity were created by the same elements and by the same forces, Neo-Confucians reinterpreted their socio-political ethic of humaneness in mystical terms of forming one body with the universe. Some even spoke of heaven and earth as their parents and the myriad entities, organic and inorganic, as their companions.

Most Neo-Confucians also declared human nature (Chinese: *xing*; Japanese:

sei), or the original psycho-physical disposition of human beings, to be good, not empty as the Buddhists asserted. Furthermore Neo-Confucians claimed that the nature (*sei*) of the universe, i.e., its moral character, was originally good. The human project, as defined by Neo-Confucians, was to preserve this original goodness by moral self-cultivation and by moral action in the world. For Confucians and Neo-Confucians alike, "the quest for self-knowledge" thus became one corresponding to "a search for an understanding of heaven" (Tu, 1976: 116). Further explaining the Confucian mentality, Roger Ames says that, for Confucians, heaven is "not a preexisting creative principle which gives birth to and nurtures a world independent of itself;" rather it is "a general designation for the phenomenal world as it emerges of its own accord" (Ames, 1987: 207).

Confucius (551–479 BC) too had respected the natural world. The *Analects* (Chinese: *Lunyu*; Japanese: *Rongo*, 6/12), the most authentic record of his thought, relates that the wise person loves water (Chinese: *shui*; Japanese: *sui*), while the humane person loves mountains (Chinese: *shan*; Japanese: *san*). This suggests that early Confucianism linked moral concerns to ecological ones. It is also artistically significant because the Chinese word for landscape painting is *shansui* (Japanese: *sansui*), denoting a pictorial representation combining water with mountains, the constituent elements in the genre. Landscapes were not, therefore, simply stylized depictions of nature; they were equally reflections of, even commentaries on, the moral consciousness of the artist and the state of morality in the world in which he lived. Confucians and Neo-Confucians, following Confucius' views on mountains and water, furthermore believed that morality involved right behavior towards the natural world and humanity.

Evidence abounds revealing a practical, even quasi-scientific, interest in the world of nature by Confucians and Neo-Confucians in Japan. By the late seventeenth century, forests throughout the archipelago had been depleted due to overexploitation resulting from a boom in the construction of castles and urban residences. Confucian scholars promptly diagnosed the environmental crisis and called for its cure. Yamaga Sokō (1622–1685), for example, argued that forests should be conserved to ensure future productivity. Sokō admonished loggers to harvest lumber only in the proper season, not to overcut, and to replant areas they had cut. One of Sokō's disciples, Tsugaru Nobumasa, daimyo of Hirosaki domain in northeastern Honshū, claimed that the three fundamental concerns of a feudal lord were (1) his family, (2) his heir, and (3) his mountain forests.

In *Daigaku wakumon* (*Dialogues on the Great Learning*), Kumazawa Banzan (1619–1691) argued that humane government involved afforestation, river dike repair, and other conservation practices that would maximize agricultural productivity. Noting the crisis at hand, his disciples declared that "mountains and rivers are the foundations of a country." Full of hope, despite the depleted forests that were all too evident, Kumazawa claimed that even bald mountains could be covered with trees again if oats and other cover crops were planted first. Ultimately, *Daigaku wakumon* advocated harnessing samurai energy for the sake of the agricultural ecosystem; Kumazawa argued that samurai should be returned to the countryside to labor as farmers rather than required to live in castle towns where they fell prey to urban vices. Kumazawa's proposals seemed so radical, however, that they were not heeded. Generations after his death, his more environmentally oriented ideas, as advocated by others, did find more favor.

Kaibara Ekken (1630–1714), influenced by the Neo-Confucian call to investigate the principles of things, and in particular the natural principles of things (*shizen no ri*), authored the first systematic botanical study in Japan, the *Yamato honzō* (*Natural History of Japan*, 1709), describing and classifying over 1550 trees, plants, flowers, birds, fish, seashells, etc., into some 37 categories. Ekken's preface declares that because heaven and earth produce and reproduce myriad life forms, scholars must study their principles. Many later studies in herbology, botany, pharmacology, and zoology were influenced by Ekken's *Yamato honzō*. He also encouraged the research of Miyazaki Antei (1623–1697), which culminated in the conservation-minded *Nōgyō zensho* (*Agricultural Encyclopedia*, 1696). This text was "the most widely used reference work in the proliferation of agronomical ideas throughout village Japan" (Najita, 1987: 46). In his introduction to Antei's text, Ekken observed that studying the *Nōgyō zensho* facilitated sage government, which nourished and educated the people, because through it people could understand how to assist heaven and earth by cultivating the productive forces of the natural world.

Given the dense population that appeared during the Tokugawa and its heavy taxation of the ecosystem, Japan could easily have become an eroded moonscape rather than a verdant archipelago (Totman, 1989). In the seventeenth century, Japan was well on its way to a state of deforestation. It was saved, in part, by the Confucian and Neo-Confucian scholars who called for systematic, scholarly solutions to the egregious ecological predicament. At the same time we cannot deny credit to Shinto and Buddhism, although admittedly they were less conspicuous in enunciating a conservation program for saving Japan's forests and thus its mountains, rivers, and fields.

Also, it must be admitted that not all Confucian scholars of the Tokugawa period were equally concerned with the natural realm. For example, Ogyū Sorai (1666–1728) acknowledged that while he knew that "the winds, clouds, thunder, and rain are the mysterious operations of heaven and earth," he declined to claim any more knowledge about the workings of the natural world. Regarding various theories concerning *yin* and *yang*, the spirits, and magical powers associated with it, Sorai insisted on his ignorance. In the end, in a manner much like the ancient Chinese realist, Xunzi (298–238 BC), Sorai declared that Confucian learning simply involved the study of the Way for the sake of governing the realm and pacifying the people; all else was idle speculation. While Maruyama Masao has lauded Sorai's supposed break with naturalistic philosophizing as an indication of the beginnings of modernity in Japan, it seems questionable whether many would endorse the assumption that this transformation is necessarily linked to a rejection of the belief that the natural, human, and ethical orders are intrinsically related. Indeed, if contemporary trends involving a heightened environmental awareness are any indication as to what is "modern", then Sorai's views seem rather dated by comparison.

A host of Confucian-minded scholars in eighteenth-century Japan, many of them associated with the Kaitokudō merchant academy in Osaka, demonstrated that admiration for the natural world did not necessarily depend on leading an agricultural life. According to Tetsuo Najita's study of these scholars, they collectively articulated two contrasting epistemologies, or theories about the nature of knowledge: (1) the historicist epistemology, claiming that the ultimate source of knowledge is history; and (2) the natural ontology, asserting that epistemological certainty derives from the natural realm alone. While Najita recognizes Kaibara Ekken and Miyazaki Antei as the "pivotal philosophers in this [latter] tradition," he adds that merchant scholars such as Nishikawa Joken (1648–1724) and Goi Ranju (1697–1762), who played "decisive roles in the intellectual development of the Kaitokudō," were equally advocates of it, extolling the natural realm as the ground of final, philosophical knowledge. Later Kaitokudō scholars associated with this tradition were Yamagato Bantō (1748–1821), an advocate of the heliocentric view of the universe, and Kaiho Seiryō (1755–1817), a proponent of mathematical thought. With these thinkers, the Neo-Confucian notion of "the investigation of things" (kakubutsu) as a means of "exhaustively understanding the principles of the natural world" (kyūri) functioned as the starting point for their own development of a natural ontology grounded in the epistemological view that the natural world itself is not just an object of knowledge but indeed the ground for the possibility of any and all knowledge.

Some late Tokugawa intellectuals were also influenced by notions of Western science as introduced to Japan via Dutch traders permitted at Nagasaki. Satō Nobuhiro (1769–1850), exposed to Western science but also influenced by Shinto, advocated techniques of agricultural management based on the scientific study of natural law. He argued that this would realize the divine aim inherent in creation. Rejecting such clever short cuts, Ninomiya Sontoku (1787–1856) insisted instead on following the natural, creative cycles of heaven in agriculture and repaying the virtue of heaven with conservation techniques that would rejuvenate the natural world. Miura Baien (1723–89), influenced by Western science, slightly modified the Neo-Confucian project of investigating things by advocating the investigation of the rational order (jōri) of heaven and earth in a disinterested, objective way. Andō Shōeki's (1703–1762) writings, particularly his Shizen shin'ei dō (The Way of Natural and Authentic Activities) extolled the "world of naturalness" (shizen no yo) over the "world of law" (hōsei), denouncing the latter in favor of a return to the former. Insofar as Shōeki idealized the state of naturalness and spontaneity, his socio-political critique of the Tokugawa order was somewhat analogous to Jean-Jacque Rousseau's Social Contract and its analysis of the state of nature as opposed to the existing political orders of Europe that supposedly enslaved humanity.

Along other lines, late-eighteenth and early-nineteenth century nativist scholars (kokugakusha) such as Motoori Norinaga (1730–1801), and especially Hirata Atsutane (1776–1843), sought, somewhat atavistically, to revive the study of ancient Japanese texts, rejecting the Chinese canon associated with Neo-Confucianism as one composed of pernicious foreign works that would cripple the originally pure, unsullied Yamato spirit. Not surprisingly, Norinaga and Atsutane also aggressively revived claims regarding the divinity of Japan, the supposed "land of the kami," and its mission as a sacred environment. Norinaga occasionally associated his thinking on these topics with his so-called "natural Shinto" (*shizen no Shintō*), which he differentiated from other brands of Shinto already extant. Although earlier scholars had declared that Shinto was "the natural way of heaven and earth," Norinaga insisted that his "natural Shinto" captured a new perspective, one conveying "a native and primordial way, coeval with heaven and earth and possessing the capacity to enable ancient man to be moral without morality" (Nosco, 1990: 169).

In part, the notion of Japan as a sacred space fed the xenophobic reaction, expressed in the slogan *sonnō jōi* (revere the emperor and expel the barbarian), directed against Westerners intent on opening Japan to international trade in the mid-nineteenth century. Revering the emperor meant, in part, revering the sacred natural realm of Japan and preventing pollution by foreigners. Despite the popularity of the *sonnō jōi* agenda among some samurai radicals in the 1850s and 1860s, Western nations, especially the United States, empowered by the industrial revolution and intent upon spreading their cultural values to all corners of the globe, could not be turned away. Consequently, the Tokugawa bakufu, a samurai regime that had provided over two centuries of uninterrupted peace for the nation, collapsed, as it was unable to fulfill its primary function, defending the realm against threats to its integrity.

MODERN VIEWS OF NATURE AND THE ENVIRONMENT

Following the Meiji Restoration of 1868, as Japan embarked upon a course of rapid modernization in the form of Westernization, concern for the natural environment lessened as the new Meiji state presided over the beginnings of industrialization by fostering polluting industries such as railroads and mining. Although it was anathema to many, the goal of the Meiji state was to match, if not surpass, the industrial prowess of the Western nations that had imposed inequitable treaties on Japan during the mid-nineteenth century. Voices of protest against the noxious side effects were either ignored or muffled until Japan had effectively modernized. Ironically, even as the Japanese lexicon assigned to the word *shizen* nuances linked to "nature" as understood in the West, Japan was simultaneously experiencing, due to its rapid industrialization, some of the most egregious instances of environmental degradation in its entire history.

In the late 1870s, widespread water pollution caused by Furukawa Ichibe's Ashio Copper Mine in Tochigi Prefecture resulted in a major ecological disaster and a political scandal that continued for decades, pitting farmers against private industry and the government. After election to the newly created Imperial Diet, Tanaka Shōzō made it his cause to end the Ashio mining operation. Furukawa replied that any harm caused was counterbalanced by contributions made to the Meiji project of *fukoku kyōhei* (enriching the nation and strengthening it militarily). When finally the poisonous byproducts of the

mining threatened Tokyo's suburbs, the imperial government, in 1898, enacted
regulations to prevent Furukawa's operation from dangerously polluting air
and water. Although notions of *nohon shugi* (the supremacy of agriculture)
circulated in the late-Meiji, these were frequently no more than ideologically
flattering currents.

It was also in the Meiji period that many progressive intellectuals such as
Fukuzawa Yukichi (1834–1901), Ueki Emori (1829–1897), and Nakae Chōmin
(1847–1901) translated Western theories of natural rights (*shizen no kenri*) into
the Japanese idiom, often as *tenpu jinken* (heavenly conferred human rights),
and advocated them as a means of furthering the liberal agenda within the
emerging Meiji polity. Fukuzawa, for example, opened his best-selling treatise,
Gakumon no susume (Encouragement of Learning) with the remark that "heaven
(nature) does not create one human being above or below another," thus
paraphrasing the Western natural rights notion that people are, by nature,
equal (Fukuzawa, 1986: 11).

While Katō Hiroyuki (1836–1916), another prominent Meiji intellectual,
advocated natural rights early on, in the late-nineteenth century he came to
embrace Social Darwinism, viewing the natural state of nations as one of
warfare wherein the fittest proved themselves on the field of battle, overcoming
weaker, inferior nations in a contest allowing only for the survival of the fittest.
However, Katō later questioned the ethical acceptability of many of these ideas
in writings such as *Shizenkai no mujun to shinka (Contradictions in the World
of Nature and Evolution*, 1906), *Shizen to rinri (Nature and Ethics*, 1912), and
*Jinsei no shizen to gohō no zento (Human Nature and the Future Prospects of
Our Country*, 1916). These texts, rather than trumpeting the workings of social
evolution, advocated leadership by an elite in ministering to the misery
produced by historical change, despite the countercurrents generated by the
vicissitudes of evolution.

Other Social Darwinists, such as Oka Asajirō (1868–1944), who more consistently defended their views, recognized the harshness of "nature's revenge"
against the advances of civilization (Thomas, 1998: 119). Nature responded to
economic development, Oka asserted, with flooding from polluted rivers and
to the advances of modern medicine with diseases such as tuberculosis, all in
"quiet revenge" against the challenges of human progress. Yet the alternative,
stopping progress, presented even worse options, since stagnation invited
destruction by a vengeful nature. In Oka's view, nature was anything but the
divine force that had sent *kamikaze* to save Japan. Instead, it was a daunting
foe that would have to be fought, although never possibly defeated.

While natural rights theory did have energetic proponents in Meiji Japan
and was part of the impetus behind the *jiyū minken undō* (People's Rights
Movement) of the late 1870s and early 1880s, the Meiji regime moved quickly
to promulgate a Prussian-style constitution in 1889, preempting appeals to
natural rights. Due to this conservative turn in late-nineteenth century Japan,
the politicization of nature at the popular level of people's rights was quashed.
On the other hand, during the same period, as the Meiji emperor increasingly
assumed the status of a divine and sacrosanct ruler, tracing his lineage to the

Sun Goddess Amaterasu, one ultimate aspect of nature, the sun, came to be intensely politicized as symbolic of the majesty, inescapable authority, and supposed benevolence of even authoritarian imperial rule.

Julia Thomas's study of Japanese nation-building in the early twentieth century, specifically in the Taishō period (1912–1926) suggests that rather than a process wherein the constructed nation was celebrated as a "conscious self-creation", Japan's "modern nationhood" was the result of an effort to "reinscribe the nation as natural", i.e., an effort at "naturalizing nationhood", and by extension of "nationalizing nature" (Thomas, 1998: 114–29). Thomas explains that the naturalized nationhood of Taishō times was a reaction against Meiji conceptions of nature, especially those related to Social Darwinism. The net effect of naturalizing nationhood, however, was to circumscribe legitimate political discourse and enhance oligarchic power. Although some critics, such as Minakata Kumagusu (1867–1941) and Yanagita Kunio (1875–1962), argued for the integrity of locally particular niches within the socio-natural world, over the efforts of the Taishō state to centralize them via a nationally-articulated conception of nature, their efforts were overwhelmed by the forces, religious and educational, promoting uniformity.

In early-twentieth century Japanese literature, *shizen shugi* (naturalism) developed not as an affirmation of ancient literary motifs, styles, or genres, but instead, like the modern nuances that were attached to the word *shizen* in Meiji times, as something influenced strongly by Western developments, in this case the European naturalist movement, especially the French movement exemplified by authors such as Flaubert, Maupassant, and Zola. Masterworks of Japanese naturalism include Tayama Katai's (1872–1930) *The Quilt* (*Futon*, 1907) and Shimazaki Tōson's (1872–1943) *The Broken Commandment* (*Hakai*, 1906). As a Japanese movement, naturalism emphasized the subordinate development of individual characters within the setting of their social environment. At the same time, it demanded not simple realism or objectivity, but genuine expressions of a character's innermost sentiments. Characteristically, naturalism did not strive after didacticism of any sort; instead, it sought to portray the human condition in its most sordid and often unmentioned details. Although soon eclipsed by the *watakushi shōsetsu* (I-novel), a more purely confessional genre that grew out of the movement, naturalism supposedly left an indelible mark on modern Japanese fiction (Kenney, 1983: 351).

The leading twentieth-century philosopher of nature was Watsuji Tetsurō (1889–1960). Watsuji's *Fūdo* (*Environment*, 1935) criticized Martin Heidegger's *Being and Time*, arguing that space, i.e., natural setting, and not time, was most crucial to human culture. Watsuji correlated the latter with three environmental zones: monsoon, desert, and pastoral. Japanese culture, he claimed, emerged from a monsoon environment where climatic vagaries produced passive, sentimental, intuitive, and temperamental traits. Watsuji's thinking also reacted against that of the most prominent philosopher of early-twentieth century Japan, Nishida Kitarō (1870–1945). Nishida claimed, in his *Art and Morality* (*Geijutsu to dōtoku*, 1923), that nature viewed at its deepest levels is nothing other than culture. Nishida's reduction of nature to culture was grounded in his view that the transcendental ego is the basis of the whole cultural world, and that it is capable of perceiving, in its depths, the worlds of mathematics, logic, art, religion, and natural phenomena. Watsuji's emphasis on the environmental impact on human culture, while seemingly common sense, was in effect a reversal of Nishida's earlier reduction of nature to culture and the transcendental ego. Tosaka Jun (1900–1945), a Marxist thinker, differed with both Watsuji and Nishida, neither reducing culture to nature, nor nature to culture or the self, but instead insisting, along more characteristically Western lines, that nature mediated the actions and work of human subjects. Tosaka's ideas, mostly developed in the militaristic 1930s, did not, however, exert a significant impact in their day.

While Watsuji's insights reflected an analytic sensitivity to the linkage between nature, the environment, and the world of human culture, other Japanese thinkers of the 1930s were more intent upon establishing, via appeal to ancient Shinto myths, the notion that Japan, as a divine, imperial country, was destined to establish itself as the leading political force throughout East Asia and the Pacific. In this context, the militaristic imperial ambitions of Japan were camouflaged by proclamations touting the creation of a "New Order," one embodied in the so-called "Greater East Asian Co-Prosperity Sphere" (*Dai Tōa kyōeiken*), a transnational facade established, supposedly, for the economic benefit of all, but serving primarily as an enormous source of raw natural resources, such as rubber, oil, metal, and wood, for Japan. The Greater East Asian Co-Prosperity Sphere, insofar as it was linked to Japan's divine mission and justified by appeal to the universality of the Sun Goddess, revealed the extent to which religious myths about the sanctity of nature could be appropriated for the purposes of economic and political domination of other lands and aggressive exploitation of their environments.

The atomic bombings of Hiroshima and Nagasaki in August of 1945 not only helped to bring World War II in Asia to a decisive close, they also served as the unforgettable beginnings of both a dedicated peace movement and a closely associated anti-nuclear movement. The latter has emphasized, as much as the inhumanity of nuclear warfare, the poisonous nature of nuclear weapons for the environment. The widespread appeal of this movement, based in part on acute Japanese consciousness of the nearly total destruction of both cites by the bombs dropped on them, later led Prime Minister Satō Eisaku and the ruling Liberal Democratic Party to formulate a national policy declaring that Japan would not produce, possess, or harbor nuclear weapons within its borders. Because of his leadership in making postwar Japan a nation devoted to peace and non-nuclear principles, Satō was awarded a Nobel Peace Prize in 1974.

Since 1945, a decreasing number of postwar Japanese have chosen to remain in the countryside close to the agricultural cycle, and thus nature. Just as the Meiji state rushed to modernize, so did Japan in the 1950s and 1960s race to renew its industrial sector, attaining heights earlier reached prior to defeat, yet largely oblivious to air and water pollution. In Minamata, Kyūshū, the Chisso Corporation was finally held responsible for mercury poisoning, over fifty years after the poisoning first occurred in 1908. By the 1960s, Tokyo had become internationally notorious for its air pollution. Citizen's protest groups rallied, especially in the 1970s and 1980s, forcing the government to regulate pollution.

Since 1970, environmentalists have argued for "environmental rights" (*kankyōken*), basing their claims on the 1947 constitution guaranteeing all Japanese the right to maintain minimum standards of a "wholesome and cultured life." Although not explicitly recognized by any courts, the notion of environmental rights has significant popular support and has been cited repeatedly in litigation. Such claims have possibly persuaded the courts to rule in favor of environmental causes and prompted legislators to enact environmental laws (*kankyōhō*) to protect the environment. Many of these were passed in 1970, during the so-called "pollution Diet". One of the more noteworthy laws endorsed the "Polluter Pays Principle", according to which the party responsible for environmental pollution is held liable for expenses required to redress damage to people and the environment (Upham, 1983: 225, 229).

In 1971, the Environment Agency (*Kankyōchō*), a government agency in the Prime Minister's Office with a director-general of cabinet rank, was established. This agency implements environmental laws, plans and coordinates government measures related to pollution control, determines environmental quality standards, and administers training programs in pollution control. Due to its efforts and those of citizen's groups throughout the nation, environmental studies programs are now part of the curriculum of the Japanese school system, beginning in elementary school and continuing through high school. At every level, students are made aware of the need to preserve the natural environment, how best to use natural resources, and of the close relationship between humanity, industry, and nature. More than traditional sources, environmental education of this sort accounts for the sensibilities of young Japanese regarding nature and the environment (Fujikura, 1983: 224; Kobayashi, 1983: 224).

Japanese environmentalists object not merely to air and water pollution, but to an array of socially objectionable forms such as noise, vibration, ground subsidence, foul odors, and construction that blocks out sunlight (Totman, 2000: 494–498). For example, local protests against construction of new routes for the Shinkansen (bullet train) typically are based on concerns about noise pollution. Heightened sensitivity to postwar abuses of nature has also led to impressive improvements in the quality of air, especially in urban areas. While automobile emissions are as strictly controlled in Japan as anywhere in the world, they remain an exceptionally unpleasant problem, especially in concentrated urban areas where the explosion of automobile traffic in the last thirty years has severely affected the quality of air.

More globally and environmentally concerned Japanese have become acutely concerned about the greenhouse effect and its consequences. Informed Japanese understand that the destruction of the ozone layer, and any rise in global temperatures that might follow, represent an extreme threat for their country, given the fact that so much of its population resides at sea level, and would be at risk if there were any significant rise in the oceans. Given Japan's vulnerability, and its significant responsibility for the problem as a major producer of

CFCs and carbon dioxide, it comes as no surprise that a major international accord related to global warming came out of the 1997 International Conference held in Kyoto.

Concomitant with the race for economic and then ecological revival, Japan has developed a distinct consciousness of itself as a "small" and/or "poor" country. Here, these attributes refer to the country's lack of natural resources and, in particular, the lack of resources necessary for the successful operation of the world's second largest industrialized economy. While Japan is rich in the variety of its resources, sometimes being called a "museum of minerals" (Hall, 1983: 352), the quantity of any particular mineral is relatively limited. Thus, for example, Japan imports 99.8 percent of its petroleum, 88 percent of its coking coal, 100 percent of its uranium, 100 percent of its nickel, 99.4 percent of its iron ore, and 76 percent of its copper. Overall, Japan is 90 percent dependent on imported raw materials. One conspicuous and disturbing consequence of Japan's relative poverty in regard to natural resources has been its enormous development of and increasing dependency on nuclear power plants as a source of inexpensive energy.

Popular culture both anticipated and echoed the movement for environmental rights. Kurosawa Akira's (1910–1998) film *Ikiru* (*To Live*, 1952) suggested that if communities were to realize improvements in their natural surroundings, they would have to organize themselves for action: little help could be expected from bureaucrats whose vision of the future extended only to the edges of the forms they rubber-stamped. Teshigahara Hiroshi's *Suna no onna* (*Woman in the Dunes*, 1964), based on the novel by Abe Kōbō, examined captive industrial workers relegated to harvesting sand, thus suggesting the barren consequences of a highly industrialized Japan, ironically enough, within a natural (although non-Japanese) setting, a seemingly endless desert. Kurosawa's *Yume* (*Dreams*, 1990) graphically criticized Japan's overreliance on nuclear power plants, a development largely of the 1970s, suggesting that an apocalyptic end of nature might be a byproduct of its neglect in favor of a neon culture devoted to consumption of electricity. More recently, Miyazaki Hayao's *Mononoke hime* (*Princess Mononoke*, 1997), an enormous box-office hit, situated the struggle for the environment and the spirits of nature within an animated work of historical fiction, elevating in the process a new hero, Prince Ashitaka of the Emishi peoples, as the leader in the cause of respect for the spiritual environment against the aggressive encroachments of those intent upon its subjugation.

One ongoing struggle between the Japanese government, its efforts to modernize the country, and environmentally-oriented citizen's groups, is centered around Narita airport. From the start, the decision, announced in 1965, to construct a new international airport for Tokyo at Tomisato, near Narita City, prompted violent protests from farmers. Various issues ranging from the imperial system to the Vietnam War affected the controversy, but one of the more popular grounds for support of the anti-airport movement was that of concern for the environment over the interests of economic growth and national prestige. By 1972, local farmers opposing the airport had been removed from the site, but activist groups still staged sabotage operations throughout the 1970s.

While farmers lost in the Narita struggle, they and agrarian interests of the countryside remain strong political forces in modern Japan, despite the increasingly small percentage of the population engaged in agriculture. The Nōkyō kumiai (Union of Agricultural Co-Operatives), boasting membership of virtually every Japanese farmer, is politically powerful due to the disproportionate electoral power held by rural elements. Despite the enormous move away from the countryside, electoral districts have not been radically redrawn. As a consequence, rural voters have an electoral clout that is the equivalent, in some cases, of five urban voters. The long dominant Liberal Democratic Party, rather than reform the system, has made it work for them, courting the farm vote and thus ensuring for themselves the magnified electoral power that has accrued to farmers. The result for consumers has been an inefficient agrarian sector that, due to protection afforded it by the government, produces some of the most expensive crops of rice in the world. While this electoral imbalance has been a continuing feature of postwar Japan, recent changes in the economy, nationally and internationally, suggest that significant reform is in the offing and that the interests of rural areas might finally fall from protectionist grace. What exactly this bodes for the environment is difficult to say, since in many cases postwar farmers have resorted to extensive use of fertilizers to produce higher-yielding crops. Increasingly, ecologically-concerned Japanese are asking for more naturally-grown produce, not so much because of traditional concerns about the sanctity of nature but because of their own concerns about the health of their families.

The agriculturally based Shinto appreciation of nature is "much diminished in modern Japan" (Earhart, 1983: 357–358). Despite this desacralization of nature, "the aesthetic appreciation of nature ... with its roots in the religious celebration of nature is still prominent." This is evident in the reading and writing of poetry and arts such as the tea ceremony (chadō), flower arranging (ikebana), cultivating dwarf trees (bonsai), and gardening as popular hobbies that reflect an alternative aestheticization of nature. Efforts to emphasize the cultural importance of nature, often spearheaded by the Japanese government, have increased markedly in the postwar period. For example in 1950, the government enacted the Cultural Properties Law providing for the designation and protection of "natural monuments" (tennen kinenbutsu), such as national parks, places of historic, scenic, or scientific interest, as well as animals, their habitats, plants, and geologic and mineral formations. These efforts by the government have provided an array of essentially secular, cultural approaches to nature, allowing Japanese to continue what appears to be a relationship with aspects of the natural world that is bound to exist in one form or another and under different ideological principles.

BIBLIOGRAPHY

Adler, Joseph A. 'Response and responsibility: Chou Tun-I and Confucian resources for environmental ethics.' In *Confucianism and Ecology: The Interrelation of Heaven, Earth, and Humans*, Mary Evelyn Tucker and John Berthrong, eds. Cambridge, Massachusetts: Harvard University Press, 1998, pp. 123–150.

Analects. D. C. Lau, trans. *Confucius: The Analects.* New York: Penguin Books, 1979.

Berthrong, John. 'Motifs for a new Confucian ecological vision.' In *Confucianism and Ecology: The Interrelation of Heaven, Earth, and Humans,* Mary Evelyn Tucker and John Berthrong, eds. Cambridge, Massachusetts: Harvard University Press, 1998, pp. 237–264.

Callicott, J. Baird and Roger T. Ames, eds. *Nature in Asian Traditions of Thought: Essays in Environmental Philosophy.* Albany: State University of New York Press, 1989.

Chapple, Christopher K. 'Animals and environment in the Buddhist birth stories.' In *Buddhism and Ecology: The Interconnection of Dharma and Deeds,* Mary Evelyn Tucker and Duncan Ryūken Williams, eds. Cambridge, Massachusetts: Harvard University Press, 1997, pp. 131–148.

Cheng, Chung-ying. 'The trinity of cosmology, ecology, and ethics in the Confucian personhood.' In *Confucianism and Ecology: The Interrelation of Heaven, Earth, and Humans.* Mary Evelyn Tucker and John Berthrong, eds. Cambridge, Massachusetts: Harvard University Press, 1998, pp. 211–236.

De Bary, William Theodore. ' "Think globally, act locally," and the contested ground between.' In *Confucianism and Ecology: The Interrelation of Heaven, Earth, and Humans,* Mary Evelyn Tucker and John Berthrong, eds. Cambridge, Massachusetts: Harvard University Press, 1998, pp. 23–36.

Earhart, H. Byron. 'Nature in Japanese religion.' In *Kodansha Encyclopedia of Japan,* vol. 5. Tokyo: Kodansha, Ltd., 1983, pp. 357–358.

Engi shiki. Felicia Gressitt Bock, trans. *Engi shiki: Procedures of the Engi Era, Books I-V.* Monumenta Nipponica monograph. Tokyo: Sophia University Press, 1970.

Fujikura, Kōichirō. 'Environmental agency.' In *Kodansha Encyclopedia of Japan,* vol. 2, Tokyo: Kodansha, Ltd., 1983, p. 224.

Fukuzawa, Yukichi. *Gakumon no susume.* Tokyo: Iwanami shoten, 1986.

Gotō, Seiko and Julia Ching. 'Confucianism and garden design: a comparison of Koishikawa Kōrakuen and Wörlitzer Park.' In *Confucianism and Ecology: The Interrelation of Heaven, Earth, and Humans,* Mary Evelyn Tucker and John Berthrong, eds. Cambridge, Massachusetts: Harvard University Press, 1998, pp. 275–292.

Habito, Ruben L.F. 'Mountains and rivers and the great earth: Zen and ecology.' In *Buddhism and Ecology: The Interconnection of Dharma and Deeds,* Mary Evelyn Tucker and Duncan Ryūken Williams, eds. Cambridge, Massachusetts: Harvard University Press, 1997, pp. 165–176.

Hall, David and Roger T. Ames. *Thinking Through Confucius.* Albany: State University of New York Press, 1987.

Hall, Robert B. 'Natural resources.' In *Kodansha Encyclopedia of Japan,* vol. 5. Tokyo: Kodansha, Ltd., 1983, pp. 352–355.

Imamichi, Tomonobu. 'Concept of nature.' In *Kodansha Encyclopedia of Japan,* vol. 5. Tokyo: Kodansha, Ltd., 1983, p. 358.

Joly, Jacques. *Le natural selon Andō Shōeki: Un type discours sur la nature et la spontanéité par un maître-confucéen de l'époque Tokugawa: Andō Shōeki (1703–1762).* Paris: Maisonneuve & Larose, 1996.

Kenney, James T. 'Naturalism.' In *Kodansha Encyclopedia of Japan,* vol. 5. Tokyo: Kodansha, Ltd., 1983, p. 351.

Kitabatake, Chikafusa. *Jinnō shōtōki,* H. Paul Varley, trans. *A Chronicle of Gods and Sovereigns: Jinnō shōtōki of Kitabatake Chikafusa.* New York: Columbia University Press, 1980.

Kobayashi, Manabu. 'Environmental education.' In *Kodansha Encyclopedia of Japan,* vol. 2. Tokyo: Kodansha, 1983, p. 224.

Kojiki, Donald L. Philippi, trans. Princeton, New Jersey: Princeton University Press, 1969.

Kraft, Kenneth. 'Nuclear ecology and engaged Buddhism.' In *Buddhism and Ecology: The Interconnection of Dharma and Deeds,* Mary Evelyn Tucker and Duncan Ryūken Williams, eds. Cambridge, Massachusetts: Harvard University Press, 1997, pp. 269–290.

Kumazawa, Banzan. *Daigaku wakumon,* Galen Fisher, trans. *Dai Gaku Wakumon: A Discussion of Public Questions in Light of the Great Learning.* Transactions of the Asiatic Society of Japan. 2nd Series, vol. 16, May, 1938.

LaFleur, William R. 'Saigyō and the Buddhist value of nature.' In *Nature in Asian Traditions of*

Thought: Essays in Environmental Philosophy, J. Baird Callicott and Roger T. Ames, eds. Albany: State University of New York Press, 1989, pp. 183–209.

Man'yōshū, Ian Hideo Levy, trans. *The Ten Thousand Leaves: A Translation of the Man'yōshū, Japan's Premier Anthology of Classical Poetry*, vol. 1. Princeton, New Jersey: Princeton University Press, 1981.

Maruyama, Masao. *Studies in the Intellectual History of Tokugawa Japan*. Mikiso Hane, trans. Princeton, New Jersey: Princeton University Press, 1974.

Miyazawa, Antei. *Nōgyō zensho*. In *Kinsei kagaku shisō*, Furushima Toshio and Aki Kōichi, eds. Nihon shisō taikei, vol. 62. Tokyo: Iwanami shoten, 1972.

Najita, Tetsuo. *Japan: The Intellectual Foundations of Modern Japanese Politics*. Chicago: University of Chicago Press, 1974.

Najita, Tetsuo. *Visions of Virtue in Tokugawa Japan: The Kaitokudō Merchant Academy of Osaka*. Chicago: University of Chicago Press, 1987.

Nihon shoki, W.G. Aston, trans. *Nihongi: Chronicles of Japan from the Earliest Times to A.D. 697*. Rutland, Vermont: Charles Tuttle Company, 1972.

Nosco, Peter. *Remembering Paradise: Nativism and Nostalgia in Eighteenth-Century Japan*. Cambridge, Massachusetts: Harvard University Press, 1990.

Odin, Steve. 'The Japanese concept of nature in relation to the environmental ethics and conservation aesthetics of Aldo Leopold.' In *Buddhism and Ecology: The Interconnection of Dharma and Deeds*, Mary Evelyn Tucker and Duncan Ryūken Williams, eds., Cambridge, Massachusetts: Harvard University Press, 1997, pp. 89–110.

Parkes, Graham. 'Voices of mountains, trees, and rivers: Kūkai, Dōgen, and a deeper ecology.' In *Buddhism and Ecology: The Interconnection of Dharma and Deeds*, Mary Evelyn Tucker and Duncan Ryūken Williams, eds. Cambridge, Massachusetts: Harvard University Press, 1997, pp. 111–128.

Piovesana, Gino K. *Recent Japanese Philosophical Thought, 1862–1996*. 3rd revised edition, including a new survey by Naoshi Yamawaki, 'The philosophical thought of Japan from 1963 to 1996.' Richmond, U.K.: Curzon Press, 1997.

Reischauer, Edwin O. *The Japanese Today: Change and Continuity*. Cambridge, Massachusetts: Harvard University Press, 1988.

Sakamaki, Shunzō. 'Shintō: Japanese ethnocentrism.' In *The Japanese Mind: Essentials of Japanese Philosophy and Culture*, Charles A. Moore, ed. Honolulu: University of Hawai'i Press, 1967, pp. 24–32.

Sakamoto, Yukio. 'On the "attainment of Buddhahood" by trees and plants.' In *Proceedings of the IXth International Congress for the History of Religions*. Tokyo: Maruzen, 1960, pp. 415–422.

Shaner, David Edward. 'The Japanese experience of nature.' In *Nature in Asian Traditions of Thought: Essays in Environmental Philosophy*, J. Baird Callicott and Roger T. Ames, eds. Albany: State University of New York Press, 1989, pp. 163–182.

Shinada, Yutaka. 'Natural monuments and protected species.' In *Kodansha Encyclopedia of Japan*, vol. 5. Tokyo: Kodansha, Ltd., 1983, p. 352.

Sponberg, Alan. 'Green Buddhism and the hierarchy of compassion.' In *Buddhism and Ecology: The Interconnection of Dharma and Deeds*, Mary Evelyn Tucker and Duncan Ryūken Williams, eds. Cambridge, Massachusetts: Harvard University Press, 1997, pp. 351–376.

Tellenbach, Hubertus and Bin Kimura. 'The Japanese concept of "nature".' In *Nature in Asian Traditions of Thought: Essays in Environmental Philosophy*, J. Baird Callicott and Roger T. Ames, eds. Albany: State University of New York Press, 1989, pp. 153–162.

Thomas, Julia Adeney. 'Naturalizing nationhood: ideology and practice in early twentieth-century Japan.' In *Japan's Competing Modernities: Issues in Culture and Democracy, 1900–1930*, Sharon A. Minichiello, ed. Honolulu: University of Hawai'i, 1998, pp. 114–132.

Totman, Conrad. *The Origins of Japan's Modern Forests: The Case of Akita*. Honolulu: Center for Asian and Pacific Studies, University of Hawai'i Press, 1985.

Totman, Conrad. *The Green Archipelago: Forestry in Preindustrial Japan*. Berkeley: University of California Press, 1989.

Totman, Conrad. *Early Modern Japan*. Berkeley: University of California Press, 1993.

Totman, Conrad. *The Lumber Industry in Early Modern Japan.* Honolulu: University of Hawai'i
     Press, 1995.
Totman, Conrad. *A History of Japan.* Oxford: Blackwell Publishers, 2000.
Tsuda, Sōkichi. 'Outlook on nature.' In *An Inquiry into the Japanese Mind as Mirrored in Literature,*
     Matsuda Fukumatsu, trans. Tokyo: Japan Society for the Promotion of Science, 1970.
Tu, Wei-ming. *Centrality and Commonality: An Essay on Chung-Yung.* Honolulu: University of
     Hawai'i Press, 1976.
Tu Wei-Ming. 'The continuity of being: Chinese visions of nature.' In *Nature in Asian Traditions of
     Thought: Essays in Environmental Philosophy,* J. Baird Callicott and Roger T. Ames, eds.
     Albany: State University of New York Press, 1989, pp. 67–78.
Tucker, Mary Evelyn. *Moral and Spiritual Cultivation in Japanese Neo-Confucianism: The Life and
     Thought of Kaibara Ekken, 1630–1714.* Albany: State University of New York Press, 1989.
Tucker, Mary Evelyn. 'The relevance of Chinese Neo-Confucianism for the reverence of nature.'
     Environmental History Review 15 (2): 55–67, 1991.
Tucker, Mary Evelyn and Dunken Ryūken Williams, eds. *Buddhism and Ecology: The
     Interconnection of Dharma and Deeds.* Cambridge, Massachusetts: Harvard University Press,
     1997.
Tucker, Mary Evelyn and John Berthrong, eds. *Confucianism and Ecology: The Interrelation of
     Heaven, Earth, and Humans.* Cambridge, Massachusetts: Harvard University Press, 1998.
Watsuji, Tetsurō. *Fūdo,* Leopold G. Scheidl, trans. *Die Geograhischen Grundlagen des Japanischen
     Wesens.* Tokyo: Kokusai bunka shinkōkai, series B, no. 35, 1937.
Upham, Frank K. 'Environmental law.' In *Kodansha Encyclopedia of Japan,* vol. 2, Tokyo:
     Kodansha, Ltd., 1983, pp. 224–225.
Upham, Frank K. 'Environmental right.' In *Kodansha Encyclopedia of Japan,* vol. 2. Tokyo:
     Kodansha, Ltd., 1983, p. 229.
Weller, Robert P. and Peter K. Bol. 'From heaven-and-earth to nature: Chinese concepts of the
     environment and their influence on policy implementation.' In *Confucianism and Ecology: The
     Interrelation of Heaven, Earth, and Humans,* Mary Evelyn Tucker and John Berthrong, eds.
     Cambridge, Massachusetts: Harvard University Press, 1998, pp. 313–342.
Williams, Duncan Ryūken. 'Animal liberation, death, and the state: rites to release animals in
     medieval Japan.' In *Buddhism and Ecology: The Interconnection of Dharma and Deeds,* Mary
     Evelyn Tucker and Duncan Ryūken Williams, eds. Cambridge, Massachusetts: Harvard
     University Press, 1997, pp. 149–164.
Yamashita, Samuel. H. *Master Sorai's Responsals: An Annotated Translation of Sorai sensei
     tōmonsho.* Honolulu: University of Hawai'i Press, 1994.

ADDITIONAL READING

Perceptions and attitudes

Asquith, Pamela and Arne Kalland, eds. *Images of Japanese Nature: Cultural Perspectives.* London:
     Curzon Press, 1997.
Berque, Augustin. *Japan, Nature, Artifice and Japanese Culture,* Ros Schwartz, trans.
     Northamptonshire, U.K.: Pilkington Books, 1997.
Bruun, Ole and Arne Kalland, eds. *Asian Perceptions of Nature: A Critical Approach.* London:
     Curzon Press, 1995.
Henshall, Kenneth and D. Bing, eds. *Japanese Perceptions of Nature and Natural Order.* Hamilton,
     New Zealand: Centre for Asian Studies, University of Waikato, for the New Zealand Asian
     Studies Society, 1992.
Morris-Suzuki, Tessa. 'Concepts of nature and technology in pre-industrial Japan.' *East Asian
     History* 1: 81–96, 1991.
Ohnuki-Tierney, Emiko. *Rice as Self: Japanese Identities Through Time.* Princeton, New Jersey:
     Princeton University Press, 1993.
Yagi,Yasuyuki. 'Mura-zakai: the Japanese village boundary and its symbolic interpretation.' *Asian
     Folklore Studies* 47(2): 137–151, 1988.

Japanese perceptions of wildlife and animals

Asquith, Pamela. 'The monkey memorial service of Japanese primatologists.' In *Japanese Culture and Behavior*, W.T. Lebra and T.S. Lebra, eds. Honolulu: University of Hawai'i Press, 1986, pp. 29–32.

Kellert, Stephen. 'Japanese perceptions of wildlife.' *Conservation Biology* 5(3): 297–308, 1991.

Kellert, Stephen. 'Attitudes, knowledge, and behaviour towards wildlife among the industrial superpowers: United States, Japan, and Germany.' *Journal of Social Issues* 49(1): 53–69, 1993.

Knight, John. 'On the extinction of the Japanese wolf.' *Asian Folklore Studies* 56(1): 129–159, 1997.

Knight, John. 'Monkeys on the move: the natural symbolism of people-macaque conflict in Japan.' *Journal of Asian Studies* 58(3): 622–647, 1999.

Animals and religion/symbolism

Asquith, Pamela. 'The Japanese idea of soul in animals and objects as evidenced by *kuyo* services.' In *Discovering Japan*, D.J. Daly and T.T. Sekine, eds. Toronto: Captus Press, 1990, pp. 181–188.

Matsuoka, Etsuko. 'The interpretation of fox possession: illness as metaphor.' *Culture, Medicine and Psychiatry* 15(4): 453–477, 1991.

Naumann, Nelly. 'Whale and fish cult in Japan: a basic feature of Ebisu worship.' *Asian Folklore Studies* 33(1): 1–15, 1974.

Ohnuki-Tierney, Emiko. *The Monkey as Mirror. Symbolic Transformations in Japanese History and Ritual.* Princeton, New Jersey: Princeton University Press, 1987.

Smyers, Karen A.. *The Fox and the Jewel. Shared and Private Meanings in Contemporary Japanese Inari Worship.* Honolulu: University of Hawai'i Press, 1999.

Environmental issues

McKean, Margaret. *Environmental Protest and Citizen Politics in Japan.* Berkeley: University of California Press, 1981.

Colligan-Taylor, Karen. *The Emergence of Environmental Literature in Japan.* New York: Garland Press, 1990.

Geomancy

Brock Johnson, Norris. 'Geomancy, sacred geometry, and the idea of a garden: Tenryu-ji temple, Kyoto, Japan.' *Journal of Garden History* 9(1): 1–19, 1989.

Kalland, Arne. 'Geomancy and town planning in a Japanese community.' *Ethnology* 35(1): 17–32, 1996.

Kalland, Arne. 'Houses, people and good fortune: Geomancy and vernacular architecture in Japan.' *Worldviews: Environment, Culture, Religion* 3: 33–50, 1999.

GRAHAM PARKES

WINDS, WATERS, AND EARTH ENERGIES:
FENGSHUI AND SENSE OF PLACE

In China, and now throughout the world, the question of *qi* and the correlative issue of *fengshui* have been encumbered with a mishmash of obtuse superstitions dictated by the most retrograde kind of charlatanism. But this does not prevent this question, if one disengages it from these accretions, from bearing vitally on the reality of being human – and indeed at the very foundations of human existence, including our own, however non-Chinese we may think ourselves.

Augustin Berque

When one reflects on the problems that beset nature and the environment at the beginning of the 21st century, it is clear that China, with its huge population and ongoing modernization and industrialization, is going to be one of the major contributors to those problems as well as an important factor in the search for solutions. What is striking about most of the current discussions of environmental questions is just how parochial are the terms in which they are conducted – presupposing a Cartesian-Newtonian view of the natural world as a mass of "dead matter in motion", deriving from natural-scientific discourse that arose in western Europe during the seventeenth century. It is worth recalling that most human beings throughout most of human history have understood the natural world in a variety of quite different ways from this.

Numerous among those human beings are the Chinese, who have pursued sophisticated scientific investigations into the natural world for millennia without its ever occurring to them that they might be investigating anything like "dead matter". But the enormous efficacy of the Cartesian-Newtonian worldview, especially in having made possible the wonders of modern technology, is now so predominant that many regard it as *true*. Without denying its efficacy in letting us manipulate the natural world toward our desired ends, we do well to consider the extent to which its prevalence encourages environmental degradation, and to entertain as plausible – and experientially accessible – a very different view deriving from the classical Chinese tradition.

Along with acupuncture, the best-known representative of Chinese science today is *fengshui* (風水). The end of the twentieth century saw an explosion of publications on the topic: a search in the Harvard libraries database for titles

185

H. Selin (ed.), Nature Across Cultures: Views of Nature and the Environment in Non-Western Cultures, 185–209.
© *2003 Kluwer Academic Publishers. Printed in Great Britain.*

containing the words "feng shui" in August 2001 yielded 72 titles, while a similar search at Amazon.com came up with 368, all published in English and most since 1990. The boom, then, has been on the popular level rather than the scholarly. A corresponding internet search yielded the addresses of an astounding "around 252,000" websites. (A slight consolation was that the very first listing happened to be for a site describing itself as "dedicated since 1995 to helping Feng Shui shed its snake-oil-and-incense image.")

A cursory perusal of this mass of resources suggests that a large number of people in the United States and Western Europe are paying fengshui "experts" large sums of money to align their expensive coffee tables with their even more expensive sofas, in the hope of bringing more wealth, and perhaps some happiness, into their already affluent households. This seems a gross perversion of the basic spirit of fengshui – which would say that happiness, and certainly some wealth, would come more easily if these people simply sold off all the furniture and other clutter that's obstructing their contact with their natural surroundings. In view of the obscurity of its origins, it is unclear how much mystification and charlatanry entered into the practice of fengshui during its early development, but I believe we can draw from its more commonsensical aspects some pointers for deepening our understanding of our relations to nature and the environment.

Fengshui is often translated as "geomancy", though the two graphs that comprise the term simply mean "wind" and "water" respectively. (The practice was originally known under the more formal name of kanyu (堪輿), meaning "canopy [of Heaven] chariot [of Earth]"). Let us ignore the mantic and divinatory aspects of the practice as distracting from its more down-to-earth applications, as well as the panoply of arcane symbolism of colors and animals and planets and stars, which the uninitiated find mystifying – and which perhaps accounts by the same token for fengshui's surge of popularity in the New Age. Much of the exoticism that so fascinates homeowners in California these days does not transfer well at all. The association of the east with the Azure Dragon and the White Tiger with the west works in China because of the directions of the Yellow Sea and Tibetan Plateau respectively. Transposed to Southern California, this schema would have the White Tiger floating in the Pacific Ocean and the Azure Dragon shivering in the High Sierra.

Astronomy and astrology do play a role in fengshui, and there is no denying that since the planets and stars that slowly circle round above our heads are part of our natural environment at every moment, they may indeed have an influence on our activities. But this is so vast a topic that it is only practical to restrict the notion of place to the sublunary realm and narrow our focus to those aspects of fengshui that might enhance our relations to this more down-to-earth sense of place.

The ambiguity of fengshui – as between a practice encouraging charlatanry based on mystification and a set of sensible recommendations grounded in sensitivity to the natural environment – is reflected in an ambivalence pervading the history of its reception in the West.

EARLY RECEPTION

The first European-language monograph devoted to fengshui is by Ernest John Eitel, a German pastor who heard "a call from the Lord to preach the gospel to the Heathen" and went to Guangdong province in China in 1861 with the Evangelical Missionary Society of Basel. Unlike many of his colleagues there, Eitel quickly became fluent in Chinese and developed great enthusiasm for the culture. After several years he transferred to the London Missionary Society, married an Englishwoman, and assumed British nationality. He was dispatched to Hong Kong in 1870, and the same year he received a doctorate from the University of Tübingen for his research on Chinese Buddhism. He gave a series of lectures on fengshui at Hong Kong City Hall, the revised texts of which were published under the title *Fengshui: The Science of Sacred Landscape in Old China* (1873). In this work, as well as in his writings on Chinese Buddhism, the author proceeds from the premise of the revealed truth of Christianity (Wong, 2000: 74–5).

Eitel begins by warning the reader that the Chinese have "made Feng-shui a black art" on a par with the arts of the astrologers and alchemists of Medieval Europe. "Practically speaking it is simply a system of superstition, supposed to teach people where and when to build a tomb or to erect a house so as to insure for those concerned everlasting prosperity and happiness," while on another level fengshui is "but another name for natural science." In the same sentence as the empiricist in Eitel deplores the "absence of practical and experimental investigation," the theologian in him admires in Chinese natural science its "spirit of sacred reverence for the divine powers of nature." While on the one hand he regards China as "childishly ignorant as regards matters of intellect," he also expresses the fervent wish that "our own men of science had preserved ... that sacred awe and trembling fear of the mysteries of the unseen ... which characterize these Chinese gropings after natural science" (Eitel, 1973: 4–7).

In spite of his ambivalence, Eitel offers an even-handed chapter on astrology which mentions the imagery associated with the four directions: the Azure Dragon with the east, the Sable Warrior (or black tortoise-and-snake) with the north, the White Tiger with the west, and the Vermilion Bird with the south. He also connects the Five Planets (Jupiter, Mars, Saturn, Venus, Mercury) with *wuxing* (五行), the Five Processes (wood, fire, soil, metal, water), and the dragon and tiger with male and female energies in the earth's crust, which he likens to positive and negative magnetic currents.[1] The most favorable site will be where the dragon and tiger energies come together (ch. 2). A chapter on numerology relates the fengshui compass to the system of divination developed in the *Yijing* (*I Ching*, or *Book of Changes*, ch. 3).

Eitel then discusses the notion of *qi* (氣) as "the breath of nature" in its alternation between expanding and reverting. "Between heaven and earth there is nothing so important, so almighty and omnipresent as this breath of nature. It enters into every stem and fibre, and through it heaven and earth and every creature live and move and have their being." There are two basic principles of siting according to fengshui: exposure to wind (*feng*) results in the breath's being dissipated from a site, while if water (*shui*) runs away from the site in a straight and rapid (rather than a slow and meandering) course, it will similarly deplete the local breath (ch. 4).

Eitel lays appropriate emphasis on fengshui's concern with the reciprocity between the human being and natural environment: "It is the boast of the Feng-shui system that it teaches man how to rule nature and his own destiny by showing him how heaven and earth rule him" (ch. 5). The chapter on "the history and literature" is disappointingly slight, perhaps reflecting the limited array of materials accessible to the author, who sums the subject up as "a strange medley of superstition, ignorance and philosophy" (ch. 6). A pejorative tone also pervades the concluding chapter, where fengshui is dismissed as a "farrago of nonsense and childish absurdities" and "the blind gropings of the Chinese mind after a system of natural science." In view of the open-minded manner in which he presents some of the more sensible features of fengshui, one suspects that the good Reverend may have protested too much at the end in order to avoid discomfiting his superiors in the London Missionary Society.

A contemporary of Eitel's from England, Edwin Joshua Dukes, wrote a book called *Everyday Life in China* that was published in London by the Religious Tract Society in 1885. The conclusion of the chapter entitled "Feng-Shui, the Biggest of all Bugbears" condemns the practice for obstructing the dissemination in China of Christianity, trade, and empirical science (Dukes, 1885: 159). In discussing the way fengshui involves the whole community (including its dead) in the planning of any new building project, the author waxes so ironical that his attempt to condemn unintentionally commends what may now be seen as its major benefits with respect to the environment.

> It will suggest itself at once to the reader that if we ignorant European outsiders were to live where we choose in China, to build as we like, to make roads and railways, to erect telegraph posts, to quarry stone wherever we saw any to our fancy, to delve recklessly into the bowels of the earth for coal, we should, in the opinion of the Chinese, be like "a maniac scattering dust" and "a fury slinging flame." We should put steeples to our churches and tall chimneys to our factories, and in doing so commit the unpardonable crime of upsetting the serenity of the spirit-world. No vengeance would be too dire to execute upon the rash mortal who could disregard the interest of his fellow-creatures in such a manner (Dukes, 1885: 151–52).

The intended irony aside, it is hard to imagine a more passionately ecological exhortation to take into account the natural surroundings and neighbors, living and dead, when deciding where and how to encroach upon the environment.

The next substantial treatment of fengshui to appear in a Western language, by J. J. M. de Groot in a chapter of the third volume of his monumental work, *The Religious System of China*, deprecates its subject throughout its 120 (large) pages. While containing more historical detail than Eitel's account, the chapter's focus is narrower: on "the part it plays in grave-building" (De Groot, 1897: 937). The author begins by characterizing fengshui as "a quasi-scientific system, supposed to teach men where and how to build graves, temples and dwellings, in order that the dead, the gods and the living may be located therein exclusively, or as far as possible, under the auspicious influences of nature" (935).

Two pages later, however, we learn this about the practice and the culture that produced it:

> Fung-shui is a mere chaos of childish absurdities and refined mysticism, cemented together, by sophistic reasonings, into a system, which is in reality a ridiculous caricature of science ... It fully shows the dense cloud of ignorance which hovers over the whole Chinese people; it exhibits in all its nakedness the low condition of their mental culture, the fact that natural philosophy in that part of the globe is a huge mount of learning without a single trace of true knowledge in it (937–38).

One will not expect to learn much from a mind as narrow as De Groot's, vast though his scholarship may be, and aside from a wealth of historical detail there is little here that was not presented in Eitel's much more concise account.

By the end of the chapter De Groot has worked himself up into paroxysms of outrage at the pernicious effects of fengshui practice: "At the outset a benumbed viper, it has, carefully fostered by the nation, developed into a horrid hydra suffocating the whole Empire in its coils and deluging it with its venom throughout its length and breadth" (1048). A far cry indeed from the beneficent Azure Dragon that Eitel sincerely strove to understand some twenty-five years earlier. And when De Groot concludes by wondering whether "foreigners may be able to shed some rays of the light of science upon the Middle Kingdom" or whether its inhabitants may be "stamped for ever with the total incapacity to rise to a higher level of mental culture," his answer is a pessimistic "no" and "yes" (1055–56).[2]

By the time De Groot delivers the American Lectures on the History of Religions in 1911, his contempt seems to have diminished somewhat, though his animosity toward unscrupulous fengshui practitioners still blinds him to the value of some of the underlying philosophy. He writes, for example, of the "philosophical nonsense of the [five] elements" (De Groot, 1912: 300). The grounds for his animosity turn out to be the way fengshui is "an obstacle to all sorts of enterprise which might be of the greatest advantage to the people: the cutting of a new road or canal, the construction of a new bridge, a railroad, tramway, or telegraph line" (316). Seen in the light of the subsequent environmental devastation wrought in the name of technological progress, an obstacle that makes us pause to take into account the features of the landscape may not necessarily be a bad thing. It was in the following year that a professorial chair for Sinology was first established at the University of Berlin, with the querulous De Groot as its first occupant.

We learn from De Groot that the predominance of ancestor reverence in ancient China, based on the belief that the spirits of the dead continue to inhabit this world in which they resided bodily, gave rise to the idea that the site and orientation of the tombs in which the dead were buried would determine the quality of their influence on their living descendants. Proper placement of the houses of the dead ensures that the qi emanating from the ancestors will enhance the energies of the living descendants and bring good fortune. Even though the living reside on the earth for far shorter periods than the corpses of the dead reside in it, it is natural to suppose that the site and orientation of our residences while alive will similarly affect the quality of our lives. For the living, then, fengshui can be said to concern "the relations to the surrounding nature, the influence of the landscape on the beauty of the buildings and the happiness of the inhabitants" (Boerschmann, 1924: viii).

These are the words of a contemporary of De Groot's, Ernst Boerschmann, whose writings evince a refreshingly different attitude toward the Chinese level of mental culture. Boerschmann went to China in 1906 for a three-year visit sponsored by the German Imperial Government, for the purpose of "an investigation of Chinese architecture and its relation to Chinese culture" (Boerschmann, 1912: 539). He is fascinated and impressed by almost all manifestations of Chinese culture he encounters in the course of his extensive travels.

> One imposing conception of the universe is the mainspring of all Chinamen, a conception so comprehensive that it is the key defining all expressions of life ... especially fine arts and architecture. They exhibit in nearly every work of art the universe and its idea. The visible forms are the reflex of the divine ... In the microcosm is recognized and revealed the macrocosm (542).

It was right at this time that De Groot published a study of this central idea in Chinese culture, according to which the macrocosm is not only represented or revealed in the microcosm, but can also be regarded as actually present in certain special or ritually defined areas. He calls this idea "universism" (De Groot, 1912).

Boerschmann was especially impressed by the aesthetic effects of fengshui practice, as evidenced by the ways the architecture is integrated into the landscape.

> The large cities and almost all others are located in most clever concord with the natural conditions to combine most advantageously the industrial interests with the most beautiful environment possible. The manner in which the Chinese artistically build their structures to harmonize with the natural environment is astonishing (Boerschmann, 1912: 572).

This admiration does not derive from a mere aestheticism, but from a view that seeks to balance "industrial interests with the most beautiful environment possible" – a view that seems to have been sadly obscured in the current rush (which builds on momentum initiated with the Revolution of 1949) to modernize the Middle Kingdom.[3]

Boerschmann also took numerous photographs of landscapes, buildings, and statuary, which he published in the magnificent volume, *Picturesque China: Architecture and Landscape* (1924). In his introductory essay he ascribes the harmony between buildings and landscape in China to the influence of fengshui, which consistently favors curved over straight lines. This is because straight lines are thought to be conduits of noxious winds and energies, *fengsha* (風煞) and *shaqi* (煞氣) (Feuchtwang, 1974: 115). The result is

> ... that feeling of restful comfort and harmony of our soul [that] arises at the sight of Chinese buildings. For we not only enjoy the unity of the extensive edifices and grounds with the immediate surroundings and nature, with which we feel ourselves a part in the picture of the buildings and the landscape. We also feel that the buildings themselves, nay, even their ornaments must somehow be imbued with nature's living spirit for them to evoke this mood of consummate peace (xiii).

To this no longer so vital spirit Boerschmann's photographs are a most handsome tribute.

It was not until the 1950s that Western scholars began serious attempts at a comprehensive understanding of Chinese science, as signaled in particular by the first volumes of Joseph Needham's monumental work *Science and Civilisation in China*. Needham devotes large sections of his second volume to the philosophies of nature found in Daoism and "the school of Naturalists" (the so-called *Yin-Yang* thinkers), the theory of *wuxing* (the five elements or five processes), symbolic correlations and correlative thinking, and the system of the *Yijing*.

We have here for the first time a comprehensive, scholarly, and open-minded account of the philosophical and scientific background from which fengshui emerged. The practice of "geomancy" itself, however, Needham relegates to a chapter on "the Pseudo-Sciences", considering it under the heading "Divination" along with such practices as scapulimancy (prediction of good or bad fortune on the basis of the behavior of ox and deer shoulder-blades when subjected to heat), astrology, chronomancy (divination to determine a favorable time for action), cheiromancy (palmistry), and the like. Operating on the straightforward definition of fengshui as "the art of adapting the residences of the living and the dead so as to cooperate and harmonise with the local currents of the cosmic breath [*qi*]," he devotes four pages to a brief history and description of the practice (Needham, 1956: 2: 359–63).[4] After remarking on its advantages and drawbacks, he concludes that "all through, [fengshui] embodied a marked aesthetic component, which accounts for the great beauty of the siting of so many farms, houses and villages throughout China" (2: 361).

It will help at this point to broaden our focus and consider some basic features of the Chinese worldviews from which fengshui arose. The major background assumption that needs to be highlighted is the understanding of the world as a dynamic play of forces, or energies, rather than an aggregate of material things, with a corresponding emphasis on "becoming" over "being". In other words, China has always inclined toward "process" rather than "substance" cosmologies.

COSMOLOGIES OF *QI*

The philosophers of the classical period in China did not begin to develop cosmologies until around the middle of the third century BCE, before which time cosmological speculation was the business of astronomers, diviners, physicians, and others at the imperial court.[5] Philosophical Daoist cosmology then sets the direction for two millennia of subsequent Chinese thought in understanding the cosmos as a field of energies known as *qi*. In a later chapter of the Daoist classic *Zhuangzi* we read: "Man's life is the assembling of *ch'i*. The assembling is deemed birth, the dispersal is deemed death ... Running through the whole world there is nothing but the one *ch'i*" (ch. 22).[6] Since breathing is a process that distinguishes the living from the dead, it was natural to think of the breath as a manifestation of the energy that animates the cosmos.

Corresponding to inhalation and exhalation are two forms of energy: "All things can now be conceived as condensing out of and dissolving into a universal *ch'i* which as Yang (陽) is pure and so free moving and active, and as Yin (陰) is impure and so inert and passive" (Graham, 1989: 328).

A *locus classicus* for this idea is the beginning of the third chapter of the syncretic Daoist text known as the *Huainanzi* (mid-second century BCE):

> The Dao began in the Nebulous Void.
> The Nebulous Void produced spacetime;
> Spacetime produced the primordial *qi*.
> A shoreline (divided) the primordial *qi*.
> That which was pure and bright spread out to form Heaven;
> The heavy and turbid congealed to form Earth ...
> The conjoined essences of Heaven and Earth produced yin and yang.
> The supercessive essences of yin and yang caused the four seasons.
> The scattered essences of the four seasons created all things.
>
> (3: 1a: 1, in Major, 1993: 62)[7]

Here qi is characterized as the source of all things, the variety among them depending on where they lie on the spectrum from the most rarefied ("pure and bright") to the most condensed ("heavy and turbid") forms of energy. A remarkably similar idea seems to have arisen independently in ancient Greek cosmology, and especially in the thought of Anaximines, for whom the "underlying nature is one and infinite [and identified] as air." In particular he writes of condensation (*puknotēs*) and rarefaction (*manotēs*) as the two basic transformations of this one "nature".

> It differs in its substantial nature by rarity and density. Being made finer it becomes fire, being made thicker it becomes wind, then cloud, then (when thickened still more) water, then earth, then stones; and the rest come into being from these ...
>
> It is always in motion: for things that change do not change unless there be movement ... The most influential components of generation are opposites, hot and cold (Kirk and Raven, 1963: 144–45).

Like the *Huainanzi*, the Daoist compilation known as the *Guanzi*, which contains passages dating from around the third century BCE and earlier, includes numerous passages concerning cosmological ideas relevant to our topic. They deal with themes such as the *wuxing*, conduct appropriate to the four seasons, and the parallels between the microcosm of the human body and the macrocosm of the landscape. In keeping with the Daoist idea that pure qi is transformed first into wind, or breath, in the realm of Heaven and into water in the realm of Earth, a chapter of the *Guanzi* begins: "Water is the original source of the myriad things, the root of all that lives, that from which beautiful and ugly, worthy and unworthy, foolish and eminent are born. Water is the blood and *qi* of Earth, like that which courses through the muscles and veins."[8] Of interest here is the naturalness with which aesthetic and moral qualities are said to derive from the natural phenomenon of water, the allusion to circulation of the blood and oxygenation of the muscles, and the correspondence between channels of energy in the body and the earth. There is an ancient parallel in Greek thought, again according to Anaximines: "As our soul, being air holds us together and controls us, so does wind [or breath] and air enclose the whole world" (Kirk and Raven, 1963: 158).

At one level the emphasis on isomorphism between human body and landscape comes simply from an appreciation of environmental influences. The chapter from the *Guanzi* just quoted ends by reiterating water as "the source of the myriad things" and then discussing the influence of water quality in the seven major states on the constitutions and characters of each area's inhabitants. "The water of Qi is forceful, swift, and twisting. Therefore its people are greedy, uncouth, and warlike. The water of Chu is gentle, yielding, and pure. Therefore its people are lighthearted, resolute, and sure of themselves" (*Guanzi*, ch. 39). The conclusion is that the sage who would transform the world has to understand the influences of bodies and flows of water on the human body.

A similar discussion of environmental influences, with an explicit focus on the importance of the local qi, is to be found in the *Huainanzi*.

> Various sorts of earth give birth, each according to its own kind.
> For this reason, the *qi* of the mountains gives birth to a preponderance of men;
> The *qi* of the low wetlands gives birth to a preponderance of women ...
> The *qi* of stone produces much strength.
> The *qi* of steep passes produces many cases of goiter ...
> All things are the same as their *qi;* all things respond to their own class.
>
> (4: 7a: 1, in Major, 1993: 167)

The list of topographical qi influences runs to fifteen items. Politically speaking (and the political is usually present in classical Daoism) the ruler will be better able to govern, like the corresponding sage in the *Guanzi*, if he understands the kinds of environmental influences operating on the constitutions of his subjects. Physiologically speaking, one is advised to pay attention to the energetic background or context of all beings with whom one comes into contact. The point is made with reference to the three facets of the individual in the *Huainanzi*'s first chapter:

> Thus, if one's physical form is placed in an unsuitable abode, it will become incapacitated;
> If one's *qi* is made to fill what it does not rightfully fill, it will be leaked;
> If one's spirit is active when it is not suited to act, it will grow dim.
> These three are what one should watch over carefully (Lau and Ames, 1998: 131–32).

Among the ten-thousand things traditionally said to have been produced by yin and yang after they separate out from the primal unity, there are five in particular that one should watch over carefully.

WUXING: THE FIVE PROCESSES

As mentioned earlier, Needham provides a wealth of information concerning early Chinese science and cosmology, but in this section I draw instead from the more recent work of A. C. Graham, because of its more philosophical orientation. A fourth-century BCE historical text known as the *Zuo Commentary* (to the *Annals* of Lu) lists six kinds of atmospheric influences or energies: "Heaven has the Six *Ch'i* ... shade [*yin*] and sunshine [*yang*], wind and rain, dark and light." Here yin and yang refer in their pre-philosophical use to the shady and sunny sides of a hill respectively (north and south sides in China, as in any place in the northern hemisphere), and thus also have to do with cold and heat as well as dark and light.

Corresponding to the Six Ch'i of Heaven are the Five Processes (*wuxing*: literally, "five goings," "transitions," "conducts," or "doings") associated with Earth: wood (木), fire (火), soil (土), metal (金), and water (水). The three terrestrial elements of the ancient Greek conception of the four elements – earth, water, and fire – are significantly different from their counterparts in China. Rather than referring to static elements that form the building blocks of the world, wuxing denotes the five primary phases of transformation through which the cosmic or telluric energies pass (Needham, 1956: 232–61; Wang, 2000: 3). An important early mention is in the *Hongfan* (*Great Plan*) chapter of the pre-Qin Dynasty *Shujing* (*Historical Classic/Book of Documents*), where the Five Processes appear at the beginning of a list of nine groups of things necessary for good government, immediately followed by the "five things to do":

> 1. The five processes: (1) Water: wetting, sinking. (2) Fire: flaming, rising. (3) Wood: bending, straight. (4) Metal: yielding to moulding. (5) Soil: permitting sowing and reaping.
>
> 2. The five things to do: (1) Demeanour: respectful. (2) Speech: accordant. (3) Looking: seeing clearly. (4) Listening: hearing clearly. (5) Thinking: understanding.[9]

Rather than enumerating abstract names of classes or properties of things, these items refer to actual, perceptible processes with dual aspects: not just wood, but wood as bending or straight; not just looking, or clear sight, but seeing clearly. There is also the Daoist idea here, reminiscent of the *Laozi*, that good governing emulates the way of nature (*tiandao:* 天道): the five natural processes are presented as basic and the five things for humans to do as secondary.

The Five Processes generate each other (*xian sheng* 相生) in cyclical sequence:

Wood generates fire, bursting into flame.
Fire generates soil, reducing wood etc. to ash.
Soil gives birth to metal, in veins of ore beneath the earth.
Metal gives birth to water, liquefying when heated.[10]
Water generates wood, nourishing the growth of plants.

Any natural environment is constituted by the ceaseless self-generation of these five phases, and insofar as we develop a sense for how each gives rise to the next, we stand to benefit, in our concern to shelter and nourish ourselves throughout our own transformations, from harmonizing these with the perpetual changes of the world.

This generating cycle of the Five Processes is derivable from, and thus correlated with, the four seasons and the four directions:

Wood	**Fire**	**Soil**	**Metal**	**Water**
Spring	Summer		Autumn	Winter
East	South	(Center)	West	North

Soil, as the most basic of the processes, is assigned the central place, or "here" position, among the four directions. Fitting it into the cycle of the four seasons is a problem, the most elegant solution to which, in view of soil's intermediary as well as central function, has been to assign it to the third lunar month of each season (Schipper, 1993: 35).

For inhabitants of the northern hemisphere, the correlation of south-north (the Chinese compass puts south at the top and north below) with summer-winter makes sense in view of the movement of the sun's position through the seasons. Since many parts of China receive more precipitation during the summer monsoon than other seasons, the correlation of these pairs with fire-water is based rather on early Chinese cosmogonies that remark the tendency of water, as yin to fire's yang, to withdraw into the cold and dark. If the cycle of the seasons begins in spring, as the day begins with the sun's rising in the east, it is as natural to correlate these as to pair autumn with the west. The correlation of the wood-metal pair with spring-autumn works insofar as "branches and leaves grow in spring and turn brittle, rigid, metallic in autumn" (Graham, 1989: 344). And at least for places nearer the center than the periphery of the Middle Kingdoms, wood is found in the forests of the east and metals in the mountains to the west.

A second cycle involving the Five Processes, in which they "conquer" or "overcome" one another (*xiang ke:* 相克), was recognized on a different basis from the directions and seasons:

Soil conquers water, by damming or absorbing it.
Wood overcomes soil, by digging with spade and plow.
Metal conquers wood, by cutting with the blade.
Fire overcomes metal, by melting it to liquid.
Water conquers fire, by extinguishing.

By contrast with the purely natural generating cycle, this one reflects the practices of hydraulic engineering and irrigation, agriculture, carpentry, and metallurgy – all activities that intervene in natural processes in such as way as to direct them toward human purposes. The Greek idea of the four elements as the essential components of a material world goes along with an understanding of gods and humans as creators and makers of things through the introduction of formative powers from outside and with a view to a pattern or paradigm external to those things (even if internal to the mind of the creator). The notion of the Five Processes, however, is part of a cosmology in which ceaseless transformations are driven by the cosmic energy that flows through the particulars according to patterns that emerge from the place of those particulars within the larger matrix. In such a world human beings thrive by creatively engaging these transformations in the appropriate ways – ways that it is the aim of Chinese medicine and fengshui practice to articulate. In any particular place, for example, one will find manifestations of several or all of the Five Processes, and the quality of the different parts of the place will depend on whether the processes are in more creative or destructive relationships.

Thanks to the general Chinese predilection for correlative thinking, wood-fire-earth-metal-water were soon correlated with a large number of other phenomena: with the five viscera (Spleen, Lungs, Heart, Kidneys, Liver), the five constituents of the body (muscles/membranes, skin, pulse/blood, bone/marrow, flesh), the five tastes (sour, bitter, sweet, acrid, salt), the five kinds of creatures (scaly, feathered, naked, hairy, shelled), the five measures (compasses, weights, plumblines, T-squares, balances), the five colors (blue-green, yellow, red, white, black), the five notes of the pentatonic scale (*que, zhi, gong, shang, yu*), the five numbers (8, 7, 5, 9, 6), and so forth.[11] In the political realm the conquering cycle was seen to be reflected in the succession of historical dynasties: the Yellow Emperor, the Xia, the Shang, the Zhou, and the dynasty to come. Since the correlations may appear to become rather tenuous when one goes beyond those with the seasons and directions, let us consider a fact that appears to set the basis of the major ones on relatively firm ground.

Graham points out the remarkable correlation between the generating and conquering cycles: "In the one required to correlate with the seasons and directions each Process is generating the immediate predecessor of the Process it conquers." This is remarkable because the conquering cycle is independent of other correlations and presumably stems rather from hands-on experience in working (with) the Five Processes. He also shows how this correlation can be subsumed under a larger pattern from the cosmology of the *Huainanzi*, and also how, when the Five Numbers are added to the correlation between the generating and conquering cycles, a "magic square" is produced in which the numbers add up to 15 in every direction (Graham, 1989: 344).[12]

Just as the Five Processes were from the outset correlated with "the five things to do", so they are also naturally applicable to energetic transformations within the microcosm of the human body.[13]

EARTH AND SYMBOLIC BODY

It is common for Daoism to extend the correlation between the body and the landscape to the country understood as the state. The *Huangdi neijing (Inner Classic of the Yellow Emperor)*, the oldest surviving Chinese medical text (from around the 3rd century BCE), has this to say about the various parts of the body: "The heart functions as the prince and governs through the *shen* ["soul"]; the lungs are liaison officers who promulgate rules and regulations; the liver is a general and devises strategies" (ch. 8, cited in Schipper, 1993: 100). Kristofer Schipper has shown how the theme of "inner landscape" runs through the entire Daoist tradition, beginning with commentaries on the *Laozi* that emphasize the parallels between the body politic and the individual's body: "The Sage's rule over the country corresponds to the rule over the body" (Schipper, 1993: 191).[14] Again the macrocosm provides patterns for microcosms.

Numerous works and commentaries in the Daoist Canon provide detailed accounts of the inner landscape in terms of ancient Chinese mythology and yin-yang and wuxing cosmology, accounts based on Daoist techniques of meditation and introspection. Although they hardly form a system, Schipper observes that, "the fundamental themes recur with surprising regularity throughout all descriptions of the inner landscape." The first part of his paraphrase of the accounts in the second-century *Laozi zhongjing* (*Laozi's Classic of the Center*) gives an idea of how rich and expansive these inner landscapes are:

> The landscape of the head consists of a high mountain, or rather a series of peaks around a central lake. The lake lies midway between the back of the skull and the point between the eyebrows. In the middle of the lake stands a palatial building, where there are eight rooms surrounding a ninth, central one. This is the Hall of Light (*mingtang*), the house of the calendar of the kings of ancient China [built in the form of a mandala of the universe]. In front of this palace and the lake around it, lies a valley (the nose). The entrance to the valley is guarded by two towers (the ears) (Schipper, 1993: 105–06).

These correlations continue through the rest of the head area with its valleys and lakes and streams and fountains; the thorax with its sun and moon, Pole Star and Big Dipper, Scarlet Palace, Yellow Court, Purple Chamber, and granary or warehouse; the abdomen with its great ocean, inverted mountain, and the Cinnabar Field – *dantian* (丹田), root of the human being and origin of the wuxing.

Few have patience or time for the practice required to discover such vast and rich inscapes, and if those who do have not undertaken Daoist practice, they are unlikely to discover such structures as the Hall of Light – but perhaps rather their non-Chinese equivalents. At any rate, to entertain in general the idea of an "imaginal" space or places within the body of the imagination, which is not so foreign to the Western traditions, would help us situate our activities more fruitfully in the natural environment insofar as our perception of it is conditioned by projections from the landscapes within.[15]

Attention to the correlations between yin and yang qi, the Five Processes, and the five viscera reminds us of our ongoing embodiment in the current physical circumstances prevailing in a particular place in the world. As the seasons proceed, the cosmic qi alternates between yang and yin: yang qi is in the ascendant from the beginning of spring and reaches its highest point in midsummer; it then diminishes as the yin qi begins to increase, which peaks at the winter solstice. It will make sense to try to harmonize the currents of qi in one's body with the larger ebbs and flows outside, and the *Inner Classic* recommends precisely this, in conjunction with four of the five viscera from the wuxing system. (The spleen, in the central position, is omitted.) A. C. Graham suggests helpfully that seasonal activities connected with crops provide a metonymic transfer to the qi associated with the corresponding viscera:

Four Seasons	Spring	Summer	Autumn	Winter
Qi	Coming to life	Growing up	Gathering in	Storing away
Five Viscera	Liver	Heart	Lungs	Kidneys

For each season, and the kind of qi that predominates in it, the text encourages certain activities on the part of the individual that will keep him in tune with the tendency (for example, "coming to life") of the natural forces of that season.

Each prescription is followed by a proscription of going against the qi of the season, on pain of harming the respective internal organ, which will in turn inhibit the tendency ("growing up", both of crops and processes internal to the body) of the upcoming season. Graham's summation of the import of the entire discourse cannot be improved on:

> The measures good for one's health ... are the measures one is moved to take when one understands how the seasons act on the body ... Man is in spontaneous interaction with things, but responds differently according to the degree of his understanding of their similarities and contrasts, connexion or isolation ... To know how things compare and connect, in particular whether in connecting they support or conflict with each other ... is to know their patterns (*li*) and the Way which unites them all (Graham, 1989: 355–56).

It is on such grounds that a fuller awareness of the relations between our own energies and those of our physical environment will be conducive to our flourishing.

Corresponding to these seasonal prescriptions and proscriptions concerning the individual's health are chapters in several classic texts – notably the calendrical chapters of the *Lu Spring and Autumn* and the "Four Seasons" chapter of the *Guanzi* – which advise the ruler how best to govern according to the season. The way *not* to do this (and this applies to the ruler of the individual body as well as to the ruler of the state) is by having in mind a fixed image of one's goals, to be pursued in resolute disregard of environment or season. This would correspond to a flouting of fengshui principles by cutting through the landscape in straight lines rather than following the curvilinear transformations of the dragon.

BRIEF HISTORY OF FENGSHUI

The historical beginnings of fengshui are shrouded – appropriately for such an enigmatic science – in mystery. Marcel Granet cites as perhaps "the first mention of beliefs that are at the origin of geomancy" a passage in the *Book of Songs* (*Shijing*; 9th to 5th centuries BCE) where the founder of a town is said to have observed "the shadows" (which would have indicated south) and also "the surrounding yin and yang" (Granet, 1926: 1: 20). Feuchtwang remarks in this connection that the application of fengshui principles to the layout of towns and cities is apparently much earlier than to houses and graves.

Another indication of an awareness of fengshui principles comes from a story concerning Meng Tian, the Qin Dynasty general who supervised the building of the first part of the Great Wall. Construction was begun in 221 BCE: "He ... built a Great Wall, constructing its defiles and passes in accordance with the configurations of the terrain. It started at Lin-t'ao and extended to Liao-tung, reaching a distance of more than ten thousand *li*."[16] Thanks to its following the contours of the land, the Wall is traditionally seen as resembling a great dragon, which is an image that comes to refer in fengshui to any important topographical formation. In 210 BCE, upon the death of his patron, the first Qin Dynasty emperor, Meng fell victim to a dastardly political plot and was ordered to commit suicide. On hearing the news, he is said to have cried out in uncomprehending lamentation over this cruel stroke of fate. Then, on reflection, he said gravely: "Indeed I have a crime for which to die ... I have made ramparts and ditches over more than ten thousand *li*, and in this distance it is impossible that I have not cut through the earth's veins [*di mo*: 地脈]: this is my crime." He thereupon swallowed a lethal dose of poison. Another feat of engineering that is more likely to have cut into the earth's veins is a road that Meng Tian built on the orders of the First Emperor a decade after the Great Wall was started, in the course of which he "made cuts through the mountains and filled in the valleys, over a distance of one thousand eight hundred *li*" (Bodde, 1940: 61, 55–56).

The *mo* of *di mo* ("earth's veins") corresponds to something in the human body like veins or arteries or pulses, but more closely to the acupuncture meridians, since no physical "envelope" for the flow is perceptible. General Meng's concern for respecting the earth is often echoed in the Ming period, which saw a special flourishing of garden culture in China. Fear of damaging the "earth's veins" fueled opposition to gypsum mining in Taihe county in the fourteenth century, and the government prohibited the digging of ponds in Nanjing "lest they damage the *qi* of the earth in the imperial capital" (Clunas, 1996: 181). The sixteenth-century author of the *Nongshuo* (*Talks on Farming*) writes eloquently of "energy arteries running within the earth" and how "earth and bone are like the arterial system of the human body which carries the energy-blood" (cited in Hay, 1985: 42). When the body is understood as an organism within the larger organism of the environment, its various energetic pulses (*mo*) correspond to the dynamic configurations referred to earlier as the earth's "lifelines" (*shi*: 勢).

In a survey of the history of fengshui the Ming Dynasty author Wang Wei writes: "The theories of the geomancers have their sources in the ancient Yin-Yang school. Although the ancients in establishing their cities and erecting their buildings always selected the sites (geomantically), the art of selecting burial sites originated with the *Burial Book* [*Zangshu*] in twenty parts, written by Guo Pu [276–324] of the Jin Dynasty" (cited in March, 1968: 261). Central to the *Zangshu*, also known as the *Book of Funerals*, is the idea that certain sites are blessed with flows of especially vitalizing energy known as *sheng qi* ("vital breath", "life energy"), which is a phase of the larger circulation of cosmic energies: "When the *ch'i* of *yin* and *yang* breathes out it is wind. When it ascends it constitutes the clouds, and when it falls it is rain. It travels on and in the ground and becomes vital *ch'i*. Vital *ch'i* travels on and in the ground and engenders the myriad things." Vital *qi* is further subject to the forces of wind and water: "The [*Burial*] *Classic* says that *ch'i* rides the wind and disperses. When bounded by water, it halts" (*Zangshu*, cited in Bennett, 1978: 9–10). With respect to a place characterized by multiply dividing streams, the *Shuilongjing* (*Water Dragon Classic*) says: "If the wind shakes the willow branches, or if the wind bends the grass whether passing over the position or not, it will mean trouble, and even meandering water will not justify the site. It will bring decay and sickness" (Feuchtwang, 1974: 139).

Water molds the natural environs from the outside in obvious ways, primarily through watercourses' cutting into the earth, but also through precipitation's sculpting the shapes of mountains over time. Winds, too, move earth, if not mountains, over the long term, in ways less obvious than waters, since the movement of air is itself invisible, becoming perceptible only through its effects on water, vegetation, and loose soil. But fengshui is concerned with winds and waters in a deeper sense too: with the invisible "breath of the earth" discussed above, and with the "flows" of qi beneath the earth that were thought to be responsible for the formation of minerals (Needham, 1970: 3: 637, 650).

The art of proper burial consists, then, in choosing a site with favorable life-breath. Guo Pu puts it succinctly by saying, *cheng sheng qi* (乘生氣), "burying [is a matter of catching] life-breath" (cited in March, 1968: 256). The understanding of qi expressed in the *Zangshu* has been aptly characterized as:

> ... 'the breath at the origin of things, forever circulating,' which flows through the whole of space, endlessly engendering all existing things, 'deploying itself continuously in the great process of the coming-to-be and transformation of the world' and 'filling every individual species through and through' (Jullien, 1995: 91–2).

The quality of a place, according to Guo Pu, depends on the local *shi*, meaning configurations of earth-energy (which François Jullien translates in this context as "lifelines"): "The vital breath circulates along the lifelines of the terrain and is concentrated at the points where they come to an end." John Hay draws a helpful distinction between *shi* (勢) as "dynamic configuration" and *xing* (形) as "form" or "shape" of concreted objects. "It is the changefulness of *shi* that is lasting, whilst the fixedness of *xing* is transient" (Hay, 1985: 53). On the fengshui understanding of a landscape, *shi* refers to both the "veins" of earth through which the qi flows and also the "skeletal structure" or "spinal column" of the terrain. In order to perceive the dynamic configurations of a landscape, it is necessary to gain some distance for a broader perspective: "The lifelines are visible from a thousand feet away, the particular configurations of the terrain from a distance of one hundred feet" (*Zangshu*, as cited in Jullien, 1995: 93). Another way, as Jullien suggests, is to consult an appropriate masterpiece of Chinese landscape painting, the primary principle for which was to "achieve the *shi*" of the landscape (99).

Another important source of ideas behind fengshui is the *Huangdi zhaijing* (*The Yellow Emperor's Siting Classic*), attributed to the fifth-century thinker Wang Wei. As the title leads one to expect, a central idea in this text is that of *zhai* (宅), meaning "site/siting" or "place/placing": "All human dwellings are at sites ... sites are the foundation of human existence" (cited in Bennett, 1978: 5). A *zhai* is therefore not merely some location in abstract space, but rather a place as defined both by a particular topography and by the kinds of human activities that take place in it.[17] Since an inhabited place is a dynamic locus of flowing energies, the *Siting Classic* addresses the temporal as well as the spatial aspects of fengshui practice: "Every year has twelve months, and each month has positions in time and space of vital and torpid ch'i [*shengqi*: 生氣 and *siqi*: 死氣]" (Bennett, 1978: 7). Human activities undertaken at times of vital qi are more likely to be successful than at times when the qi is torpid. The quality of qi also varies within smaller cycles, such as the diurnal. Manfred Porkert characterizes *shengqi* as "the quality of energy during the *yang* hours of the rising sun (midnight to noon) ... [which has] a quickening and invigorating effect on active enterprises," and the opposite for the *yin* hours from noon to midnight (Porkert, 1974: 172–73).

The *Siting Classic* also asserts an isomorphism between a *zhai* and a human body, a correlation that is central to Daoist thinking about the relations of the human being to the landscape. "The forms and configurations are considered to be the body; water and underground springs are the blood and veins; the earth is the skin; foliage is the hair; dwellings are the clothes; door and gate are the hat and belt" (Bennett, 1978: 13).

In spite of the prevalence of all these ideas, it was not until the Tang dynasty (618–906) that fengshui theories and practices began to be synthesized and formalized into a distinct school named as such.[18] Wang Wei's historical survey distinguishes two schools, one using the "Jiangxi method" and the other the "Ancestral Hall" or "Fujian method", both of which he claims are derived from Guo Pu (March, 1968: 261). The latter (founded a century or two later) is also known as the "compass" school, and it emphasizes the importance of the Eight Trigrams of the *Yijing* and the Five Planets, as well as the indispensability of the geomantic compass. Since this school's methods are more abstract and pseudo-scientific, it seems to have been infected by more charlatanry than its counterpart and is thus of less interest in our present context.[19]

The school employing the Jiangxi method is also known as the "form and configuration" (*xing shi*) school, because of its concern with intuiting the configurations (*shi*) of qi from the shapes or forms (*xing*) of the landscape. It was apparently founded by an imperial fengshui master by the name of Yang Yunsong (*ca.* 840–888). In its consideration of mountains and watercourses, the Form and Configuration School lays particular emphasis on the motif of the dragon as a pattern to be found in "all topographical formations" (Feuchtwang, 1974: 141), as indicated by the titles of such important treatises by the founder as the *Hanlongjing* (*Classic on Arousing the Dragon*) and the *Yilongjing* (*Classic on Approximating the Dragon*).

It may, however, be better to talk of the dragon as an image for all *vital* topographical formations, since some places may simply be "dead" in terms of the flows of qi. As a later author, Shen Hao (17th century), puts it:

Surely nothing but the writhings of the magic dragon is an adequate figure of the mountain-ridges' permutations. What does not resemble the permutations of the magic dragon does not realize the subtle geomantic essence. Therefore it is said: if it has permutations, call it dragon; if it has none, call it barren mountain (cited in March, 1968: 256–57).

This is not the Azure Dragon as an image for one of the four directions, which is culturally and geographically specific to China, but "a universal symbol of the powers of nature" and especially of "the power of [self-]transformation" (Feuchtwang, 1974: 149–50). Since Chinese culture is based on the premise that all events in the world are continually transforming themselves, the dragon is its perfect emblem. It is also an archetypal image common to an amazingly wide range of cultures and mythologies, in some of which its meaning is negative by contrast with its generally auspicious quality in China (Hay, 1994).

To fully appreciate sciences based on becoming rather than being, on energy rather than matter, we need to reorient our ways of perceiving our environment. In introducing his discussion of the notion of *shi* as "lifelines" in the earth, François Jullien recommends that we stop regarding nature as "an object of science" in the Western sense. "Rather, we should here perceive nature intuitively, through the sensibility of our bodies and their activity, as the single common principle within and outside us that operates throughout reality and explains how the world is animated and functions. Let us imagine a new 'physics' and stop thinking of nature abstractly" (Jullien, 1995: 91). If this idea seems too exotic – a physics associated with a transformation in our experience – we might recall that there exists something similar in the Western tradition, with the Stoics and Epicureans. As Pierre Hadot puts it, "Contemplation of the physical world and imagination of the infinite are important elements of Epicurean physics. Both can bring about a complete change in our way of looking at things. The closed universe is infinitely dilated, and we derive from this spectacle a unique spiritual pleasure" (Hadot, 1995: 87–8).

Just as the breath that animates the human body through bringing oxygen to the blood is invisible (the misty exhalations in cold weather being water vapor rather than air), so the cosmic breath animating the body of the earth cannot be seen, although it can be felt or otherwise sensed.[20] The science of acupuncture, which is closely related to fengshui, has been slow to gain acceptance as valid on the part of practitioners of Western medicine. The main reason is that its background assumptions are so different, and there is also that fact that the meridians (*jingluo:* 經絡) through which the currents of qi flow through the human body are invisible, and Western researchers were looking for literal conduits such as veins or nerves.

Corresponding difficulties arise with the idea of energies flowing through the earth (in the broad sense, including watercourses, vegetation, and the other processes of metal and fire) along "lifelines" that cannot be directly seen but can be intuitively discerned by the well-trained practitioner. This notion should, however, be less mysterious to Western physicists since the discovery of the earth's magnetic field (the magnetosphere) whose energies flow along lines that are similarly invisible. Indeed the Chinese appear to have been the first to understand the phenomenon of magnetism, with 3rd century texts referring to the "south-controlling spoon" – a piece of lodestone carved into the shape of the Northern Dipper (*Ursa Major*). These Chinese discoveries naturally took place within the context of fengshui (Needham, 1969: 71ff; 1962: 4/1, 229–334).

I should like to conclude with some reflections on how, through the medium of practice in the arts, one might come to an experiential understanding of the reasonable ideas and features of fengshui as a basis for their application to everyday life.

RELEVANT ARTS, FINE AND MARTIAL

We saw that Guo Pu understands the lifelines of the land as both its "skeletal structure" and its "veins", and when the Form and Configuration School of fengshui applies the image of the dragon to any important topographical formation, it is on the assumption that the dragon is animated by qi flowing through its bones as well as its veins. This idea is associated with the distinction between the "mountain" (*shan:* 山) and "waters" (*shui:* 水) qi animating a landscape (*shanshui*), with peaks and ridges as skeletal structure and watercourses as veins, and it is also exemplified in one of China's greatest contributions to the world's art, landscape painting (Feuchtwang, 1974: 141–48). François Jullien quotes and comments upon the tenth-century aesthetician Jing Hao on this topic:

> Under the painter's brush, as in nature, 'the aspects of mountains and waters are born from the interaction of vital breath and the given layout to which that force imparts dynamism.' In China, the purpose of painting is to rediscover the elemental and continuous course of the cosmic pulsation through the figurative representation of a landscape (Jullien, 1995: 94, also 95–102).

Thus an excellent way of learning to discern the lifelines in a landscape is to study Chinese landscape paintings – which is in itself one of life's great pleasures.

Another great Chinese contribution to the world of the arts, again associated with fengshui, is the art of the garden. We saw earlier that fengshui allows for certain kinds of human intervention in order to improve, when necessary, the conditions of a particular site – though the tragic figure of Meng Tian stands as a reminder of the dangers of too deep an intervention. As an important adjunct to the dwellings of the living in China – of the well-to-do in particular – the garden provides an opportunity to put the principles of fengshui into practice.

The classic garden manual, the seventeenth-century *Yuanye* (*Craft of Gardens*) by Ji Cheng, talks of the way a well constructed garden, where one has "visualized the balustrades as if they were in a painting," can "flood the heart with intoxication" and create "a pure atmosphere around our tables and seats [so that] the common dust of the world is far from our souls" (Ji, 1988: 43–44). Chinese gardens would often be laid out in such a way as to "borrow" features of the neighboring landscape, by offering views from within the garden of a local mountain or lake. A basic premise of the Chinese garden is the microcosm/macrocosm correlation we saw between the body and landscape; the well designed garden sets up a pattern of energies that corresponds to the dynamic configurations of a larger landscape – and indeed energies just as powerful thanks to the amplifying effect of miniaturization.[21]

As the sterling work of John Hay in this field has shown, the most important components of the Chinese garden are generally the rocks: "Rocks were both the frame of the garden and the foci of visual attention" (Hay, 1985: 17). As forces of nature, rocks are powerful configurations of qi, not only as "bones of the earth" analogous to the human frame, but also as "kernels" of energy. "The essential energy of soil forms rock ... Rocks are kernels of energy: the generation of rock from energy is like the body's arterial system producing nails and teeth ... The earth has the famous mountains for its support ... Rocks are its bones."[22] The garden is then above all an environment of energies in which the human body can restore itself and renew its strength, insofar as its own minerality (minerals make up a good part of our dry weight) is surrounded and enhanced by that of the rocks.

Certain rocks are regarded as especially powerful configurations of qi because they are microcosms of mountains, smaller configurations of the huge telluric forces that thrust land up into the sky. Others, such as the greatly prized Taihu rocks, formed of limestone deposits from the floor of Lake Tai that are three hundred million years old, are more zoomorphic, resembling enormous sponges of billowing stone (Hay, 1985: 36). These rocks exemplify the three foremost qualities according to Chinese aesthetics of stone: leanness, surface texture, and foraminate structure – all of which help reveal to the outside the vast interior forces that formed the rock. The properly arranged garden is a prime example of the benefits of fengshui. Simply to be in the presence of such magnificent kernels of energy, let alone to contemplate them aesthetically, brings tonic strength to the human organism.

For those of us who do not have the luxury of a garden, there are other ways of cultivating an appreciation for the relations between the body's energy configurations and those of the environment – through practices such as *qigong* (氣功): literally, "energy work", a self-healing art that combines movement and meditation, and *taiji quan* (太極拳), "Great Ultimate bare-hands-combat-technique", one of the softer martial arts to have been developed in China. Anyone who is skeptical about the existence of the qi that flows through and energizes the body need only try some qigong exercises (for example, the set known as the *ba duan jin* (八段錦), or "eight sections of brocade", often used as a warm-up for *taiji* practice) in order to experience the flow of qi in his or her own person. As in yoga, the breath (itself a manifestation of the cosmic breath in the human body) is central to taiji practice. The practitioner is encouraged to breathe smoothly and deeply from the *dantian* (the "cinnabar field" discussed above, located a couple of inches below the navel) throughout all movements.

Taiji quan is believed to have originated in the Southern Song Dynasty (12th to 13th centuries), though an alternative – less well-attested – tradition traces the practice back to the time of the Six Dynasties (3rd to 6th centuries). It incorporates numerous ideas from Daoism, although the first part of its name refers all the way back to the "Great Treatise" appended to the *Yijing*, where the Taiji (Great Ultimate, Primal Beginning) is said to "engender the two primary forces" of yin and yang (Wilhelm, 1967: 318). Various accounts explain how yin and yang generate the wuxing, the five viscera, and so forth – accounts that harmonize with all that has been stated above concerning the correlations between the body and the environment (Despeux, 1981: 16–21, 40–58; Needham, 1962: 2: 460–64).

In accordance with the ancient Chinese conception of the human being as inhabiting the space between the earth and the heavens, taiji encourages an awareness of our bodily activities in this milieu and a sense that all our movements depend on our current configuration of energies in relation to the forces of heaven and earth. Entertaining the image of the head's being suspended from the heavens by an invisible cord encourages a sense for the way our energy maintains the upright posture, while the injunction to let the shoulders and elbows relax and drop down positions us with respect to gravity and the earth. The practice also invokes our relations to other inhabitants of the milieu between earth and heaven insofar as many of the movements are named after animals. The Yang style, for instance, includes movements such as: grasp bird's tail, crane spreads wings, carry tiger to mountain, fend off monkey, raise hand pat horse, part wild horse's mane, snake creeps down, golden cock stands on one leg, step back ride tiger, and so on.

While the natural spontaneity of the animal is taken as a model, the means of emulating it are anything but natural; one has to practice the moves over and over, employing thought, imagination, and memory, for a long time before spontaneity comes. Subsequent practice brings about precisely what fengshui encourages – a greater awareness of the relations between one's activities and the configurations of the surroundings, whether natural or built.

An appreciation and/or practice of such arts helps us understand how ideas from fengshui may be integrated into contemporary worldviews. They are admittedly incompatible with Cartesian-Newtonian understandings of the cosmos – to which we are in any case no longer obliged to subscribe. The Cartesian perspective is singularly parochial, and however much it might further our domination of nature through technology, this is no reason to think it will help us flourish in other ways. The sensible core of fengshui seems by contrast quite compatible with the physics and biology of the 21st century, and the corresponding modes of experience are at any rate accessible to anyone nowadays with an open mind and opened senses.

Unless one has the luxury of choosing a place to live and building a dwelling there, there is no opportunity for engaging in the practice of fengshui as siting or participating in the architect's decisions concerning the orientation of the house in relation to prevailing sun, wind, rain, and terrain. But one can always change the ways a living space is configured in a residence already built, by the simple expedient of becoming more aware of one's activities in relation to the surroundings. It is thus possible to improve the positioning of bed, dining table, chairs, work table, etc., without going to the expense of hiring an expert consultant. A little common sense combined with heightened sensitivity to place goes a long way. Opportunities to set up one's work space will generally be more limited than in the home, but improvement is usually possible nonetheless.

One might take encouragement from those writers on fengshui like Shen Hao who emphasize intuition. Coming upon a place that is right, he writes,

> One's eyes are opened; if one sits or lies, one's heart is joyful. Here the breath gathers, and the essence collects. Light shines in the middle, and the magic goes out on all sides. Above or below, to right or left, it is not like this ... Try to understand! it is hard to describe (cited in March, 1968: 259).

Among contemporary scholars, no one has put it better than François Jullien, who paraphrases Guo Pu as follows:

> Not only my own being, as I experience it intuitively, but the entire landscape that surrounds me as well, is continuously flooded by subterranean circulating energy ... The most glorious sites will be those where it is most densely accumulated, where the circulation of the breath is

most intense, its transformations most profound ... By rooting one's dwelling here rather than elsewhere, one locks into the very vitality of the world, taps the energy of things more directly (Jullien, 1995: 92).

If in response to such an expression of the more sensible tenets of fengshui we could become suspicious of the modern conceit that we can be at home anywhere, regardless of the telluric and atmospheric forces around us, we would surely witness a greater flourishing of human lives and the natural environment on which they depend.

ACKNOWLEDGMENT

I am grateful to Roger Ames, Augustin Berque, Marty Heitz, Hans-Georg Möller, and Hai-ming Wen for their helpful comments on an earlier draft.

NOTES

[1]   For an early classic correlation among the Five Processes, four directions and center, the five planets, and five symbolic animals (with a yellow dragon corresponding to earth in the central position), see the third chapter of the *Huainanzi* (Major, 1993: 70–73).

[2]   As for the level of Chinese mental culture, we might note in an aside the number of discoveries and inventions that originated in China and East Asia: printing, gunpowder, the magnetic compass, "mechanical clockwork, the casting of iron, stirrups and efficient horse-harness ... segmental-arch bridges and pound-locks on canals, the stern-post rudder, fore-and-aft sailing, quantitative cartography" (Needham, 1954: 1: 15; 1969: 11).

[3]   For an account of the extent of the assault on the environment during the revolution, see Shapiro, 2001.

[4]   The quote is cited from the article by H. Chatley in Samuel Couling's *Encyclopaedia Sinica* (1917).

[5]   See Graham, 1986: 325. The next few paragraphs are based on the section "The Cosmologists" in his *Disputers of the Tao*, 315–70.

[6]   Chinese transliteration has changed from using a system called Wade-Giles (*ch'i*) to the one preferred today, Pinyin (*qi*). In cases where the quotations come from older sources, I have retained the Wade-Giles transliteration.

[7]   My colleagues Roger Ames and Hans-Georg Möller have warned in private communications that this translation may be misleading and contradict the picture I am otherwise trying to convey of early *qi*-based cosmologies. For the term *sheng*, "was born out of" would be better than "began" and "gave birth to" or "vitalized" preferable to "produced". The ideas to be avoided are those of an independent "producer" and any kind of creation *ex nihilo:* it is rather a matter of ceaseless "taking shape" or "vitalizing" from formless "chaos". See Ames and Hall, 2001: 19–30, and Möller, 2001: 114–29.

[8]   *Guanzi* ch. 39, cited in Graham, 1989: 356. Translation modified, by emending the first word, *di* (earth) to *shui* (water), on the basis of the reasons given by Rickett in the commentary to his translation (Rickett, 1998: 2: 98–100).

[9]   Graham, 1989: 326, except that for the sake of the structure I have used Needham's translation for the characterization of soil in the last sentence of the Five Processes section (Needham, 1956: 243).

[10]   Needham (1956: 2: 255) suggests as another basis for this generation the way metal mirrors exposed to the night air attract or secrete "sacred dew".

[11]   See the table in Needham, 1956: 2: 262–63, who notes that over a hundred groups of correlations(!) are mentioned in the extant texts as a whole (2: 264). The situation is further complicated by the fact that some of the authors of the many texts in which these further correlations appear arrange them differently. See, for instance, John Major's commentary on the Five Processes in the *Huainanzi* (Major, 1993: 185–89) and the section "Han Cosmologies" in Hall and Ames, 1995:

256–68. The initial capitals of the names of the five viscera remind us that the Chinese terms as used in medicine refer to the entire functions of the physical organ and its associated meridian.

¹² For a more detailed treatment of this issue, see Major, 1984.

¹³ For the role of the "Five Evolutive Phases" in Chinese medicine, see Porkert (1974: 43–54).

¹⁴ For the full argument, see especially the chapters "The Inner Landscape" (pp. 100–12) and "Keeping the One" (pp. 130–59).

¹⁵ "Imaginal" is a coinage – to avoid the ontologically denigratory connotations of "imaginary" – of the great scholar of Islamic philosophy Henry Corbin, in his works on the archetypal imagination (such as *Creative Imagination in the Sufism of Ibn Arabi*). Nietzsche is one of the few Western philosophers to have been interested in the idea of inner landscapes: see Parkes, 1994, chapter 4, "Land- and Seascapes of the Interior".

¹⁶ Cited in Bodde, 1940: 54.

¹⁷ Western philosophy has generally devoted far more thought to abstract space than to lived place. For a judicious restoration of this imbalance, see Casey, 1997.

¹⁸ For example, the *Guanshi dili zhimeng (Mr. Guan's Geographical Indicator)* attributed to the third-century author Guan Lo, the *Zhangshu (Book of Funerals)* attributed to Guo Pu, and Wang Wei's *Huangdi zhaijing (The Yellow Emperor's Siting Classic)*; see Needham, 1956: 360.

¹⁹ A comprehensive account of the practices of the Fujian School, with fascinating descriptions of the geomancer's compass, is to be found in Feuchtwang, 1974: 18–95.

²⁰ Needham (1959: 3: 469) notes the parallel between the idea of qi animating the earth and Aristotle's notion of the two terrestrial emanations (*anathumiasis*), one aqueous (*atmidodestera*) and one gaseous (*pneumatodestera*) (*Meteorologica*, I, iv [341b6 ff.]).

²¹ The classic work on this topic is Stein, 1990: see, especially, part one: "Trees, Stones, and Landscapes in Containers", and "Survey of Themes". For a brief history of the remarkable petromania and litholatry (worshiping rocks and stones) that have characterized the Chinese tradition, see the first two sections of my essay in Berthier, 2000.

²² From the 86-page entry on "rock" in the eighteenth-century encyclopedia entitled *Classified Contents of the Mirror of Profound Depths*, cited in Hay, 1985: 52. See also Needham (1959: 3: 637) who cites the late sixteenth-century *Great Pharmacopoeia (Bencao Gangmu)*: "Stone is the kernel of qi and bone of the earth."

BIBLIOGRAPHY

Ames, Roger T. and David L. Hall. *Focusing the Familiar: A Translation and Philosophical Interpretation of the Zhongyong*. Honolulu: University of Hawaii Press, 2001.

Bennett, Steven J. 'Patterns of the sky and earth: a Chinese science of applied cosmology.' *Chinese Science 3*: 1–26, 1978.

Berque, Augustin. *Écoumène: Introduction à l'étude des milieux humains*. Paris: Belin, 2000.

Berthier, François. *Reading Zen in the Rocks: The Japanese Dry Landscape Garden*, Graham Parkes, trans. Chicago: University of Chicago Press, 2000.

Bodde, Derk. *Statesman, Patriot, and General in Ancient China*. New Haven, Connecticut: American Oriental Society, 1940.

Boerschmann, Ernst. *Chinese Architecture and its Relation to Chinese Culture*. Washington, DC: Smithsonian Institute Report 1911 (published 1912), pp. 539–577.

Boerschmann, Ernst. *Picturesque China, Architecture and Landscape; A Journey Through Twelve Provinces*. London: Unwin, 1924.

Casey, Edward S. *Getting Back into Place: Toward a Renewed Understanding of the Place-World*. Bloomington: Indiana University Press, 1993.

Casey, Edward S. *The Fate of Place: A Philosophical History*. Berkeley: University of California Press, 1997.

Clunas, Craig. *Fruitful Sites: Garden Culture in Ming Dynasty China*. Durham, North Carolina: Duke University Press, 1996.

Corbin, Henry. *Creative Imagination in the Sufism of Ibn Arabi*, Ralph Manheim, trans. Princeton, New Jersey: Princeton University Press, 1969.

Despeux, Catherine. *Taiji quan: art martial, technique de longue vie*. Paris: Éditions de la Maisnie, 1981.

Dukes, Edwin Joshua. *Everyday Life in China; or, Scenes Along River and Road in Fuh-Kien*. London: Religious Tract Society, 1885.

Elvin, Mark, ed. *Sediments of Time: Environment and Society in Chinese History*. Cambridge: Cambridge University Press, 1998.

Eitel, Ernest J. *Feng-Shui: or, The Rudiments of Natural Science in China*. Cambridge: Cockaygne, 1973.

Feuchtwang, Stephan D.R. *An Anthropological Analysis of Chinese Geomancy*. Ventiane, Laos: Vithagna, 1974.

Graham, A. C. *Chuang-tzu: The Seven Inner Chapters and Other Writings from the Book Chuang-tzu*. London: Allen & Unwin, 1981.

Graham, A.C. *Yin-Yang and the Nature of Correlative Thinking*. Singapore: Institute of East Asian Philosophies, 1986.

Graham, A. C. *Disputers of the Tao: Philosophical Argument in Ancient China*. LaSalle, Illinois: Open Court, 1989.

Granet, Marcel. *Danses et légendes de la Chine ancienne*. 2 vols. Paris: Alcan, 1926.

de Groot, J.J.M. *The Religious System of China*, vol. 3. Leiden: Brill, 1897.

de Groot, J.J.M. *Religion in China: Universism: A Key to the Study of Taoism and Confucianism*. New York: G. P. Putnam's Sons, 1912.

Hadot, Pierre. *Philosophy as a Way of Life*, Michael Chase, trans. Oxford: Blackwell, 1985.

Hall, David L. and Roger T. Ames. *Anticipating China: Thinking through the Narratives of Chinese and Western Culture*. Albany: State University of New York Press, 1995.

Hay, John. *Kernels of Energy, Bones of Earth: The Rock in Chinese Art*. New York: China Institute in America, 1985.

Hay, John. 'Structure and aesthetic criteria in Chinese rocks and art.' *Res* 13: 6–22, 1987.

Hay, John. 'The Persistent Dragon (Lung).' In *The Power of Culture: Studies in Chinese Cultural History*, Willard J. Peterson, et al., eds. Hong Kong: The Chinese University Press, 1994.

Ji Cheng. *The Craft of Gardens*, Alison Hardie, trans. New Haven, Connecticut: Yale University Press, 1988.

Jullien, François. *The Propensity of Things: Toward a History of Efficacy in China*, Janet Lloyd, trans. New York: Zone Books, 1995.

Lau, D. C. and Roger T. Ames, trans. *Yuan Dao: Tracing Dao to its Source* [first chapter of the *Huainanzi*]. New York: Ballantine, 1998.

Major, John S. 'The five phases, magic squares, and schematic cosmography.' In *Explorations in Early Chinese Cosmology*, Henry Rosemont, Jr., ed. Chico, California: Scholars Press, 1984, pp. 133–66.

Major, John S. *Heaven and Earth in Early Han Thought: Chapters Three, Four and Five of the Huainanzi*. Albany: State University of New York Press, 1993.

March, Andrew L. 'An appreciation of Chinese geomancy.' *Journal of Asian Studies* 25: 253–67, 1968.

Morris, Edwin, T. *The Gardens of China: History, Art, and Meanings*. New York: Scribners, 1983.

Möller, Hans-Georg. *In der Mitte des Kreises: Daoistisches Denken*. Frankfurt: Insel Verlag, 2001.

Needham, Joseph. *The Grand Titration: Science and Society in East and West*. London: Allen & Unwin, 1969.

Needham, Joseph, with Wang Ling. *Science and Civilisation in China*. Vol. 1: *Introductory Orientations*. Cambridge: Cambridge University Press, 1954.

Needham, Joseph. *Science and Civilisation in China*. Vol. 2: *History of Scientific Thought*. Cambridge: Cambridge University Press, 1956.

Needham, Joseph. *Science and Civilisation in China*. Vol. 3: *Mathematics and the Sciences of the Heavens and the Earth*. Cambridge: Cambridge University Press, 1970.

Needham, Joseph. *Science and Civilisation in China*. Vol. 4, part 1: *Physics and Physical Technology*. Cambridge: Cambridge University Press, 1962.

Parkes, Graham. *Composing the Soul: Reaches of Nietzsche's Psychology*. Chicago: University of Chicago Press, 1994.

Parkes, Graham. 'The role of rock in the Japanese dry landscape garden.' In *Reading Zen in the*

Rocks: The Japanese Dry Landscape Garden, François Berthier, ed. Chicago: University of Chicago Press, 2000, pp. 85–155.

Pennick, Nigel. *The Ancient Science of Geomancy*. London: Thames and Hudson, 1979.

Porkert, Manfred. *The Theoretical Foundations of Chinese Medicine*. Cambridge, Massachusetts: MIT Press, 1974.

Rickett, W. Allyn, trans. *Guanzi: Political, Economic, and Philosophical Essays from Early China*. 2 vols. Princeton, New Jersey: Princeton University Press, 1985.

Schipper, Kristofer. *The Taoist Body*. Berkeley: University of California Press, 1993.

Shapiro, Judith. *Mao's War against Nature: Politics and the Environment in Revolutionary China*. Cambridge: Cambridge University Press, 2001.

Sivin, Nathan. *Science in Ancient China: Researches and Reflections*. Brookfield, Vermont: Variorum, 1995.

Skinner, Stephen. *The Living Earth Manual of Feng-Shui: Chinese Geomancy*. London: Routledge & Kegan Paul, 1982.

Stein, Rolf A. *The World in Miniature: Container Gardens and Dwellings in Far Eastern Religious Thought*, Phyllis Brooks, trans. Stanford: Stanford University Press, 1990.

Twitchett, Dennis and John K. Fairbank, eds. *The Cambridge History of China*, vol. 1. Cambridge: Cambridge University Press, 1978.

Wang, Aihe. *Cosmology and Political Culture in Early China*. Cambridge: Cambridge University Press, 2000.

Wilhelm, Richard. *The I Ching: or Book of Changes*, Cary F. Baynes, trans. Princeton, New Jersey: Princeton University Press, 1967.

Wong, Timothy M.K. 'The limits of ambiguity in the German identity in nineteenth century Hong Kong, with special reference to Ernest John Eitel (1938–1908).' In *Sino-German Relations Since 1800: Multidisciplinary Explorations*, Ricardo K. S. Mak and Danny S. L. Paau, eds. Frankfurt: Peter Lang, 2000, pp. 73–91.

JOHN D. KESBY

THE PERCEPTION OF NATURE AND THE ENVIRONMENT IN SUB-SAHARAN AFRICA

The sub-Saharan Africa major cultural region, as here defined, includes the whole of the Sahel belt immediately south of the Sahara, and all of the area of the African landmass south from there to the extreme southern coast, excluding the Ethiopian Highlands and deserts of the Horn. These highlands and deserts are inhabited by peoples who have strong cultural ties to the peoples of North Africa, Arabia and, beyond that, to the rest of the North African and Near Eastern major cultural region. Without trying to characterise the sub-Saharan peoples as a group, they share a single cultural background, albeit a complex one, and they share a great deal in common, despite some striking differences, in their modes of subsistence and the habitats in which they live. In broad terms, when Europeans began to document first the coasts and then the interior of the area south of the Sahara, there were four main habitats: lowland wet forest, highland wet forest, dry woodland and savanna, and desert. Not surprisingly, these different groups meant that there were different animals and plants available to different peoples, depending on where they lived, so that animals familiar to some peoples were unknown to others.

At the time of the first encounters between Europeans and sub-Saharan peoples, a minority of ethnic groups within the major region were hunter-gatherers who cultivated no crops and whose only domestic animals were dogs. However, few, if any, of these groups were not in contact with cultivators of crops, or with pastoralists, or with both; and they therefore had access to foods other than those hunted and gathered. As just indicated, some peoples were exclusive pastoralists, relying for their basic production on their herds of cattle and flocks of goats and sheep. Like the hunter-gatherers, these pastoralist peoples were a minority among the sub-Saharans, and probably all of them traded with cultivators, or with hunter-gatherers, or with both, for food and other commodities which they did not produce themselves. Certainly, the vast majority of African sub-Saharan societies and peoples were cultivators of various types. Some, in areas with relatively reliable rainfall, or groundwater, were intensive in their approach; a larger number used more extensive methods and did not put as much effort into cultivation as did the others. Throughout

211

H. Selin (ed.), Nature Across Cultures: Views of Nature and the Environment in Non-Western Cultures, 211–228.
© 2003 Kluwer Academic Publishers. Printed in Great Britain.

the area of the savanna and dry woodland vegetation types, recurrent droughts were always a probability and discouraged more than a limited investment in the sowing of crops.

Vegetation was one of the variables affecting African peoples, their mode of subsistence was a second, and the third was the presence or absence of social classes. Some of the societies had no class systems: individuals were distinguished by age, by sex, by ability and by personal characteristics, but no one inherited social status from his parents. All of the hunter-gatherers, all of the pastoralists, and some of the cultivators were classless, but the remaining cultivators lived in kingdoms with a royal family, aristocracy and thoroughgoing class system. The distribution of these two types of societies did not follow a random pattern; they were not a spotty patchwork quilt of irregularly scattered forms, incoherently distributed among each other, with no apparent regular order. However, there is no evidence now to enable us to untangle the process by which, during late prehistory, the long and serpentine band of kingdoms came into existence, in the process separating the stateless (and classless) societies into three distinct blocs.

One further criterion by which the sub-Saharan societies can be distinguished from each other is that of literacy. By AD 1800 some of the kingdoms were literate, in the sense that they had upper classes with members who could read and write. Not all the kingdoms were literate, however, and those that were occurred in two zones, both of them on the borders of the major region. One of these was in and near the Sahel belt, from the big bend of the River Niger to Lake Chad, and thence out to the east to Darfur. The other was on the Indian Ocean coast, where the Swahili-speaking city-states used the Arabic script which had been brought there, presumably by sea, in the course of trading activities between that coast and the Near East. Arabic script was equally the basis for the literacy of the Sahel group of states. The name "Sahel" is derived from the Arabic word for "coast", and, although it is not a seacoast, the zone has long been a sort of southern "coast" for the caravans of camels arriving in the Northern Savannah of Africa from North Africa. The Swahili (the same word basically as Sahel) city-states are on a true seacoast; and these two "coasts" of sub-Saharan Africa are the zones where the culture of North Africa and the Near East has most powerfully irradiated that of the sub-Saharan peoples.

As a result it is not just the Arabic language and script which arrived there, but also Islam, a substantial volume of trade goods, and a strong tendency for the formation of urban centres. The city-states of the Indian Ocean coast have already been mentioned. Another distinctive complex of cities, west of Lake Chad, are those of the Hausa people. Indeed West Africa as a whole seems to have had an elaborate network of urban centres, with their accompanying markets, long before the Portuguese started feeling their way by sea along the Guinea Coast. It is not clear why West Africa, but not most of Central, East and Southern Africa, had such an elaborate commercial development so comparatively early, but it is likely that, like the Hausa kingdoms and cities, the whole area had been affected by the centuries of trading by camel across the

Sahara. There is, however, no evidence that these urban centres, or those of the east coast, were so thoroughly "urbanised" that their populations had lost touch with the key features of the landscape, of farming or of animals and plants, a process associated with the populations of the huge conurbations of Europe. This process by which the basic ethnic culture of a people is lost, through industrialisation and urbanisation, produces the unfamiliarity with plants and animals that is characteristic of the big city populations of northwest Europe or of North America. In these places the ability to identify particular species is stimulated by books and television and is linked with a conscious conservationist backlash. This process is now represented south of the Sahara in the sprawling complex of the urban growth of Gauteng province, in South Africa, but it is new in Africa, and it was not there in 1800.

The Gauteng industrial area is part of a relatively late sequence of developments which set in during the nineteenth century, as European trade goods and African slave traders penetrated the interior. The Christian missions, based in Europe, became bolder, and reached inland from the coasts, and European governments, after 1885 and somewhat reluctantly, established administrations over the area. These late processes had the effect of bringing sub-Saharan populations into much closer contact with Europeans, generating new elites in each territory, with educations of European type, and detonating also new urban areas and a vast increase in urban populations.

THE COGNITIVE SYSTEMS

This sketch of cultural features and processes provides the framework for systems of cognition within the sub-Saharan area. It is vital to emphasise at this stage that an introduction to the sub-Saharan knowledge of nature and the environment is essentially premature, because the detailed research upon which it must be based has begun in only a few localities, and any general models that might be suggested have not been tested over most of the area. It is true that there is a considerable volume of fragmentary information, such as bird names from South Africa and plant names for the Gikuyu and Mbeere, of East Africa, but there is a very limited amount of information on the *systems* of cognition which underlie such vocabularies. The account which is given here is therefore over-bold, because it is pushing ahead of the evidence, although that evidence so far does not falsify the model which is used here. It may also help readers to know that the model is strongly affected by the author's experience with the Rangi people of East Africa, with their neighbours, and with some reference to Swahili, now the lingua franca of a large area of eastern Africa. A thorough testing of what is summarised here therefore is needed in all the sub-regions of the sub-Saharan major region; but this testing has not yet been carried out. Morris' detailed information from the peoples living west of Lake Malawi does not contradict the model, but this area is not far removed from East Africa. However, the claim that underpins this discussion is that the observations made here are substantially reliable for the sub-Saharan region as a whole, and the first of these observations is securely reliable.

wow!

Sub-Saharan languages do not have a word for "nature", nor do they have one for "the environment". These two words have come into English, and their cognates into other European languages, by processes which are embedded in Euroamerican cultural history, and they are specific to it. Basically, Natura was a goddess in early historic, and probably later prehistoric, Italy. After the end of the Roman Empire in the west she evolved into an abstraction in the philosophical theology of disputes within the universities of Western Europe. She was now the basic framework of unbreakable rules which governed the Universe, as created by Almighty God. The theologians' sense was picked up by secular writers, and in the process of the Enlightenment and the Romantic Movement, during the eighteenth and nineteenth centuries, "nature" came to mean the whole ordered pattern of physical reality, minus people. However, there was always an ambiguity here, since people partake of nature, being ordered and physical. When Carl Linné produced his book *Systema Naturae* in Sweden, during the eighteenth century, he included people, under the newly coined name of *Homo sapiens*, in his group Primates, which were part of Mammals, and firmly embedded in Natura.

"The environment" emerged into prominence during Europe's nineteenth century, and meant basically not the whole of "nature" but that part of it which closely surrounded people; indeed it could be used of the surroundings of any living organism, for instance a species of large cat or of tree. However, "the environment" did not include people; they were not part of it.

Turning back to the sub-Saharan knowledge systems, they have a considerable amount in common with those of Europeans, but have not, as far as we now know, undergone the philosophical and theological elaboration which characterises the literate classes in Europe. In many ways they are, if anything, more comprehensible to most people, including those who are not Africans. Within sub-Saharan societies there are two very important oppositions. One is between the inhabited village or settlement of the human community and the wild places, such as the forest, "the bush" or the hills. The other is between the human sphere of activity and knowledge and the sphere of the superhuman powers, including God, the gods and the different kinds of spirits. These two oppositional dyads are neither logically nor practically opposed to each other, and by following the ways in which African peoples spell out in experience the overlays between the various features of the two it is possible for an outside observer to see that they are covering the ground implied in "nature" and "the environment", but without the ambiguities involved in a long history of philosophical development, as in Europe.

It is worth emphasising that within all the African cognitional schemes, the presence of God and of metaphysical realities is assumed. The physical aspect of daily human experience is assumed to be a kind of "skin" on the surface of the metaphysical. The two are in intimate and inseparable contact, and the detailed events of the physical are interpreted by way of the underpinning connections of the metaphysical. Thus when diviners use their various methods to read the significance of illness or other trouble, they are referring back at all times from the observable symptoms to unseen and non-physical, forces "below the surface". This intimate link between the observable and the non-observable is by no means peculiar to sub-Saharan peoples, and can be paralleled all over the world, for example among the Amerindian peoples of North and South America.

However, Amerindians and sub-Saharans differ appreciably in culture, and their differences in food, music and languages, for example, raise the question of whether there is a distinctively African mode of cognition. The quick answer to the question is that, on the basis of what has been gathered for comparison so far, it is not possible to discern any pattern of perception which is peculiar to the sub-Saharans. This may be the result of inadequate and patchy research, rather than the absence of such a pattern. However, as emphasised already, there do seem to be similarities between basic cognitional features among Amerindians and Africans, and a more prolonged survey reveals the same basic contrasting dyads among peoples in other major cultural regions. What is, however, quite definitely distinctive about sub-Saharan peoples is the co-ordinated assemblage of detail of what it is that they know. Their interest in the sun, the moon, the hills and the rivers may not be distinctive, but the animals and the plants which they know form an idiosyncratic pattern because they are precisely those of the Afrotropical region and of its botanical counterpart. Nowhere else on earth has this precise combination of living forms. Thus sub-Saharans share their experience of some genera and species with a limited range of other major regions. Huge evergreen fig trees (*Ficus* spp.) are prominent in the landscapes of the Indian and Southeast Asian regions as well as the sub-Saharan, and the very conspicuous baobabs (*Adansonia digitata*) of tropical African landscapes are paralleled by members of the same genus of trees in Madagascar and in Australia. However, some species are much more restricted. Aardvaarks (*Orycteropus afer*) are distinctively Afrotropical, while spotted hyenas (*Crocuta crocuta*) and ostriches (*Struthio camelus*) extend somewhat beyond the borders of the region, but are named in one African language after another, presumably because they are so distinctive.

For sub-Saharan peoples, animals and plants are two of the implicit major categories which seem to be recognised everywhere as making up the world about them. These implicit categories are:

1. sun, moon and stars;
2. hills, valleys, lakes, rivers and coasts (the basis of the inorganic landscape);
3. soils and vegetation;
4. plants and fungi; and
5. animals.

These implicit categories must be inferred from the vocabularies of African languages, since they are not explicitly named, and their existence is therefore open to challenge. But they do seem to occupy the place in African knowledge systems which is occupied in literary English by the explicit categories "nature" and "the environment".

Unsurprisingly, the sun and the moon figure prominently in African knowledge and symbolic representation, because the sun defines the year and the moon the months (Remember that the English "month" is derived from the word "moon"). The stars also attract some attention, especially the Pleiades, which, because of their movements, are identified with seasons. (Even in the equatorial wet forest the people have some sense of the changing seasons). However, in most areas the stars do not receive many individual names. Despite their great importance as part of the general framework of knowledge, further-more, the sun and the moon are not characters in myths over most of the area, the conspicuous exception being the Southwestern region, where the Hottentot (Khoi) and Bushman (San) groups of peoples have a background of telling stories in which the sun and the moon are portrayed as behaving rather like human beings, a feature of storytelling in some other parts of the world. This peculiarity of the Southwestern peoples is one of several which suggest that they are distinct from the majority of the peoples of the region. Indeed it seems likely that at some stage in late prehistory there were two major regions south of the Sahara and not just one. At that stage, we can hypothesise, there was a Southern African major cultural region, with Khoisan languages and without cultivated crops. It extended from the southern coast northwards to the south-ern edge of the Central African Wet Forest, wherever that threshold may have been at the time, and to about the centre of what is now Tanzania. To the north of the border zone was the Equatorial African major region; with the passage of time Bantu languages, knowledge of iron smelting, the rule of kings and the cultivation of crops moved into the Southern area from the Equatorial. If this was so, it is realistic to expect some detailed features of cognition to be different among the Southwesterners from those of their counterparts to the east and north. However, the evidence available to test this hypothesis is very thin.

LANDSCAPES, SOILS AND CLIMATE

Much more abundant is the detail on contemporary African knowledge of the landscape about them. When we turn from celestial to terrestrial features, it is necessary to emphasise that at the time of their first contact with Europeans each African people was devoid of a sense of a landmass called Africa, or any other single name, and, in more cultural terms, devoid of any sense of a sub-Saharan region. The "world" experienced by each people consisted of them-selves, neighbouring peoples and those areas which people from their area reached in their travels. These worlds must have been of differing sizes, because, for instance, in West Africa the large towns with their markets and the long-distance trade routes, together with networks of specialised trading families, must have meant that some individuals had a knowledge of an area the size of West Africa, plus the western part of the Sahara, and of at least parts of North Africa, linked southward by the camel caravans. By contrast, in the Congo Basin, there must have been villages along the river whose world was long and narrow and followed the main trading route along the widest part of the principal river course. In other parts of the basin, for example along the Kwilu River, the distances usually covered were apparently much shorter, and the experienced world was correspondingly smaller.

However, at the same time as emphasising that African worlds were of limited size, it is vital to appreciate that they were very detailed. People who did not know what happened on the nearest ocean coast, which could be hundreds of miles away, knew exactly what was happening in their own settlement and in nearby settlements. They knew the scenery of their area; the sources of water at different seasons, vital in the driest areas, such as the Kalahari Desert; the soils and what could be grown on them, where that was relevant, for instance, not in the Kalahari; the uses of the woods of various plants, and the medicinal uses to which plants could be put; as well as a host of technical processes, such as, in many areas, iron smelting. It is very easy for people who live in huge industrial conurbations to underestimate the technical proficiency, local knowledge and sheer self-sufficiency of local communities in Africa and in other non-industrial areas. As noted earlier, it is these conurbational populations, such as those in New York City or London, who have lost touch with farming, with birds and other animals and with plants. They have also lost contact with other people. A human individual only has a limited ability to absorb knowledge; beyond each individual's limit, they just stop. Each of us risks overload. Londoners and New Yorkers pick up information from all over the world, by way of television and the Internet, but their knowledge is patchy. They often do not know the people next door, or what is happening in the next street, and they hear about the riot on the other side of town – on the television. In place of this kind of knowledge, the people of any African village had a limited but very detailed and intense knowledge, centred on the small area where they lived.

In order to illustrate the kind of landscape knowledge which sub-Saharan peoples had, it is useful to take just three examples from different parts of the area. In the Central African Wet Forest the Fang people, in what is now Gabon and nearby, recognised basically two kinds of country: the wet forest itself and clearings made by people within it, that is by the Fang and neighbouring peoples. It is very likely that centuries ago the vegetation of equatorial wet forest was already being changed in composition because of a long-term process of making small clearings for cultivation. It is unlikely that Fang had a detailed narrative of how the vegetation had been changed, but they did have a refined appreciation of differences in the composition of different tracts of vegetation, which reflected the fact that tree species were not spread evenly through the Fang area or nearby.

By contrast, on the Northern Savannah, where it stretched west of Lake Chad, pastoral Fulani in some areas had no experience of wet forest whatever but recognised the differences between dry woodland, dense thicket, grassland, parkland (with scattered trees among the grassland) and seasonal swamp, under water in the wet season and the soil dried hard in the dry season. On the East African Plateau, the Maasai, another group of pastoralists, recognised the same variety of vegetation types as the pastoral Fulani but had more opportunity than they did to encounter highland wet forest, grazing their herds and flocks well up in the hills to the edge of the forest. In West Africa, because of the comparative scarcity of highland areas, the Fulani have had contact with this type of forest only in restricted areas, for instance the Futa Jalon.

As with the vegetation, so the soils were distinguished from each other within sub-Saharan classifications. One such example among the Rangi people, in what is now Tanzania, involves the recognition of three types of soil within the territory. One distinctive type is that of areas of seasonal swamp, which have just been mentioned among the Fulani and Maasai. In these areas, known as *mbuga* in Swahili and *nyika* in Rangi, the soil is very dark, sticky when it is wet, and baking hard when it dries. For this type of soil the Rangi use their term for "dark", and for the two soil types of dry well drained country they use their terms for "pale" and "reddish". The first of these types is pale grey or pale yellow, the second orange or chestnut, strongly coloured by iron oxides. Apparently, it was on these two types that the Rangi grew their crops prior to the second half of the nineteenth century, and even by the mid-twentieth they were still cultivating the *nyika* only for crops which arrived from the coast after *ca.* 1850, that is, for bananas, sugar cane and sweet potatoes. The value of the *nyika* levels during the twentieth century has been mostly for the dry-season grazing of cattle, sheep and goats, and presumably at the beginning of the nineteenth century they were used for nothing else.

To the east of the Rangi are the Maasai, already mentioned; to the southwest are the Sandawe, in country which is drier and more subject to famine-causing droughts than is Rangi country. Within Sandawe country, Newman has identified at least seven named soil types, each broadly identifiable with a vegetation type, and recognised by the Sandawe to offer different potentialities for crop production (Newman, 1970: 82–98).

Returning to their neighbours the Rangi, it is striking that their main soil types are named after words for colour which are the only words for colour in the Rangi language, and at first glance this might seem very significant. However, many African languages have only three words for colour, and those words regularly refer to the same groupings of colour: dark, pale and "reddish", i.e. orange-red-brown (i.e. bright brown). Furthermore, the same pattern of naming recurs in various parts of the world, far from Africa, for instance, in New Guinea, Australia and the Americas. And it seems that it is convenient that Rangi vocabulary lends itself so well to local soil types, but that this is not the basis for that vocabulary. A more probable correlation, valid it seems across the world, is: dark: pale: "reddish":: night: day: blood. However, colour symbolism draws us away from the immediate issue of soils and vegetation. It is not necessary in the context of these features to elaborate the processes of symbolization, or the significance of colour symbols in particular, but within African systems of knowledge, vegetation and soils flow seamlessly into other images, including a variety of symbols.[1]

The refinement and good sense of African reactions to the conditions of the landscapes in which they live is forcefully, but disturbingly, illustrated by a sequence of events in East Africa, during the years after 1945. As the fighting in Europe and the Far East came to an end, the British Government responded to the large-scale food shortage of the period with a series of initiatives, one of which is usually known in the documents of the time as the Groundnut Scheme. Westminster and Whitehall aimed at a massive boost to production of vegetable oils by growing large acreages of groundnuts in Tanganyika Territory, now incorporated into Tanzania. The scheme was to be developed rapidly, using machinery to clear woody vegetation and to work the soil. The planners in different parts of the territory identified three large areas, and these were then organised as the key centres of production. One important criterion for the choice of the three areas of the scheme was that they should be large but with few or no human inhabitants, because it was likely, in the light of administrative experience in tropical Africa, and elsewhere, that moving numbers of people would lead to obstructive resentment, to protests, and probably therefore to delays.

Within Tanganyika Territory, apparently not a single administrative officer within the civil service, and not a single technical officer, supported the scheme, which seemed to them to lack detailed local knowledge. The planners went ahead, ignoring the misgivings of the people on the spot. Despite the heavy investment, the three areas of the scheme never developed into long-term groundnut-producing localities, such as those already by then in existence in Senegal and in the north of Nigeria. The boost in global vegetable oil production did not occur, and the venture cost the British tax-payers an estimated 81 million pounds, in dead loss, and at the money values of *ca.* 1950. The Tanganyika Groundnut Scheme should be a compulsory course in any degree for development planning or tropical agriculture – but it is not. The opposition of British administrators and technical staff is very significant, but in the context of soils, vegetation and weather conditions, it is also necessary to consider the reactions and the local knowledge of the groups of people who lived near to the three large tracts designated for the scheme. One of these tracts was in the Kongwa area of Gogo country. To a European travelling through East Africa, the tract is not obviously very different from the rest of the area inhabited by the Gogo people – flat, dry, poor soils, tough woody vegetation, leafless through the long dry season. In the years after 1960, after the Groundnut Scheme was officially over, Peter Rigby, then living with the Gogo of their central area (roughly the Dodoma District), asked some people near the old Kongwa tract why so few people, even by Gogo country's sparse standards, had lived there. They replied that they knew that stretch of country well, and that the rainfall there was more than usually unreliable, that cultivated crops and grazing for the cattle and flocks were at risk, and that the name for the area in the Gogo language was equivalent to the English "drought country". But apparently no one who favoured the Groundnut Scheme had ever asked any of the Gogo of that locality![2]

This example, from an area just to the south of the areas of the Sandawe, Rangi and Maasai peoples, emphasises better than many more general examples the amount of detail which has entered into the local knowledge of African peoples, in this case in assessing the links between expected weather conditions and the productive returns on crops and livestock. That same attention to detail is particularly well demonstrated by the number of names in African languages for animals and plants. What needs emphasising is the very large number of names in common use in any one language, and that, although some individuals were better informed than others, at the time of first contact with Europeans it was normal for all the individuals of a community to be able to identify a substantial number of types of "folk species", not necessarily the same as the species of formal European taxonomy.

PLANTS AND FUNGI

In any particular area of Africa a European naturalist of the present time would recognise more species of living forms than would the local people; this is because these people have not been through the Linnean revolution in natural history of the eighteenth century, nor through the events of its aftermath. Even so, for someone who is not familiar with plants and animals, the sub-Saharan vocabularies can be daunting. Apparently, on the basis of the peoples who have been adequately documented, each language carries an implicit distinction between plants and animals, but without a word for each distinct kingdom. Basically, both types of organisms are alive, unlike, say, rocks or hills, but whereas plants stay still, animals move about. Within the plants, there is a much more explicit distinction between large plants, of the kind which in English are called trees, shrubs and large herbs, and small plants, mostly herbaceous, which form the layer of vegetation which is underfoot for people and for domestic livestock. Among these small plants are the important vegetation-forming family of the grasses, in the narrowest sense, that is the Gramineae (or Poaceae). An important distinction for African peoples, as elsewhere, is that between plants, large or small, which are domesticated and cultivated and those which are wild. A further distinction which is probably general within African languages, but needs some further investigation, is the implicit recognition, probably everywhere, of the category which European naturalists would call fungi. Names for fungus folk species are frequent in sub-Saharan languages, and some fungi are readily eaten, but an explicit category parallel to "fungi" is not a regular feature of the languages, while at the same time the fungi do not fit readily into the categories for "larger plants" and "smaller plants". For these Cecil Brown, in his survey of universal life form categories in various languages, has coined the terms "grerbs", that is grasses and/or herbs. "Fungi" seem to be some sort of implicit category among the "living forms which do not move".

Within these comprehensive groupings are the folk species, which each do have a name. There does not seem to be any great mystery about how each "species" is identified within a particular language. Are the plants large or small? This criterion has been dealt with already. Are they green in part, or in whole? This seems to be important for separating off the fungi, although some flowering plants lack chlorophyll and are therefore not green. It is worth asking how often these are classified with fungi. What colour are the flowers? What are the fruits like? What shape and size are the leaves? Are they edible? Are they particularly desirable, because in demand, and at the same time not particularly abundant? What can you do with their wood or other parts of the plant? Do they have a striking taste? Do they have a striking smell? These criteria are hardly unfamiliar in Europe, or Borneo, or Middle America. They are the stuff of folk classification and of its systematic counterpart.

An example of the importance of scent is incense, which has played a part in ritual in, for example, the Near East, India and China; plants have also been burnt as incense in sub-Saharan societies. A slightly different emphasis is the use of scented plants to attract bees to settle in a hive, the local people claiming that they are attracted by the scent. Thus, in East Africa, both the Gikuyu and Rangi peoples have used some species of Labiatae (or Lumiaceae, a family with numerous aromatic species) for the purpose of attracting swarming bees. The link with incense, if any, is obscure.

Scent, however, is not the only symbolic aspect of plants. Visual impact is also important. Conspicuous in some of the numerous dry landscapes of Africa, and the metaphysical symbolic landscapes too, are the very large fig trees, of the genus *Ficus*, with a variety of species, for example *F. fischeri* and *F. sycamorus*. During the dry season, when most trees and shrubs drop their leaves, the individuals of these species are easily seen at a distance, not only because they are huge, but because they keep their dense crowns of glossy dark green leaves. Local populations see them as symbols of vitality and health, of continuing life in the midst of prevailing death. They are thus the tropical equivalent of the holly and ivy of Europe, and of the evergreen "Christmas tree", except that in Europe the sterility is that of winter and not of the dry season.

Although some plants are symbolic, there are great numbers of species which are not. At the same time, they are not eaten or apparently used for anything, but are recognised because they are distinctive. They are recognisable because they are a familiar part of the habitat within which a particular people or sub-group live. However, a very large proportion of plants within the sub-Saharan area are seen there to have medicinal value. There is nothing peculiar about this, because it is a feature of human groups everywhere. Indeed what would be exceptional would be to find a people somewhere who attached little or no medicinal value to plants. When collecting details about plants and their uses among the southern Gikuyu, Leakey discovered that some names covered a pair or more of species (as seen from a European naturalists' point of view), which formed a single folk species because they were similar in their medicinal qualities (as seen from the local people's point of view). The Gikuyu speaking to Leakey registered that in terms of appearance, and of the observable criteria, which formed the basis of identification for most plants, the various individuals of such folk species were not all similar to each other. However, in these cases the medicinal usage was an over-riding factor in the naming and identification (Leakey, 1977: 1286).[3]

Such merging of visually distinct plants is not unusual, however, probably because medicinal use of plants is so important everywhere that people live. An example from feudal and early modern Europe is the name *Saxifraga* (English: saxifrage), which was applied to a genus of flowering plants, *Saxifraga*, as now generally recognised by botanists, and to a very different form, now recognised as a species by botanists under the name *Pimpinella saxifraga*. These two types of saxifrages look very different, and belong to two distinct botanical families, *Pimpinella* to Umbelliferae (or Apiaceae), and *Saxifraga* to Saxifragaceae. What they have in common, in the medicinal values of European culture, is that they were believed to disperse kidney stones and believed to do so because their roots were able to break stones and rocks as they were growing (*Saxifraga* means "breaking rocks" in academic Latin). Similar imagery, pharmacy and therapy occur not only in Europe and Africa but also in the contexts of other cultural traditions of healing.

The healing powers of plants south of the Sahara were widely known in each community, and lay people with no specialist knowledge practised some remedies. They were the equivalent of the "old wives' remedies" of the English. However, much of this herbal pharmaceutical knowledge in Africa, as apparently everywhere else, was available only to specialists – the diviners, that is the people, usually men rather than women, who appear in Hollywood films of Africa as "witch doctors", a term best avoided. Because of the esoteric character of much of this knowledge, and because of the vast extent of herbal pharmaceutical knowledge, it is necessary to restrict coverage of the theme here. Otherwise, it will dominate too much of a survey devoted to the whole of knowledge of the habitat, not just a part which the majority of people everywhere clearly find endlessly absorbing.

ANIMALS

The medicinal substances derived from animals are fewer, but the animals seem to be seen as forming a more complex pattern than do plants and fungi, and they also play a much more prominent role in storytelling. The degree of refinement in the identification of animals which is current among African peoples is well illustrated by the entry into European knowledge of the short-necked member of the family Giraffidae which is now known in English as the okapi. In the years around 1900 Harry Johnston (Sir Harry Johnston as he became in the process) spent a number of years near the lake of Malawi, on the East African Plateau, and in the Congo Basin. It was in this last area, in the equatorial wet forest there, that he encountered the Mbuti people, pygmies who are hunter-gatherers. Johnston was a prominent example of the late Victorian and Edwardian adventurers who attracted the attention of Sir Arthur Conan Doyle, the novelist. He was a naturalist and linguist, interested in anthropology as well as in administration, who was in his element among the Mbuti, whose vast knowledge of animals and plants, and ability to guide him through the wet forest, were a huge advantage. In the midst of his wealth of experience, however, Johnston found one feature of Mbuti knowledge puzzling. They showed him the skins of a large mammal, which they knew as *okapi*; the skins were purple-brown with zebra-like striping on the hindquarters, and looked as if they belonged to an animal related to the zebras and quaggas. On the other hand, the hoof prints which the Mbuti showed him in the wet soil of the forest floor were cloven, unlike the horse-like undivided hoof of the zebras, and resembling those of a large antelope, such as an eland. Johnston, by his own account, wondered if on this issue the Mbuti were making a complex mistake; but the group of Mbuti men now set off through the forest with Johnston, to show him one of the elusive animals, and eventually there, in a small clearing, was an okapi. Presumably, Johnston was the first European ever to see one; and the European natural history books speak readily of the "discovery" of the okapi, ignoring implicitly the Mbuti hunters who took Johnston on that dramatic excursion. Today those same books call the species *Okapia johnstoni*, a reminder of the way in which the "global" system of taxonomy tends to ignore the local vernacular systems of knowledge on which it is built, although the European taxonomists have borrowed a local name for the genus.

The Mbuti justify a small digression. Unlike most African peoples, but like the okapi, they are mammals of the equatorial wet forest, and this has had a striking effect on their symbolism. A number of African peoples use the sky or the sun as a symbol for God, but the Mbuti are in their element in the twilight under the forest canopy. They even suffer from sunburn if they spend too long in the sunlight of the clearings. For them, therefore, the appropriate image of God is not the sky or the sun, but the forest itself.

Returning to the okapi, if one asks how the Mbuti knew that these animals were indeed that, rather than something else, the quick answer is that they look like okapi, or, as the birdwatchers say, that they had the "jizz" of okapi. If we analyse out what criteria lie behind that "jizz", we find that, as with plants and fungi, there are a number of features which are, apparently, unconsciously checked when individuals are recognising folk species. Basic are size, shape, colour, and distinctive behaviour. Their smell is a less general feature in identification than is the case among plants; many of them do not wait to be sniffed. However, the black-and-white species of Mustelidae, and notably the zorilla, or African skunk, are fetid enough on occasion to be noticed. Another of these striking black-and-whites, the ratel or honey badger, is noted among many African peoples as a plucky fighter, tenaciously holding its end up, and sometimes, in self-defence, biting off the genitals of large male mammals, including humans.

Examining those vocabularies of animal names available, it is clear that a high proportion of them are among the birds and mammals, just as among the plants and fungi the majority of names are for flowering plants. This is also the case in European languages, and very widely across the world, away, that is, from unusual marginal areas, such as Polynesian islands, with their notable shortage of mammals and abundance of marine fish species. Away from the birds and mammals, in the sub-Saharan area, the naming is less often at species level. Among the reptiles, the black cobra, the puff adder and the agama lizard are likely to be named, but then, among invertebrates, there is one word for snail, one word for slug, one word for millipede, one for centipede, one for scorpion, one for spider, one for solifuge, or sun spider, one for dragonfly, and one for earwig. On the other hand, beetles (*Coleoptera*) and bugs (*Hemiptera Heteroptera*), being more diverse in appearance, are divided into various folk species. Cicadas, crickets and grasshoppers are unusual insects because they make sounds, and, like some species of birds, they are often identified by the sounds which they make. Like some species of birds also, some of these insects have names which are onomatopoetic.

Apparently, within sub-Saharan Africa and in other areas across the world, the animal folk species are grouped into physical "levels": sky, middle level, below. There is also a further category – the waters – which is perceived either as part of "below" or separately, as a category equal in value to the other three. Just as the three physical levels are a regular feature of human knowledge systems, so are particular animals linked symbolically to each level. In broadly global terms, the key animals at each level are birds of prey with the "above", big cats with the middle level, and snakes with the "below". In some regions, for instance in India and the Far East, snakes are linked equally with the waters and with the "below", an indication that in those areas the waters and the "below" merge readily into each other. However, in the sub-Saharan area this is much less clear. Crocodiles and hippopotami seem to have some significance as symbols of the waters, and in marine contexts the sea cows (*Sirenia*) also seem to have this role, manatees on the Atlantic coast, dugongs on that of the Indian Ocean. None of these, however, have the emotional impact of snakes, not usually linked by most African populations to the waters. One further complication is worth noting, and that is that pythons, which are very distinctive in appearance and not readily confused with other snakes, are seen as more auspicious and less sinister or threatening than cobras or a host of small non-poisonous species.

The "below" is not the sphere of human activity, but the middle level distinctively is, and its animal symbols are perceptibly closer to people than are those of the "below" or the sky. Pre-eminently, they are lions or leopards. They are always leopards in the lowland wet forest areas, where lions are absent; and usually lions elsewhere, apparently because they are bigger. Apparently too, no African languages (and perhaps no languages) have a word strictly translating the naturalists' technical form "mammals", but within sub-Saharan knowledge systems there does seem to be an implicit recognition of a category "animals like us". This does not include bats, which are of the sky, nor dolphins and whales, which are of the waters, and a sort of fishes. Among these "animals like us" are the domesticated forms, pre-eminently dogs, cattle, goats, sheep and donkeys. These are implicitly set apart from the wild forms, and chickens are perhaps linked to them, rather than to wild birds, because they are domestic, do not fly much and are hardly of the sky. However, they represent a problematic category of ambiguous character, joined by more recently acquired turkeys and ducks, and by the ambiguous flightless ostrich. Problematic too, but in a very different sense, is the villain of the middle 'level', malicious, sinister, bold and intimately linked with witches. This is the spotted hyena (*Crocuta crocuta*), an "animal like us", but feared and hated. The striped hyena (*Hyaena hyaena*) is also present in parts of sub-Saharan Africa and is readily linked to the spotted hyena, but it is of a much more retiring temperament and overshadowed in African feelings and symbols by its larger and bolder relative. Sub-Saharan peoples have never acquired the image of "the cowardly hyena", which Europeans built up, on the basis of the behaviour of the striped hyena – in India. It is more clearly the bold and sinister spotted hyena with takes us off into the large and important theme of witchcraft.

Without too long a digression, it is useful to give a working definition of witchcraft: it is the practice of harming members of one's own moral community by the use of superhuman power. To put the issue another way: if you use that same superhuman power to harm the foreign enemies of your community, that is not witchcraft; it is patriotism. In other words, witchcraft is not defined by the power which you use but by the moral assessment of the use which you make of that power. Witchcraft itself is one aspect of an orderly universe, with regular connections of cause-and-effect, pervaded by justice. Within sub-Saharan systems of knowledge, in which everything is made by God, it follows that God must have made witchcraft and witches. However, few sub-Saharans carry this argument through to asking the question, "Is God therefore evil?" In Africa, as elsewhere, the problem of evil arises. This could be the starting point for another book.

In the "sky" or at the "above" level, there is no figure of generally widespread witchcrafty evil, such as the spotted hyena, but there are the owls, which in Africa, as elsewhere, are disturbing because they are well informed and therefore sometimes tell people news which they do not want to hear, especially news of impending death. Furthermore, in some areas, as among the Mambila people and their neighbours in Cameroon, owls are directly linked to witches, because witches frequently turn into owls. By day, however, the place of the nocturnal owls is taken by eagles, vultures and other falconiform birds of prey, and also by other highflying birds, notably storks and cranes.

Without attempting a lengthy analysis of the way in which these animals fit into this pattern of "tiers" or "levels", the underlying theme is that each of the symbolic forms is a flashpoint at which the hidden metaphysical aspect of reality connects with physical "skin", which is more readily perceived by human observers. Thus spotted hyenas provide a sort of window into the activities of witches; and owls provide a similar window, and also one into the possibilities of clairvoyance, of seeing what is normally, in the mundane world, unseeable to human beings. A particularly vivid and compressed symbol is provided by pangolins, mammals with a scale-like body covering, and hence their English name of scaly anteaters. In terms of animal symbols pangolins appear to cross the invisible boundaries between levels; they are "animals like us", in the way in which they give birth and feed their young on milk, they are scaly, and hence, like snakes or fish, belong to the "below"/waters level; and in some areas there are species which climb trees, and hence belong also to the "above"/sky "level". An example of the tree-climbing triple level symbol occurs among the Sherbro people, who live on a small island off the coast of West Africa. This joining of the animal symbols of the different levels is not unique to Africa. Another example is the feathered snake, symbol of the god Quetzalcoatl, in Mesoamerica.

In contrast with the animals of the levels, and their fusions, are the animals that play the role of Trickster. There is nothing peculiar about the presence of Trickster stories south of the Sahara, since they are found throughout human cultures, but African peoples have a very large number of animal stories involving the Trickster, and a distinctive assemblage of species to play the role. Trickster is an amoral adventurer, always male, who tries to outwit his opponents, but is frequently outwitted by them. He represents the, apparently deliberate, perversity of human experience, the inexorable law that things always go wrong; he provides one partial solution for "the problem of evil". Within the sub-Saharan area Trickster is played variously in different areas by jackal, hare, water chevrotain (*Hyemoschus aquaticus*), dikdik and some other small antelopes, tortoise and spider. All are small animals, or at least relatively small, and none of the large animals of the area ever play the role, although they are often the butts of Trickster's schemes. Furthermore, none of the animals symbolising the "levels" are ever Trickster. However, like these animals, the Trickster species provide a window into the metaphysical world below the physical "skin".

Although not all animals which are known to African peoples are powerfully symbolic, many certainly are, and European naturalists working their way through the natural histories of African peoples often try to separate the observable appearance and behaviour of animals from the "resonances" of those animals which are immediately perceived by any African observer. For instance, when an individual Rangi sees a spotted hyena he sees not only a picture in a textbook or a sequence from a natural history film, but he also sees obscenity, evil, witchcraft and incest. Similarly snakes or chameleons detonate a whole set of moral metaphysical responses. There is a difference between the European naturalist and the Rangi in their perception of animals.[4] Even Europeans with very prim and proper modern educations still respond to elephants, lions, hyenas and snakes in ways which are symbolic, metaphysical and emotional, and not just in the ways prescribed by formal education. Most Africans do so in a less complicated or guilt-ridden way. In most African languages, there is no word that can be translated "science", none which can be translated "religion", just as there are none for "nature" or for "the environment", at least none with the same meanings as in post-modern European languages. However, the situation is becoming increasingly complex and internally contradictory. Some processes of formal education are Muslim in character and have their historical and cultural roots in the Near East, but many more, in government and mission schools, are rooted in Enlightenment modern Western Europe. It follows that large numbers of people south of the Sahara are increasingly affected by the nuances of "science" and of the Linnean revolution.

Even so, it is vital not to stretch this contrast of worldviews too far. Carl Linné's systematic natural history was rooted in the Swedish folk natural history which he absorbed as a boy, but he moved away from the levels of that natural history: bats moved away from the sky/above/birds category and became mammals that fly. Porpoises, dolphins and whales, at the same time, moved away from the waters/fish category, and became mammals specialised for swimming.

However, he was not only splitting existing categories, but also grouping them: people became mammals too, like the dolphins and the bats. They joined chimpanzees and gorillas, and numerous other species, in a grouping called Primates, and were treated as part of the Animal Kingdom, like other animals. Hitherto, European folk natural history had treated people as very distinct from animals; and the sub-Saharan approach has been basically similar, in seeing people as uniquely privileged observers of other living forms, observers who name them. Even so, the embedding of people in the world of animals was also present. Over the whole of the East African Plateau, and indeed over most of the sub-Saharan area, the local peoples would not eat chimpanzees, gorillas, baboons or monkeys, depending upon which species were present in their particular area. Behind this aversion apparently lay the recognition that these Primates are very like people. In other words, some animals are "like us", and among them Primates are "very like us". There was a kind of implicit recognition of the Primates, but with no name for them. Significantly, in parts of the Central African Wet Forest, the local people *did* eat chimpanzees and gorillas, but in those areas they also ate people (well, foreigners, anyway). The refinements of sub-Saharan animal classification go far beyond the scope of a brief introduction. But it is sufficiently clear that African peoples classified themselves as cognitionally superior to other animals, even "animals like us" and that they saw themselves as having clear affinities with God.

NOTES

[1] Any reader who would like to digress into these other byways could read Turner's (1967) account of ritual symbols in the Ndembu sub-group of Lunda in the Southern Savanna sub-region.

[2] I owe this account of the Gogo comments to the late Peter Rigby.

[3] Leakey himself readily uses the term 'magical', which is best avoided because it has no precise technical meaning.

[4] Even a small dull-coloured beetle or plant bug is significant, because it is one of the less prominent representatives of the 'below'.

BIBLIOGRAPHY

Brown, C.H. 'Folk zoological life-forms: their universality and growth.' *American Anthropologist* 76: 325–7, 1979.

Evans-Pritchard, E.E. *Witchcraft, Oracles and Magic Among the Azande.* Oxford: Clarendon Press, 1937.

Jacobson-Widding, A. *Red-White-Black as a Mode of Thought.* Stockholm: Almquist and Wiksell, 1979.

Kesby, John D. 'The Rangi classification of animals and plants.' In *Classifications in Their Social Context,* Roy F. Ellen and David Reason, eds. London, New York, San Francisco: Academic Press, 1979, pp. 33–56.

Kesby, John D. *Rangi Natural History: The Taxonomic Procedures of an African People.* 3 vols. New Haven, Connecticut: Human Relations Area Files, 1986.

Kingdon, Jonathan. *East African Mammals*. 3 vols., 7 parts. London, New York, San Francisco: Academic Press, 1971–82.

Kingdon, Jonathan. *The Kingdon Field Guide to African Mammals*. San Diego: Academic Press, 1997.

Leakey, L.S.B. *The Southern Kikuyu before 1903*. 3 vols. London, New York, San Francisco: Academic Press, 1977.

Middleton, John, and E.H. Winter, eds. *Witchcraft and Sorcery in East Africa*. London: Routledge and Kegan Paul, 1963.

Morris, Brian. 'Chewa conceptions of disease: symptoms and etiologies.' *Society of Malawi Journal* 38(1): 14–36, 1985.

Morris, Brian. 'Additional note on the herbalist associations.' *Society of Malawi Journal* 38(1): 37–43, 1985.

Morris, Brian. 'Medicines and herbalism in Malawi.' *Society of Malawi Journal* 42(2): 34–54, 1989.

Morris, Brian. 'Animals as meat and meat as food: reflections on meat eating in southern Malawi.' *Food and Foodways* 6(11): 19–41, 1994.

Morris, Brian. 'Woodland and village: reflections on the 'animal estate' in rural Malawi.' *Journal of the Royal Anthropological Institute* 1(2): 301–316, 1995.

Morris, Brian. *The Power of Animals: an Ethnography*. Oxford: Berg, 1998.

Morris, Brian. *Animals and Ancestors: an Ethnography*. Oxford: Berg, 2000.

Newman, James L. *The Ecological Basis for Subsistence Change Among the Sandawe of Tanzania*. Washington, DC: National Academy of Sciences, 1970.

Oliver-Bever, Bep. *Medicinal Plants in Tropical West Africa*. Cambridge: Cambridge University Press, 1986.

Riley, Bernard W. and David Brokensha. *The Mbeere in Kenya*. 2 vols. Lanham, New York,: University Press of America, 1988.

Roberts, A. *Birds of South Africa*. Cape Town: John Voelcker Bird Book Fund, 1940, and revisions.

Turner, Victor W. *The Forest of Symbols: Aspects of Ndembu Ritual*. Ithaca, New York: Cornell University Press, 1967.

Watt, John Mitchell and Maria Gerdina Breyer-Brandwijk. *Medicinal and Poisonous Plants of Southern and Eastern Africa*. Edinburgh and London: E. and S. Livingstone, 1962.

J.L. KOHEN

KNOWING COUNTRY: INDIGENOUS AUSTRALIANS AND THE LAND

I feel it with my body
With my blood
Feeling all these trees
All this country
When the winds blow you can feel it
Same for country ...
You feel it
You can look
But feeling ...
That makes you.

Bill Neidjie*

In these words, Bill Neidjie, a traditional owner of part of Kakadu National Park in the Northern Territory, clearly illustrates the relationship between Aboriginal people and the land. It is by "feeling" the land that a person is "made", or really exists. He is talking about what he feels when he is in his "country", a term used by Aboriginal people to describe those geographic areas and the landscapes within them with which they have inherited rights and responsibilities. Aboriginal Australians have a special relationship with the land and with everything that exists upon the land.

An Aboriginal person looking at a landscape will be a part of that landscape, not just an external viewer. If the land is the place where he or she was born, then there is a special relationship between the person and the place. If the land is not a person's own "country", there may be links to the land through extended family networks. Even if the land is not their "country", there will be respect shown towards the land and the people who are traditional owners of the land. Even today in urban Australia, if an Aboriginal person is speaking before an audience, it is customary first to acknowledge the traditional owners of that place.

Amongst the Miriwoong people of the East Kimberley region of Western

---

* Cited in Birckhead, De Lacey and Smith, 1993.

H. Selin (ed.), Nature Across Cultures: Views of Nature and the Environment in Non-Western Cultures, 229–243.

west

Australia, the links to the land are passed down through the women. The strongest links are with the mother's country, and with the mother's mother's country, although connections still exist with the father's country and the grandfather's country. In much of coastal southeastern Australia, the link to country is passed down through the father, and the father's father; links to the mother's country are less important.

Traditional Aboriginal society is divided into several levels of organization. At the broadest level, people may speak different languages, but they can have close ceremonial and spiritual links, perhaps sharing dreaming stories. For example, Aboriginal people from the Kimberley in the northwest of Western Australia to Alice Springs in Central Australia share a creation story which extends across many language groups over thousands of kilometres.

The tribe is generally defined in terms of people who share a common language, although several dialects of the language may be spoken. In areas where resources are plentiful, the population density before white settlement was high, and several different language groups may exist within a few hundred square kilometres. For example, five distinct languages, Darug, Kuringgai, Dharawal, Gundungurra and Darkingjung, were all spoken within a radius of 100 km of Sydney (Kohen, 1995a). In contrast, in the arid environments of Central Australia, a single language such as Aranda or Warlpiri may extend over hundreds of thousands of square kilometres (Blake, 1981; Dixon, 1980).

The land owning unit in Aboriginal societies is generally the clan. A clan consists of a group of biologically related families, numbering from perhaps as few as twenty people to over one hundred, who traditionally live on and use the land for both spiritual and economic activities. Indeed, it is often not possible to separate the two, for when hunting and gathering takes place, religious prohibitions and totemic beliefs will always be considered.

Kinship, or "relatedness", extends beyond biological links, and involves the classification of individuals into like social categories. In this way, all Aboriginal people will be related to each other. In southern Australia, it is still common for elders in the Aboriginal community to be given the classificatory title of "uncle" or "auntie", even though no biological relationship actually exists. In the same way, it is possible for a person to be "related" to an animal or plant. If a person has a kangaroo for a totem, then often the kangaroo is seen as a brother or sister, and might not be harmed in any way, although this is not always the case with totems and cannot be assumed *a priori*.

Oral traditions are vitally important. Although paintings and engravings often depict creation events, without a written language Aboriginal cultural traditions are usually passed on by stories, song and dance. Important events in the creation of the landscape are often depicted in song and dance, and only those people who have a right to them know these stories. An Aboriginal custodian will know how to "sing" a place, to tell its story by repeating a special song which relates to its creation and significance. Knowing how to sing a place is an affirmation of ownership and responsibility for that place.

As hunters and gatherers, Aboriginal people have always needed to utilize the plants and animals which exist on their land. In order to maintain these resources, cultural and spiritual prohibitions have evolved which ensure that there will always be enough food to maintain the population. In addition, active management of the land to sustain the plants and animals is needed, and this is referred to as "caring for country" (Young *et al.*, 1991). Aboriginal people believe that setting fire to the country will maintain and improve it.

TRADITIONAL ABORIGINAL USE OF FIRE

Aboriginal people have interacted and continue to interact with natural environments right across Australia (Bowman, 1998, 2000; Kohen, 1995b). In Arnhem Land in the Northern Territory, several rock shelter occupation sites have been lived in for 50,000 years (Roberts *et al.*, 1990). In the Kimberley Region of Western Australia, recent archaeological evidence suggests that Aboriginal people have lived there for over 42,000 years (O'Connor, 1999), and even in the far south in Tasmania, prehistoric sites have been dated to 35,000 years ago (Mulvaney and Kamminga, 1999). Over this history, they exploited the resources which were available to them in much the same way that hunters and gatherers had been doing in other parts of the world and developed land management practices which allowed them to persist with the longest isolated cultural tradition in the world.

One of the technologies which may well have had a major impact on the environment is the use of fire. An understanding of the way Aboriginal people use fire, both in the past and in the present, is critically important, because the use of fire as a land management tool is closely linked to beliefs about the responsibility to care for "country", as it is in many parts of the world. Aboriginal people will often say that they are "cleaning up the country" when they burn it. They see fire as a way to maintain their relationships with the land by caring for it. Fire is a form of housekeeping, designed to remove unwanted understorey vegetation and to promote grass (Young *et al.*, 1991).

Some researchers, like David Horton from the Australian Institute of Aboriginal and Torres Strait Islander Studies, suggest that, "Aboriginal use of fire had little impact on the environment and ... the patterns of distribution of plants and animals which obtained 200 years ago would have been essentially the same whether or not Aborigines had previously been living here" (Horton, 1982: 237; see also Horton, 2000). Other people have misunderstood or misrepresented Aboriginal fire regimes. For example, Barnett (1998) suggested that Aboriginal fire management in Kakadu National Park is totally destructive. He failed to recognise that the diversity of ecosystems present in Kakadu is the result of active management by the Aboriginal custodians. In many parts of Australia, Aboriginal traditional knowledge about appropriate fire regimes is being used to good effect in the management of National Parks (Birckhead *et al.*, 1993; Creagh, 1992; Rose, 1995).

The archaeologist Rhys Jones first proposed the idea of "firestick farming", the implication being that Aboriginal people unintentionally and intentionally modified the environment by the use of fire (Jones, 1969). Certainly Aborigines had been observed using fire to burn large tracts of land since the first European settlements, and it was clear that fire was an important tool. However, Jones was one of the first to suggest that this burning was controlled or directed. He saw fire as important in increasing the productivity of the land, by replacing mature forests with open woodlands and grasslands.

Based on the use of fire by contemporary Aboriginal communities, particularly in northern and Central Australia, it is now recognised that Aboriginal people used that fire primarily to clear the underbrush to make travel easier, hunt large game, facilitate the gathering of small game (burning possums out of hollow trees) and drive away snakes from camping spots.

However, there were other beneficial consequences. Fire would recycle nutrients and promote new growth, which would subsequently attract herbivores like kangaroos and wallabies. Many plants were favoured by regular low intensity burning. A burning pattern consisting of frequent, low intensity fires removes the woody understorey plants, and allows many of the grasses, yams, orchids and lilies to flourish (Pain, 1988). Orchids and lilies often possess underground storage organs or tubers, which were eaten by the Aborigines.

There may also be negative impacts caused by Aboriginal burning. It has been suggested as the cause of hill slope instability (Hughes and Sullivan, 1981), valley infilling, sedimentation of the estuaries (Hughes and Sullivan, 1986), reduction of the margins of the rainforest, and changes in ground cover which affected the distribution of small terrestrial animals (Kohen, 1995b). Added to these are the likely impacts on arboreal mammals and the larger terrestrial animals. Clearly, fire at least has the potential to change dramatically both the physical and the biological environment in which Aboriginal people exist.

One phrase which has been coined to describe traditional Aboriginal burning practices is "peripatetic pyromania" (see Jones, 1969), which gives the image of people wandering aimlessly across a landscape setting fire to the bush wherever they went. The evidence suggests this was not the case. The burning generally took place at appropriate times of the year and when weather conditions were right. This ensured that there was a low intensity burn and therefore little danger of a crown fire developing. Traditional burning patterns maximised the species diversity in any particular area, because burning tended to leave a mosaic of vegetation which had been burned at different times. Aboriginal people burned in country which they knew and understood, as were the consequences of burning.

This regular firing favoured not only fire-tolerant or fire-resistant plants, but also encouraged those animals which were favoured by more open country. The actual species which were encouraged varied from place to place, depending on the climate, vegetation and rainfall pattern. Aboriginal burning, in many areas at least, did impact on the "natural" ecosystem, producing a range of vegetation associations which would maximise productivity in terms of the food requirements of the Aborigines. What Aboriginal people were trying to achieve was a balance between the need to burn some areas to promote certain resources and the need to protect other areas where particular plant foods grew, perhaps in areas like rainforests and wet sclerophyll[1] forests. Some fires were clearly designed to act as firebreaks, to prevent wildfires burning valuable fire-sensitive resources. Rather than suggesting that Aborigines burned the landscape, perhaps it is better to say that they managed the landscape, and that fire was one of the tools which they used.

FIRE AND CONTEMPORARY ABORIGINAL COMMUNITIES

One of the difficulties in understanding the consequences of Aboriginal burning is that at all locations where Aboriginal communities still practice burning today, their patterns of burning have changed. In some regions Aboriginal people no longer have access to all of their territory, they now have much greater mobility due to the use of cars and trucks, they are much more selective in what they hunt, and it is much easier to shoot a kangaroo with a rifle than to spear it. Fires now tend to be lit in proximity to roads and tracks and to be of much greater spatial extent than the previously controlled small scale burns.

In the East Kimberley, Aboriginal fire sometimes gets out of control. Fire is occasionally used as a means of satisfying "payback" responsibilities, and extensive wildfires have resulted from fires lit beside the road to light the way for people travelling on foot at night. Indeed, Head (1994) argues that the East Kimberley fire regimes should not be seen as a continuation of "traditional" Aboriginal burning practices, in contrast to, say, Arnhem Land in the Northern Territory. This is because the pastoral invasion was experienced by the parents of people alive today, so there has been dislocation from country. Aboriginal people are much more sedentary on one hand (living in communities and at outstations), but also more mobile (using vehicles and boats) on the other. She argues that the vegetation associations which will change with increased Aboriginal burning are those which are not fire adapted, and she points to the differences in vegetation associations across the country. With such variation, it would be unusual if there were no variations in the response of vegetation to fire. She points out that in recent times, Aborigines have been maintaining both fire-dependent and fire-resistant communities.

Even burning undertaken on behalf of the managers of cattle stations by Aboriginal workers is often of doubtful or limited value. This is either because the reasons for the burn are not clearly enunciated or because the person lighting the fire simply considers it an "appropriate" time and place to burn, without any valid cultural reasons, either from an Aboriginal perspective or from a European land manager's.

On the other hand, recent ethnographic accounts in Arnhem Land suggest that current burning regimes there produce few accidental fires (Russell-Smith et al., 1997). Pyne (1991) describes the burning practices of the Gunei tribe in detail and suggests that burning actually starts at special sites while the rainy season is still in progress. It escalates as drying spreads, and it culminates at the end of the Dry with a conflagration of those places destined for burning but not yet fired. He clearly suggests that there is controlled, planned, directed burning.

The reasons for the different stages are also suggested, although Aboriginal informants would not necessarily concur with the proposed reasons that they traditionally burn in a particular fashion or in a particular sequence. Protective early burning acts like a firebreak to prevent damage to the rainforests (Bowman, 2000). The burning regime of different habitats is varied, with lowland sites generally burnt every year and other areas burnt on a cycle of two, three or even four years. It is the local knowledge of "country" which results in the ability to use fire as an effective management tool.

Such burning practices may result in both increases and decreases of some animal species in the short term. Waterfowl thrive in sedges which regenerate shortly after burning, but snakes have nowhere to hide and disappear into the surrounding areas. To counter the claim that many of the results of Aboriginal burning are accidental, there are many accounts in the literature which demonstrate conclusively that Aboriginal people were using fire in a well-planned manner. Pyne (1991: 123) suggests that the people in Arnhem Land "... exercise control by timing the fires with diurnal wind shifts, by relying on the evening humidity, and by exploiting topographic features like cliffs and streams and old burns ... [T]he fires were sequential, and burning a composite of practices in a mosaic of environments that extended over nine to ten months. Most of the grasslands and savannahs burned; portions of the floodplains burned twice; the woodlands and forests burned on the order of a fourth to a half their area".

Implicit in this statement is the view that some areas are intentionally burned and some are left unburned; this is called mosaic burning. Pyne goes on to suggest, "What distinguished the Aboriginal regime was the amount of burning that occurred in the midseason, the regularity of the annual burning cycle, the biological timing of the fires, the insistence that some areas never be fired and that others be fired as often as possible". In many ways, it is the strong seasonality in the tropical north which determines when fires will be lit. In other parts of Australia, where rainfall is not divided into a consistent wet and dry season, other patterns developed.

In the desert areas, other factors may influence burning patterns. Rainfall, or lack of it, is sporadic and unpredictable and can override any man-made firing regime. If there is no standing crop then there is nothing to burn, so there are no fires. If the land is underwater, there are no fires. In contrast to the tropical north, there is no defined wet and dry season on which the burning pattern can be based. Under these circumstances, when conditions did allow an area to burn, the fire could burn over much wider distances. It is possible to view Central Australia as a region where Aboriginal burning of particularly valuable areas may indeed have protected these areas from more intense wildfires which could otherwise have been initiated by lightning strikes and which could have caused depletion of resources.

The impact of the cessation of Aboriginal burning in the arid zone became clear in the 1920s and 1930s, when traditional people began to move from the deserts into the missions and reserves. Without regular burning, large wildfires broke out in the 1920s, 1950s and again in the 1970s, initiated by lightning strikes. Because there was such an accumulation of fuel, because of the cessation of regular burning, when fire did break out it was unusually large. It resulted in large-scale extinction of the medium sized animals, those with body weights between forty-five grams and 5 kilograms. In those places where traditional burning regimes remained, such as in the Gibson Desert where the Pintubi people still hunted and gathered, the medium sized animals survived. Admittedly, by the 1920s other changes had occurred in central Australia. Exotic plants and animals had become established, and Europeans had begun to impact on the margins of the deserts with their land management practices.

In 1974 and 1975, 80% of Uluru National Park in the Northern Territory was burnt by a fire, an event which simply could not have happened if traditional Aboriginal burning practices had been allowed to continue. Now, these traditional burning practices are being incorporated into the park's plan of management (ANPWS, 1991).

While Aboriginal people in northern Australia used fire as a tool for increasing the productivity of their environment, Europeans in southern Australia see fire as a threat. Without regular low intensity burning, leaf litter accumulates, and crown fires can result, destroying everything in their path. Europeans fear fire because it can destroy their houses and their crops, and it can destroy them. Yet the environment which was so attractive to them for grazing sheep and cattle was created by fire. It has been suggested that the European settlement of Tasmania followed almost exactly those areas which the Palawa or Tasmanian Aborigines had regularly burned (Pyne, 1991).

If fire is rare in a community, then a fire may result in the loss of species, but where fire is common, and where most species are fire-adapted, most species will have effective strategies for survival and will therefore not be adversely impacted in the long term, although of course there may be short term reductions in the number of individuals within the community. However, even in communities which are adapted to fire, adaption is to a particular fire frequency-intensity regime. A change in the regime can also result in a reduction in the number of species. A good example is a plant which is killed by fire and which produces short-lived seed. Successive fires can eliminate these species from a community. This is the kind of problem which managers of National Parks are trying to cope with at the moment. If they do not burn the parks, some species will become extinct; if they do, others may become extinct.

The strong interrelationship between plants, animals and fire is apparent in Australia. Such relationships evolved because the Australian vegetation was exposed to fire long before Aborigines arrived on the continent. Plants and animals in dry sclerophyll communities co-exist in dynamic equilibrium because the structure, size and distribution of populations are constantly changing. The creation of vegetation mosaic based on frequency of firing by Aboriginal people therefore must be seen as a mechanism for increasing the range of habitats, the range of stages of regeneration after burning, and therefore the range of micro-environments. In terms of the overall species diversity, such a pattern would increase the total number of species because it maximises the habitat range. Some species will be disadvantaged, and the populations will decline, and others will be advantaged, but the diversity will be maintained.

If Aborigines wanted to maximise the productivity of the landscape, what strategy should they adopt? The animals in the sclerophyll forests are adapted to fires and will survive fire. Woodlands and forests can be burnt regularly without any long-term decline in the fauna, and if they are burnt every 1–3 years the kangaroo and wallaby populations will be kept near the sustainable maximum level. On the other hand, it would be unwise to burn wet sclerophyll and rainforests, because the animals in those vegetation associations tend to be specialist, not well adapted to fire. If you want to protect those animals, you do not burn the rainforests.

With European settlement, Aboriginal people were displaced and traditional Aboriginal burning regimes changed. Fires were no longer routine. Vast areas of Australia were no longer subject to annual low intensity fires. Even when Aboriginal people had access to parts of their land, the spatial extent and frequency of fires changed. The cultural need to burn was important, but because of restricted access, some burning took place in the wrong season. Some areas were left unburnt, allowing an accumulation of fuel, and when a fire did occur it was often a crown fire, destroying the forests. European land management practices could not tolerate the possibility that crown fires might develop and destroy the infrastructure that was necessary to run a property. A new tool was introduced – hazard reduction burning. Areas of land were intentionally burnt to reduce the accumulation of fuel and to ensure that a crown fire did not develop.

The consequences for the fauna of changing fire regimes when Aboriginal burning ceased are also apparent. The shift from summer fires to winter fires in the arid spinifex (spikey grass) country changed the nature of the vegetation and consequently the fauna. Small mammals have been impacted most severely. As long as the fires are not too frequent in the woodlands and dry sclerophyll forests, and provided there is a mosaic of vegetation maintained by varying the frequency of fires in different areas, the burning pattern in these areas would not seem to be critical. However, the fire sensitive animals are in the tall forests, wet sclerophyll forests, and rainforests, and an increase in the fire frequency in these areas is likely to have an impact on the specialised animals which occupy those specialised niches.

Why do Aboriginal people use fire? Fire would not be routinely and regularly used unless there was a particular need for it. The Gunei in Arnhem Land do not burn for nine or ten months of the year because they are pyromaniacs; they burn to increase the productivity of the landscape. Fire is used for immediate hunting, to promote food plants and animals in the future, and to make travel easier. These things are only necessary if food is scarce or people need to travel over long distances.

The intensive use of fire, so ubiquitous in Aboriginal culture, tells us that Aboriginal populations were exploiting the landscape to the best of their ability, supporting the greatest human population density, ensuring a constant supply of food both in the present and in the future, and facilitating movement across the landscape to allow these other things to happen.

ABORIGINAL CONSERVATION MECHANISMS

While it has been suggested that Aboriginal people can and did cause localized extinctions of some animals under some conditions, the evidence is at best circumstantial (Abbott, 1980; Flannery, 1994). As with any group of people dependent on renewable resources, Aboriginal people must have mechanisms in place which ensure that individual resources are not over-exploited. This can be done in many ways, and Aboriginal people use a combination of practical skills, totemic restrictions and religious beliefs to ensure that valuable resources are sustained.

When women are gathering yams, it is usual for the top of the yam to be broken off and replanted, ensuring the resource will once again be available the following year. Similarly, seeds of fruits are dispersed around major camp-sites to increase the local availability in the future. In Central Australia, trading networks were established which included exchange of valuable resources including seeds of food plants (Kimber, 1984). The knowledge of how to germinate seeds to produce additional food was passed down from generation to generation in dreaming stories.

An important conservation measure is the system of totems which determines a person's relationship with all other people and animals. If a man has a kangaroo for a totem, he will not hunt kangaroo, so at any time, a proportion of the population will not be hunting kangaroo. Similarly, other totemic animals will not be hunted. This reduces the likelihood that a particular animal or plant will be over-exploited. Because animals like kangaroos and possums are common totems, the hunting of these animals will be minimized, although under some conditions one can eat certain parts of a person's totemic animal.

Cultural prohibitions also exist against hunting and eating either male or female animals at certain times of the year, or when an individual is passing through a particular stage of life. There are restrictions on what foods can be hunted and eaten by young children, pregnant women, boys undergoing initiation (and in many cases also their parents), and adult men and women during ceremonies. In southeastern Australia, only initiated men could eat female emu, while in the Sydney region stingrays and sharks were not eaten at all (Kohen, 1993).

Amongst the Miriwoong people of the Eastern Kimberley, crayfish and crayfish eggs were actively transported and released into new waterholes to increase their range and availability. This process seems to have operated over much of inland Australia. Female crayfish carrying eggs were released, and only males were eaten. For some species of shellfish, there were prohibitions against the collection of juvenile individuals. In Arnhem Land, if children collected shellfish which were too small, they were told to return them to the water (Meehan, 1982). If a resource did become scarce, then an alternative source of food was found until the population could recover.

There were also areas which were rarely visited and which acted as reserves for some animal species. Often these were places of special religious significance, which would only be visited on special occasions. The knowledge of the location

and importance of these special sites was retained by the traditional owners
and custodians of the land.

Since the arrival of Europeans in Australia, Aboriginal people have been
dispossessed from their traditional lands and their culture forced to adapt to
at least some aspects of the European lifestyle. However, in many parts of
Australia, notably Arnhem Land, Central Australia, the Kimberley, Cape York
and the Western Desert, Aboriginal people have retained control over large
areas of land. In many cases, they still use their land to obtain food. However,
they have adopted many modern conveniences in order to improve the effectiveness of their hunting and gathering, and they often mix European goods
and traditional foods. The conservation value of land which is controlled by
Aboriginal people is well recognised by conservationists and politicians alike.

A few hundred kilometres to the south of Derby in Western Australia, the
Martujarra people of the Little Sandy Desert maintain a far greater use of
traditional plant foods than many other Aboriginal communities. Martujarra
women dig tubers from the ground using both a traditional digging stick and
a shovel, and Cyperus bulbs are similarly dug up and stored in plastic dishes.
Veth and Walsh (1988) suggest that tubers of Ipomoea are a favourite "bush
tucker" of people who live in Jigalong. Acacia seeds, particularly green ones,
and grass seeds also still form part of the traditional diet. Twelve different
fruits are eaten, and the Santalum and Solanum species are common in spinifex
areas which have been burnt less than 5 years ago, so fire is still an important
component of their socio-economic system. The Martujarra retain the social
context of gathering and use a balance of "bush" foods and bought foods, both
of which are available in the vicinity of their community. In order to maintain
this lifestyle, they must continue to burn the environment in the way their
ancestors did.

Contemporary communities use land for a variety of purposes, including
hunting and gathering, pastoralism, tourism and cultural practices. Aboriginal
culture has always been dynamic, incorporating new technology and new
beliefs, while maintaining an underlying "Aboriginality" (Young et al., 1991).
The holistic nature of Aboriginal society is reflected in the way in which the
relationship with the land is part of the economic, social and spiritual makeup.
It is only by maintaining the spiritual links to the land that the land will be
productive. If social and religious customs are not observed, there are likely to
be adverse economic consequences. Food animals may become scarce, or water
holes may dry out. Strang (1997: 84) argues correctly that within Aboriginal
land use patterns, "economic interactions are never wholly divorced from social
and spiritual interactions".

The view that Aboriginal people traditionally lived a harmonious existence,
with little impact on the environment, is simplistic. However, in the modern
world, traditional owners of country could view what may be seen as a dramatic
environmental disaster in an entirely different light. For example, the Kuki-

Yalanji people of northeast Queensland strongly supported the construction of a road through the Daintree Rainforest, which was part of their country (Anderson, 1989). White conservationists saw this as something contrary to what they perceived as the "proper" protection of country which should have been given. However, access to spiritual sites and areas rich in resources may well have more than compensated for the potential loss of land resulting from the construction of the road. Indeed, the possibility of generating income from increased tourism, especially if part of the tourist activities included exposure to Kuki-Yalanji culture, is beneficial to the Aboriginal community as a whole, as well as to those visitors who learn about the culture. The value of cultural tourism has long been recognised as being of important economic and social value for contemporary Aboriginal communities (Altman, 1988). The Tiwi people of Melville and Bathurst Islands, and the Aranda and the Pitjantatjara of the central desert, are involved in cultural tourism which incorporates non-exploitative use of wildlife (Collins et al., 1996).

The changes which have taken place in hunting and gathering over the last 200 years have meant that many traditional animal foods are often no longer available. Legislation may restrict the hunting of animals, or, as was the case with the dugong hunters at Cape York and in the Torres Strait, the indigenous communities may stop hunting voluntarily because the population was declining dramatically (Marsh, 1996; van Tiggelen, 1996). In this case, the community decided that the continued existence of a significant totemic and spiritual animal was more important than the economic benefit gained by eating dugong meat.

European land managers, through joint management programs of National Parks, have used traditional Aboriginal knowledge. In the Northern Territory, both Kakadu in the tropical north and Uluru Kata Tjuta in the Central Australian desert are leased back to the Federal Government, even though the traditional owners were granted inalienable freehold title (Nutting, 1994). There is clear recognition that Aboriginal ecological knowledge can contribute to the conservation of rare and endangered species. It is also a de facto recognition that Aboriginal owners of country are the best managers of that country, although often an "environment first" approach is adopted by non-Aboriginal managers (Creagh, 1992).

Aboriginal people have begun to regain control over their land, particularly in Northern and Central Australia. Much of the land has been overgrazed and abused by European land managers. As a consequence, Archer et al. (1998) suggest that a number of problems may arise. On the positive side, traditional fire regimes may assist in the conservation of rare and endangered species, but on the negative side, Aboriginal people may not have the skills and resources necessary to rehabilitate degraded land.

From a non-Aboriginal perspective, the use of introduced species as food resources may lead to conflict with conservationists who wish to eradicate them, and there will be opposition to the continued hunting of rare and endangered species. In addition, the use of outstations has changed the density and distribution of Aboriginal populations, which, along with access to modern technology and greater mobility, may affect predation rates. It is also likely that social imperatives to "clean up country" may be stronger than ecological ones to protect fire-sensitive food products. Langton (1998) recognises that these are all becoming important issues for northern Australian Aboriginal communities.

Clearly there may be significant differences between the western conservation ethic and Aboriginal cultural perspectives. Aboriginal communities manage their resources through social and spiritual controls, but the methods and outcomes may well be different. Aboriginal people know the land where they belong. Land is a culturally defined resource. It plays an integral role in affirming Aboriginal identity through its use as the basis for subsistence, cultural practices, tourism and conservation (Altman, 1988; Altman and Finlayson, 1992).

The concept of Native Title, and the legislative framework which was adopted by the Australian Government in 1993, goes some way to defining in a European legal sense what "country" is to traditional owners. It recognises that Aboriginal people may retain rights of access for ceremonies, hunting and gathering on land over which native title rights have not been extinguished. Under the legislation, certain conditions apply, and for an Aboriginal family or group to successfully claim native title rights they must be able to demonstrate that they are the descendants of people who were traditional owners, that they have maintained continuous association with the land, and that they continue to maintain "traditional" activities on the land. This may include fishing, hunting, collecting plants or pigments, or carrying out ceremonies.

Regardless of whether the legal recognition of native title rights is granted, Aboriginal people will continue to assert their connection to country. The focus of Aboriginal existence is the relationship with the land and the local knowledge of the land. For indigenous Australians, the land has always belonged to them, and they are obligated to look after it.

NOTES

[1] Sclerophyll plants are those which have evolved leathery evergreen foliage and are adapted to survive arid conditions.

BIBLIOGRAPHY

Abbott, I. 'Aboriginal Man as an exterminator of wallaby and kangaroo populations on islands round Australia.' *Oecologia* 44: 347–354, 1980.

Altman, J.C. *Aborigines, Tourism and Development: The Northern Territory Experience.* Darwin: ANU North Australian Research Unit Monograph, 1988.

Altman, J.C. and J. Finlayson. *Aborigines, Tourism and Sustainable Development.* Canberra: Centre for Aboriginal Economic Policy Research, 1992.

Anderson, C. 'Aborigines and conservationism: The Daintree-Bloomfield road.' *Australian Journal of Social Issues* 24: 214–227, 1989.

ANPWS. *Sharing the Park. Anangu Initiatives in Ayers Rock Tourism.* Canberra: Australian National Parks and Wildlife Service, 1991.

Archer, M., I. Burnley, J. Dodson, R. Harding, L. Head and L. Murphy. *From Plesiosaurs to People: 100 Million Years of Australian Environmental History.* Australia: State of the Environment

Technical Paper Series (Portrait of Australia). Canberra: Department of the Environment, 1998.

Barnett, D. '"Fire-stick farmers" are killing Kakadu.' *Financial Review*, 22 January, 1998.

Benson, D. and J. Howell. *Taken for Granted. The Bushland of Sydney and its Suburbs*. Sydney: Kangaroo Press in association with the Royal Botanic Gardens, 1990.

Birckhead, J., T. De Lacy and L. Smith, eds. *Aboriginal Involvement in Parks and Protected Areas*. Canberra: Australian Institute of Aboriginal and Torres Strait Islander Studies Report Series, Aboriginal Studies Press, 1993.

Blake, B. *Australian Aboriginal Languages*. Sydney: Angus and Robertson, 1981.

Bowman, D.M.J.S. 'Tansley Review No. 101. The impact of Aboriginal landscape burning on the Australian biota.' *New Phytologist* 140: 385–410, 1998.

Bowman, D.M.J.S. *Australian Rainforests. Islands of Green in a Land of Fire*. Cambridge: Cambridge University Press, 2000.

Catling, P.C. 'Ecological effects of prescribed burning practices on the mammals of southeastern Australia.' In *Conservation of Australia's Forest Fauna*, D. Lunney, ed. Mosman: Royal Zoological Society of NSW, 1991, pp. 354–363.

Collins, J., N. Klomp and J. Birckhead. 'Aboriginal use of wildlife: past, present and future.' In *Sustainable Use of Wildlife by Aboriginal Peoples and Torres Strait Islanders*, M. Bomford, and J. Caughley, eds. Canberra: Australian Government Printing Service, 1996, pp. 14–36.

Coombs, H.C., J. Dargavel, J. Kesteven, H. Ross, D.I. Smith and E. Young. *The Promise of the Land: Sustainable Use by Aboriginal Communities*. Canberra: Australian National University, 1990.

Creagh, C. 'Looking after the land at Uluru.' *Ecos* 71, Autumn 1992.

Cribb, A.B. and J.W. Cribb. *Wild Food in Australia*. Sydney: Collins, 1974.

De Graaf, M. 'Aboriginal use of fire.' In *Report on the Use of Fire in National Parks and Reserves*, R.E. Fox, ed. Darwin: Department of Northern Territory, 1976, pp. 14–20.

Dixon, R. *Languages of Australia*. Cambridge: Cambridge University Press, 1980.

Dodson, J., ed. *The Naive Lands. Prehistory and Environmental Change in Australia and the Southwest Pacific*. Melbourne: Longman Cheshire, 1992.

Ellis, B. 'Rethinking the paradigm: cultural heritage management in Queensland.' *Ngulaig* 10: 1–25, 1994.

Flannery, T. *The Future Eaters. An Ecological History of the Australasian Lands and People*. Sydney: Reed Books, 1994.

Gamble, C. 'The artificial wilderness.' *New Scientist* 10 April 1986, pp. 50–54.

Gill, A.M. 'Fire and the Australian flora. A review.' *Australian Forestry* 37: 4–25, 1974.

Gill, A.M., R.H. Groves and I.R. Noble, eds. *Fire and the Australian Biota*. Canberra: Australian Academy of Science, 1981.

Hallam, S. *Fire and Hearth*. Canberra: Australian Institute of Aboriginal Studies, 1975.

Head, L. 'Landscapes socialised by fire: post-contact changes in Aboriginal fire use in Northern Australia, and implications for prehistory.' *Archaeology in Oceania* 29(3): 172–181, 1994.

Head, L. and R. Fullagar. '"We all la one land": Pastoral excisions and Aboriginal resource use.' *Australian Aboriginal Studies* 1: 39–52, 1991.

Horton, D. 'The burning question: Aborigines, fire and Australian ecosystems.' *Mankind* 13(2): 237–251, 1982.

Horton, D. *The Pure State of Nature: Sacred Cows, Destructive Myths and the Environment*. Sydney: Allen and Unwin, 2000.

Hughes P.J. and M.E. Sullivan. 'Aboriginal burning and Late Holocene geomorphic events in eastern N.S.W.' *Search* 12: 277–78, 1981.

Hughes, P.J. and M.E. Sullivan. 'Aboriginal landscape (effects of Aboriginal occupation and fires on erosion).' In *Australian Soils: The Human Impact*, J.S. Russell and R.F. Isbell, eds. St. Lucia: University of Queensland Press, 1986, pp. 117–33.

Isaacs, J. *Bush Food. Aboriginal Food and Herbal Medicine*. Sydney: Ure Smith Press, 1987.

Jones, R. 'Fire stick farming.' *Australian Natural History* 16: 224–228, 1969.

Jones, R. 'Landscape of the mind: Aboriginal perceptions of the natural world.' In *The Humanities and the Australian Environment*, D.J. Mulvaney, ed. Canberra: Australian Academy of the Humanities, 1991, pp. 21–48.

Kimber, R.G. 'Resource use and management in Central Australia.' *Australian Aboriginal Studies* 1984(2): 12–23, 1984.

Kohen, J.L. *The Darug and Their Neighbours. The Traditional Aboriginal Owners of the Sydney Region.* Blacktown: Darug Link in association with Blacktown and District Historical Society, 1993.

Kohen, J.L. 'Mapping Aboriginal linguistic and clan boundaries in the Sydney region.' *The Globe* 41: 32–39, 1995a.

Kohen, J.L. *Aboriginal Environmental Impacts.* Kensington: University of New South Wales Press, 1995b.

Kohen, J.L. and A.J. Downing. 'Aboriginal use of plants on the western Cumberland Plain.' *Sydney Basin Naturalist* 1: 1–8, 1992.

Langton, M. *Burning Questions: Emerging Environmental Issues for Indigenous Peoples in Northern Australia.* Darwin: Centre for Indigenous Natural and Cultural Resource Management, Northern Territory University, 1998.

Lassak, E.V. and T. McCarthy. *Australian Medicinal Plants.* Melbourne: Mandarin Australia, 1990.

Low, T. *Wild Food Plants of Australia.* Sydney: Angus and Robertson, 1988.

Low, T. *Bush Medicine. A Pharmacopoeia of Natural Remedies.* Sydney: Angus and Robertson, 1990.

Mangglamarra, G., A.A. Burbidge, and P.J. Fuller. 'Wunambal words for rainforest and other Kimberley plants and animals.' In *Kimberley Rainforests of Australia*, N.L. McKenzie, R.B. Johnston and P.G. Kendrick, eds. Chipping Norton, New South Wales: Surrey Beatty and Sons, 1991, pp. 413–421.

Marsh, H. 'Progress towards the sustainable use of dugongs by indigenous peoples in Queensland.' In *Sustainable Use of Wildlife by Aboriginal Peoples and Torres Strait Islanders*, M. Bomford and J. Caughley, eds. Canberra: Australian Government Publishing Service, 1996, pp. 139–151.

Martin, R. 'Koala politics.' *Nature Australia*, Spring 1997, p. 80.

Meehan, B. *Shell Bed to Shell Midden.* Canberra: Australian Institute of Aboriginal Studies, 1982.

Mulvaney, D.J. and J. Kamminga. *Prehistory of Australia.* Sydney: Allen and Unwin, 1999.

Mulvaney, D.J. and J.P. White, eds. *Australians to 1788.* Sydney: Fairfax, Syme and Weldon, 1987.

Nutting, M. 'Competing interest or common ground? Aboriginal participation in management of protected areas.' *Habitat (Australia)* 22(1): 30–7, 1994.

O'Connor, S. '30,000 years of Aboriginal occupation: Kimberley, North West Australia.' *Terra Australis* 14. Canberra: Department of Archaeology and Natural History and Centre for Archaeological Research, The Australian National University, 1999, p 5.

Pain, S. 'The healthiest restaurant in Australia.' *New Scientist* 18 August 1988, pp. 42–7.

Pyne, S.J. *Burning Bush. A Fire History of Australia.* New York: Henry Holt and Co., 1991.

Roberts, R.G., R. Jones and M.A. Smith. 'Thermoluminescence dating of a 50,000 year old human occupation site in northeastern Australia.' *Nature* 345: 153–6, 1990.

Rose, B. *Land Management Issues: Attitudes and Perceptions Amongst Aboriginal People of Central Australia.* Alice Springs: Central Land Council, 1995.

Russell-Smith, J. 'A record of change: studies of Holocene vegetation history in the South Alligator Region, Northern Territory.' *Proceedings of the Ecological Society of Australia* 13: 191–202, 1985.

Russell-Smith, J., D. Lucas, M. Gapindi, B. Gunbunuka, N. Kapirigi, G. Namingum, K. Lucas, P. Giuliani and G. Chaloupka. 'Aboriginal resource utilisation and fire management practice in Western Arnhem Land, Monsoonal Northern Australia-notes for prehistory, lessons for the future.' *Human Ecology* 25(2): 159–195, 1997.

Specht, R.L. 'Responses to fires in heathlands and related shrublands.' In *Fire and the Australian Biota*, A.M. Gill, R.H. Groves and I.R. Noble, eds. Canberra: Australian Academy of Science, 1981, pp. 395–415.

Strang, V. *Uncommon Ground: Cultural Landscapes and Environmental Values.* Oxford: Berg, 1997.

van Tiggelen, J. 'Survival in the Strait. Is the dugong at risk of over-hunting?' *Ecos* 86: 12–17, 1996.

Veth, P.M. and F.J. Walsh. 'The concept of "staple" plant foods in the Western Desert region of Western Australia.' *Australian Aboriginal Studies* 1988(2): 19–25, 1988.

Walker, D. 'The development of resilience in burned vegetation.' In *The Plant Community as a*

Working Mechanism, E.I. Newman, ed. Oxford: Blackwell Scientific Publications, 1982, pp. 27–43.

Whelan, R.J. 'Patterns of recruitment to plant populations after fire in Western Australia and Florida.' *Proceedings of the Ecological Society of Australia* 14: 169–178, 1988.

White, M.A. *The Greening of Gondwana*. Sydney: Reed Books, 1986.

Wilson, G., A. McNee and P. Platts. *Wild Animal Resources: Their Use by Aboriginal Communities*. Canberra: Bureau of Rural Resources, Australian Government Publishing Services, 1992.

Yapp, G.A. 'Wilderness in Kakadu National Park: Aboriginal and other interests.' *Natural Resources Journal* 29: 171–184, Winter 1989.

Young, E., H. Ross, J. Johnson, and J. Kevesten. *Caring for Country: Aborigines and Land Management*. Canberra: Australian National Parks and Wildlife Service, 1991.

EDVARD HVIDING

BOTH SIDES OF THE BEACH: KNOWLEDGES OF NATURE IN OCEANIA

NATURE AND CULTURE: DIVERSITY IN A MARITIME CONTINENT

In this essay I provide some insights into how the peoples of Oceania relate to their environments of land and sea. For the reader interested in following up these questions, I provide extensive references to ethnographies and comparative studies from most corners of this far-flung region, as well as to related discussions of theoretical relevance. A note on regional definitions is in order. "The peoples of Oceania" for the present purposes includes the Pacific Islanders on both sides of the Equator (among them also the peoples of the large island of New Guinea), as well as the indigenous peoples of Australia and Aotearoa New Zealand. This use of "Oceania" rather than the more restricted (but, admittedly, very evocative) "South Pacific" follows present-day usages by indigenous scholars (notably Hau'ofa 1993, 1998), and as a definition it accommodates the indigenous peoples of Australia. For reasons of space Australia is sparsely treated in this essay, yet Aboriginal Australian views of the environment have a significant place in the overall comparative perspective. [Editor's note: See the chapter by J.L. Kohen on Indigenous Australians and the Land.]

Using "Oceania" also avoids the logical inconsistency implied by the fact that numerous "South Pacific Islands" are actually located north of the Equator (i.e., the Micronesian islands and the Hawaiian archipelago) – and it conveniently restricts the view to a "Pacific" that does not include the Asian and American Pacific coasts. Thus the focus of this essay is firmly on the islands of the tropical Pacific (Figure 1), scattered unevenly across the ocean between the Tropic of Cancer in the north, the Tropic of Capricorn in the south, and from the continental landmass of New Guinea in the west to tiny, outlying Easter Island in the east. I warn the reader, though, that I shall use the terms "Oceania", "Pacific", South Pacific" and "Pacific islands" with only subtle contextual indications as to the intended scope of the terms.

Attempting a regional overview of systems of knowledge of nature in Oceania is no small order, as the areas in question cover almost half the globe and are home to more than a quarter of the world's languages (1300–1500, depending

H. Selin (ed.), Nature Across Cultures: Views of Nature and the Environment in Non-Western Cultures, 245–275.
© 2003 *Kluwer Academic Publishers. Printed in Great Britain.*

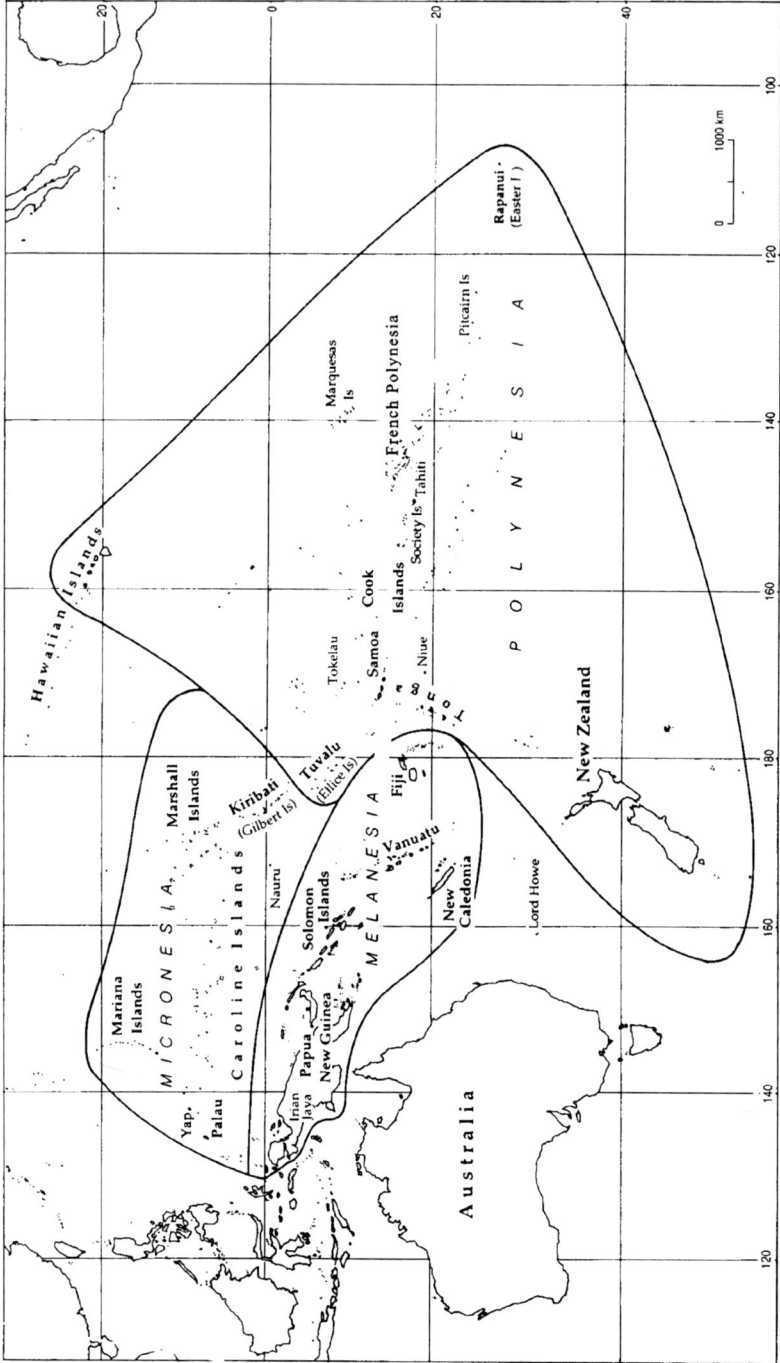

Figure 1 Oceania, showing regional subdivisions (Map adapted from original in Denoon, 1997).

on linguistic definitions; see Oliver, 1989: 66–77). Each of these languages conveys its speakers' knowledge of nature (that nature, however, showing less variation between islands and archipelagos). Linguistic diversity in Oceania is often held to represent cultural diversity, reaching staggering levels in places like the island nation of Vanuatu where a population of around 200,000 speak 110 different languages. The most widespread of Oceania's major language groups, the Austronesian, is also the world's most widely distributed in that its languages are spoken by people distributed over "two-thirds of [the world's] circumference from Madagascar to Easter Island" (Spriggs, 1997: 4). With all the non-Austronesian languages and an overwhelming majority of the Oceanic Austronesian languages, the Western Pacific region defined as Melanesia and including New Guinea, Solomon Islands, Vanuatu and New Caledonia is home to "a quarter of the world's languages, spoken, over a land area much less than a hundredth of the inhabitable land surface of the world, by a fraction of the world's population that is too small to bother calculating" (Laycock, 1982: 33).

All this is summed up in Oliver's statement that, "[b]etween thirteen hundred and fourteen hundred 'languages' were being spoken in Oceania just before Europeans arrived and introduced a few more of their own" (Oliver, 1989: 66). The quotation marks used by Oliver direct our attention to a particular pattern of linguistic diversity in Oceania: dialect areas are contiguous, and distinctiveness of languages is not necessarily defined by mutual incomprehensibility. It is clear that the languages of Oceania reflect a complex cultural history of migrations, a "history of movement" wherein "[a]lmost all the people, plants, and animals in the Pacific originated from Asia in the distant past" (Crocombe, 1983: 3; see Denoon, 1997 for overviews). It is noteworthy that the migrations of fauna and flora from Asia into the Pacific do not all belong to the realm of ancient natural history. In fact the more recent human migrants (particularly the "Lapita peoples" and the early Polynesian voyagers) had a penchant for bringing useful plants and animals to the new islands they settled, thereby transforming the ecology to greater or lesser degree (Rappaport, 1979).

The Island Pacific was populated through successive waves of migration from mainland Southeast Asia through the Philippine and Indonesian archipelagos. While the Australian and non-Austronesian languages reflect human migrations at least 50,000 years ago from Indonesia into New Guinea and subsequently into Australia, the Austronesian languages mirror the rapid sea migrations from ca. 5,000 years ago of so-called "Lapita peoples", journeying from the Southeast Asian archipelagos into coastal Melanesia and expanding into the rest of the Pacific, where they formed the basis for the Polynesian settlement of every major island in the eastern Pacific (Kirch, 1997 and 2000; Spriggs, 1997). This is demonstrated not least by linguistic evidence, showing a gradual diversification as well as a number of retained similarities along the route of Austronesian migration from South China and Southeast Asia and out through the Island Pacific and into Southeast Asia. For example, taking into account slight phonetic variation, the basic word for fish is *ika*, the basic word for eye is *mata*, and for the numeral five *lima*, as far apart as in Java and

Hawai'i and in hundreds of Austronesian languages dispersed throughout the region between these extremes.

Anthropologists, geographers, archaeologists and linguists (as well as politicians) divide the islands of the tropical Pacific into the three main cultural regions of Melanesia, Polynesia and Micronesia (Figure 1). Seventeenth and eighteenth century European explorers invented these divisions. Despite considerable cultural-historical overlap in assumed boundaries and contestations of the analytical usefulness of the distinctions (e.g., Thomas, 1989), this threefold regional division has remained in use.

Melanesia includes New Guinea, Solomon Islands, Vanuatu, New Caledonia and, by most definitions, Fiji. While the western half of New Guinea is under Indonesian domination, the eastern half and adjacent coastal islands comprise the nation state of Papua New Guinea (independent since 1975). Solomon Islands, Vanuatu and Fiji are also independent states (established in 1978, 1980 and 1970), while New Caledonia remains under French colonial rule. Melanesia derives its name from Greek *melas* (i.e., "black islands", a term dating from early European exploration of the Pacific) – referring either to the dark skin of the islands' inhabitants or to the dark impression made by the large mountainous islands on the horizon under typically grey and rainy skies. The region has a total population of about eight million, most of whom live on the island of New Guinea.

I have mentioned that a quarter of the world's languages are spoken as vernaculars in Melanesia. Neighbouring groups may speak entirely unrelated languages and follow different principles of social and political organization, although they rely on mostly uniform ecosystems characterized by fringing reefs, lagoons and estuarine mangrove swamps along the coasts and steep, forested slopes and river valleys in the interior. These environments promote lifestyles focussed on fishing, trade and shifting agriculture on the coast and its low hinterlands and more intensive agriculture (sometimes irrigated) and hunting in the interior. Until the end of the 20th century, when Asian logging companies escalated their operations in Papua New Guinea and Solomon Islands (Barlow and Winduo, 1997), Melanesia had some of the last undisturbed rainforests of the world. This is a region of high terrestrial and marine biodiversity, with a striking diversity also in the ways in which the environment is perceived, utilized and managed from the perspectives of a multitude of indigenous cultures.[1]

The island societies of Polynesia (derived from Greek "many islands") are far more culturally uniform, and the 36 or so languages of the region are all closely related, having a shared ancestral language (Kirch, 2000: 212ff). Around half a million people (excluding New Zealand and Hawai'i) live on a total land area only about half that of, say, Solomon Islands but are spread out for more than 10,000 kilometres along and (mostly) south of the Equator, in the "Polynesian Triangle" defined by Hawai'i, New Zealand and Easter Island. Many important Polynesian islands are high and volcanic, often connected economically and politically with coral atoll satellites. Irrigated agriculture, whether rain-fed or based on groundwater (as on atolls) is common throughout the region. There are strong ties between most Polynesian archipelagos, relating to common histories of migration and settlement.

Similarly, social relations and cultural affinities are strong within and among most island groups of Micronesia (from Greek "small islands"), although a greater number of mutually unintelligible but related languages are spoken there. The inhabited islands of Micronesia are scattered along and north of the equator for more than 4,000 kilometres, and inter-island travel by sailing canoe has continued to this day in remoter parts of the region.

This summary of the conventionally classified cultural regions of the tropical Pacific presents several contradictions to assumptions that cultural difference is generated by spatial distance, isolation, and ecological variation. For example, Melanesia, the region with the least distance between islands, with the most compact archipelagos and with the largest landmasses, also has the far greatest cultural diversity. Polynesia, the most remote of the regions, shows much cultural similarity throughout its dispersed scatters of oceanic islands. In terms of cultural and, especially, linguistic difference, virtually any limited portion of New Guinea shows far greater variation than the entire Polynesian region. Attempts have been made to correlate variation in social and political organization in Oceania with types of ecosystem and associated regimes of production (Sahlins, 1958, 1963). But arguing, for example, that ranked, stratified Polynesian societies are produced by the ecological circumstances of high islands that permit complex irrigated agriculture would seem belied by the fact that perhaps the most stratified Polynesian society of all is found on Tongatapu, a low coralline island with less than abundant water. And an attempt to correlate the egalitarian social systems of Melanesia and the stratified systems of Polynesia with ecological difference believed to produce different production regimes is disproved by the fact that ecological conditions on the larger islands of Polynesia and Melanesia are much the same and by the prominence of stratified societies and irrigated agriculture in parts of island Melanesia. In Oceania (as elsewhere) there is no one-sided causative relation between environment and society – but still, the question may be asked whether living by the sea on an oceanic island promotes a certain view of the world.

WHOSE NATURES, WHOSE KNOWLEDGES?

The following questions guide my further discussion: how are the natural environments of Oceania viewed, known and used by their human inhabitants? How have these peoples – so often viewed as in harmony with, even "one with", nature – engaged actively and productively with lands, forests, reefs and seas through the generations? What is "nature" in Oceania's cultures?

The pervasive distinction between "nature" and "culture" in Western views has deep roots in Enlightenment philosophy. From Descartes' seventeenth-century theses on a fundamental dualism between the mental and material, a dichotomy could be derived between nature and culture, in which the culture of humans enabled them to dominate nature. In this context, and in line with earlier views held by Aristotle about the natural world as a hierarchical given, "nature" was created as a culturally specific Western notion – an objective material domain forming the basis for the development of "exact" natural science.[2] Any images and interpretations of the natural environment are, however, simplified, ordered representations of patterns in observed environmental phenomena (see Ellen, 1982: 205). It comes as no surprise that there should be multiple orderings of the "reality" constituted by those phenomena. The intensive study of local perceptions and classifications of the natural environment is often referred to as "ethnobiology" (see Berlin, 1992) – or "ethnoecology" when focusing on perceived linkages between organisms and environment.

A number of anthropologists working in the Pacific region, mainly in New Guinea, have made influential contributions to debates about the reality of "nature". Roy Rappaport has argued for a distinction between a "cognized environment", being the sum of environmental phenomena perceived and interpreted by a human population, and an "operational environment", the sum of ecological relations in which organisms, or entire human populations, are involved (Rappaport, 1979). Thus the "multiple orderings of reality" just mentioned could refer to different cognized environments defined by the peoples of Oceania and by Western scientists, respectively. No cognized environment can possibly include all aspects of the operational; even Western science, aiming to unravel "everything", is one way of constructing a cognized environment. What, then, about the assumed "natural given" of "operational environments"? Roy Wagner (1975: 142) has argued that "... although we allow ... that other cultures comprise sets of artefacts and images which differ in style from our own, we tend to superimpose them on the same reality – nature as we perceive it". In discussing a New Guinea society in which a concept of nature seems patently absent, Marilyn Strathern draws on Wagner and states, "... there is no such thing as nature or culture. Each is a highly relativized concept whose ultimate signification must be derived from its place within a specific metaphysics" (Strathern, 1980: 177).

Some arguments to be brought along in the present discussion of Oceanic views of nature, are (1) that the Western scientific way of looking at, describing, and defining nature is not necessarily an all-encompassing truth against which non-Western views may be compared and evaluated; and (2) that non-Western peoples relate to their environments from beliefs and assumptions that may, or may not, be compatible with Western scientific ones. The latter point is important: although non-Western views of nature may seem exotic and strange to the Western observer, they are not therefore by definition incompatible with Western science. For Oceania, the rapidly expanding attention given in recent decades to sciences of Pacific Islands peoples reflects a recognition of many traditional forms of knowledge as science, empirically-based yet founded in distinctly different worldviews (Morrison, Geraghty and Crowl, 1994; Clarke, 1990; Hviding, 2003; for case studies see Clarke and Thaman, 1993; Johannes, 1981; Majnep and Bulmer, 1977).

With these caveats in mind, let me embark on a survey of observable aspects of Oceania as a diverse ecological environment for human life, from the deserts of Australia, through the rugged rainforest lands of New Guinea, to a multitude of oceanic island environments. I pay particular attention to the latter, chiefly owing to the uniqueness in global terms of the Pacific islands as a setting for human life: a vast and varied number of small land masses in an ocean spanning half the world, widely dispersed yet in terms of ecology and the activities of their human inhabitants integrated with the surrounding seas as well as with other islands nearby and beyond the horizon.

While Australia and the huge island of New Guinea are integral parts of Oceania but must be seen as continents in their own right, the islands of the Pacific ocean are of four main types (Kirch, 2000: 47–50; Oliver, 1989: 3–26). The large mountainous islands of the Western Pacific archipelagos, sometimes called "continental", have river valleys, fertile soils and rainforests and are rimmed with mangrove estuaries, lagoons and coral reefs. Second, high volcanic islands, found throughout the tropical Pacific, vary greatly in size and in the degree to which they have rivers and surrounding reefs. Third, there are the coral atolls, long narrow strips of low land surrounding a central lagoon; and fourth, there are the raised atolls (referred to by the Polynesian word *makatea*), rather dry islands with no lagoons, little or no fresh water, and sheer cliff coasts. These island types offer very different opportunities for human settlement and economic activity.

In general, biodiversity decreases from west to east in the Pacific, but seen as a whole "the region has the world's highest proportions of endangered species, and probably endemic species, per unit of land area or inhabitant" (Dahl, 1986: 2). The region's reefs and seas have the world's greatest numbers of marine animal and plant species, and the lands have unique examples of distinct forest types and high levels of endemic plants and animals (Randall, Allen, and Steene, 1990; Mueller-Dombois and Fosberg, 1998; Paijmans, 1975; Dahl, 1986). One notable example concerns the "compression" of tropical rainforest types in the Melanesian islands. For example, a walk of a few hours from coast to mountain ridges in the Solomon Islands covers an extensive range of basically Southeast Asian forest types that could only be encountered during several days, maybe weeks, of walking inland in New Guinea or Borneo. In the high islands of the Solomons, mossy montane forest occurs at elevations of around 800m, while such vegetation is only found above 2,000m in the continental island of New Guinea (Whitmore, 1998: 171–172). Forest types on a typical island in the tropical Pacific include Indo-Pacific beach flora (often replaced by planted coconuts), secondary forest interspersed with agricultural areas, lowland rainforest on limestone and/or volcanic soils, and – in the high islands of the Western Pacific – the landwards and seawards extremes of montane cloud forest and estuarine mangrove and swamp forest. Beyond the seashore the high islands of the Western Pacific often have seagrass beds, extensive lagoons with small low islands, and outer reefs where a special saltwater-resistant vegetation grows above the high-water mark, while atolls and smaller volcanic islands have fewer transitional zones between land and sea.

Birds dominate the land fauna of the Pacific islands. Many birds are hunted for food and as a source of decorative feathers, and studies into the knowledge and uses of the avian fauna in the islands of the Western Pacific show how birds of the rainforest have profound cultural importance in everyday life, religion and ritual.[3] Even where there are fewer bird species, winged creatures are important. For atoll dwellers and sea-oriented peoples of high islands, seabirds have special cultural significance as partners of fishermen (who rely on the birds to locate schooling fish) and as helpers of navigators.

Unlike birds, land mammals are not prominent in nature's Pacific repertoire, although Australia and New Guinea are remarkable for the presence of many endemic marsupials – some of which, like the spiny, ant-eating egg-laying echidna, are seen as very strange creatures indeed, not only by Western biologists, but also by the people who live with them, such as the Mountain Ok of Inner New Guinea (Jorgensen, 1991). The many marsupials of New Guinea provide not only valued protein, but also much food for thought – both to Western science (Flannery, 1989) and in indigenous worldviews (Barth, 1975 and 1987). In contrast to New Guinea and Australia, the islands of the Pacific have an indigenous mammal fauna restricted to rats, fruit bats (a delicacy in many island societies, and invariably classed with birds), and that notable marine animal, the large dugong (or "sea cow") that roams the seagrass beds of tropical Western Oceania and is a prestigious item among coastal peoples. Smaller marine mammals, like porpoises and dolphins, are culturally important in parts of coastal Melanesia, but are generally classed with fish (Takekawa, 2000). Two domesticated mammals are important in cultural repertoires throughout the Pacific. Pigs are especially important as prestige food and exchange objects in Melanesia, where the large islands also have considerable feral populations. The dog (including Australia's dingo) was, like the pig, brought into the region thousands of years ago by settlers from Southeast Asia. It has traditionally been both a hunting companion (where feral pigs occur) and, in islands without game, a valued source of meat.[4]

Reptiles are abundant throughout Oceania, but mostly in the form of small lizards. Snakes are limited to the Western Pacific, where there are also large monitor lizards often imbued with religious importance. The large islands at the Western perimeter of the South Pacific (as well as Australia) are the haunt of estuarine (or saltwater) crocodiles, huge and dangerous animals best avoided in everyday life. They are important in Oceanic religion and folklore – also in places far beyond their usual distribution, as Best demonstrated (1988) in an interesting study of probable crocodile sightings and beliefs about dragon-like beasts in such dispersed locations as New Zealand, Kiribati, Fiji and Samoa.

Whereas the land environments of Oceania have some very special types of forest composition and many unique species of plant, bird and other animals, it is the marine environment that holds the record for biodiversity. Furthermore, while humans have altered the terrestrial environments and ecology of most Pacific islands (through the introduction and extinction of species and the alteration of forest ecosystems), the marine surroundings of those islands have not been as drastically altered. In a Oceanic world in which seafood in the form of fish, shells and crabs (supplemented with rather more special creatures such as turtles and dugong) is people's main source of protein, fishing and marine gathering have remained important and meaningful activities, and the knowledge underpinning this field of food production has been maintained. Religion and folklore throughout the Pacific islands also abound with some of the more remarkable inhabitants of reefs and seas, such as sharks, turtles, dugong, dolphins, tuna and other fishes (Figure 2) and giant clams. Mythology, sacrifice, totemism and initiation in Pacific islands societies place marine animals at centre stage, and in contemporary Oceanic identity politics and creative writing, the general fact of living by and with the sea is a crucial focus of what it means to be a person of Oceania (Hau'ofa, 1993, 1998; Diaz and Kauanui, 2001). Epeli Hau'ofa has suggested "the development of a substantial regional identity that is anchored in our common inheritance of a very considerable portion of the Earth's largest body of water, the Pacific Ocean" and quoted Pacific scholar and poet Teresia Teaiwa: "We sweat and cry salt water, so we know / that the ocean is really in our blood" (Hau'ofa, 1998: 392).

While the large "continental" islands of the Western Pacific archipelagos, and a number of high volcanic islands further out in the Pacific, are ecologically rich and complex, have abundant freshwater, and are often endowed with marine resources as bountiful as those of the land, atolls and raised islands are impoverished in comparison. They generally lack freshwater – on atolls this is supplied through wells from a so-called "freshwater lens", a thin underground layer of fresh water that floats on the heavier salt water in the interior of the coral. They have poor limestone soils, and while raised atolls have no lagoons

Figure 2 A good catch of skipjack tuna is landed at Chea village in the Marovo Lagoon of Solomon Islands. Nowadays caught by trolling from motorized fibreglass canoes, the tuna remains a sacred animal in Marovo. When caught it must be handled in prescribed respectful ways and always be landed with its tail facing the sea whence it came (Photo by Edvard Hviding, 1987).

and often lack fringing coral reefs, even atoll lagoons may be shallow, land-locked and resource-poor compared to the deeper, more open lagoons along many high island shores.

Thus human life on "a South Sea Island" may have more of the hardships of water shortage, resource scarcity and exposure to harsh weather than the idyllic stereotype would imply.[5] Some "South Pacific" environments are indeed quite severe contexts for human activity. Most island dwellers of the tropical Pacific are likely to have some perception of the finiteness of resources, especially since land (apart from on very large islands with low population density) is obviously limited (Crocombe, 1987), and since the open ocean beyond the reefs is a barren world in immediate terms of human sustenance. The huge stocks of pelagic fish (particularly tuna) that roam the Pacific were, and are, beyond the reach of most village-style food-gathering activities, which tend to focus closely on the regionally varying, but always diverse, offerings of the environments of land and nearshore reefs and seas. In terms of the combined ingenuity of their indigenous systems of agroforestry and reef-and-lagoon fishing, and the local knowledge bases of these activities, Pacific islanders on the whole are probably unmatched in the world (Barrau, 1958; Clarke, 1971; Clarke and Thaman, 1993; Denoon and Snowden, 1982; Hviding, 1995, 1996a; Hviding and Bayliss-Smith, 2000; Johannes, 1981; Kwa'ioloa and Burt, 2001; Sillitoe, 1996).

The fact that the island types described above in many cases co-exist in one and the same archipelago also makes for much interaction among the inhabitants of islands of different ecological type. Throughout the Pacific, contact rather than isolation has been the rule as islanders engaged in economic specialization, inter-island trade and exchange systems and military alliances. Regional systems like the "Tongan maritime empire" (Kirch, 1984: 217–242), the "Yapese empire" (Hage and Harary, 1996: 30–35) and the kula exchanges of the Massim islands off eastern New Guinea (Malinowski, 1922; Leach and Leach, 1983) were founded on their participants' command over long-distance maritime travel. Navigation, thus, is another field of environmental knowledge where Pacific Islanders are hardly surpassed; numerous classic studies attest to the complexity of traditional navigation of sailing canoes between distant islands, based on knowledge of stars, sky, sea swells, currents and winds (Gladwin, 1970; Lewis, 1972; Feinberg, 1988). Maritime travel has remained a focus of everyday life in most Pacific islands societies, even if Yamaha outboard motors have become more ubiquitous than sails woven of pandanus leaves (see Feinberg, 1995). In recent decades long-distance navigation by star courses has also been a key focus in Pacific movements towards indigenous cultural and political self-determination, exemplified most strikingly by the successful voyages of the Hawaiian canoe replica Hokule'a to far corners of the Pacific (Finney, 1994; Kyselka, 1978).[6]

The typical approach taken by Pacific Islanders to the island environment, then, is characterized on the one hand by detailed knowledge of and intense engagement with the land and its associated reefs and inshore seas, and on the other by a fundamental outwards-looking view of the world as not confined to the home island but connected across the ocean with other natures and cultures. This twin foundation of Pacific islanders' worldviews – in intensive uses of the local and extensive overseas contacts – implies that island landscapes as context for human activity are not confined to single islands. While the islanders of the Pacific probably know and use a greater percentage of their island flora than most rainforest peoples do (Henderson and Hancock, 1988; Whistler, 1992), and while they probably know and use a greater variety of species of marine animals than most other fishing peoples (e.g., Johannes, 1981), this sometimes encyclopaedic knowledge of local nature coexists with awareness of worlds beyond the home shores. The double groundedness in the local and the widely regional runs as a continuous thread in Oceanic cultural history (see Kirch, 2000; Waddell, Naidu and Hau'ofa, 1993; Diaz and Kauanui, 2001).

IN THE SEA OF ISLANDS

For those who live on the coast, the sea does not divide, nor is it a deterrent to contact. In an influential paper entitled "Our Sea of Islands", the Tongan anthropologist and writer Epeli Hau'ofa (1993) presents an alternative model of Oceania. His model emphasizes inter-island relations and contact in past, present and future and squarely confronts the tendency to stress the smallness and remoteness of the islands in the Pacific by focusing on their appearance as small dry surfaces in a vast ocean far from the centres of power. Oceania must no longer be seen as "islands in a far sea", Hau'ofa concludes, but as "a sea of islands".

Let us shift briefly to another time and another Pacific place. In his comparative analysis of the epistemologies of scientific practice and discovery, Bruno Latour (1987: 215ff) recounts what happened on 17 July 1787 when the French explorer Lapérouse landed in his ship *L'Astrolabe* on an unknown area of land in the far north Pacific, called "Segalien" or "Sakhalin" in older sources. Lapérouse's immediate question was whether this was an island or a peninsula – not least since "a fierce dispute had ensued among European geographers" (Latour, 1987: 215) as to the accuracy of old maps and records that variously showed Sakhalin as an island and as a peninsular outcrop of Asia. When Lapérouse landed, a few "savages" came to the beach and exchanged salmon for pieces of iron, and Lapérouse was astonished but pleased when it turned out that "not only did [the 'savages'] seem to be sure that Sakhalin was an island, but they also appeared to understand the navigators' interest in this question and what it was to draw a map of the land viewed from above" (Latour, 1987: 216). Lapérouse and his expedition never returned to France. The two ships were wrecked in 1788 on the reefs of Vanikoro in the eastern Solomon Islands. But in his account of the expedition, based on journals sent home at intervals, Lapérouse noted for this Sakhalin incident how "the ease with which they had guessed our intentions led me to believe that the art of writing was not unknown to them" (Valentin, 1969: 78).

Islanders indeed know that they live on islands – a recognition that has implications both regarding the finiteness of land, the absoluteness of coasts, and the openness of overseas linkages to other islands and other peoples. It is important to note that in Oceania, the coral reefs that line the coasts are usually seen as connected to the land. Thus both agricultural land, hunting areas and fishing grounds are generally subject to limitations on entry and exploitation – the inshore seas of Oceania are not commons with open access leading to the "tragedy" (Hardin, 1968) of unregulated overharvesting. But the widespread existence of ownership, limited access and internal resource use regulations does not imply that all Pacific Islands peoples have at all times practised wise and sustainable resource management. Although for the region in general, the repertoire of traditional conservation practices is large and varied – especially concerning fisheries regulations which in many ways predate modern Western fisheries management measures (Johannes, 1978; Ruddle and Akimichi, 1984; Ruddle and Johannes, 1990; South, Goulet, Tuqiri and Church, 1994; Ruddle, Hviding and Johannes, 1992) – there are numerous historical examples from the Pacific of resource depletion and environmental degradation (Rappaport, 1979; see also Kirch, 2000). Sometimes, such degradation was a direct result of overfishing and intentional burning of vegetation; in other cases it was an indirect consequence of human-introduced animals like rats, dogs and goats.

But examples of depletion and environmental degradation do not modify the fact that there were, and are, many traditional practices of resource management in Oceania that are conscious conservation measures. Often connected to religious observances and taboos, such practices appear designed to maximize long-term yields of fishing grounds, garden lands and hunting areas. Many measures seem to reflect a view of natural resources as being limited and depletable; these are facts of life on small islands surrounded by reefs but encompassed by more or less barren seas. Further, resources may be limited both for ecological and political reasons. Inter-group warfare between and within islands in the Pacific region often created inequalities in territorial divisions that forced people to make the most out of what resources they possessed. One example of this can be taken from the Marovo Lagoon of Solomon Islands, where I have spent a total of some three years living on a small, densely populated island (an extinct volcanic peak) whose original rain-forest has been transformed, through the gradual substitution of trees for over two centuries, into a dense sustainable agroforest with almost no plants not directly useful for food, medicine, building materials and other purposes (Hviding and Bayliss-Smith, 2000: 128–130). Being "salt-water" people owning sea and reefs but little land, this particular group transformed its negligible landholdings into a fully domesticated cultural landscape, much of which to outsiders still looks like untouched rainforest.

Whereas for Pacific Islanders, the open seas are barren in terms of exploitable resources (apart from the seasonal abundance of large schools of tuna closer to shore), they provide another valued item – access to the surrounding world. Oceanic seas are linkages between distant populations, and with effective maritime technology and navigation, "... [o]cean spaces ... become highways rather than barriers" (Lewis, 1972: 15). Most peoples of the tropical Pacific have not been isolated island dwellers, but rather frequent ocean travellers, possessing sophisticated maritime technology and navigation practices.

The navigation expert David Lewis has reconstructed inter-island patterns of close contact in the pre-colonial Pacific (Lewis, 1972), and his results present a number of striking facts that contradict a notion that island cultures should be characterised by isolation. First, it is possible to sail between almost all the inhabited islands of the South Pacific from Southeast Asia without once making an open-sea crossing longer than 310 nautical miles. The only exceptions are Easter Island, Hawai'i, and New Zealand. Second, the 23 zones of regular, close contact identified by Lewis include virtually every inhabited island in the tropical Pacific and show much overlap (again with the above-mentioned exceptions). Third, Lewis has gone back to early sources, in this case Captain Cook's journals, to identify the approximate bounds of the world known by Tupaia, a Polynesian navigator and priest who Cook met in 1771 when the latter visited Tahiti on his first voyage. Tupaia's world as communicated to the British included "... every major group in Polynesia except Hawaii and New Zealand, and it extended for 2600 [nautical] miles from the Marquesas in the east to Rotuma and Fiji in the west, equivalent to the span of the Atlantic" (Lewis, 1972: 17n). Tupaia was to be important for Cook's further discoveries. He decided to come along in Cook's ship the *Endeavour* for the journey from Tahiti and provided exact navigational directions for a large number of hitherto "undiscovered" islands to the west and southwest. In most locations Tupaia seemed to know the chiefs and he acted as interpreter and cultural broker for the European explorers (Beaglehole, 1974, chapters 8–9). Although Tupaia did not know how to get as far as New Zealand, once the ship arrived there he established working relations between the British and the Maoris. Tupaia died of malaria in Batavia on the *Endeavour*'s way back to Europe. His importance for the first European discoveries of the South Pacific has not been widely publicized by imperial historians and can only be elucidated by a close reading of Cook's journals.

Traditional Pacific Island navigators command high respect in their societies (Gladwin, 1970). They are the recognized experts in dealing with the outside world, and at some stage in ancient times, the expertise of the navigator was crucial for securing the survival of groups of migrants at sea, in search of new homelands that the navigator as often as not would not have visited before. Many aspects of Oceanic navigation, such as interpreting seabird behaviour, ocean swells and cloud patterns, serve as well to find land not previously known – and much settling of new islands in Polynesia may have been in response to overpopulation (or conflict) that forced people to set off in search of unknown new homelands (Irwin, 1992; Lewis, 1972; Gladwin, 1970). The traditional Polynesian navigators were no less heroes in their own society than discoverers like Marco Polo or Vasco da Gama were in medieval Europe.

BOTH SIDES OF THE BEACH

In the Pacific Islands, the close interrelationships between land and sea, mountain and shore, and river and lagoon form particularly important components of indigenous worldviews and politics. The beach, with its characteristic fringe of salt-tolerant (and endlessly useful) coconut palms, is the zone of confluence between land and sea as tides ebb and flow. It is the scene for human arrivals and departures, and to Pacific islanders it represents movement and connectedness to worlds beyond the local. It is also the place at which one stands, with one's back to the land, to scan the horizon for seabirds indicating the whereabouts of schooling tuna or other fish or for signs of approaching bad weather. Yet at the same time the land immediately behind the beach of a Pacific island offers a tightly packed diversity of conditions for human life: sheltered land for settlement, soil for growing root crops and leaf vegetables, forest dense with plants useful for food, medicine, house construction, boat building and more, and historical sites of former settlement, burials and so forth, establishing continuity in human presence (Figure 3).

A well-documented characteristic of Pacific Islanders' perceptions of their environment is a widespread lack of dichotomy between sea and land. This applies to everyday resource use, where fishing, marine gathering and agricultural activity go together with seafood and root crops to form complementary, interrelated and interdependent domains in work and diet. Normally, the food systems of coastal villagers in the Pacific combine resources of land and sea in such ways,[7] and so the preferred traditional diet in most islands of Oceania consists of root crop staples and marine protein in the form of fish and shellfish.

Figure 3 Village shore, Marovo Lagoon, Solomon Islands. The beach – backed by forest and fronted by reefs – is where land and sea, as well as the local and the outside world, meet and merge; it is a focus of everyday life and history for people of the Pacific Islands (Photo by Edvard Hviding, 1987).

Miriam Kahn, in her study of food and symbolism among the coastal Wamira in the Milne Bay region of Papua New Guinea, says, "Wamirans divide their food world into two categories: *tia* (animal food) and *lam* (vegetable food) ... The category of *tia* ... has fish as its most stable element" (Kahn, 1986: 46–47). Beyond the village level, and particularly in the Melanesian archipelagos of the Western Pacific, many trade systems have been based on the bartering of agricultural products for fish and other fruits of the sea. Such trade was often linked closely with ritual forms of exchange, and in parts of Melanesia the flow of produce between land and sea has remained crucial for the maintenance of regional political and social relationships. Often, ecological and economic linkages between land and sea have a parallel in indigenous ideas about dependencies between, for example, "fish and taro", "bush people and saltwater people", or simply "up and down".[8] In Fiji, the complementarity is framed directly as being between people of sea and people of land (e.g., Sahlins, 1985: 99–102), so that "everyone is always 'land' or 'sea' in relation to others" (Toren, 1995: 171). Through such paired concepts, Pacific islanders' own views tend to emphasize distinction, yet also mutual dependency and trade, between differently endowed groups (for example taro cultivators living uphill and coastal fishing people), and in a sense the ecological aspects of local and regional economic and social relationships are explicitly recognized.

Pervasive integration of sea and land is also reflected in the traditional resource management systems of Oceania. As a rule, localized groups whose members are related through kinship and co-residence control the access to and utilization of resources in territorial holdings that embrace land, reefs and sea, usually subsumed in a single indigenous term (Ruddle, Hviding and Johannes, 1992; see also Sahlins, 1958). Thus, traditional resource use and management throughout the Pacific islands take an integrated view of terrestrial and marine resources. This pattern is also seen among the Melanesian Torres Strait Islanders where land boundaries extend over the reefs into the sea (Johannes and MacFarlane, 1991), and in the coastal resource tenure systems of indigenous Australians (Peterson and Rigsby, 1998). The peoples of Oceania held such views and practised resource management long before the Western world realized the importance of what is now called "integrated coastal zone management" (Johannes, 1978). Indeed, in a number of locations in island Melanesia, recent resistance against foreign mining and logging companies has had its foundations precisely in the explicit concern that river-carried pollution from mining and other large-scale land transformations will be detrimental to the marine environments of reefs and lagoons (Hviding, 1996a: 340–343).

Generally, then, a lack of a cognitive dichotomy between land and sea and the resources there characterizes views of "nature" among the peoples of Oceania. Land and sea resources are managed by broadly similar mechanisms and viewed as different – but related and equivalent – types of a main category of owned territory that forms the basis of a group's existence. Many of these customary resource management systems remain active and are increasingly used to meet new challenges of resource development and conservation (Ruddle, Hviding and Johannes, 1992; Hviding, 1998b). In some island societies, each local group controls a territory that embraces the entire available range of resource zones, from inland mountains through coastal lowlands, out to (and including) lagoon and reefs.[9] In other societies, particularly in Melanesia, distinct groups of "bush people" and "saltwater people" live adjacent to each other, with the former owning and managing the land and forest resources and the latter owning the lagoon and reefs and managing marine resources. In such situations, the two types of groups tend to be linked through trade systems, sometimes including exchange of use rights in forest and fishing grounds.

That land and sea in Oceania are seen as sub-groups of the same overall category stands in striking contrast to many other regions, where land and sea are cognitively dichotomized and where the use of the two main types of resources is guided by widely different management ideologies, often as regulated land versus unregulated sea (see Durrenberger and Pálsson, 1987). The Oceanic version is one in which fishing grounds are classified in local languages as something akin to "sea land" (see Lingenfelter, 1975 for Yap). In Hawai'i and many other parts of Polynesia, customary systems of rights to productive resources imply that the people of a given locality have access to all important resource types. In Hawai'i, each defined area (or, in technical terms, "corporate estate") included zones of both land and sea, usually in the form of a wedge-shaped territory (ahupua'a) extending from the central mountain cone of the islands through the coastal lowlands into the sea, at least a couple of kilometres from the beach. Main districts and their subsections were all divided according to such criteria, following a pattern of "overlapping stewardship" (Sahlins, 1958: 14). Each single territorial division thus defined contained within its boundaries hunting areas, forest, dryland and irrigated agricultural zones, swamp, beach, reefs and open-sea fishing grounds. Such tenure systems, building on the principle of access to all important resource zones, have also been documented for Fiji (the vanua, see Thompson, 1949; Ravuvu, 1983), from Yap (the tabinaw, see Lingenfelter, 1975), and other places. These examples represent a general pattern of integrated lands and seas; yet the concepts within which this view is locally grounded varies in semantic content. While the Hawaiian ahupua'a means "altar of the pig", the Yapese tabinaw refers to the household, and Fiji's vanua is an all-embracing "land". Yet all these concepts are core symbols within their respective cultural contexts.

I shall turn to how people living around a coral lagoon in the Melanesian Western Pacific relate to the environments of sea, coral reef, and rainforest on which they depend for their material and spiritual sustenance. The Marovo Lagoon, located in the New Georgia area of Western Province, Solomon Islands, is an ecologically diverse environment dominated by 700 square kilometres of lagoon, delimited by a long chain of raised barrier reef and backed by high volcanic islands with rainforest. In 2001 around 11,000 people lived in some 50 villages, mainly on the lagoon coasts of the high islands (Figure 4). Household-based production remains centred on shifting cultivation of root crops (mainly sweet potatoes), reef and lagoon fishing and a small but diverse cash sector (Hviding, 1996a). Adherence to Christian churches is universal throughout the area. By early 1995 the Marovo Lagoon with its "natural and

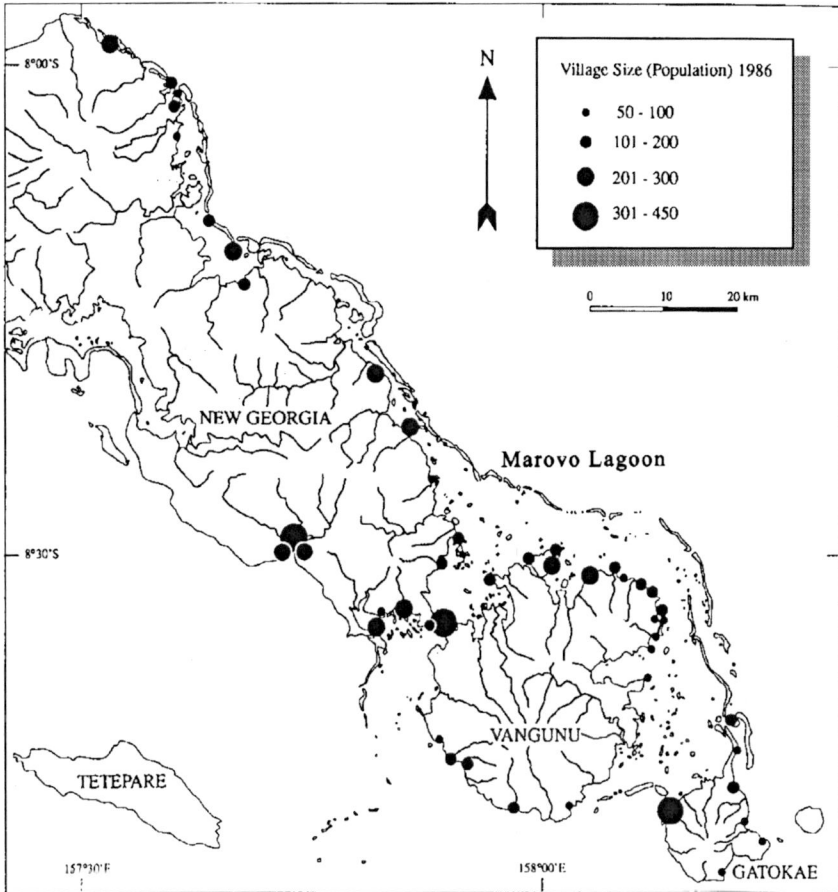

Figure 4 Map of the Marovo Lagoon area of New Georgia, Solomon Islands, showing settlement patterns as of the 1986 national census. Although the population has increased by some 60%, the relative sizes of the villages remain similar. Note that numerous extended-family "hamlets" with less than fifty inhabitants are not shown (Map by University of Bergen).

cultural wonders" was listed as a UNESCO World Heritage Site; since then a complex multitude of Asian logging and fishing companies and Western conservationists have engaged with the local land- and sea-holding groups in a sometimes bewildering variety of encounters and projects (Hviding and Bayliss-Smith, 2000; Hviding, 1998a).

Let me initially give a closer examination of indigenous views of the lands and seas under customary ownership, as subsumed in the Marovo concept of *puava* (Figure 5), in a restricted sense meaning "earth, ground". The basic relationships between people, land and sea in Marovo are between kin-based groups (*butubutu*) and delimited territories (*puava*) of land and – usually – sea and reefs. A land- and sea-holding butubutu bears the same name as its puava,

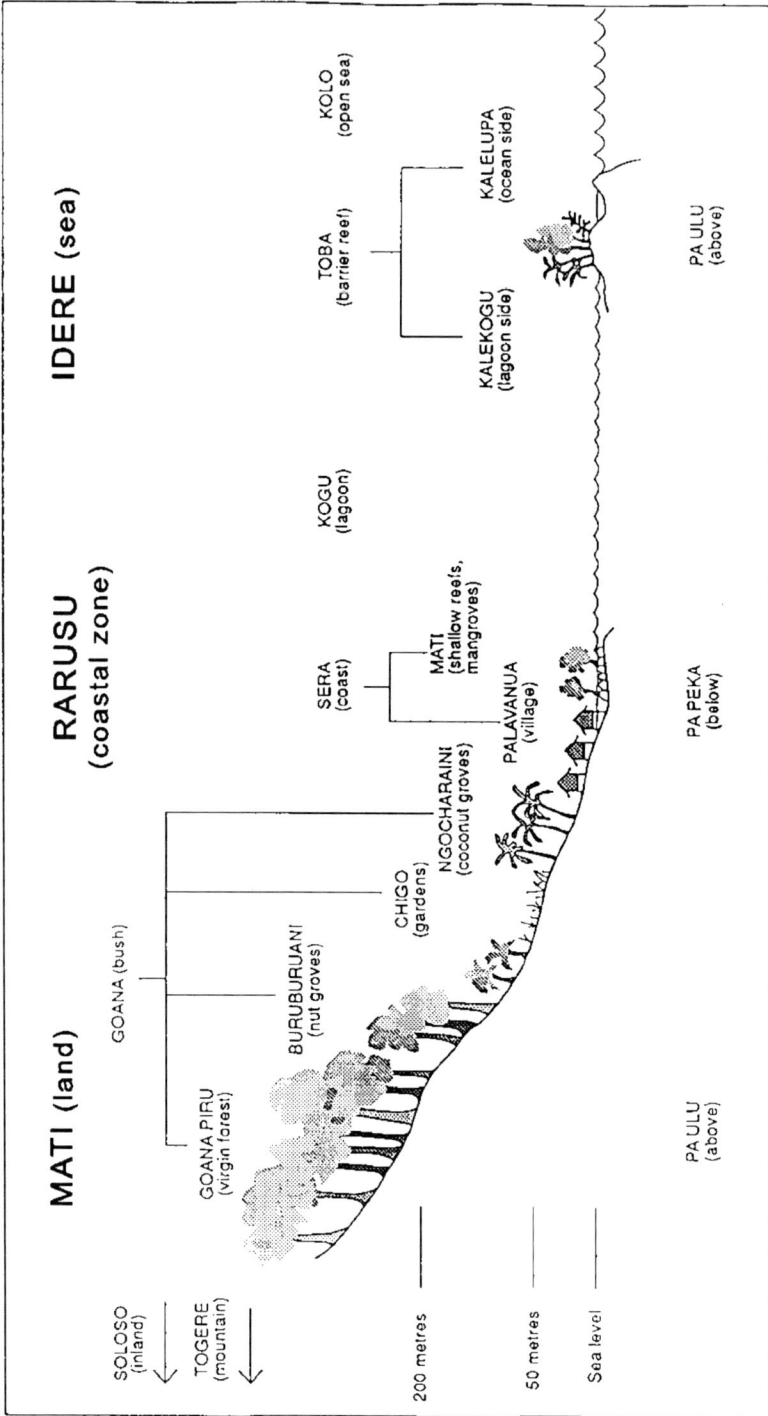

Figure 5 The *puava* concept of Marovo, Solomon Islands: indigenous classification of a continuum of land-and-sea environment, held as group territory (University of Bergen).

and there is a saying which sums up this foundational relationship between people and environment: "*Kino pa Marovo ieni soto pa tututi oro soto pa puava*" ("Life of the Marovo people is joined to genealogy and land"). Two other standard expressions in Marovo convey a mutualist (see Ingold, 1992) definition of butubutu-puava relationships: "If there is no *puava*, there can be no *butubutu*"; "The *butubutu* guards the *puava*". This mutualism between defined units of people and of environment is seen as constitutive of the continued lives of both. A butubutu derives its identity from the puava it holds as inalienable ancestral title, and the puava cannot be sold or otherwise transferred to non-members (a principle also enshrined in Solomon Islands national legislation concerning customary tenure).

In the classification of a typical *puava*, "land" (*mati*) and "sea" (*idere*) are the foundational pairs of complementary zones. The land is further divided into "forest" or "bush" (*goana*) and "coast" or "beach" (*sera*). *Sera* is the zone of human settlement and also, on seashore and in mangroves (referred to in a wider sense as *rarusu*, "coastal zone"), the transitional area where land and sea merge through the flow of tides and rivers. Surprisingly, the word *mati* (land) re-emerges here as the term for nearshore shallow reefs and mangrove areas, with reference to the anomalous position of these zones in terms of "sea" (*idere*). Thus *sera* is a central focus of the puava; it is also invariably referred to as being located "below" (*pa peka*) relative to the forest (which is elevated also literally) and the sea, both of which are "above" (*pa ulu*). From the coast or village, one moves "up" to the forest, but also "up" to the outer lagoon and open sea. Conversely, when located "up" in one of these more remote zones, one moves "down" to the central coast again. The overall zone of forest or bush (*goana*) escapes the classificatory ambiguity of the coast and is straightfor-wardly divided into coconut groves (*ngocharaini*), swidden gardens (*chigo*) some of which at any time are under fallow, special cultivated groves (*buruburuani*) of nut trees, and successively more remote zones named after typical vegetation or topography and culminating in *goana piru* (wild forest), that is assumed never to have been significantly disturbed by human activity and that, unless logged by recent commercial operations, remains as hunting ground. A special category of *goana piru* is *goana pa togere* (mountain forest), an environment of upland ridges and peaks whose vegetation is regarded as inhospitable to humans and game animals alike. In the Marovo language the concept *soloso* refers to these far inner lands of mountains, limestone pinnacles and steep ravines, as well as being a term for "large mountainous island" and indeed, in an extended sense, for "the world". In the latter sense, this reflects some long-standing views of a sizeable volcanic island and its reef-and-lagoon surround-ings as being a foundation of localized existence for people, plants and animals.

Much more can be said about the details of environmental classification from a Marovo perspective. As an area of extraordinary combined marine and terrestrial biodiversity the Marovo Lagoon is the focus of a similarly rich body of indigenous knowledge of these environments and their bountiful offerings to human existence. Although the scale of biodiversity and knowledge involved may be greater, the Marovo example well demonstrates typical patterns for

Island Oceania in terms of the integration between land and sea in human life-worlds – on "both sides of the beach". Marovo repertoires for talking about land, reefs and sea and the non-human occupants of these environments include a basic vocabulary of names for some 500 plants, 60 birds, 350 fishes, and 100 marine shells. Added to this are some 50 distinct terms for forest types, land topography and freshwater systems, and more than 80 separate terms for reef types and underwater topography, as well as innumerable designations for the habitual whereabouts of specific species of marine and terrestrial life (Hviding, 1995). Reef-related terms, for example, have much specific information value for fishing, as they differentiate between a great number of distinct forms of fish habitats requiring one or another of 60 common fish-catching techniques, most of which are still practised around the Marovo Lagoon (Hviding, 1996a: 210–224). Whereas *saghauru* is a generic term for "reef" (as is *goana* for "forest"), knowledge of the great repertoire of named reef zones and bottom types converge with knowledge about the migration patterns and seasonal spawning aggregations of important food fishes – not unlike the way in which the complementary repertoire of land-related terms for forest zones and topography guides the hunter and crop cultivator in their search for game, wild honey, medicinal plants and good plots for new gardens.[10]

KNOWING NATURE: SCIENCES OF PACIFIC ISLANDS PEOPLES

It has been noted by some commentators, writing from Melanesian ethnography, how a certain obsession with observation and empirical validation in search of a verifiable "truth" seems to be widespread in the region. As Tom Ernst comments in a discussion of "empirical attitudes" among the Onabasulu of the Great Papuan Plateau, "The empirical attitudes of Melanesians are striking ... The importance of 'seeing' or experiencing something in determining its 'reality' is ubiquitous" (Ernst, 1991: 200, 203). Such arguments subvert more popular but misinformed notions that New Guinea, for example, is a place of superstition and extraordinarily strange religious beliefs where local knowledge is anything but scientific. In the Marovo Lagoon, I have met with many thinkers who readily discuss their own theories of ecology, prehistory and linguistics, in contexts of migration, isolation and dispersal. I have listened to long accounts of how certain bird species used not to be present locally, but arrived in the area at the time of ancestor so-and-so and since then have established themselves in certain ways and become useful for people in other ways. Likewise, the people of Marovo hypothesize how skin colour and other morphological characteristics indicate that their own ancestors probably came from the northwest, whereas the slightly different characteristics of a neighbouring group would seem to indicate influence from another direction. Similar discussions are carried out on the long-term dispersion of different languages in the surrounding archipelago, including the replacement of a couple of languages now no longer remembered, and how totally unrelated languages are able to co-exist side-by-side among people who otherwise seem very similar. Although these islanders do not use concepts like species colonization, pheno-

type or genotype, or diffusion, issues are analogous and point to their fundamental preoccupation not with isolation, but with inter-island, inter-cultural
contact and influence.

Akimichi (1978: 323–324), in his analysis of fishing-related knowledge (which
he labelled "ethno-ichthyology") among the people of Lau Lagoon on Malaita,
Solomon Islands, emphasizes the breadth of what might narrowly seem to be
practical knowledge.

> When, where and how different fish behave are cognitively distinguished, and thereby the Lau
> make decisions about where and how to catch them ... . [T]he range of attributes, natural and
> supernatural, with which fish are vested in Lau culture, relate to both the cognitive and the
> practical realms and in no sense are limited to taxonomic criteria.

Such detailed, sophisticated environmental knowledge possessed by indigenous peoples has received much recent attention by outside observers and is
interestingly documented by a burgeoning research literature dealing particularly with traditional knowledge of tropical marine environments in the Pacific
region (see, e.g., Johannes, 1978, 1981; Ruddle and Johannes, 1985, 1990; Gray
and Zann, 1988). Much of this literature deals primarily with the knowledge
of important food fishes and their patterns of behaviour. Indigenous knowledge
in a number of these documented cases – such as that concerning lunar cycles in
spawning aggregations of reef fishes in Palau and Solomon Islands (Johannes,
1981; Johannes and Hviding, 2000) – shows a high level of detail and breadth
of coverage of species, and is founded in distinct observational and hypothesiz-
ing procedures.

In his recent book, *Knowledges*, Peter Worsley (1997) examines a global
range of knowledge including biological classification among Australian
Aboriginals, navigation technology in the Island Pacific, medical practice in
Central Africa, and more, including consumer culture and nationalist beliefs in
the Western world. Worsley's aim is to confound the belief that Western science
possesses "exact knowledge" in a global-hegemonic sense and to argue instead
for a pluralized view of comparable forms of knowledge with much in common.
He starts out by asking why anthropologists and others interested in the
knowledge systems of "other peoples" have since the 1950s persistently referred
to non-Western knowledge of the environment as "ethnoscience". This is the
heart of the matter. In a normal anthropological context "ethnoscience" refers
also to the rigorous methodological approach taken in cognitive studies of
systems of classification and taxonomic structures found in "other cultures"
(e.g., Berlin, 1992). From this emerges ethno-biology, ethno-botany, ethnoecology, and ethno-medicine, and also – as listed alphabetically to the point
of parody in a recent dictionary of anthropology (Seymour-Smith, 1986) –
ethno-mathematics, ethno-musicology, ethno-psychology, ethno-pharma-
cology, ethno-philosophy, and ethno-psychiatry. "Ethno-", thus, in most cases
is used to prefix names of disciplines that according to Western epistemology
are forms of objective science based on the rigours of hypothetical-deductive
method (see Popper, 1980). Used as prefix, "ethno-" indicates a field of "native"
knowledge relative to a canonical counterpart within non-"ethno-" Western
science; it indicates that the specific field of knowledge is that of the observed

rather than of the observer (Conklin, 1957), and it reinforces the widely-accepted notion that "we see [objective] nature in terms of [subjective] cultural images" (Ellen, 1982: 206). [Editor's note: see Ellen's essay in this volume.] A somewhat different quest for making indigenous knowledge compatible with Western science is seen in some of the recent attention given to "traditional ecological knowledge", a sometimes romanticized concept invoked in discussions of traditional conservation (cf. examples and debates in McCay and Acheson, 1987; and Dudgeon and Berkes' essay in this volume).

The push for a pluralized view of kinds of knowledge is not an abstract intellectual concern; the issue has bearings for today's debates about interdisciplinary research, indigenous epistemology and decolonized methodologies (see Smith, 1999; Gegeo, 1998; Hviding, 2003). Academic and political attention has increasingly risen around the fact that Oceania's peoples have their own sciences. A few years ago the huge diversity of knowledges held by Pacific Islanders was the subject of a large conference at the University of the South Pacific called "Science of Pacific Islands Peoples". The editorial preface to the four volumes of conference proceedings (Morrison, Geraghty and Crowl, 1994) was free from "ethno-" prefixes, instead noting how the volumes are "a unique collection of traditional scientific and technical knowledge from the Pacific Islands", thus addressing an urgent need to "give due attention to Pacific knowledge of plants and animals, astronomy, medicine, agriculture, navigation, boat-building, fishing, and other fields of knowledge as well as to the conceptual and linguistic ways of organizing this knowledge" (Morrison, Geraghty and Crowl, 1994: vii).

With this background, I return again to the Marovo Lagoon. Ontological premises prevailing in Marovo hold that the organisms and non-living components of the environment do not constitute a distinct realm of nature or natural environment separate from culture or human society. Marovo people do not view reefs, sea and forest and the living things therein as "an environment of neutral objects" (Ingold, 1992: 53; see Hviding, 1996b, for further discussion). In Marovo, as among the Baktaman of inner New Guinea, "... one is prepared to be one with an environment in which all places, species and processes are understood as being basically of one unitary kind, agreeable to man" (Barth, 1975: 195).

In the epistemology that prevails in Marovo, several successive states apply to the acquisition and validation of "knowledge" (*inatei*): From hearing about something (*avosoa*), a state of knowing (*atei*) is obtained. Previous and subsequent knowledge as well as the social context of its transmission determine whether or not this knowing entails believing (*va tutuana*, literally "imbue with truth"), a state that through repeated verifying instances of seeing (*omia*) is transformed into trusting (*norua*, literally "be convinced of efficacy") and the accumulated state of being wise (*tetei*). An example from a potentially dangerous environment illustrates these processes and premises. Six days every week, an average of 200–300 men of Marovo spend most or part of the daylight time diving on the deep outer reefs facing the ocean, for spearfishing and for collecting commercial shells. This presence of people in seas populated by fairly large numbers of potentially dangerous sharks very rarely leads to attacks by sharks on swimming humans. It is frequently pointed out that the only divers who have ever been killed by sharks in Marovo came from groups that do not stand in a totemic relationship to these man-eaters. For a number of historical reasons, shark totemism (entailing a prohibition against harming, provoking, killing and, most of all, eating sharks) is only associated with and practised by a limited number of the localized butubutu of Marovo, especially those with a maritime-oriented history and ancestral territorial holdings of mainly reefs and sea. The observation that sharks tend to attack mainly people from non-totemist, land-oriented groups is regarded by Marovo's fishermen as a valida-tion of the belief that ancestrally-imposed respect shown to the shark will in return give protection from attacks in the present. Remembered instances of fatal attacks are considered to be the test or trial (*chinangava*) of the ancestrally-derived belief in the efficacy of shark totemism, and from believing (*va tutuana*) in this notion one becomes convinced and elevates it to the level of trusting (*norua*).

Marovo people's relationships with the environment also involve manipula-tion of the latter through often standardized and widely-known acts of interven-tion that should be seen as belonging to the category of "magic". These acts appear in daily life as highly pragmatic, observable tools in handling problems posed by the environment – in a Pacific Islands' context applied not least to the role of weather in maritime travel. If heavy breakers make it impossible to launch a canoe from village beaches on the open-sea coasts, certain elders may be summoned to carry out the act of *va bule* (to calm). This act involves calming the sea for a brief moment (enough to launch the canoe) by chanting a spell and throwing a knotted length of a creeping beach plant into the waves. Practical magic is also used during sea voyages. If a feared *ivori* wind (a form of small tornado) is seen approaching during canoe travel in stormy weather, most adult men or women know how to perform an act that will sever the black column of spinning clouds and send the two parts disappearing down into the sea and back up into the clouds, respectively. This act, *seke ivori* (literally "give the death blow to the *ivori* wind") involves reciting a brief spell while making circular motions with a bushknife (invariably carried in canoes) or alternatively hammering together two stones (less often carried at sea). These acts are integral to Marovo people's practical engagement with the environment and to their own perception of constraints posed by the environ-ment on human activity. The efficacy of practical magic to overcome what would in Western scientific terms be considered natural constraints is verified regularly by use.

SACRED ENVIRONMENTS: SPIRITS, SONGS AND STONES

The Marovo *puava*, the Hawaiian *ahupua'a*, the Yap *tabinaw*, and the Fijian *vanua* as discussed earlier are all variations on a common theme where a defined social group is associated with and controls a section of the environment embracing upland forest, lowland garden areas, coastal settlement sites and

beaches, nearshore reefs and inshore fishing grounds. While Oceania's commu-
nally-owned lands and seas provide the contexts for everyday life and the
resources for sustenance, they also constitute cultural land- and seascapes
inscribed with markers of identity, history and belonging. Studies of Aboriginal
Australian cosmologies (Morphy, 1991; Myers, 1991; Stanner, 1966) show how
the utilitarian value of the land (and seas, cf. Davis, 1989) encompasses far
more than material benefit. The environment also yields spiritual sustenance
to the people associated with it, through traditions that relate the creation of
all kinds of topographical features (rivers, rocks, and so forth) and other
attributes of the environment to the activities and travels of ancestral beings,
during the mythical time of "dreaming". Although often with less emphasis on
direct ancestral creation of landscape features, the pattern is echoed in inland
New Guinea, Island Melanesia, and into every corner of the Pacific.[11]
Throughout Oceania, spiritual sustenance of local human life-worlds is provided for by a multitude of sacred sites and associated stories, songs and poems.
The Oceanic land- and seascape is a spiritual (or cultural) environment as well
as a physical (or natural) one, and sacred and otherwise significant sites – for
example, man-made stone monoliths deep in the rainforest, submerged reefs
kilometres offshore, and any manner of smaller sites including individual stones
and trees – are sprinkled throughout islands and coasts.

In the Pacific Islands the connections between ancestral and contemporary
environments are focused on the twin concepts of *mana* and *tapu*, two terms
that occur in some form or another in all Austronesian languages and cultural
repertoires.[12] Mana has most often been described as representing a spiritual
power that pervades human existence and is manifested in deities, sacred objects
(including stones, trees and animals) and persons of divine association, usually
high chiefs. Mana is linked to procreative power, abundance, and success, and
is also present in everyday life as a state of efficacy shown by, for example, a
well-growing garden or a large fish catch. The associated term *tapu* generally
stands for ritually sanctioned prohibitions against certain acts including contact
with sacred persons and things. Although these two concepts are complex and
varied, it may be said in the simplest manner that the connection between
mana and tapu in Oceanic religions necessitates that certain rules of tapu must
be upheld in order for states of mana to be attained and maintained. More
often than not, this involves the everyday observance of certain relationships
of reverence to places in forest and on reefs perceived to be the abode of
powerful spiritual beings. The near-universal introduction of Christianity to
Island Oceania has not led to the demise of the mana-tapu mindset; instead
the terms have been adapted into local-language expressions for "blessed" and
"holy", respectively.

PARADISE CONTESTED

In contrast to Oceania's grand scale in terms of cultural diversity and geographical extension, few of the world's human inhabitants live there. The region
usually receives scant attention in Western media; its role in world political

economy is miniscule compared to that taken by its powerful Asian and American neighbour countries on the western and eastern Pacific Rims. In world history Oceania has largely figured as a remote frontier (the last part of the world to be "discovered" and colonized by European imperial powers), as a source of exotica (a last vestige of "primitive peoples"), and – in more popular Western images – as the location of "Paradise"; the South Seas of sailors and adventurers. In the twenty-first century global tourism still has a definite place for the South Pacific. White beaches, turquoise lagoon waters, swaying coconut palms, and smiling young women and men with flowers in their hair bring promises, on the internet and in glossy brochures, of a "timeless tranquillity" to be experienced in "Treasure Islands", "Happy Isles", "Islands Lost in Time", or whatever other connotation of tropical paradise getaway that may be invoked by the global tourism industry in close cooperation with the regional Tourism Council of the South Pacific (see the website http://www.tcsp.com) and national offices of tourism.

From a certain point of view, the small island nations of the South Pacific have little beyond their tropical nature with which to attract tourists. This is not a part of the world where tourists seek ancient monuments and ruins of past civilizations. Although almost every island has old stone structures that mark the sites of former sacred grounds, fortifications and settlements, it is the natural environments of forest-clad, rugged volcanic islands and coral reef lagoons that catch the eye of world tourism. And in a manner strikingly similar to that of previous centuries of discovery, the human inhabitants of the islands still seem to be viewed largely as integral to that nature, in a time when some of the archipelagos marketed as tranquil paradises are nations racked by coups and civil war (Fiji, Solomon Islands) and others (such as French Polynesia, Hawai'i, and New Caledonia) remain colonized. Historian K.R. Howe has astute observations on early European searches for Oceania's paradises:

> Those [Spanish] travellers who began to enter the Oceanic world in the sixteenth century did so with a well-established mental landscape of the tropical island paradise – sweet airs, glorious abundance of flora and fauna, running fresh water, riches, and their human inhabitants living in a natural innocence and ready for co-option in imperial designs ... The discovery of the earthly Pacific paradise and the noble savage by the likes of Banks and Bougainville some 250 years after [the Spanish explorers] was but a rerun of a very old Western theme (Howe, 2000: 13–14).

It would seem that this Western theme is still being rerun. Today's tourists – including the growing number of "eco-tourists" – are invited to a paradise with a happy coexistence between a "glorious abundance of flora and fauna" and "human inhabitants living in a natural innocence" (Douglas, 1996; Hviding and Bayliss-Smith, 2000). A perhaps more dispassionate, but still celebratory, line of argument is taken by natural scientists in reviews of the unique aspects of biodiversity in the South Pacific. Against such persistent views of a biologically rich and ecologically harmonious island world, the Pacific battles of World War II, nuclear testing on remote atolls by France and the USA, logging of the rainforests of the Western Pacific, and the sinking of atolls as sea levels rise figure starkly as rapacious inroads into pristine natures and cultures by a violent and greedy world. The South Pacific of today has indeed become a powerful symbol of the developed world's capacity to destroy. Western conservationists have in recent decades been flocking to the islands of the Pacific to further their causes. Not a few of them have met with indigenous peoples who have not necessarily shared their agendas but who have been more concerned about maintaining nature as a source for local existence in the widest sense – which may well imply that useful secondary forest is much preferable to untouched, pristine rainforest (see Hviding, 1998a). It goes without saying that people in Oceania are not necessarily conservationists in a Western sense.

Images and realities of the South Pacific were becoming engaged in global agendas and processes of discovery and imperialism long before "globalization" became a catchword. Yet these images and realities – deeply founded in both Western and Pacific conceptions of "nature" – have in recent decades been increasingly entangled in the global processes of environmental degradation and destruction, as well as in the global moralities of environmental conservation. These many mutual entanglements of Oceania and the Western world, as well as the extreme vulnerability of fragile island environments to imperialist intrusions, are beautifully and playfully – but seriously – highlighted in a recent song by a prominent voice in contemporary Oceanic scholarship and art, invoking the atoll locations of French and American nuclear testing:

Bad Coconuts
(Lyrics by Teresia Teaiwa;
music by Richard Hamasaki and Doug Matsuoka)

An apple a day
Keeps the doctor away
An apple a day
Keeps the doctor away

But a coconut a day
Will kill you
A coconut a day
Will kill you
A coconut a day
Will kill you

If you live on Moruroa
If you visit Fangataufa
Return to Enewetak
Resettle Bikini

A coconut a day
Will kill you
A coconut a day
Will kill you
A coconut a day
Will kill you

For the inhabitants of Oceania, coconuts and other living things that to outsiders may seem insignificant are by no means small and simple. They remain large in terms of importance for the continuity of everyday lives in the Pacific Islands, and they underpin, in material and symbolic ways, knowledges of nature in Oceania.

NOTES

[1]   For Papua New Guinea, see Morauta, Pernetta and Heaney, 1982.
[2]   For detailed discussions, see, e.g., Serpell, 1996.
[3]   For the Kalam of New Guinea, see Majnep and Bulmer, 1977; Feld, 1982 for the Kaluli of New Guinea; and Haussmann, 1994 for Guadalcanal in the Solomons.
[4]   See Titcomb, 1969 for the dog in ancient Oceania and Rappaport, 1968 for a classic study of the significance of pigs in New Guinea.
[5]   See Alkire, 1987 for an ecologically-oriented review of Pacific atoll societies.
[6]   See also the website of the Polynesian Voyaging Society at http: //pvs.hawaii.org and a notable film by Diaz [1997] for similar movements on Guam.
[7]   See Pollock, 1992 for a review of Polynesia.
[8]   See Hviding, 1996a for Solomon Islands examples.
[9]   See Sahlins, 1958 for a comparative survey of Polynesia.
[10]   See Hviding, 1996a for details on the knowledge and practice of Marovo fishing; see Hviding and Bayliss-Smith, 2000 for analysis of Marovo rainforest uses including agroforestry.
[11]   See, e.g., for the Foi and Umeda of New Guinea, Weiner, 1991 and Gell, 1995. For the Solomon Islands, see Hviding, 1996a and Keesing, 1982. For Vanuatu, see Bonnemaison, 1994; for Fiji, Ravuvu, 1983 and Toren, 1995; and for Palau, see Parmentier, 1987.
[12]   See Shore, 1989 for a comprehensive review.

BIBLIOGRAPHY

Akimichi, Tomoya. 'The ecological aspects of Lau (Solomon Islands) ethnoichthyology.' *Journal of the Polynesian Society* 87: 301–326, 1978.
Alkire, William H. *Coral Islanders*. Arlington Heights, Illinois: AMH Publishing Corporation, 1978.
Barlow, Kathy, and Stephen Winduo, eds. 'Logging the Western Pacific: Perspectives from Papua New Guinea, Solomon Islands, and Vanuatu.' Special Issue of *The Contemporary Pacific* 9(1), Spring, 1997.
Barrau, Jacques. *Subsistence Agriculture in Melanesia*. Bulletin 219. Honolulu: Bernice P. Bishop Museum, 1958.
Barth, Fredrik. *Ritual and Knowledge among the Baktaman of New Guinea*. Oslo/New Haven, Connecticut: Norwegian University Press/Yale University Press, 1975.
Barth, Fredrik. *Cosmologies in the Making: A Generative Approach to Cultural Variation in Inner New Guinea*. Cambridge: Cambridge University Press, 1987.
Beaglehole, J.C. *The Life of Captain James Cook*. London: Adam & Charles Black, 1974.
Berlin, Brent. *Ethnobiological Classification: Principles of Categorization of Plants and Animals in Traditional Societies*. Princeton, New Jersey: Princeton University Press, 1992.
Best, Simon. 'Here be dragons.' *Journal of the Polynesian Society* 97: 239–259, 1988.
Bonnemaison, Joël. *The Tree and the Canoe: History and Ethnogeography of Tanna*. Honolulu: University of Hawaii Press, 1994.
Clarke, William C. *People and Place: An Ecology of a New Guinean Community*. Berkeley: University of California Press, 1971.

Clarke, William C. 'Learning from the past: traditional knowledge and sustainable development.' *The Contemporary Pacific* 2: 233–253, 1990.

Clarke, William C. and R.R. Thaman, eds. *Agroforestry in the Pacific Islands: Systems for Sustainability*. Tokyo, London, Paris: United Nations University Press, 1993.

Conklin, Harold. *Hanunóo Agriculture*. Rome: FAO (Food and Agricultural Organization of the United Nations), 1957.

Crocombe, Ron G. *The South Pacific: An Introduction*. Auckland: Longman Paul Ltd., 1983.

Crocombe, Ron G., ed. *Land Tenure in the Pacific*. 3rd ed. Suva, Fiji: Institute of Pacific Studies, University of the South Pacific, 1987.

Dahl, A.L. *Review of the Protected Areas System in Oceania*. United Nations Environment Programme (UNEP)/International Union for the Conservation of Nature and Natural Resources (IUCN). Cambridge: IUCN Conservation Monitoring Centre, 1986.

Davis, Stephen. *Man of all Seasons: An Aboriginal Perspective of the Natural Environment*. Sydney: Angus & Robertson, 1989.

Denoon, Donald, ed. *The Cambridge History of the Pacific Islanders*. Cambridge: Cambridge University Press, 1997.

Denoon, Donald, and Catherine Snowden, eds. *A Time to Plant and a Time to Uproot: a History of Agriculture in Papua New Guinea*. Port Moresby: Department of Primary Industries/Institute of Papua New Guinea Studies, 1982.

Diaz, Vicente M., director. *Sacred Vessels: Navigating Tradition and Identity in Micronesia*. A Moving Islands Production. Documentary film, 1997.

Diaz, Vicente M. and J. Kehaulani Kauanui, eds. *Native Pacific Cultural Studies on the Edge*. Special Issue of *The Contemporary Pacific* 13(2), Fall 2001.

Douglas, Ngaire. *They Came for Savages: 100 Years of Tourism in Melanesia*. Lismore, New South Wales: Southern Cross University Press, 1996.

Durrenberger, E. Paul and Gísli Pálsson. 'Ownership at sea: fishing territories and access to sea resources.' *American Ethnologist* 14(3): 508–522, 1987.

Ellen, Roy F. *Environment, Subsistence and System: The Ecology of Small-Scale Social Formations*. Cambridge: Cambridge University Press, 1982.

Ernst, Thomas M. 'Empirical attitudes among the Onabasulu.' In *Man and a Half: Essays in Pacific Anthropology and Ethnobiology in Honour of Ralph Bulmer*, Andrew Pawley, ed. Auckland: The Polynesian Society, 1991, pp. 199–207.

Feinberg, Richard. *Polynesian Seafaring and Navigation: Ocean Travel in Anutan Culture and Society*. Kent, Ohio: Kent State University Press, 1988.

Feinberg, Richard, ed. *Seafaring in the Modern Pacific Islands: Studies in Continuity and Change*. DeKalb: Northern Illinois University Press, 1995.

Feld, Steven. *Sound and Sentiment: Birds, Weeping, Poetics and Song in Kaluli Expression*. Philadelphia: University of Pennsylvania Press, 1982.

Finney, Ben (with Merlene Among *et al.*). *Voyage of Rediscovery: A Cultural Odyssey through Polynesia*. Berkeley: University of California Press, 1994.

Flannery, Tim. *Mammals of New Guinea*. Carina, Queensland: Robert Brown, 1989.

Gegeo, David Welchman. 'Indigenous knowledge and empowerment: rural development examined from within.' *The Contemporary Pacific* 10: 289–315, 1998.

Gell, Alfred. 'The language of the forest: landscape and phonological iconism in Umeda.' In *The Anthropology of Landscape*, Eric Hirsch and Michael O'Hanlon, eds. Oxford: Clarendon Press, 1995, pp. 232–254.

Gladwin, Thomas. *East is a Big Bird: Navigation and Logic on Puluwat Atoll*. Cambridge, Massachusetts: Harvard University Press, 1970.

Gray, F. and L. Zann, eds. *Traditional Knowledge of the Marine Environment in Northern Australia*. Townsville: Great Barrier Reef Marine Park Authority, Workshop Series No. 8, 1988.

Hage, Per and Frank Harary. *Island Networks*. Cambridge: Cambridge University Press, 1996.

Hardin, Garrett. 'The tragedy of the commons.' *Science* 162: 1243–1248, 1968.

Hau'ofa, Epeli. 'Our sea of islands.' In *A New Oceania: Rediscovering Our Sea of Islands*, Eric Waddell, Vijay Naidu and Epeli Hau'ofa, eds., 2–16. Suva, Fiji: University of the South Pacific, 1993, pp. 2–16 (Reprinted in *The Contemporary Pacific* 6: 148–161, 1994.)

Hau'ofa, Epeli. 'The ocean in us.' *The Contemporary Pacific* 10: 391–410, 1998.

Haussmann, Eberhard. *Vögel in der Kultur der Mbirao. Ergebnisse ethnornitologischer und ethnotaxonomischer Untersuchungen in einer Ethnie der Salomonen.* Mundus Reihe Ethnologie, Band 83. Bonn: Holos Verlag, 1994.

Henderson, C.P., and I.R. Hancock. *A Guide to the Useful Plants of Solomon Islands.* Honiara: Research Department, Ministry of Agriculture and Lands, 1988.

Howe, K.R. *Nature, Culture, and History: The 'Knowing' of Oceania.* Honolulu: University of Hawai'i Press, 2000.

Hviding, Edvard. *Kiladi oro vivineidi tongania tingitonga pu ko pa idere oro goana pa Marovo/Of Reef and Rainforest: A Dictionary of Environment and Resources in Marovo Lagoon.* Bergen: Centre for Development Studies, University of Bergen, in cooperation with Western Province Division of Culture, Gizo, Solomon Islands, 1995.

Hviding, Edvard. *Guardians of Marovo Lagoon: Practice, Place, and Politics in Maritime Melanesia.* Pacific Islands Monograph Series, 14. Honolulu: University of Hawai'i Press, 1996a.

Hviding, Edvard. 'Nature, culture, magic, science: on meta-languages for comparison in cultural ecology.' In *Nature and Society: Anthropological Perspectives*, Philippe Descola and Gísli Pálsson, eds. London: Routledge, 1996b, pp. 165–184.

Hviding, Edvard. 'Western movements in non-Western worlds: towards an anthropology of uncertain encounters.' *Journal of the Finnish Anthropological Society* 23(3): 30–51, 1998a.

Hviding, Edvard. 'Contextual flexibility: present status and future of customary marine tenure in Solomon Islands.' *Ocean and Coastal Management* 40: 253–269, 1998b.

Hviding, Edvard. 'Between knowledges: Pacific studies and academic disciplines.' *The Contemporary Pacific* 15: 43–73, 2003.

Hviding, Edvard and Tim Bayliss-Smith. *Islands of Rainforest: Agroforestry, Logging and Ecotourism in Solomon Islands.* Aldershot, UK: Ashgate, 2000.

Ingold, Tim. 'Culture and the perception of the environment.' In *Bush Base, Forest Farm: Culture, Environment and Development*, Elizabeth Croll and David Parkin, eds. London: Routledge, 1992, pp. 39–56.

Irwin, Geoffrey. *The Prehistoric Exploration and Colonisation of the Pacific.* Cambridge: Cambridge University Press, 1992.

Johannes, R.E. 'Traditional marine conservation methods in Oceania and their demise.' *Annual Review of Ecology and Systematics* 9: 349–364, 1978.

Johannes, R.E. *Words of the Lagoon: Fishing and Marine Lore in the Palau District of Micronesia.* Berkeley: University of California Press, 1981.

Johannes, R.E. and Edvard Hviding. 'Traditional knowledge possessed by the fishers of Marovo Lagoon, Solomon Islands, concerning fish aggregating behavior.' *Traditional Marine Resource Management and Knowledge Bulletin* 12: 22–29, 2000 (also at http: //www.spc.in/coastfish).

Johannes, R.E. and J.W. MacFarlane. *Traditional Fishing in the Torres Strait Islands.* Hobart: CSIRO Division of Fisheries, 1991.

Jorgensen, Dan. 'Echidna and *kuyaam*: classifications and anomalous animals in Telefolmin.' *Journal of the Polynesian Society* 100: 365–380, 1991.

Kahn, Miriam. *Always Hungry, Never Greedy: Food and the Expression of Gender in a Melanesian Society.* Cambridge: Cambridge University Press, 1986.

Keesing, Roger M. *Kwaio Religion: The Living and the Dead in a Solomon Island Society.* New York: Columbia University Press, 1982.

Kirch, Patrick Vinton. *The Evolution of the Polynesian Chiefdoms.* Cambridge: Cambridge University Press, 1984.

Kirch, Patrick Vinton. *The Lapita Peoples: Ancestors of the Oceanic World.* Oxford: Blackwell, 1997.

Kirch, Patrick Vinton. *On the Road of the Winds: An Archaeological History of the Pacific Islands Before European Contact.* Berkeley: University of California Press, 2000.

Kwa'ioloa, Michael and Ben Burt. *Our Forest of Kwara'ae.* London: British Museum Press, 2001.

Kyselka, Will. *An Ocean in Mind.* Honolulu: University of Hawaii Press, 1978.

Latour, Bruno. *Science in Action.* Cambridge, Massachusetts: Harvard University Press, 1987.

Laycock, D.C. 'Melanesian linguistic diversity: a Melanesian choice?' In *Melanesia: Beyond*

Diversity, R.J. May and H. Nelson, eds. Canberra: Research School of Pacific Studies, Australian National University, 1982, pp. 33–38.

Leach, Jerry W. and Edmund Leach, eds. *The Kula: New Perspectives on Massim Exchange.* Cambridge: Cambridge University Press, 1983.

Lewis, David H. *We, The Navigators: The Ancient Art of Landfinding in the Pacific.* Canberra: Australian National University Press, 1972.

Lingenfelter, Sherwood G. *Yap: Political Leadership and Culture Change in an Island Society.* Honolulu: University Press of Hawaii, 1975.

McCay, Bonnie J. and James M. Acheson, eds. *The Question of the Commons: The Culture and Ecology of Communal Resources.* Tucson: University of Arizona Press, 1987.

Majnep, Ian Saem and Ralph Bulmer. *Mnmon Yad Kalam Yakt: Birds of My Kalam Country.* Auckland and Oxford: Auckland University Press and Oxford University Press, 1977.

Malinowski, Bronislaw. *Argonauts of the Western Pacific: An Account of Native Enterprise and Adventure in the Archipelagoes of Melanesian New Guinea.* London: Routledge & Kegan Paul, 1922.

Morphy, Howard. *Ancestral Connections: Art and an Aboriginal System of Knowledge.* Chicago: University of Chicago Press, 1991.

Morauta, Louise, John Pernetta and William Heaney, eds. *Traditional Conservation in Papua New Guinea.* IASER Monograph 16. Boroko, Papua New Guinea: Institute for Applied Social and Economic Research, 1982.

Morrison, John, Paul Geraghty and Linda Crowl, eds. *Science of Pacific Island Peoples*, 4 vols. Suva, Fiji: Institute of Pacific Studies, University of the South Pacific, 1994.

Mueller-Dombois, Dieter and F. Raymond Fosberg. *Vegetation of the Tropical Pacific Islands.* New York: Springer, 1998.

Myers, Fred R. *Pintupi Country, Pintupi Self: Sentiment, Place, and Politics among Western Desert Aborigines.* Berkeley: University of California Press, 1991.

Oliver, Douglas H. *Oceania: The Native Cultures of Australia and the Pacific Islands.* 2 vols. Honolulu: University of Hawaii Press, 1989.

Paijmans, K., ed. *New Guinea Vegetation.* Canberra: Australian National University Press, 1975.

Parmentier, Richard. *The Sacred Remains: Myth, History and Polity in Belau.* Chicago: University of Chicago Press, 1987.

Peterson, Nicholas and Bruce Rigsby, eds. *Customary Marine Tenure in Australia.* Oceania Monograph No 48, Sydney: University of Sydney, 1998.

Pollock, Nancy. *These Roots Remain: Food Habits in the Islands of the Central and Eastern Pacific since Western Contact.* Honolulu: University of Hawaii Press/Laie, Hawai'i: The Institute for Polynesian Studies, 1992.

Popper, Karl R. *The Logic of Scientific Discovery.* London: Unwin Hyman, 1959, reprinted 1980.

Randall, John E., Gerald R. Allen and Roger C. Steene. *Fishes of the Great Barrier Reef and Coral Sea.* Honolulu: University of Hawaii Press, 1990.

Rappaport, Roy A. *Pigs for the Ancestors: Ritual in the Ecology of a New Guinea People.* New Haven: Yale University Press, 1968.

Rappaport, Roy A. 'Aspects of man's influence upon island ecosystems: alteration and control.' In *Ecology, Meaning, and Religion*, by R.A. Rappaport. Berkeley: North Atlantic Books, 1979, pp. 1–26.

Ravuvu, Asesela. *Vaka i Taukei: The Fijian Way of Life.* Suva, Fiji: Institute of Pacific Studies, University of the South Pacific, 1983.

Ruddle, Kenneth, and Tomoya Akimichi, eds. *Maritime Institutions in the Western Pacific.* Osaka: National Museum of Ethnology, 1984.

Ruddle, Kenneth, Edvard Hviding and R.E. Johannes. 'Marine resources management in the context of customary tenure.' *Marine Resource Economics* 7: 249–273, 1992.

Ruddle, Kenneth and R.E. Johannes, eds. *Traditional Marine Resource Management in the Pacific Basin: An Anthology.* Jakarta: UNESCO, Regional Office for Science and Technology in Southeast Asia, 1990.

Sahlins, Marshall D. *Social Stratification in Polynesia.* Seattle: University of Washington Press, 1958.

Sahlins, Marshall D. 'Poor man, rich man, big man, Chief: Political types in Melanesia and Polynesia.' *Comparative Studies in Society and History* 5: 285–303, 1963.

Sahlins, Marshall D. *Islands of History*. Chicago: University of Chicago Press, 1985.

Serpell, James. *In the Company of Animals: A Study of Human-Animal Relationships*. Cambridge: Cambridge University Press, 1996.

Seymour-Smith, Charlotte. *Macmillan Dictionary of Anthropology*. London: Macmillan Press, 1986.

Shore, Bradd. *'Mana* and *Tapu.'* In *Developments in Polynesian Ethnology*, Alan Howard and Robert Borofsky, eds. Honolulu: University of Hawaii Press, 1989, pp. 137–173.

Sillitoe, Paul. *A Place Against Time: Land and Environment in the Papua New Guinea Highlands*. Amsterdam: Harwood Academic Publishers, 1996.

Smith, Linda Tuhiwai. *Decolonizing Methodologies: Research and Indigenous Peoples*. London and New York: Zed Press, 1999.

South, G. Robin, Denis Goulet, Seremaia Tuqiri, and Marguerite Church, eds. *Traditional Marine Tenure and Sustainable Management of Marine Resources in Asia and the Pacific*. Suva, Fiji: International Ocean Institute – South Pacific, University of the South Pacific, 1994.

Spriggs, Matthew. *The Island Melanesians*. Oxford: Blackwell Publishers, 1997.

Stanner, W.E.H. *On Aboriginal Religion*. Oceania Monographs 11. Sydney: Oceania Publications, 1966.

Strathern, Marilyn. 'No nature, no culture: the Hagen case.' In *Nature, Culture and Gender*, C. MacCormack and M. Strathern, eds. Cambridge: Cambridge University Press, 1980, pp. 174–222.

Takekawa, Daisuke. 'Hunting method and the ecological knowledge of dolphins among the Fanalei villagers of Malaita, Solomon Islands.' *SPC Traditional Marine Resource Management and Knowledge Information Bulletin* No. 12: 3–11, 2000 (also at http: //www.spc.in/coastfish).

Thomas, Nicholas. 'The force of ethnology: origins and significance of the Melanesia/Polynesia division.' *Current Anthropology* 30: 27–42, 1989.

Thompson, Laura. 'The relations of men, animals and plants in an island community (Fiji).' *American Anthropologist* 51: 253–267, 1949.

Titcomb, Margaret (with Mary Kawena Pukui). *Dog and Man in the Ancient Pacific*. Bernice P. Bishop Museum Special Publication 59. Honolulu: B.P. Bishop Museum, 1969.

Toren, Christina. 'Seeing the ancestral sites: transformations in Fijian notions of the land.' In *The Anthropology of Landscape*, Eric Hirsch and Michael O'Hanlon, eds. Oxford: Clarendon Press, 1995, pp. 163–183.

Valentin, F., ed. *Voyages and Adventures of La Pérouse*. Translated from the French by Julius S. Gassner. Honolulu: University of Hawaii Press, 1969 [orig. 1875].

Waddell, Eric, Vijay Naidu, and Epeli Hau'ofa, eds. *A New Oceania: Rediscovering Our Sea of Islands*. Suva, Fiji: School of Social and Economic Development, University of the South Pacific, 1993.

Wagner, Roy. *The Invention of Culture*. Chicago: University of Chicago Press, 1975.

Weiner, James F. *The Empty Place: Poetry, Space, and Being among the Foi of Papua New Guinea*. Bloomington and Indianapolis: Indiana University Press, 1991.

Whistler, W. Arthur. *Polynesian Herbal Medicine*. Lawai, Kauai, Hawaii: National Tropical Botanical Garden, 1992.

Whitmore, T.C. *An Introduction to Tropical Rain Forests*. 2nd ed. Oxford: Oxford University Press, 1998.

Worsley, Peter. *Knowledges: What Different Peoples Make of the World*. London: Profile Books, 1997.

WILLIAM BALÉE

NATIVE VIEWS OF THE ENVIRONMENT IN AMAZONIA

Amazonia is that region drained by the Amazon River and its tributaries together with adjacent lowlands. It represents about 4% of the earth's land surface. It is roughly the size of the contiguous 48 U.S. states or the island continent of Australia. But Amazonia far exceeds both those comparable regions in terms of the biotic and linguistic diversity of its varied landscapes. Landscapes refer not to nature paintings or manicured lawns and gardens in Euro-American tradition, but rather to distinctive juxtapositions of living species in spatial contexts that are verifiable empirically by soil, light, temperature, and historical conditions. In Amazonia, swidden fields produced by slash-and-burn horticulture, anthropogenic old-growth fallow forests of the terra firma, and dooryard gardens found in the context of Native American villages and camps, among other definitive zones, constitute landscapes in this sense. The environment to native Amazonians is virtually never a monolithic entity, encompassing ecosystems and ecospheres; rather, almost everywhere it represents a heterogeneous association of landscapes, each having a different character and different index of its essence. Ample diversity of native languages also exists: Amazonia has about 300 languages in 170 different family groupings.

The examination of Amazonian environmental knowledge may be facilitated by recognition of four working principles:

1. Amazonian languages and the extensive vocabularies of which these are comprised indicate that people recognize biotic diversity that is intrinsic to the region.
2. Environmental knowledge by Amazonian peoples tends to be restricted to local settings.
3. Amazonian languages and cultures probably encode a tremendous amount of accurate, empirically and independently verifiable knowledge about the environment.
4. Amazonian peoples since prehistory have transformed landscapes by redistributing suites of species across the region, enriching the biota in local and regional contexts and reordering the slope and contents of the surface of the land in diverse locales.

277

H. Selin (ed.), *Nature Across Cultures: Views of Nature and the Environment in Non-Western Cultures*, 277–288.
© 2003 *Kluwer Academic Publishers. Printed in Great Britain.*

Concomitantly, Amazonian peoples of today know that they and their anteced-
ents have impacted and diversified Amazonian landscapes in such ways.

THE RECOGNITION OF DIVERSITY

The first principle to be apprehended in examining how native Amazonian
peoples view the environment seems to be that their languages encode and
reflect, to a greater or lesser extent, the high diversity of landscapes together
with the living and non-living material contents of these that surround their
effective living spaces. Amazonia contains a greater diversity of species of plants
and animals than any other land region of comparable size in the world,
including at least 30,000 plant species. Excluding South America, no single,
entire continent boasts more than 30,000 plant species (Lizarralde, 2001: 266).
When insects are added to the list, millions of animal species may be found in
Amazonia, and biological systematists are far from being able to identify and
classify them all. Marked richness in fish, avian, and mammalian species has
been noted as well (Lizarralde, 2001). The biological diversity of Amazonia is
sometimes contrasted with the seeming monotony and homogeneity – in terms
of high heat and humidity, low saturation of sunlight on the forest floor,
architecture of massive trees, and so on – of the forests that harbor such
diversity. Native Amazonian groups as well as traditional peasant societies
(*caboclos* in Lusophone Amazonia, *ribereños* in Hispanic Amazonia), however,
recognize and name distinctively numerous types of forest habitats (Fleck and
Harder, 2000; Frechione *et al.*, 1989; Shepard, Jr. *et al.*, 2001). And the native
classifications of plant and animal species thus far described indicate a critical
and deep knowledge of diversity.

In some cases, native Amazonians classify a living species known to science
into more than one folk species, a phenomenon known as "overdifferentiation"
(Berlin, 1992: 18–19). Overdifferentiation most often occurs with domesticated
plants. Manioc (cassava) [*Manihot esculenta*], a tuber crop that is a staple
source of carbohydrates in many native Amazonian societies, seems to have
had its origins in Amazonia in the early to middle Holocene (i.e., seven to nine
thousand years ago [Piperno *et al.*, 2000]). Although manioc is recognized as
constituting but one species scientifically, it is typically subdivided into between
15 and 137 folk species in diverse Amazonian cultures, with the average number
of named folk species per language being 22 (Balée, 1994; Chernela, 1986;
Descola, 1994; Grenand, 1980). Overdifferentiation occasionally occurs also
with a nondomesticated species. In the Ka'apor language of eastern Amazonia,
the crabwood tree (*Carapa guianensis*), a forest hardwood in the mahogany
family, is subdivided into "red" and "white" crabwood folk species, depending
on the color of the heartwood and whether the individual is found growing in
seasonally flooded forest or on high, well-drained ground – the ones with red
heartwoods grow in the terra firma, whereas the ones with white heartwoods
are found in lower elevations (Balée, 1994). The Achuar people of the Peruvian
Amazon recognize twelve folk species of felines, but fewer than half these folk
species are distinguished as separate taxa in systematic zoology (Descola, 1994:

83). The converse process of underdifferentiation, in which a biological species is grouped under one term with other biological species, seems to be less common cross-culturally, especially with regard to higher plants and vertebrates. The numbers of folk taxa in Amazonian languages for the plant and animal kingdoms, while high both when compared to classification of environments with less biotic diversity and when juxtaposed to folk Euro-American terminologies for their own natural biotas, are nevertheless not in the aggregate higher than the millions of species worldwide that are potentially recordable by science.

LIMITS ON SYSTEMS OF CLASSIFICATION

A second principle in understanding native Amazonian views of the environment is that while there are multiple environments to inhabit, and individual groups of people have over time traveled across many of them in migrations and wanderings (Vidal, 2000), the classification of landscapes and biotic forms is limited to localities that are familiar and more or less immediate. Folk genera of plants or animals (in folk English, monomial labels for taxa like "oak", "rabbit", and "wren") normally do not exceed 500 terms in any language, other than the language of scientific taxonomy (Berlin, 1992: 96–101). But sometimes overdifferentiated taxa are, in scientific retrospect, worth the validating of new species. The Kayapó people of central Amazonian Brazil recognize 56 folk species of bees of which 11 were found to be either unknown or new to science and not yet described systematically (Posey, 1983). Likewise, a new species of capuchin (organ-grinder) monkey was recognized by science after the Ka'apor of eastern Amazonia (unrelated linguistically to the Kayapó) had already encoded it as a separate folk species (see Queiroz, 1992). In virtually every Amazonian language, a large percentage of vocabulary is devoted to plants and animals, and this reflects both the biotic richness of the region as well as empirical knowledge of that richness, ingrained in speech and cognition over long periods of historical time.

To some extent, throughout the Amazon, real creatures may be imbued with human qualities. Other human groups may be conceptualized as having animal attributes, and some humans are believed to be able to take on certain animal forms, while some animals may assume human forms ("shape-shifting") (Descola, 1996; Slater, 1994; Viveiros de Castro, 1998). A common shape-shifter to peasants who dwell along the Amazon River in Brazil, known as caboclos, is the gray dolphin (*Sotalea fluviatilis*). The dolphin is believed to assume the form of a strange but handsome youth at ceremonial occasions such as the Feasts of Saint John and Saint Peter in June. The dolphin, known as the *boto tucuxi*, traditionally is assigned the paternity of infants born to young, unwed mothers, who may have ventured too far from the party, and too close to the river, on such occasions (Wagley, 1976). Much of the environmental knowledge and lore possessed by Amazonian peasants is probably syncretic, that is, a product of the collision between Native American and Iberian systems of knowledge (Pace, 1998: 80). Certain surviving native systems of knowledge ascribe earthly life in most respects to some creatures that are otherwise otherworldly, as with a blue-boned, kinkajou-like animal known in the folklore of the Ka'apor people. The same people recognize an anaconda-like snake of one kilometer in length, known as the Mayu, to be at once both a fearful spiritual and physical entity, capable of shifting form from a snake into a seductive woman who is imbued with dark and potent magical abilities, though the Mayu and the kinkajou-like creature cannot be shown to have any independent empirical existence at all. In Amazonia, one notes that some human-like forms with human-like personalities (though lacking sometimes in speech) are believed to be "mothers of the game animals", i.e., protectors of supplies of edible meat from the forest, as seen among the Munduruku of north-central Amazonia (Murphy, 1960), the Achuar (Descola, 1994), and many other groups. Certain food taboos and food avoidances among diverse Amazonian native groups, as on raptors, capybaras, and sloths, are very widespread and seem to constitute a common, ancient foundation of religious and spiritual knowledge that affects the distribution of the local biota. Religious knowledge is integrated with environmental knowledge into a single epistemological frame of reference in Amazonia. Some mythic heroes in one form or another are believed to have created various features of landscapes known to people in the present as their homeland (Vidal, 2000), much like the Dreaming Beings of Australian Aboriginal lore.

NATIVE KNOWLEDGE OF THE ENVIRONMENT

For the most part, species and landscapes are not mere products of mind; they are real entities with finite boundaries that can be described cross-culturally. In other words, native Amazonians "know" a lot about the environment, which is a third principle in understanding their views of it. The empirical landscapes of Amazonia provide the observer with a wealth of biotic and abiotic minutiae, a wealth that is premised on the staggering diversity of rainforests. The rainforest as a habitat only seems monotonous to naive, not native, observers. Similarly, the semideserts and deserts of southern Africa and Western Australia often presented to Western visitors bleak and uninhabitable landscapes (see commentaries in Tonkinson, 1991: 32–34 and Lee, 1993). But for the hunter-gatherers, those regions contained much natural wealth and beauty together with resources for survival. Native Amazonians tend to consider forests and diversity home. In fact, native Amazonians typically divide the "forest" into more than one type in their local milieus. Remarkably, the Matses people of Amazonian Peru recognize 178 habitats in their area alone (Fleck and Harder, 2000), a number that greatly surpasses in diversity the previously existing scientific classification of their rainforest. But in retrospect, each Matses rainforest habitat is different in terms of the empirically verifiable distribution of indicator plant and animal species (Fleck and Harder, 2000). In similar fashion, and with some overlap among the various categories, the Matsigenka people, also of the Peruvian Amazon, identify 69 habitats defined in terms of vegetation, 29 habitats defined in terms of abiotic factors, and 7 habitats understood in terms of faunal indicator species (Shepard, Jr. et al., 2001).

No single Amazonian classification of plants, animals, and landscapes may be taken as a model to represent the others. For the 300 or so native languages spoken in Amazonia, there are at least 300 or so ethnobiological and ethnoecological systems of classification. One of the main differences concerns relations with the environment. The few hunter-gatherers (nonagricultural people) seem to have fewer folk species names (in folk English, binomial labels for taxa, like "live oak," "jack rabbit," and "house wren") than horticultural peoples, especially concerning the domain of plants. This is partly because horticulturalists interact with and manage more plant species than hunting-and-gathering peoples (Balée, 1994). The hunting-and-gathering Guajá people of eastern Amazonia, however, seem more keenly aware of plants that are edible to particular game species than do the neighboring, horticultural Ka'apor people (Cormier, 2000). Apart from this difference in mode of production (i.e., food production vs. food collecting), native Amazonian societies tend to exhibit broad similarities in terms of how they name the biota. In general, wild plants are often named for domesticates, but not vice-versa, unless the plant being named is recently introduced. In the Achuar language, the wild star apple tree of the forest (*Pouteria caimito*) is called *yaas-numi*, "star apple-tree", a name that is clearly derived from the name of the domesticated and closely related star apple tree proper (*Chrysophyllum caimito*), which is simply called *yaas*, "star apple" (Descola, 1994: 79). "False yam" (*kara-ran*) in Ka'apor, a name for a wild yam (*Dioscorea* sp.), is derived from *kara*, "yam", which is the name of the domesticated yam (*Dioscorea trifida*). These derivations and others like them in many other languages suggest a recognition by native peoples of a conceptual dichotomy between domesticated and nondomesticated plants, though statistical patterns in the nomenclature related to the morphology of terms permit recognition of a third category – semi-domesticated biota (Balée and Moore, 1994). These naming patterns mean that agriculture, artificial selection, and altered genotypes of organisms that have been long subjected to human management (and are therefore quite familiar) are cognitively embedded in concepts of folk classification. So, too, is the temporal and spatial diversification of landscapes.

DIVERSIFICATION OF LANDSCAPES

The Amazon rainforests have long been thought to be pristine and virgin. Recently, however, many scholars have questioned that assumption, based on their growing awareness that the population of Amazonia before the arrival of the Europeans was much larger and that agriculture was much older and more widespread than had been earlier thought. This led to the formation of anthropogenic forests and landscapes in the present (Balée, 1989; Denevan, 1992; Dean, 1995). Agriculture, which in Amazonia traditionally involves the swidden (slash-and-burn) method, has evidently resulted in an abundance of forest types in the regrowths that follow after a few years of burning. It seems that beyond about forty years and up to about 200 years after the initial planting of a swidden field, if forest regrowth occurs, that forest tends to be high in diversity yet different in terms of the species present from surrounding, seemingly untouched pristine forest (Balée, 1989, 1994). These old forests, called fallows, have traditionally been classified as high forest (pristine forest on well-drained ground) by Western researchers. The new evidence suggests that these forests are of a different type, and that they would not exist had it not been because human agricultural activities changed edaphic (soil) and biotic conditions to favor them. These fallow forests are called *taper* in the Ka'apor language, whereas high forests are called *ka'a-te*, and the Ka'apor distinguish them by numerous indicator species present in one but not in the other type (Balée, 1994).

Clearly Amazonian people since prehistory have impacted regional and local flora and fauna through the processes associated with domestication. Such processes have also influenced landscapes. In just a few cases, massive reconstitution of Amazonian landscapes took place in prehistory. Perhaps the most dramatic example of such manipulation is in the Baurés area of northern Bolivia, where prehistoric peoples built hydraulic earthworks stretching across more than five hundred square kilometers. These earthworks, which are etched on the land surface by linear causeways and canals, seem to have functioned as enormous fish weirs, permitting ancient people to manage potentially huge populations of fish for their sustenance. In a sense, they were ancient fish farms, to date unknown from elsewhere in the Amazon region (Erickson, 2000a). In the rainy season today, areas of artificial ponds in Baurés yield up fish including armored catfish, cichlids, and characins (such as piranhas), as well as snails. In the past, according to Clark Erickson, "Artificial ponds provided a way to store live fish and snails until needed" (Erickson, 2000a: 191). The ancient earthworks of Baurés, built some five hundred to two thousand years ago, have been called the "Nazca Lines of the Amazon."[1] Probably such manipulations of the existing landscape were enabled by relatively complex sociopolitical entities, though not as complex as a civilization, which in South America is known only from the Andes.

Early Amazonian peoples domesticated several species seen pantropically today, including manioc, papaya, cashew, peanut, pineapple, arrow cane (in the grass family), tobacco, annatto (a dye tree), guava, rope plant (in the bromeliad family), cocoyam (an edible aroid, like taro), and certain chili peppers. In addition, they cultivated and managed many species of trees, such as Brazil nut, peach palm (with a large edible fruit), hog plum (which has a tart, vitamin-C rich fruit), soursop, and many others. Probably ancient Amazonians changed the genotypes of some of these species, in a sense bringing them within the realm of domestication, but the loss of Amazonian peoples to disease and depopulation, following the European Conquest, may have occasioned a loss in the domesticated crop diversity that apparently existed on the eve of contact (Clement, 1999a, b; Peters, 2000). In other words, a large crop inventory could have required substantial numbers of people to manage and protect it. Early Amazonian peoples appear to have contributed to diversity by spreading domesticated crops around the basin and by creating new agricultural landscapes. These new landscapes, because of the increased availability of solar radiation and probably higher nutrient content in the soil, when compared to pristine, dense forest, would have favored the introduction and growth of tree species that formerly had been limited to much more narrowly restricted habitats, such as swamp and riverine forests. Ancient Amazonian peoples also built elevated landscapes that favored the expansion of species and enrichment of local biota.

In Amazonian Bolivia, the habitat of the Sirionó people, who speak a language in the Tupi family of languages, is about two-thirds a grassy, seasonally flooded savanna, not entirely unlike the Everglades of South Florida; the rest, in contrast, consists of patches of forest on the terra firma. Amazonian people typically do not build settlements and sleep in marshland. The Sirionó traditionally camped in the forest patches, on the high, well-drained forest islands that dot the savanna. These forest islands, indeed, are substantially elevated above the surrounding savanna by a few to many meters, and the highest ones (up to 18 meters in height) never flood. They are anywhere from a few acres to several square miles in extent. If one takes a historical and ecological perspective, these islands are, in fact, artificial mounds that were deliberately constructed by native peoples, probably of Arawakan linguistic affiliation, from about the beginning of the Common Era to about A.D. 1200 (Erickson, 1995, 2000b). The Sirionó arrived in the area, probably migrating from the south out of what is now Paraguay, during the 1600s, perhaps earlier. Although the Sirionó people did not build the mounds, they do know that the mounds of their habitat are anthropogenic phenomena. The mounds are rich in artifacts such as potsherds and charcoal, moreover, that independently supply evidence for former, permanent human occupation of some definite magnitude and sophistication. The soil contents of one of these mounds was shown to contain 13% pure ceramics – thus, these forests have been appropriately called "ceramic forests" (Langstroth, 1996). Because of all the burning for cooking and ceramic production that must have occurred in the past, the soils of the mounds are very fertile, much more so than those of the surrounding savanna. Fallow forests elsewhere in Amazonia also tend to have richer soils than surrounding high forests; some of these soils are known as *terra preta do índio*, a charcoal- black anthrosol (man-made soil) which exhibits high nutrient content, relatively high pH, and ample exchangeable bases such that it is one of the most fertile soil formations known. It is sometimes taken to be evidence of relatively long human occupations in Amazonian forests in the remote past, occupations that represent pre-European settlement patterns never seen since the conquest (Woods and McCann, 1999). And the mound forests of the eastern Bolivian Amazon are richer in tree species and animal life than the surrounding savanna. Likewise, fallow forests elsewhere in the lowland rainforest, formerly thought to be essentially homogeneous as a forest type, often harbor a wealth of animal species attracted to the many fruit trees that occur in a higher density in these forests than in high forests that have not been subjected (in recent memory at least) to any sort of agricultural manipulation. The mound forests of the Bolivian Amazon, like the many fallows throughout Amazonia generally, are in a sense native orchards (Balée, 2000), and thus are they understood and categorized by native societies.

The Kayapó people of the well-drained savannas of central Amazonian Brazil also have forest islands in their environment. They refer to these generically as *apêtê*. Recent researchers argued that these forest islands were anthropogenic phenomena that had been "planted" by the Kayapó (Anderson and Posey, 1989). Another view, however, argued that they were relic forests of Pleistocene climate changes that had nothing to do with human intervention or creation (Parker, 1992). Africanists have since reported that forest islands found dotting the well-drained savannas of Guinea (West Africa) are anthropogenic (Fairhead and Leach, 1996) and would not have existed had it not been because of plantings of fruit trees and the management of these over time by successive generations of human inhabitants. Indeed, the mechanisms involved and the anthropogenic evidence adduced for the forest islands in Guinea constitute very similar findings from another continent to the arguments that had been made in favor of anthropogenic *apêtê*. Much of the debate on the *apêtê* hinges on whether the Kayapó themselves consider the tree species within these forests to have been planted or to have been "plantable", based on the principal interview language of Portuguese. In fact, multiple Kayapó words for "planting" do not correspond to any English or Portuguese term (Posey, 1998: 112). In the Ka'apor language (again, unrelated to Kayapó) also, a single word corresponds to English "to plant" or Portuguese "plantar". To be sure, the Portuguese and English terms have many meanings. In Portuguese, one can accomplish the planting, sowing (*semear*), and cultivating (*cultivar*) of a crop, such as wheat or barley, by the verb "plantar". One can also "plant a swidden field" (*plantar uma roça*) (Holanda Ferreira, n.d.: 1098). In Ka'apor, all these operations are denoted by distinct, non-interchangeable terms, as shown in Table 1.

The four terms in Ka'apor from Table 1 could be used as glosses for English "to plant" or "plantar" in appropriate, specified contexts. Some of the Ka'apor terms listed in Table 1, while not interchangeable with any of the others, and which derive from fundamentally different roots and concepts, may also have many meanings. *Yitim* refers not just to digging a hole in the ground and inserting seeds or cuttings therein; it also can mean burying any item, including a human corpse. *Omor* refers not only to sowing domesticated seeds (as with *pitim-ha'ĩ-omor*, "tobacco-seeds-sow" or, idiomatically in English, "sowing of tobacco seeds"); it is used also to refer to the casual tossing away of seeds from

Table 1 Concepts of "planting" in Ka'apor

English "Planting" Concepts	Ka'apor Correspondences
"plant a swidden"	*kupiša-moú*
"plant" (a seed or cutting)	*yitim*
"sow" (seeds only)	*omor*
"cultivate" (or "raise to maturity")	*mu-tiha*

non-domesticated fruits that may germinate and then grow into future, fruit-producing trees. And *omor* denotes tossing in general, as in "tossing out garbage or waste" or "tossing or throwing a ball." One "cultivates" (*mu-tiha*) domesticated plants; with the same verb, one "rears" children and "raises" pets. An important step toward determining the nuances in concepts of domestication and cultivation in native Amazonia involves securing an adequate understanding of native terms related to (and revelatory of) these concepts.

HUMAN KNOWLEDGE AND LANDSCAPE DIVERSITY

The fourth and final principle underlying native Amazonian views of the environment is that people recognize that human beings, past and present, have contributed to landscape diversity. Even notions of spontaneity of volunteer plants must be hedged against native constructs of the landscape. In Ka'apor, *yeye-tu-hem*, "by itself-[particle]-it is born" (or idiomatically in English "it germinates spontaneously"), an adjectival verb phrase used to describe volunteer species in successional growth of swidden fields, may not capture the entire cultural essence of a plant species. Sometimes, it is clear than an individual of a volunteer species was "planted" by a nonhuman animal. "Deer manioc" (*arapuha-mani'i*) is sometimes said to have been "planted by deer" (*arapuha a'e yitim*), in the sense of its seeds having been interred by the deer (perhaps through feces).[2] Deer manioc is also found in old swiddens and fallows that appear in the traditional Western classification of forests to have been high forest. This is also true of many fruit trees that grow in fallows. While people may not have intended the fallow to be a repository of edible fruits, when they placed settlements within their cleared swidden fields, they also ate fruits from other fallow forests, the seeds of which they discarded (*ha'i-omor*), or sowed, about the settlement and in the garbage pits (middens) that surrounded it, though this sowing was less deliberate perhaps than their planting of manioc cuttings and seeds of other domesticated plants. Some of the seeds casually tossed away germinated and became trees, such as non-domesticated cacao, papaya, star apple, and a host of other fruit trees more commonly found in fallow today than in high forest. Some of these fruit trees are especially attractive to game animals. Capuchin monkeys spread the seeds of non-domesticated cacao throughout the area of the fallow as they consume the sweet, edible pulp that surrounds the seeds, and the Ka'apor as well as other Amazonian peoples know this dispersal mechanism. The Ka'apor say that the monkeys "sowed" (*omor*) the seeds of non-domesticated cacao in old fallows. Although it is perhaps too early to say whether the Kayapó people, far to the south of the Ka'apor, "planted" the forest islands of their central Brazilian landscape, it must be remembered that there are many ways to "plant" a forest.

*   *   *

Four principles based on empirical research to date can be adduced in approaching how native peoples of Amazonia view the environment. First, languages encode biotic and abiotic diversity to a greater or lesser extent, seemingly greater in the case of Amazonian languages. Second, classifications of biota and landscapes are confined to local, essentially familiar environments. Third, native Amazonians know a great deal about the environments they inhabit. And fourth, native Amazonians recognize and encode linguistically the fact that people, past and present, have affected the distribution of the biota and the formation of the landscape. Much of the knowledge that informs these views has derived from about eleven thousand years of human occupation in Amazonia (Roosevelt, 1998), if not more. It remains to be seen what proportion of that knowledge, and the knowledge acquired during the subsequent development of agriculture, has survived into the post-modern world of today and what prospects such knowledge may have for persisting in the day-to-day life of Amazonian societies.

NOTES

[1] The Nazca lines lie on the desert plains of south-central Peru. The lines represent figures such as birds and other animals, geometric figures, and sometimes enclosures. There are many hypotheses for their existence. Some researchers believe they had to do with water supply, some that they were purely for religious reasons. See the article, 'Nazca Lines' in The *Encyclopaedia of the History of Science, Technology, and Medicine in Non-Western Cultures*, edited by Helaine Selin, Kluwer, 1997, pp. 777–780.
[2] They believe that deer eat parts of wild manioc plants, and research shows that they definitely eat parts of domesticated ones.

BIBLIOGRAPHY

Anderson, Anthony B. and Darrell A. Posey. 'Management of a tropical scrub savanna by the Gorotire Kayapó of Brazil.' *Resource Management in Amazonia: Indigenous and Folk Strategies. Advances in Economic Botany 7*, Darrell A. Posey and William Balée, eds. Bronx, New York: New York Botanical Garden, 1989, pp. 159–173.

Balée, William. 'The culture of Amazonian forests.' In *Resource Management in Amazonia: Indigenous and Folk Strategies. Advances in Economic Botany 7*, Darrell A. Posey and William Balée, eds. Bronx, New York: New York Botanical Garden, 1989, pp. 1–21.

Balée, William. *Footprints of the Forest: Ka'apor Ethnobotany – The Historical Ecology of Plant Utilization by an Amazonian People*. New York: Columbia University Press, 1994.

Balée, William. 'Elevating the Amazonian landscape.' *Forum for Applied Research and Public Policy* 15(3): 28–32, 2000.

Balée, William and Denny Moore. 'Language, culture, and environment: Tupí-Guaraní plant names over time.' In *Amazonian Indians from Prehistory to the Present: Anthropological Perspectives*, Anna C. Roosevelt, ed. Tucson: University of Arizona Press, 1994, pp. 363–380.

Berlin, Brent. *Ethnobiological Classification: Principles of Categorization of Plants and Animals in Traditional Societies*. Princeton, New Jersey: Princeton University Press, 1992.

Chernela, Janet M. 'Os cultivares de mandioca (tucano).' In *SUMA: Etnologica Brasileira*, vol. 1, *Etnobiologia*, Berta Ribeiro, ed. Rio de Janeiro: Vozes, 1986, pp. 151–158.

Clement, Charles. '1492 and the loss of Amazonian crop genetic resources. I. The relation between domestication and human population decline.' *Economic Botany* 53: 188–199, 1999a.

Clement, Charles. '1492 and the loss of Amazonian crop genetic resources. II. Crop biogeography at contact.' *Economic Botany* 53: 203–216, 1999b.

Cormier, Loretta A. *The Ethnoprimatology of the Guajá Indians of Maranhão, Brazil*. Ph.D. dissertation. New Orleans: Tulane University, 2000.

Dean, Warren. *With Broadax and Firebrand: The Destruction of the Brazilian Atlantic Forest*. Berkeley: University of California Press, 1995.

Denevan, William M. 'The pristine myth: The landscape of the Americas in 1492.' *Annals of the Association of American Geographers* 82(3): 369–385, 1992.

Descola, Philippe. *In the Society of Nature: A Native Ecology in Amazonia*, N. Scott, trans. Cambridge: Cambridge University Press, 1994.

Descola, Philippe. 'Constructing natures: Symbolic ecology and social practice.' In *Nature and Society: Anthropological Perspectives*, Philippe Descola and G. Pálsson, eds. London: Routledge, 1996, pp. 82–102.

Erickson, Clark L. 'Archaeological methods for the study of ancient landscapes of the Llanos de Mojos in the Bolivian Amazon.' In *Archaeology in the Lowland American Tropics*, Peter W. Stahl, ed. Cambridge: Cambridge University Press, 1995, pp. 66–95.

Erickson, Clark L. 'An artificial landscape-scale fishery in the Bolivian Amazon.' *Nature* 408: 190–193, 2000a.

Erickson, Clark L. 'Lomas de ocupación en los Llanos de Moxos.' In *Arqueologia de las Tierras Bajas*, Alicia Durán Coirolo and Roberto Bracco Boksar, eds. Montevideo: Min. de Educación y Cultura, Comisión Nacional de Arqueologia, 2000b, pp. 207–226.

Fairhead, James and Melissa Leach. *Misreading the African Landscape: Society and Ecology in a Forest-Savanna Mosaic*. Cambridge: Cambridge University Press, 1996.

Fleck, David.W. and John D. Harder. 'Matses Indian rainforest habitat classification and mammalian diversity in Amazonian Peru.' *Journal of Ethnobiology* 20(1): 1–36, 2000.

Frechione, John, Darrell A. Posey, and Luiz Francelino da Silva. 'The perception of ecological zones and natural resources in the Brazilian Amazon: An ethnoecology of Lake Coari.' In *Resource Management in Amazonia: Indigenous and Folk Strategies. Advances in Economic Botany* 7, Darrell A. Posey and William Balée, eds. Bronx, New York: New York Botanical Garden, 1989, pp. 260–282.

Grenand, Pierre. *Introduction à l'étude de l'univers Wayãpi: Ethnoécologie des indiens du Haut-Oyapock (Guyane Française)*. Paris: SELAF, 1980.

Holanda Ferreira, A.B. de. *Novo Dicionário da Lingua Portuguesa*. 14th ed. Rio de Janeiro: Editora Nova Fronteira, n.d.

Langstroth, Roberto. *Forest Islands in an Amazonian Savanna of Northeastern Bolivia*. Ph.D. dissertation. Madison: University of Wisconsin, 1996.

Lee, Richard B. *The Dobe Ju/'hoansi*. Fort Worth, Texas: Harcourt Brace College Publishers, 1984, reprinted 1993.

Lizarralde, Manuel. 'Biodiversity and loss of indigenous languages and knowledge in South America.' In *On Biocultural Diversity: Linking Language, Knowledge, and the Environment*, Luisa Maffi, ed. Washington, DC: Smithsonian Institution Press, 2001, pp. 265–281.

Murphy, Robert F. *Headhunter's Heritage: Social and Economic Change among the Mundurucú Indians*. Berkeley: University of California Press, 1960.

Pace, Richard. *The Struggle for Amazon Town: Gurupá Revisited*. Boulder, Colorado: Lynne Rienner, 1998.

Parker, Eugene. 'Forest islands and Kayapó resource management in Amazonia: A reappraisal of the *apêtê*.' *American Anthropologist* 94(2): 406–428, 1992.

Peters, Charles M. 'Pre-Columbian silviculture and indigenous management of neotropical forests.' In *Imperfect Balance: Landscape Transformations in the Pre-Columbian Americas*, David L. Lentz, ed. New York: Columbia University Press, 2000, pp. 203–223.

Piperno, D.R., A.J. Ranere, I. Holst, and P. Hansell. 'Starch grains reveal early root crop horticulture in the Panamian tropical forest.' *Nature* 407: 894–897, 2000.

Posey, Darrell A. 'Keeping of stingless bees by the Kayapó Indians of Brazil.' *Journal of Ethnobiology* 3(1): 63–73, 1983.

Posey, Darrell A. 'Diachronic ecotones and anthropogenic landscapes in Amazonia: Contesting the consciousness of conservation.' In *Advances in Historical Ecology*, William Balée, ed. New York: Columbia University Press, 1998, pp. 104–118.

Queiroz, Helder Lima de. 'A new species of Capuchin monkey, genus *Cebus* Erxleben, 1777 (Cebidae: Primates) from eastern Brazilian Amazonia.' *Goeldiana (Zoologia)* 15: 1–13, 1992.

Roosevelt, Anna C. 'Ancient and modern hunter-gatherers of lowland South America: an evolutionary problem.' In *Advances in Historical Ecology*, William Balée, ed. New York: Columbia University Press, 1998, pp. 190–212.

Shepard, Glenn H., Jr., Douglas W. Yu, Manuel Lizarralde, and Mateo Italiano. 'Rain forest habitat classification among the Matsigenka of the Peruvian Amazon.' *Journal of Ethnobiology* 21(1): 1–38, 2001.

Slater, Candace. *Dance of the Dolphin: Transformation and Disenchantment in the Amazonian Imagination.* Chicago: University of Chicago Press, 1994.

Tonkinson, Robert. *The Mardu Aborigines: Living the Dream in Australia's Desert.* Fort Worth, Texas: Holt, Rinehart and Winston, 1991.

Vidal, S.M. 'Kuwe Duwa Kalumi: The Arawak sacred routes of migration, trade, and resistance.' *Ethnohistory* 47(3–4): 635–667, 2000.

Viveiros de Castro, Eduardo. 'Cosmological deixis and Amerindian perspectivism.' *Journal of the Royal Anthropological Institute*, n.s. 4(3): 469–488, 1998.

Wagley, Charles. *Amazon Town: A Study of Man in the Tropics.* New York: Oxford University Press, 1953, reprinted 1976.

Woods, William I. and Joseph M. McCann. 'The anthropogenic origin and persistence of Amazonian dark earths.' *Yearbook of the Conference of Latin Americanist Geographers* 25: 7–14, 1999.

DAVID L. BROWMAN

CENTRAL ANDEAN VIEWS OF NATURE AND THE ENVIRONMENT

The geographic region under discussion here is the central part of the Andes, the area of the former Inca Empire. Here the political reorganization and settlement policies of the Inca, along with an already presumed mutual cultural base, led to a wide region which shared a number of concepts about the natural ways of the world. The particular focus of this paper is the high plateau area of the central Andes, the *puna* and *altiplano*, which includes the Titicaca Basin, the world's largest high elevation lake, at just over 3,800 meters or about 12,500 feet. This region was not only the original homeland of the Incas but also the homeland of the preceding sister states from which the Inca evolved, the Quechua-speaking Wari conquest state of most of Peru, and the Aymara-speaking Tiwanaku federation of Bolivia and adjacent portions of Chile and extreme southern Peru.

The altiplano and puna are part of an essentially treeless grassland (Browman, 1997), which extends along the high plateaus and basins of the Andes, with variations in rainfall defining the Colombian-Ecuadorian *paramo* at the north end, the Peruvian puna and Bolivian altiplano in the central portions, and the Argentina salt puna at the south end. Rainfall at the north end often is greater than 600 mm per year, while it decreases to less than 100 mm per year at the southern end of this zone. The semi-arid high elevation grasslands, the Peruvian puna and the Bolivian altiplano cover nearly 10,000,000 hectares, comprising nearly a quarter of the land surface of the Central Andean region; the puna and altiplano range in elevation from roughly 3,000 meters to 4,800 meters (10,000 feet to 15,000 feet). Any human adaptation to this region then must deal not only with the issue of lack or uncertainty of precipitation, but also with the problems of high elevation and marked diurnal temperature changes which are characteristic of such mountainous regions. Reduced partial pressure, a lengthy dry season, irregular precipitation, low temperatures with frequent frosts and freezes (only one month a year is totally frost free), pronounced diurnal temperature variations (20°C a day being typical in some zones), rugged topography, and poorly developed soils result in a variety of stresses and risks.

H. Selin (ed.), Nature Across Cultures: Views of Nature and the Environment in Non-Western Cultures, 289–310.
© 2003 *Kluwer Academic Publishers. Printed in Great Britain.*

The Andean plateau grasslands have been modified from the theoretical pristine state – that abstract state in which there would have been no grazing selection on the plant communities by the domestic livestock, and in which there would have been no habitat disruption by the humans' cultivating plants and selecting for specific useful economic species (for example, the need for fuel for new steam driven plants in the 19th century led to the near extinction of the cushion plant or *yareta*.)

Local adaptation to the environment of these high plateaus involves a variety of approaches to respond to extant conditions and mechanisms to modify the physical landscape. I start here with selections from meteorological commentary, as weather conditions are among the environmental parameters cited most often in my experience.

LOCAL METEOROLOGICAL PREDICTORS

Local folklore includes a number of references to both short and long term meteorological predictors. In part my intent in this article is to examine some of these predictors, to ascertain from a Western scientific basis what validity they may have and how they may function. The Central Andean plateau is a hostile and hard environment, and reading weather omens correctly has allowed the inhabitants to survive and prosper. Rainfall is particularly important, so there are many signs that predict whether the precipitation will be adequate, abundant, excessive, or deficient. Following is a list of some of the more typical events employed by Andean inhabitants to prognosticate future weather conditions.

Long term predictors

Stars

Stars in the Milky Way "river", the home of major Central Andean constellations, predict whether the year will be rainy or dry and are particularly important for foretelling the onset of the El Niño. If the dark stellar cloud constellations are particularly bright, it will be a year of drought or a bad year; if they are obscure, it will be a rainy year. If the principal stars of these constellations first appear vividly at dawn or dusk, it will be a good year (with rain). One constellation in the Milky Way, known in the western world as the Pleiades, is particularly important for long-term prognostication. In a good year, the stars shine brightly, but in a bad year they are misty or vague. Common dates of observation are the Austral winter solstice (June 21), San Juan (June 24) or Trinity Sunday in June, for constellations such as the Pleiades, or the feast days of the crucifix (September 14), St. Michael's (September 29), or the Rosary (October 7), for other constellations in the Milky Way. (Abercrombie, 1998: 496, n.9; Urton, 1981: 98, 116, 120–121, 173; van den Berg, 1989: 19; Yampara, 1993: 183). This practice has a great time depth; a 16th century document reports the Incas' similarly observing the Pleiades for weather and crop prediction (Urton, 1981: 200).

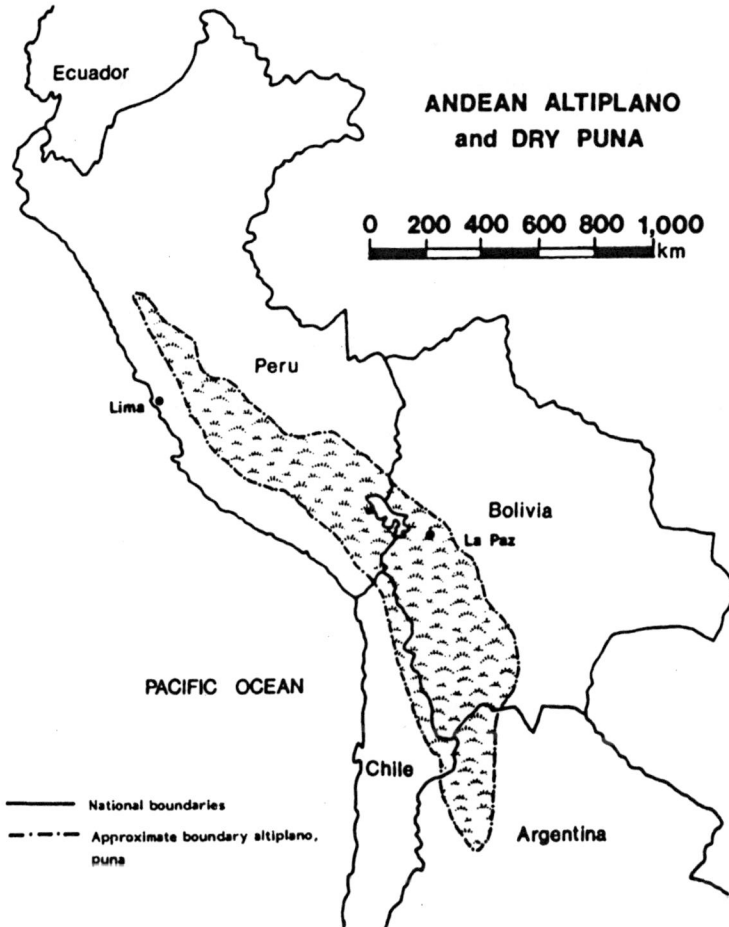

Figure 1   Map of Peru and Bolivia showing the location of the altiplano and dry puna. From Browman, David L., "Andean arid land pastoralism and development." *Mountain Research and Development* 3(3): 242, 1983.

San Juan, Independence, and San Andres days

Predictions are often made at the "agricultural New Year", which, depending on the community, ranges from June 24 to August 1 and also during the beginning of the rainy season (September through November), regarding precipitation amounts: drought, normal, or excessive.

If the ground is dewy on San Juan day (June 24), or if the smoke from the kitchen fires hangs near the earth, it will be a good agricultural year; conversely if it is dry that morning, or if the smoke disperses into the atmosphere, it will be a bad year.

If the first three days of August are rainy or cloudy, rains will come early, and it will be a good year; conversely if these days are clear and frosty, then it will be a bad year. If it is cloudy or rainy on San Andres day, November 30, it will be a good year; if it is clear and frosty, then a bad year. Local markets often adjust "futures" on crops based on predictors of this date, lowering the prices if it rains, as abundance is predicted, but raising prices if it is frosty, as future scarcity is predicted (Ayala Loayza, 1990: 246; Enriquez, 1987: 8; LaBarre, 1948: 174–5).

Lake vegetation

If water plants (such as *Elodea, Myriophyllum,* and *Potamogeton* spp.) grow well, it will be a good year for agriculture; if the lake shores are black with decaying algae and water plants, it will be a bad year. If there are many water plants at sowing, there will be frosts and it will be a bad agricultural year. However if the plant harvest is good, the fish harvest will be bad, and vice versa (LaBarre, 1984: 176; van den Berg, 1989: 22; Vellard, 1981: 170).

Waterfowl

If wild waterfowl build nests on the ground or only low to the ground, it will be a dry year, but if they build tall nests or nest in high places it will be a wet year (Ayala Loayza, 1990: 234; van den Berg, 1989: 24).

Land plants

If the firewood (*yareta, tola*) flower early and well, in June, it will be a good year; if they do not flower, it will be a bad year (van den Berg, 1989: 22; Yampara, 1993: 184).

Animals

If the foxes can be heard howling in September or October, and the skunks are out, it will be a good year, and it is time to begin planting. In a bad year they will not appear. If the fox howl is throaty, it will be a good year; if the fox howl sounds thin and high, it will be bad. If there are many guinea pigs, viscacha and other rodents, it will be a good year. If the birds sing loudly at planting time, it will be a good year (Ayala Loayza, 1990: 236–7; van den Berg, 1989: 23; Yampara, 1993: 184).

Winds

If the winds begin in August and early September, it will be a good year (adequate rain); but in a poor year (lack of rain), the winds do not begin until October (Yampara, 1993: 183).

Snows

If it snows in the late dry season (July, August, September), it will be a good year; if the first snows are not until All Saints' Day, it will be a bad agricultural year (Yampara, 1993: 183).

Short term predictors

Winds

Proliferation of small blackbirds indicates that strong winds from the east or south are about to occur, accompanied by bad frosts or heavy freezes (Enriquez, 1987: 6).

Winds from the Northeast and North will bring rain; winds from the West will bring drought and may bring hail. Winds from the East may bring hail, and winds from the South bring frost and drought (van den Berg, 1989: 20; Vellard, 1981: 170).

Clouds and mist

If the morning fog is thin or the rainbow high, it will not rain, but if the morning fog is thick or the rainbow low, then it will rain that day (LaBarre, 1948: 175–6; Tschopik, 1946: 567). If the clouds are white, it is good, as they bring rain; if the clouds turn yellow, it is bad, as they will bring lightning and hail (Bolin, 1998: 50).

Animals

If a variety of songbirds sing, or take up energetic flying, their activities announce pending rainfall that day. Frogs croaking excessively also foretell rainfall (Ayala Loayza, 1990: 234–5; LaBarre, 1948: 175–6; Tschopik, 1946: 567; van den Berg, 1989: 25).

Moon

If the moon is yellowish, it will rain soon; if the moon is white, there will be no rain (Ayala Loayza, 1990: 235, 246).

The short-term meteorological predictors are fairly straightforward. A yellow moon means haze or atmospheric moisture; a clear moon indicates dry skies and high pressure. The location of fog and clouds in the sky is also a clear indicator of whether the clouds are likely to produce precipitation. Bad storms are frequently associated with clouds so thick as to modify apparent colors. In the Andes a yellowish cast signifies bad weather, while the folklore where I live in Missouri is that bad weather (especially tornadoes or severe hail) is foretold by a yellowish green color. Around the world, many birds, animals, and insects react to changes in humidity, which are often associated with the onset of rain. The specific directions of the winds are geographically determined for the Central Andes by local topographic situations, as we shall see below. Again wind direction is a very common and straightforward predictor; for example, people in New England forecast bad weather if winds come from the Northeast (they call the storm a Nor'easter), and so on.

It is the long-term predictors that are of more interest to us here. Several of these are concerned with predicting the onset of the rainy season, its duration and intensity. Some of them have proven to be very accurate, but at first blush do not seem to have any underlying scientific rationale. For example, why should it matter what the foxes sound like? Investigation shows that high humidity in the area causes the howls to sound throatier; dry air makes them sound higher. Thus there is a western scientific relationship after all between how the foxes sound and good agricultural years (which require adequate rainfall). So what about the stars: how and why should it be possible to predict the amount of rainfall, or the onset of the El Niño, by observing the Pleiades in the dry season? As we shall see later on, there is a scientific basis for this prediction as well.

CLIMATIC VARIATIONS: METEOROLOGICAL EXPLANATIONS

The El Niño/Southern Oscillation (ENSO), the Quasi-Biennial Oscillation, the Intertropical Convergence Zone (ITCZ), and the South Pacific Convergence Zone all play a part in determining the long-term patterns (annual as contrasted to day-to-day) of precipitation and temperature in the altiplano and puna. Of particular importance for periods of severe weather anomalies are the cycles of the as yet poorly understood Quasi-Biennial Oscillation, which is thought to have a cycle of roughly 2.4 to 2.8 years, and the ENSO, which has an average estimated cycle of roughly 3.8 years. The Quasi-Biennial Oscillation has its largest magnitude from 20°N to 20°S latitude, which includes the bulk of the Central Andean region. When maxima or minima of these two oscillations coincide, the result is anomalous weather patterns. In the case of the maxima, there are generally abnormal droughts in Northeast Brazil and the altiplano, with rainfall as much as 70% below normal, and excessive rains and flooding along the northern reaches of the Pacific Coast.

While such major ENSO events do not occur often, being recorded only four times for the 20th century, they do occur at least once in the average lifetime of a Central Andean resident and are a source of environmental risk that the inhabitants must development management options to deal with in order to survive. In addition, however, even the minor ENSO events may have a measurable impact on regional precipitation, so that the Central Andean residents have developed a variety of means to try to identify their onset. *Risk* is sometimes characterized as a situation where the probability of loss is known, while *uncertainty* is limned as the situation where the probability of loss is unknown. Inasmuch as the Andean residents have lived with the environmental hazards for millennia, and have collected a good deal of folk wisdom regarding the probabilities of loss, *risk* in fact is used deliberately. While productive risks involve many social, political, and economic factors, our essay will be limited primarily to the risks related to climatic variation and physical environmental elements.

The ITCZ, in an average year, usually ranges between the Equator in winter and perhaps 12° North latitude in the summer. There are several circulation patterns influenced by variations in the ITCZ, but two are of particular importance for what the Central Andean inhabitants experience (Philander, 1990). The first is the Walker circulation, which has a zonal, or east-west circulation pattern, along the ITCZ. The Walker circulation ascends over the Pacific, with

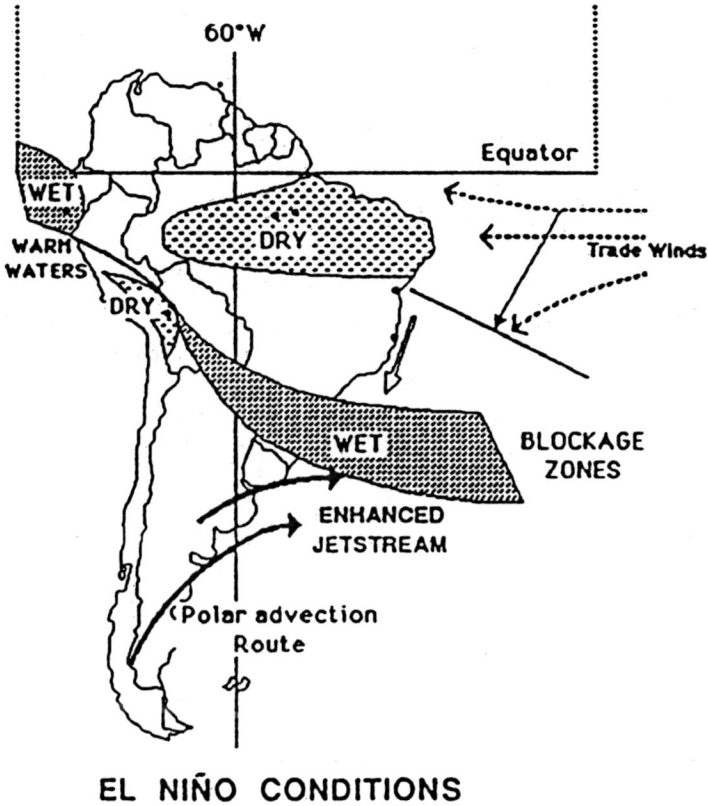

Figure 2 Moisture patterns (and blocking conditions) during an ENSO anomaly. From Martin *et al.*, 1992: 100, Figure 1. Used with permission of Louis Martin and Elsevier Publishers.

airflow from east to west. Relatively cool trade winds form over the cold upwelling currents of the Pacific coast of the central Andes. As these trade winds blow westward, the air is warmed and moistened through convection. Moist air rises where the sea surface temperatures are high in the western Pacific. As it rises, the absorbed moisture is released as precipitation. This flow then circles back eastward as cooler air higher in the troposphere. The dry air of this circulation patterns subsides where the surface water is cold, along the Pacific shores of South America. It thus descends as cool dry air over the western flanks of the central Andes, contributing both to the aridity on the coast and the winter dry season in the high plateau grasslands. The ITCZ tends to stay over the warmest sea surface temperature areas; hence it usually moves north in the winter and south in the summer, reaching its maximum southern extent usually about April.

The Hadley cells are the second of these circulation patterns; they are meridional or north-south circulation pattern phenomena. Two East Pacific Hadley meridional cells, fueled by moist air, rise in the ITCZ, each diverging poleward (one thus having a northward trajectory and the other a southern trajectory). These circulation patterns descend over regions north and south of the ITCZ. The intensity of these two cells changes depending on the season; the more intense cell has a subsiding motion in the winter hemisphere where the latitudinal thermal gradients are larger.

The variations in these two circulation patterns (Hadley and Walker) define most of the precipitation behavior over South America. Thus, during ENSO conditions, the warm western Pacific waters intrude eastward. Walker circulation weakens, as there is less energy differential between the sea's surface temperatures of the eastern and western Pacific. The ITCZ moves anomalously southward, and the Southern Pacific Convergence Zone moves further northeastward than usual. When the Walker circulation is weak, reduction in equatorial upwelling thus leads to enhanced convection and precipitation over the eastern Pacific and along the Pacific coast of the central Andean region.

Hadley circulation is thus intensified, and has a much larger poleward flux. There is a general increase in temperatures in the tropical belt or zone. The Southern Oscillation or southward displacement action of the ITCZ, coupled with the enhanced Hadley circulation activity, results in enhanced precipitation south of the Equator, usually along the southern Ecuadorian and northern Peruvian coasts, resulting in the typical El Niño phenomenon so well known in Andean literature.

For the puna and altiplano, these circulation events have teleconnections. When the Southern Oscillation is strong, there is widespread dryness over the Southern Andes, due to the anti-cyclonic strengthening of winds over north central Chile and central Argentina. In contrast, we should observe that the anti-ENSO, La Niña, or negative Southern Oscillation, results in the northward displacement of the ITCZ and brings abnormally high rainfalls to the southern Andes.

Teleconnections to circulation patterns over the Atlantic are important here. A weakening of the Pacific Walker cell (usually associated with the beginning of the ENSO) is associated with the strengthening of the South Atlantic Ocean Walker cell. This cell rises over Amazonia, with the low pressure convective zone over the western Amazon, thus contributing to heavy precipitation there; the returning subsiding branch of this circulation occurs to the east, generally over Northeast Brazil. Thus in an ENSO event, the Atlantic Walker cell intensifies and extends westward, while the Pacific Walker cell recedes. The Hadley cell circulation associated with the Atlantic Walker cell pattern also is intensified during an ENSO. The resultant high pressure over Northeast Brazil, along with the decrease in convection moisture, results in severe droughts in portions of northern Amazonia.

Precipitation in the altiplano and puna is fed by advective warm humid air from the Amazonian lowlands to the east, rather than by local convection. Continental warm air of these flows during the austral summer is responsible for the occurrence of precipitation in the altiplano and puna. More than 80% of the average yearly precipitation occurs when the ITCZ is in its southern displacement during the austral summer. The Atlantic trade winds which give rise to precipitation in the Titicaca basin and over much of the altiplano and puna must first cross the Amazon basin; thus anomalies in conditions in Amazonia will be reflected by anomalies in precipitation over the puna and altiplano grasslands. Hence the disruptions in the Atlantic Walker and Hadley circulation patterns over Amazonia during an ENSO event result in a decrease in warm humid air flow from that region over the Andean peaks. Mild ENSO events may be marked by significant decreases in precipitation, and severe ENSO events are marked by Atlantic circulation pattern-derived droughts both in Northeast Brazil and over the altiplano. It is evident that the ability to predict an ENSO event would be of significant importance to the Central Andean dweller, allowing early adjustments to his or her lifeways to counter the adverse impacts of this situation.

In the normal year, the austral winter months are marked by strong cyclonic wind circulations which originate in the South Pole region and transport cold dry air along the backbone of the Andes and its eastern slopes northward. On occasion, because of regional high pressure anomalies, these airflows occur in the fall planting season or the spring ripening period and thus can cause major calamities among plant-based agricultural villages impacted by the anomalous occurrence of extremely cold air patterns associated with these circulations. During the winter or dry season, higher wind velocities are common in the puna and altiplano, in themselves creating a greater environmental hazard. But on irregular occasions, polar advections result in extremely strong frigid winds which can lower the temperature along the eastern slopes and backbone of the Andes by as much as 20°C in a single day. In most cases in the Central Andes, these polar intrusions are of short enough duration not to cause dramatic losses, although clearly a drop in temperature of this magnitude will impact both plants and animals adversely. There may be some linkages with the ENSO meteorological events, as in the winter following the 1982–1983 major ENSO. That winter was particularly severe, particularly in June and July, with severe strong advective polar winds occurring 50% more frequently than normally, exacerbating losses from the ENSO drought and flooding.

The puna and altiplano are in a region close enough to the Equator that ordinarily the inhabitants do not need to worry about snows and winter climatic events typical of higher latitudes. The austral winter is the dry season, so little if any precipitation falls. Temperatures fall rapidly in the high elevations to below the freezing point, but also as well in the thinner air, warm up very quickly the following day. Thus until recently, snowfalls were an exceeding rare occurrence. Lasting snowfalls occurred so infrequently, as a rule, that Andean livestock herders did not put up hay or have barns or shelters in which to shelter their animals. Folk knowledge said that severe snows just do not (or should not) occur during the dry season. My Aymara informants in the altiplano, in fact, claimed that it "never" used to snow during the dry season, until the recent past, when now it is not at all infrequent for snows to occur during the "dry" season. While my informants blamed space exploration or a variety of other recent western technological activities for this shift, meteorologists indicate that the root cause is the deforestation of Amazonia, and thus the survival of some Atlantic weather fronts much further westward, still with significant moisture content as they now rise up the east flanks and cross the altiplano and puna. While snowfalls are known to occur during the wet season, the wet season is summer, and the snows rapidly melt. In the dry seasons, however, because of the ambient cold temperatures, snowfalls can remain for extended periods, covering forage for days at a time and leaving herds with no shelter into which to retreat. Not unexpectedly, high mortality has resulted.

METEOROLOGICAL AND ENVIRONMENTAL RISK IMPACT MANAGEMENT

Water management and planting several crop varieties are major productive techniques employed by Central Andean populations to deal with frost and altitude risk. Frost risk is greater on the flat areas or bottom lands than lower adjacent slopes (even though higher in elevation) because cold air, being denser, sinks. Typical households will seek to secure agricultural plots in several different topographical localities to help manage environmental risk, thus having both valley bottom and hill slope fields, fields of both sandy and clayey soils, fields on slopes with morning sun and also on slopes with afternoon sun, and so on. Thus rather than "economy of scale", with the most efficient land-holding being a single large plot, as would be typical in modern agronomists' recommendations, the high altitude farmers find it most rational to have their landholding scattered.

To avoid frost danger, the Central Andean farmers exploit the small microclimate variations these scattered plots provide. Frost and altitude risk can also be dealt with by selection of crops. For example, the native bitter potatoes, (*amarga* or *luki* varieties, with higher solanine content) which require more processing time to render them palatable for human consumption, also are more frost resistant and have greater pest resistance. Thus the shorter processing time of "regular" or "non-bitter" potato is offset by the double benefit of lower losses due to nematodes and other pests, and the lower losses from frost damage. In addition, farmers may shield more sensitive plants by building low walls around fields to protect them from wind stress, as well as planting frost-resistant species around the more sensitive plants.

The average number of frost-free days ranges from more than 140–150 days in the northern reaches (and also in the lake-effect zone adjacent to Lake Titicaca), to less than 90–100 frost-free days in the southern sectors (Enriquez, 1987). Total losses of crops due to frost and freezes have been regularly mentioned in reports since Colonial times and are in the memory of current farmers. While the standard rule of thumb is that there is a 0.5°C decrease in average annual temperature at this latitude for each 100 meters in increase in elevation of the field, so that planting fields at higher elevations might seem contra-indicated, in fact planting fields at different elevations may be a frost management technique, a means to reduce risk of loss of plants by unseasonal frosts which impact valley bottom fields more. Such a practice has the additional benefit of having the harvest spread out over an extended period of time. About two weeks more maturation time is required for every 100 meters in increase

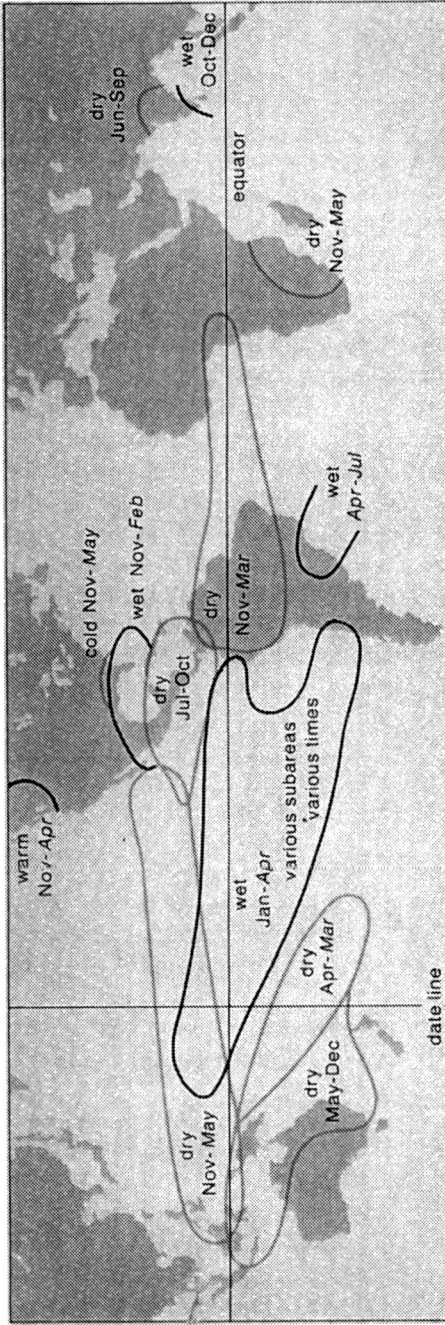

Figure 1. An occurrence of the El Niño–Southern Oscillation phenomenon involves such large-scale changes in climate as drought in normally productive agricultural regions and heavy rains in normally dry regions. Areas where cold and dry climatic anomalies typically appear are enclosed in colored lines; areas of wet and warm anomalies are enclosed in black lines. Duration of the anomalies is expressed in months; roman type indicates a month during the year of the major warming of the waters along the coasts of Ecuador and Peru; italics indicate a month during the following year.

Figure 3   Global ENSO wet and dry pattern. From Rasmussen, 1985: 169, Figure 1.

in elevation (Browman, 1987a: 175). By planting at different elevations, the farmer can reap an added benefit by avoiding the risk of higher labor costs required by needing to secure labor to bring in harvests from several fields which all mature at the same time.

Droughts are a regular event; one Aymara community in my research estimated that one year in seven was a complete agricultural failure and that yields would be sub-standard in two additional years as well. Some other estimates include a catastrophic drought 1 year in 4 (Montes de Oca, 1992), or a drought every 3 to 5 years (Claverias and Manrique, 1983). Records for this century bear out this conventional wisdom, as meteorological stations recorded significant droughts in the region from 18 to 35 percent of the years recorded for the 20th century, that is, from 1 in 6 to 1 in 3 years (Browman, 1987a: 175). Long-term droughts also occur, with more devastating impacts: the four year-drought that ended in 1943 resulted in *totora* (local cattail plants) beds drying up in Lake Titicaca and lake levels dropping so low that fish spawning grounds in these beds were impacted, with a major collapse of fisheries.

Irrigating fields in the puna is an obvious risk management technique for handling drought and is used as such. This approach also has the benefit of modifying altitude constraints. For example, crops can be planted at higher elevations with irrigation, and also may mature up to one month earlier than non-irrigated fields, allowing the cultivator to reduce or spread out labor inputs.

Environmental risk management may involve local technological investments to reduce productive risk by creating more local optimal environments to enhance the land's carrying capacity. For cultivators, increased productive capacity may include various land and water management systems, such as raised or ridged fields (*camellones*), terrace systems (*andenes*), specialized water containment networks (*qochas*), and sectorial fallowing systems. For herders it generally involves movement in special pasturing strategies, as well as stocking options.

The *qochas, andenes* and *camellones* are three different systems for water and frost management on agricultural lands. *Andenes* are the terraces from which the Andes received their name. *Camellones* refer to several types of raised or ridged fields systems, usually with parallel furrows and collector and drainage canals. *Qochas* refer to a water management system that is based on the construction of small depressions, which may resemble small ponds in the wet season, with water storage and drainage in these depressions also managed by a radial system of furrows and canals.

In some years, with high precipitation, the Andean root crops suffer high losses in the poorly drained clay soils. The drainage features of these three systems prevent tuber crops from rotting. In other years, drought ensues, and the water captured can be employed to "splash irrigate" the plants. Drainage/ponding features are not just important for water management in terms of irrigation. In the high frost-risk puna and altiplano, water's heat retention properties are significant in the reduction of frost damage and loss. Drainage canals and features are placed either at right angles or parallel to the daily course of the sun from east to west. Both of these alignments maximize heat capture by the water, allowing the water features to serve as solar heat sinks. To accentuate this effect, wide shallow flat bottom canals of width equal to or greater than the adjacent raised field platform are constructed to maximize efficiency in heat capture (Erickson, 1988, 1994). Radiation of this energy at night keeps the local microenvironment a degree or so warmer than the surrounding flat lands, reducing or eliminating frost damage.

Algae and water plants growing in these water management features are employed as green manure for the adjacent furrows, thus enhancing agricultural productivity (up to two times greater than adjacent dryland fields). Many of the water management features also contain exploitable fish populations and attract waterfowls, providing additional subsistence resources. However, during the process of "modernizing" or "westernizing" Andean agricultural production, the bulk of these millennium old raised field systems (with the earliest examples dated between 3,000 and 4,000 years ago in this region) have been destroyed by plowing or otherwise abandoned. The use of the raised and ridged fields systems to manage rainfall, to provide drainage, to moderate frost damage, and to provide green manure, has been for the most part lost in the current time period.

Sectorial fallowing is a means of risk reduction by allowing a field to regenerate fertility naturally during a fallowing period, without the need of capital inputs such as chemical fertilizers and pesticides. In addition to natural regeneration of fertility, insect pest vectors are reduced by extending the fallow period to sufficient lengths that the insect population diminishes or disappears, owing to lack of proper host. Recent intensification of agricultural production has led to abandonment of sectorial fallowing systems, increasing productive risk from insect pest vectors and in terms of the need for more chemicals and fertilizers.

Andean herders are very concerned with climatic variations and climatic risks. Among the most frequently cited causes for human illnesses are cold weather and winds; for animal species precipitation (either excess or lack) is the most frequently cited folklore factor in animal disease. To insure adequate grass production, regardless of precipitation variations, herders have tried various means of forage enhancement. One is by creating locally permanent marshy areas. Plants growing in these boggy or marshy zones are the preferred forage species for the native herd animals, the alpacas and llamas. Thus throughout the puna and altiplano, herders block small streams or excavate shallow pond areas to create what are termed *bofedales, pantones, oqhas, wayllas*, and also *qochas* (Browman, 1983: 243). If the local terrain is not appropriate for creating standing waters in ponds or marshes, the herders employ seasonal irrigation to help maintain year-round production of forage grasses.

Maintenance of the grasslands by reduction of less palatable and less nutritious (for the livestock) woody and shrubby species is usually accomplished by burning. Aymara herders around Lake Titicaca frequently practice annual burning of grazing areas. Modern livestock management specialists argue that used at lengthy intervals, say 5 years or more, fire is an appropriate management tool to eliminate undesirable woody shrubs; they frown, however, on annual burning, because it has a propensity to reduce the amount of available nutrients over time.

Herders traditionally preferred the native livestock, llamas and alpacas, over the introduced sheep, goats, hogs and cattle. The current market prices, however, make it much more rewarding to shift to these introduced European species. The herders know that the introduced species cannot eat the same forage as llamas and alpacas but must have special feed; these European animals are not yet acclimatized and adapted to the Andean environment. The native camelids can metabolize the high cellulose pasture in the highlands with a 10 to 25 percent greater efficiency (depending on the introduced species being compared [Browman, 1987b: 126]). The native animals, having evolved in this region, also have substantially lower mortality rates due to environmental stress, particularly during periods of drought. In addition the softer footpads of the plantigrade native livestock distribute the weight more evenly, unlike the sharper hoofs of the sheep, goats, and cattle. This feature of the introduced livestock causes heavy trampling damage to grass root systems in the high elevation environments.

WEATHER PREDICTION

Celestial visibility

For many years, ethnographers working in the Central Andes reported that it was typical for the local inhabitants to observe constellations to predict the onset of the ENSO, as noted above. Among the constellations of importance is what the western world calls the Pleiades; their appearance on or near the austral winter solstice determines when to initiate the planting season. Because the ethnohistoric record showed Inca populations employing the Pleiades in this fashion, it is presumed that this practice has a time depth of a millennium or more. In earlier times, the austral winter solstice was more frequently mentioned as a pivotal observation date, but as the Andean populace became Christianized, this period of observation was often transferred to the feast day of San Juan, June 24. Traits that were most frequently mentioned were the first observation of the helical rise of the constellation, its perceived brightness, and the apparent size of the stars in the constellation.

Most western observers, from what seemed like solid scientific knowledge, assumed that the first rising of the stars, and certainly their perceived size and brightness, should have been empirically stable, and this folk knowledge was generally written off as mere superstition. However, after the 1982–83 ENSO, when several ethnographic observers noted that the Andean *yatiris* (those who know) seemed to be making amazingly accurate forecasts about the occurrence of the ENSO that year, as well as precipitation forecasts for the coming agricultural year in general, they focused new attention on this issue.

When these *yatiris* or other local indigenous experts recorded the stars in the Pleiades as larger and brighter, a larger harvest was predicted for the growing season. The occurrence of larger and brighter stars were also said to be indicators of the need to plant the crops earlier. If the stars were recorded as being dimmer and smaller, then the harvest would be, if you will, dimmer and smaller, and the beginning of the planting season was set at a later date.

If in fact there is a strong ENSO event, rainfall in the Central Andes is adversely impacted during the growing season. Very strong events can result in near-drought conditions, but even milder ENSO events can result in much reduced rainfall. Potatoes, *oca*, *ulluco*, and other root crops are major cultigens in the altiplano and puna. If the tuber is planted too late, when the soils are waterlogged, the seed rots. If it is planted too early, when there is a moisture deficit in the soil, the plant may wither and die in the worse case scenario, but even in the best case scenario, the lack of moisture retards root formation early on and later reduces the number of shoots and stems per plant, thus reducing the overall tuber production. Particularly for the tuber farmer, it is of critical importance to know when to plant, but it is nearly as important that the indigenous grain crops such as quinoa and cañihua be planted at the correct time in the rain cycle.

Western ethnographers noted that in fact the observations of star size and brightness reported by the *yatiris* did seem to occur, but most knew of no empirical basis for this to occur, and many, as I did, chalked it up to auto-suggestion. It has been known for some time that moisture in higher levels of the stratosphere results in the masking or diminishment of visibility; faint stars will not be seen, and bright stars seem dimmer and smaller. But since our Central Andean observations were made in the middle of the austral winter dry season, these factors did not seem to lend themselves initially to an appropriate explanation. After all, it was the middle of the dry season, and not a cloud was to be seen in the skies.

However, Dr. Benjamin Orlove was convinced that the amazing accuracy of predictions based on the presumed appearance, brightness/dimness, and size of the stars in the Pleiades must have some empirical basis. After perusing the ethnographic literature, and finding more than a dozen well-confirmed recent reports by fellow ethnographers, he recruited two researchers from the Lamont-Doherty observatory of Columbia University to work with him on the project (Orlove *et al.*, 2000). The astronomers sought to test the phenomenon by observing the relative atmospheric transparency in terms of high cloud formation, employing data from the International Satellite Cloud Climatology Project, and also to assess the intensity of winds, in case high altitude turbulence also was a factor in the apparent visibility and magnitude of the stars as perceived by the human eye.

The recently available satellite imagery allowed the team to map out the high altitude moisture content in June over a series of years. More severe ENSO conditions in the altiplano are often associated with temperature anomalies as well as moisture anomalies; in such events, the growing season tends to be warmer and drier than the average year. Research of the imagery, plus other meteorological data bases, indicated that tropic cirrus cloud cover increased in warm ENSO years over the Central and South Central Andes. During the dry season, this higher moisture amount is significant, but to the human eye, "subvisual". That is, there was the formation of a high-thin cloud system derived from higher moisture contents that was essentially too small to be identified by the naked human eye.

The evident solution was that during significant ENSO anomalies, this increased water vapor level in the atmosphere effectively reduced visibility of the Pleiades, making the stars seem dimmer, and in extreme cases, perhaps even handicapping the observation of the actual first day of their helical rising. Because atmospheric conditions change day to day, the question was raised whether a single observation on the austral winter solstice, or on San Juan feast day, would be accurate. The literature sources, however, indicated that the prognostication was the sum of a series of observations made on several nights for a week or more surrounding those dates.

While it has recently been well documented that major ENSO anomalies are associated with drought conditions in the central Andes, it was not immediately evident from the literature whether such situations were always associated with low rainfall anomalies at the beginning of the planting season (mid-September to mid-October). Reviewing the meteorological data for the target period, the two researchers from the Lamont-Doherty observatory found that there was a positive correlation with low rainfall and higher temperature anomalies at the beginning of the planting season during the occurrence of significant ENSO events. In addition, these variations also correlated with the actual harvest of tubers as reported by the agricultural agencies in the impacted areas.

Thus: the predictions based on the averaged naked-eye observations by the native elders and farmers, taken over a series of days around the austral winters solstice, do in fact have an empirical basis. During the onset of significant ENSO events, more high elevation moisture overflows the Andes from the Atlantic windflows. Although not visible to the naked eye, the high elevation moisture does impact the apparent brightness and other visibility factors for the earth-bound observers. Apparently dimmer and smaller star sizes do correlate with the beginning of a significant ENSO; a significant ENSO does correlate with reduced moisture in the beginning of the planting season in mid-September through October. Thus this Central Andean age-old folklore that the dates of the planting season are to be determined by the brightness of the Pleiades around the austral winter solstice or San Juan's feast day has a solid scientific basis.

Lake Titicaca and the puna/altiplano

Accurate weather prediction is also important for the exploitation of the resources from Lake Titicaca – resources which include harvesting plants such as the totora reeds as well as fishing. For the 20th century, long-term records have been kept at the major lake ports of Puno, Peru and Guaqui, Bolivia. There is a periodicity of lake level changes correlated to meteorological phenomena: a cycle of *ca.* 2.8 years, thought to be associated with the ENSO, and a longer one of around 11 years, associated with sunspot activity (Kunzel and Kessler, 1986).

The average annual variation of lake levels is usually about 0.70 meters from summer to winter. In years of drought, the level of the lake can increase as little as 0.15 meters. However, in years of major rainfall, the rise from one season to the next can be quite dramatic: in 1935–1936, there was a 1.79 meter rise, in 1984–1985, a 2.20 meter rise, and several sources report more than a 3 meter rise in the period between September 1985 and April 1986. In the generally flat terrain of the altiplano, this resulted in thousands of hectares being flooded, in some cases for years; the excess rainfall also resulted in waterlogged agricultural fields and major crop failure, with rotting of plant roots. In addition, there were also high animal losses of sheep, llamas, and alpacas, because of water-soaked newborns perishing in the cold night winds, and high cattle losses because of the asphyxiation of the totora plants due to rising lake levels, resulting in major losses of crucial fodder. Access to the totora beds was a critical factor in periods of drought as well; herders without access to lake *totorales* suffered losses of 50% of livestock or more in many cases in the crisis period of 1982–1983.

The greatest variation for that period was between the low measurement of December 1943 and the high of April 1986, which was recorded as reaching 6.35 meters at Guaqui and 6.37 meters at Puno, or 20.9 feet (Browman, 1993; Morlon, 1996). Because the altiplano is a largely flat basin, a change in lake level of 2.5 meters would only change the lake volume *ca.* 2.5%, but the area covered by lake waters could change up to 25%. Thus it is clear that within a single resident's lifetime, the configuration of Lake Titicaca can change dramatically. Such changes can be a mixed blessing. In the ENSO-caused drought of 1982–83, the lake level fell 1.6 meters in 12 months, and there were staggering losses in surrounding agricultural zones. An estimated 50% of natural pasture was lost. This resulted in an animal mortality level of 20–30% of the herds, but even among the surviving animals there was an estimated 40% meat loss and up to 30% wool loss. In addition, potato and quinoa crop losses were in the 60–70% range (Claverias and Manrique, 1983). On the other hand, the lake level drop resulted in exposing rich new agricultural lands, which were exploited with great success the next growing season. And even though the drought resulted in the loss of herds and crops, the lake provided insurance resources, in terms of totora stems and roots, fish, and water birds to eat.

As noted, one of the particularly important resources from Lake Titicaca is the local cattail or *totora*. This plant is used in thatching roofs, in constructing balsa fishing boats, in providing fodder for the cattle industry, in creating spawning grounds for fish and nesting areas for waterfowl, and in providing a root and new growth stem for human consumption. Although not a domesticated plant, by strict definition, the totora are extensively managed by the Titicaca basin inhabitants.

New totora beds are usually planted in September and October. If the totora is planted too early, the lake could fall too much, and the sun dry up the new plants when they are short and vulnerable; if the totora is planted too late, they run the risk of early rains' making the water too turbid, causing young plants to rot or resulting in new plants' uprooting and floating to the surface.

By planting in September or October, the locals are able to monitor the lake level when the dry season is far enough advanced to permit a precise estimate of the minimum level that the lake will reach. If the lake is high, totora is planted at the shallow, onshore end of totora beds, to compensate for the plants that drown at the outer edge. If the lake level is low, they plant at the deeper, offshore end, to compensate for loss of totora from plants that have dried out at the inner edge. Totora must be planted most intensively in dry years, because without their efforts the totora will not come back as fully in years of higher water. If the water is too deep to plant by hand, they push roots in with a pole or drop clumps of roots into water weighted with stones.

Dried totora loses weight but does not shrink in volume. It is used to thatch roofs, to make mats and mattresses, to make storage containers, to make *balsa* boats, and to make the "floating islands" of Puno Bay. Fresh green totora is used to feed llamas and alpacas, and more recently, to feed horses, mules, and cattle. Totora is harvested by the *pichu*: 1 *pichu* makes 3 standard reed mats, while a *balsa* requires 30 to 33 *pichu* and must be replaced twice a year as it slowly waterlogs (Levieil, 1987).

Totora reed beds are usually held as private property, but *llachu*, principally the floating water weeds (*Elodea, Myriophyllum, Potamogeton* spp.) but which also may include smaller water algae and plants (*Chara, Cladophora* and *Nostoc* spp.), are open to all shore dwellers, and fishing is a communally-held right (Levieil *et al.*, 1989; Levieil and Orlove, 1990). The *llachu* is a highly productive resource, yielding 4–5 times more dry matter than totora (Dejoux and Iltis, 1992). In addition to feed, *llachu* is used as a fertilizer, and when dried, also as a fuel.

Villages (rather than individuals) own fishing territories. Shoreside communities tolerate some outside fishing as social risk insurance, to make "friends" for periods of future lacks. The local native species are primarily killifish (*Orestias* sp.) which live in the totorales and submerged macrophytes and also some catfish (*Trichomycterus* sp.) which habituate the submerged macrophytes or river mouths. Because the killifish tend to migrate little horizontally, it is possible to "own" a fish population; hence individual families own fish weirs, landing spots, and fishing zones (Levieil, 1987). However, larger totorale fishing areas are controlled and managed by the adjacent shoreside villages, and the community controls fishing rights in the zone. This small-scale management area is argued to result in conservation, administrative efficiency, economic efficiency, and social equity. The drought of 1983, however, resulted in overfishing, because the fish were constrained to a smaller area and were easier to catch. There was a reduced feeding and breeding area, and more shore dwellers activated social ties with the fishing communities, to turn to fishing as an emergency subsistence strategy.

Winds play a critical part for both fishermen and farmers. Fishing is mainly a winter or dry season occupation, and perceived Lake Titicaca water colors are employed as weather predictors. If the lake color is blue, good weather will ensue and fishing is safe. If there is no wind, with the lake a dead calm, the water in the distance can sometimes appear white. If it is windy, the lake surface seems to turn green, with darker green indicating a stronger wind. Strong winds are also signaled by the appearance of black water in the distance, invariably in the direction from which the wind will start blowing. When wind dies down, black turns to gray and then to blue. Where there are major stream inlets into lake, the lake is sometimes seen as yellow or red due to sediment carried, so that the direction and strength of wind-driven currents in the lake can be inferred from the density of color (Orlove, 2002).

Farmers want a calm day for thatching roofs with totora reeds, and they need a day of steady firm winds for threshing and winnowing quinoa and other grains. While the wind direction is seen as critically important, the winds are often named not for the specific cardinal direction but for the geographic feature from the direction they come (usually a geographic feature proximal to that village). Winds are also paired left and right; stronger winds are seen as coming from the right or male side (usually correlated with the East and South). Thus any village may have 6, 8, or 12 named winds, but always as a left and right pair. As noted earlier in the section on short-term meteorological predictors, winds from the South and East bring hail and strong freezes. This is in part because, as reported in our discussion of the ENSO, occasional polar advections result in extremely strong frigid winds which can lower the temperature along the eastern slopes and backbone of the Andes by as much as 20°C in a single day.

NUTRITIONAL ISSUES AND GEOPHAGY

The Central Andean diet has been analyzed by a variety of researchers looking at the standard western perception of comestible items, and often it has been viewed as being deficient in iron, calcium, and Vitamins A and C. One of the issues that these earlier studies overlooked is the fact that the floating water plants (the *llachu*) from Lake Titicaca and other highland lakes and ponds are dried and traded in blocks and are a good source of Vitamins A and C, iron and manganese, as well as an excellent source of protein. Some varieties of the float duckweeds and algae have up to 45–65% protein dry weight (Dejoux and Iltis, 1992; Morlon, 1996).

One of the better sources of dietary minerals is the comestible clays that are consumed with dehydrated, freeze-dried potatoes (*chuño*), often two or three times a week. The Central Andean inhabitants employ more than 30 named earths in comestible and medicinal concoctions, with uses primarily in one of four areas: (a) phyllosilicate clays, used to detoxify plants for human consumption; (b) comestible salts – calcium and sodium salt earths; (c) sulfur minerals, mainly for medicinal purposes; and (d) metallic mineral earths (mainly copper and iron compounds) for various medicinal purposes. Some of these earths also function as colorants, mordants, acarcides, and fertilizers (Browman and Gunderson, 1993). Phyllosillicate clays are good at absorbing a number of phytotoxins, especially the various glycoalkaloids in tubers, the alkaloids of chenopod grains, and the steroid and cardiac glycosides found in other high Andean domesticates; they also contain a number of bioavailable minerals.

The practice of consuming earths (geophagy) has been argued (Johns, 1990) to be a more effective means than cooking of dealing with heat stable and water insoluble compounds such as these phytotoxins.

It is important to detoxify the plants that form the basic diet of the indigenous inhabitants, particularly because of the altitude factor. The chenopod grains quinoa and cañihua were, until replaced by rice in recent times, the primary grains eaten by the Central Andean groups. These chenopods have high amounts of the organic alkaloid, or glucoside, saponin, which destroys red blood cells, causing hypoxia at these elevations. Cañihua, which tolerates higher amounts of soluble salts and is more resistant to frost and hence a very desirable domestic plant for the salt puna and southern altiplano area, also has increased amounts of glucoside saponin. While much of the saponin is in the outer coat which is winnowed away, there still remain sufficient amounts to make the grain unpalatable to humans until it is removed by washing and leaching.

The legumes contain phaseolunatin, a cyanogenetic glucoside, which interferes with calcium metabolism and can lead to anemia. The native lupine tarwi has lupinine alkaloids, which cause respiratory distress. The native potatoes have high levels of solanine, which interferes with breathing (in severe cases possibly leading to respiratory paralysis), and also have inhibitors which interfere with digestive enzymes necessary to break down protein. Companion domestic tubers such as oca have substantial quantities of oxalate, which interferes with calcium metabolism (already a problem), and so on. The freeze-drying process (basically employing freezing to rupture cells so that the water soluble toxins can be leached out, and then drying the tuber in the sun and wind, where heat labile toxins are destroyed) is thus supplemented by the geophagic practice of consuming phyllosilicate clays, as the anions of these clays have an affinity for the complex biochemical phytotoxin substances and take them out of solution as insoluble precipitates.

Thus a practice that earlier was thought to be merely a cultural anomaly now has been shown to have physiological bases and to be a necessary component of the means by which the human populations deal with the problems of the Central Andean environment.

*   *   *

I have elected to employ the Titicaca basin as the principal proxy for discussing the Central Andean perceptions of the environment for a number of reasons. On one hand, a focus on Lake Titicaca allows us to investigate both land-based and water-based indigenous environmental concepts. But more importantly, it is issues that occur in this area, such as the ability to detect the ENSO by looking at the stars, or the consumption of clays with meals – both cultural features which had earlier been derided or written off as obscure cultural anomalies – that I wanted to use to illustrate their meteorological and physiological bases. We have been too hasty in the past to write such traits off; they inform us of the unsuspected wisdom that we often overlook in cultures which view the world from different eyes that Westerners do.

BIBLIOGRAPHY

Abercrombie, Thomas A. *Pathways of Memory and Power: Ethnography and History among an Andean People*. Madison: University of Wisconsin Press, 1998.

Ayala Loayza, Juan Luis. *Insurgencia de los Yatiris: manifestaciones culturales del hombre andina*. Puno: CONCYTEC, 1990.

Bastien, Joseph W. and John M. Donahue, eds. *Health in the Andes*. Washington DC: American Anthropological Association, 1981.

Bolin, Inge. *Rituals of Respect: The Secret of Survival in the High Peruvian Andes*. Austin: University of Texas Press, 1998.

Browman, David L. 'Andean arid land pastoralism and development.' *Mountain Research and Development* 3(3): 241–252, 1983.

Browman, David L. 'Agro-pastoral risk management in the Central Andes.' *Research in Economic Anthropology* 8: 171–200, 1987a.

Browman, David L., ed. *Arid Land Use Strategies and Risk Management in the Andes: A Regional Anthropological Perspective*. Boulder: Westview Press, 1987b.

Browman, David L. 'Climatic influences in the Titicaca Basin cultural sequence.' Paper presented at the 13th International Congress of Anthropological and Ethnological Sciences, Mexico City, August 4, 1993.

Browman, David L. 'Environment and nature: South America – The Andes.' In *Encyclopaedia of the History of Science, Technology, and Medicine in Non-Western Cultures*, Helaine Selin, ed. Dordrecht, The Netherlands: Kluwer Academic Publishers, 1997, pp. 307–309.

Browman, David L. and James N. Gundersen. 'Altiplano comestible earths: prehistoric and historic geophagy of highland Peru and Bolivia.' *Geoarchaeology* 8(5): 413–425, 1993.

Claverias H., Ricardo, and Jorge Manrique M., eds. *La Sequia en Puno: Alternativas Institucionales, Tecnológicas y Populares*. Puno: Universidad Nacional Tecnica del Altiplano, 1983.

Dejoux, Claude and Andre Iltis, eds. *Lake Titicaca: A Synthesis of Limnological Knowledge*. Monographiae Biologicae, Vol. 68. Dordrecht, The Netherlands: Kluwer Academic Publishers, 1992.

Enriquez, Porfirio. 'Indicadores andinos que anucian heladas.' *Boletin del Instituto Estudios Aymaras* (2nd series) 2(26): 4–15, 1987.

Erickson, Clark L. *An Archaeological Investigation of Raised Field Agriculture in the Lake Titicaca Basin of Peru*. Ph.D. dissertation. Urbana: University of Illinois, 1988.

Erickson, Clark L. 'Methodological considerations in the study of ancient Andean field systems.' In *The Archaeology of Garden and Field*, Naomi F. Miller and Kathryn L. Gleason, eds. Philadelphia: University of Pennsylvania Press, 1994, pp. 111–152.

Granadino, Cecilia and Cronwell Jara Jimenez. *Las Ranas, Embajadoras de la Lluvia y otros relatos*. Lima: MINKA, 1996.

Johns, Timothy A. *With Bitter Herbs They Shall Eat It: Chemical Ecology and the Origins of Human Diet and Medicine*. Tucson: University of Arizona Press, 1990.

Kunzel, F. and A. Kessler. 'Investigation of level changes in Lake Titicaca by maximum entropy spectral analysis.' *Archives for Meteorology, Geophysics, and Bioclimatology, Series B: Theoretical and Applied Climatology* 36: 219–227, 1986.

LaBarre, Weston. *The Aymara Indians of the Lake Titicaca Plateau, Bolivia*. Memoir 48. Menasha, Wisconsin: American Anthropological Association, 1948.

Levieil, Dominique Philippe. *Territorial Use-Rights in Fishing (TURFs) and the Management of Small-Scale Fisheries: The Case of Lake Titicaca (Peru)*. Ph.D. dissertation. Vancouver: University of British Columbia, 1987.

Levieil, Dominique P., Q.C. Cutipa, C.G. Goyzueta, and F.P. Paz. 'The socio-economic importance of macrophyte extraction in Puno Bay.' In *Pollution in Lake Titicaca, Peru: Training, Research and Management*, T.G. Northcote, et al., eds. Vancouver: Wastewater Research Centre, University of British Columbia, 1989, pp. 155–175.

Levieil, Dominique P. and Benjamin Orlove. 'Local control of aquatic resources: community and ecology in Lake Titicaca, Peru.' *American Anthropologist* 92(2): 362–382, 1990.

Martin, Louis, *et al.* 'Enregistrements de conditions de type El Nino, en Amérique du Sud, au cours des 7,000 dernières années.' *Comptes Rendus de l'Académie des Sciences, Paris.* Série II: *Mécanique, Physique Chimie* 315(1): 97–192, 1992.

Morlon, Pierre, ed. *Comprender la agricultura campesina en los Andes Centrales, Peru-Bolivia.* Lima: Institut Français d'Études Andines, and Cuzco: Centro de Estudios Regionales Andinos Bartolome de Las Casas, 1996.

Orlove, Benjamin S. *Lines in the Water: Nature and Culture at Lake Titicaca.* Berkeley: University of California Press, 2002.

Orlove, Benjamin S., John C.H. Chiang, and Mark A. Cane. 'Forecasting Andean rainfall and crop yield from the influence of El Niño on Pleiades visibility.' *Nature* 403: 68–71, 2000.

Philander, S. George. *El Niño, La Niña and the Southern Oscillation.* San Diego: Academic Press, 1990.

Rasmusson, Eugene M. 'El Niño and variations in climate.' *American Scientist* 73(2): 168–177, 1985.

Tschopik, Harry, Jr. *The Aymara of Chucuito, Peru.* Part 1. *Magic.* New York: Anthropological Papers of the American Museum of Natural History, vol. 44, part 2, 1951, pp. 135–320.

Tschopik, Harry, J. 'The Aymara.' In *Handbook of South American Indians*, vol. 2, Julian H. Steward, ed. Washington DC: Bureau of American Ethnology Bulletin 143(2), 1946, pp. 501–574.

Urton, Gary. *At the Crossroads of the Earth and the Sky: An Andean Cosmology.* Austin: University of Texas Press, 1981.

Van den Berg, Hans. *La tierra no da asi no mas: los ritos agricolas en las religion de los Aymara-Cristianos.* Amsterdam: Centro de Estudios y Documentacion Latinoamericanos, 1989.

Vellard, Jehan A. *El Hombre y Los Andes.* Buenos Aires: Ediciones Culturales Argentinas, Ministerio de Cultura y Educacion, 1981.

Yampara Huarachi, Simon. 'Economia comunitaria andina.' In *La Cosmovision Aymara*, Hans van den Berg and Norbert Schiffers, eds. La Paz: Hisbol, Biblioteca Andina No. 14, 1993, pp. 143–186.

ELLEN BIELAWSKI

NATURE DOESN'T COME AS CLEAN AS WE CAN THINK IT": DENE, INUIT, SCIENTISTS, NATURE AND ENVIRONMENT IN THE CANADIAN NORTH

"And that is all I know about your white men who once came to our land, and perished; whom our fathers met but could not help to live."

Qaqortingneq (in Petrone, 1988: 32)

The indigenous people of northern North America who survived the past 400 years are in the process of recovering their own terms for themselves. The most inclusive terms that accurately apply to them are plural words that translate loosely as "The People". In the Subarctic forest live the Dene (*den*-ay) known in the literature and academically as Athapaskans, one of many groups of North American "Indians". On the treeless coasts and tundra live the people whose names for themselves are all variants of the word Inuit *(in*-you-eet)*:* Inuvialuit, Inupiat, Cup'ik, Yup'ik. Europeans called them Eskimos, and thus they are known in European literature and science. Northern Dene live largely in the Subarctic Boreal Forest of North America (other Athapaskans include Navajos, Apaches, and smaller tribes scattered along the Pacific coast of the United States). Inuit live along the Arctic coasts, on both sides of the Bering Strait and in Greenland as well as in North America. Inuit across the circumpolar world – Greenland, Canada, Alaska, and Siberia – are represented internationally through the Inuit Circumpolar Conference. In this text I refer to all of them as "Inuit" although they prefer and use regional variations. The regional names occur in the text, as does the term "Eskimo" when I am citing written work where "Eskimo" occurs, or when I refer to one specific group among the circumpolar Inuit. All members of the Athapaskan linguistic stock call themselves some form of the word "Dene" and that is how I refer to them. Again, quoted literature may include the terms Athapaskan or Indian.

Summarizing views of nature and environment as they are expressed by members of Inuit and Dene cultures, adapted to an area sweeping from the western tip of Siberia to the east coast of Greenland, and from Thule in Northwest Greenland to the northern parts of Canada's prairie provinces, is

311

H. Selin (ed.), Nature Across Cultures: Views of Nature and the Environment in Non-Western Cultures, 311–327.
© 2003 *Kluwer Academic Publishers. Printed in Great Britain.*

an enterprise doomed to misconception if not inaccuracy. With that caveat, I include many references to locally specific studies in this attempt at a broad overview. The following is the briefest of glimpses into not one, but two complex and enduring human traditions in nature, each with its own unique subtlety.

THEIR LAND, WATERS, COAST, SEA, ICE AND SKY

There are exceptions, but generally Inuit live along the Arctic coast of North America from the Bering Strait in Alaska to the coasts of Hudson Bay and Labrador in eastern Arctic Canada. Sea ice was, and is, an enormous part of their environment (Nelson, 1969). When the ice was not frozen solid, it filled the Inuit world with the ice floe edge, an area of rich biodiversity, and all the subsistence and navigational challenges that moving ice in a frigid sea presents. To a lesser extent, Inuit lived terrestrially, always exploiting caribou and riverine fish on tundra well away from the coast. But only the Caribou Inuit of interior Arctic Canada, and the Nunamiut of the Brooks Range in Alaska, lived year-round away from the Arctic Ocean's shore. Yup'ik, Cup'ik and Aleutiiq peoples lived along the coasts of the Aleutian Islands and southern Alaska, on a very cold but usually ice-free ocean.

Dene lived primarily inland, at and within the treeline (the jagged, ever-shifting and ephemeral boundary between Arctic tundra and Subarctic taiga) and within the huge boreal forest that covers northern North America from the southern edge of Alaska's Brooks Range to the southern shore of Hudson Bay. The enormous lakes and river systems of the Canadian Shield country and the long river drainages across interior Alaska were as important to Dene as the sea ice and sea were to Inuit.

Even though Inuit are primarily coastal peoples who relied heavily on marine mammals for food and fuel, both Inuit and Dene relied on caribou. Dene hunted caribou year-round. Even Inuit with a reliable supply of marine mammal meat required caribou hides, essential for winter clothing, and antler for tools. Dene and Inuit territory overlapped along the southern tundra and northern treeline as each group followed the annual caribou migration from winter forest foraging to summer calving grounds on the tundra.

Consequently, and in the way of people everywhere, Dene and Inuit make up part of each other's world. Both groups, no matter how infrequent their immediate contacts with each other, were probably aware of the other and their differences since each first occupied its traditional territory. Contact between the two groups preceded contact with Europeans by many centuries. It included trade, sometimes through long-standing and ritualized trading partnerships (Burch, 1979; Burch and Correll, 1972). Dene in Alaska relied on marine mammal oil while Inuit relied on interior products such as caribou hides, moose and mountain goat. European explorers frequently described Inuit and Dene relationships as antagonistic. Samuel Hearne named a much frequented fishing spot on the Coppermine River "Bloody Falls" after the Dene with whom he was traveling killed several Inuit there in 1771. Such enmity, while deep and long-standing, was not the only thread in the fabric of relationship between Inuit and Dene. Archaeological evidence shows that Dene ancestors occupied the Subarctic forest from about 8000 years ago. Inuit ancestors probably occupied the Bering Strait region by at least 3000 years ago. Direct ancestors to contemporary Inuit occupied Arctic Alaska, the Canadian Arctic coasts, and Greenland beginning about 1000 years ago.

CATEGORIES OF LANGUAGE AND THOUGHT

Inuit and Dene views of nature are not accurately expressed in English.[1] The most obvious example is that in western science, the term "environment" typically excludes people. For both Dene and Inuit, however, environment includes people, land, animals, air, insects, water, fish, birds, plants, rocks, and everything else we can perceive or imagine. Environment also includes relationships among all of these things we can label. Relationships among things western scientists perceive as living might be termed, scientifically, an ecosystem. But would a western scientist include social or spiritual relationships among all things in an ecosystem? Dene and Inuit do. Their indigenous environment includes spiritual knowledge as well as knowledge of where and when and how life and death happens. It recognizes relationships between the living and the dead. It recognizes that knowledge and spirit are carried on from generation to generation among species in the land and water they inhabit.

There are many other English words that do not translate cleanly into Inuit and Dene languages, just as the comparable ideas do not translate well into English. Unfortunately, the use of these terms has been largely taken for granted in the conduct of science on Inuit and Dene lands. These terms may lead directly to mistaken perceptions and misunderstanding. These include "research", "knowledge", "past", "present" (Sharp, 1988: xviii) "land", "water" and a cornucopia of terms captured under the English label "environmental management" (Bielawski, 1993: 8–9).

Given these difficulties, I caution the reader to recognize the conceptual limits of western thought in considering any indigenous peoples' views of nature.

THEORY

It is obvious to Dene and Inuit that their views of what people from the western scientific tradition call "nature" and "environment" are different from western scientific conceptions. The scope of difference was so great to Europeans who first encountered Inuit and Dene, and described them to the educated classes of Europe, that for a very long time, the western scientific tradition believed that Inuit and Dene views of nature and environment were in no way systematic and especially not scientific. Macdonald's synthesis of Inuit astronomy (2000) begins with an excellent review showing the disparity between what Inuit elders know and what classic ethnographic sources recorded about Inuit astronomy.

... These very real practical drawbacks to stargazing in Arctic regions during winter are easily overlooked. Birket-Smith does exactly this when he tries to contrast the apparent paucity of star lore among the Inuit with his assumption that Arctic winters provide optimum conditions for celestial viewing: 'The astronomical knowledge of the Caribou Eskimos is not great in consideration of the opportunity which the open prospect and the long, starlit, arctic nights give for observation. Probably the explanation is that the stars, apart from serving as a guide and a certain approximate division of the night, have no significance to the Eskimos' (Birket-Smith, 1929: 56, MacDonald, 2000: 12).

Further, from MacDonald (2000: 15–16): "Birket-Smith complains that 'in many cases it is impossible to get statements from one's informants to agree, because very often the names [of stars] are quite confused'."

Contrast that statement with this: "On the basis of interviews and observations made with Igloolik elders, the claims of ... Birket-Smith are not entirely supported. At Igloolik the 'common' names given to particular stars by various Inuit elders are essentially consistent and, differences of regional nomenclature aside, there was usually agreement on the nonliterary Inuktitut designations for all major identified stars and constellations" (MacDonald, 2000: 16).

Last, MacDonald writes, "The point is that one finds all too often in the writings of anthropologists, missionaries, and explorers, from the earliest times and over the entire Arctic area, vague, unsubstantiated, or incomplete references to stars and other celestial or environmental phenomena of significance to Inuit" (2000: 3).

Although studies as comprehensive as MacDonald's for astronomy do not exist for Inuit or Dene knowledge as it pertains to most of the western science disciplines, similar disparity between what ethnographers recorded and what Inuit and Dene actually know, or knew, are the rule rather than the exception in the literature. Often, in fact, the classic sources are silent on scientific subjects or classify them as myth. This has two results: the epistemology of indigenous knowledge is obscured, and the content of indigenous knowledge is recorded as lacking, or reduced to the status of myth as western science has treated it. Again, citing Laurens Van der Post, MacDonald makes this point concisely:

The word 'myth' in common usage was the label applied to what the rationalist in command of the day dismissed as illusion, nonexistent, apocryphal or some other of the proliferating breed of reductive words the cerebral norms of our time produce for dismissing the existence of any invisible and non-conceptual forms of reality (van der Post, 1975, in MacDonald, 2000: 18).

Similarly, in his study of Koyukon Dene natural history, Richard Nelson writes that one purpose of his work is

... to show how real and tangible the Koyukon belief in nature is. Because this belief differs vastly from our own, we may have difficulty appreciating its power and substantiality for those who are its inheritors. It lies beyond our emotional grasp, so we are inclined to pass it off as quaint folklore or mere fantasy. I hope the chapters that follow will make this Koyukon view of nature more concrete for persons who have learned to see a different one (1983: xv).

Unfortunately, the generalizations of early ethnographers about the paucity of Inuit and Dene knowledge pertinent to scientific disciplines were largely brought forward to near the end of the 20th century.

Certain epistemological questions are present for the observer and/or describer of indigenous views of nature. It is essential to examine one's theoretical framework and assumptions before proceeding with comparisons between western science and indigenous knowledge.

My early research on this topic was as an attempt to understand, then interpret, the indigenous knowledge of northern peoples accurately, and in ways mutually comprehensible to, Inuit, Dene, and scientists who study the environment Dene and Inuit have occupied for thousands of years. I approached comparing indigenous knowledge and science through philosophy of science precisely because scientific research in the north will benefit if researchers grasp indigenous knowledge in terms they can all understand and respect within their explanatory frameworks. I emphasize that validity in terms of western science and philosophy is not required to acknowledge or validate indigenous knowledge in and of itself.

I cannot emphasize enough that Dene and Inuit epistemologies differ as much from each other as western science does from indigenous knowledge. Again, specialists in aspects of Dene and Inuit cultures illustrate this conclusively. From the Kaska Dene, John Honigmann describes a "relativistic philosophy":

> One characteristic of epistemological thinking stands out clearly, the relativistic nature of truth. The Indian does not regard his thinking in absolutist, or universal, terms of validity ... implying the realization that white and other people have conceptual systems which differ from that held by the Kaska. Knowledge to the Indian is derived from experience and tradition. Since not all peoples have the same experience nor identical traditions, there follows the readily accepted assumption that the world contains different kinds of knowledge and different truths (1949: 215–216).

In contrast, "... Eskimos take a far more absolutist and ethnocentric view, which generally classifies their own way as correct or 'genuine' and their ways as inferior or invalid ... I was once assured that Eskimos are the only 'real' human beings, and all other people are 'something else, I don't know what'" (Nelson, 1973: 289).

When considering indigenous knowledge and western science, persons who are trained in the western scientific tradition must examine their theoretical assumptions about rationality and relativism. Indigenous knowledge did not develop in a context requiring comparison with the parameters of science but compares well when challenged with these parameters. Several studies show that Inuit knowledge is consensual, replicable, generalizable, incorporating and to some extent experimental and predictive (Bielawski, 1992, 1996; Denny, 1986). I have not seen evidence that Inuit controlled conditions for experiment, in contrast to scientific method. Nor did they, over time, increase accuracy in measurement.

Realism allows that both science and indigenous knowledge contribute to human understanding of the Dene and Inuit environments. A realist approach requires that one accept the natural world as real and holds that the objects of nature exist in and of themselves, were here before science, and will remain regardless of the activities of inquiry directed toward them. This position takes science seriously as a form of knowledge different from the indigenous knowledge of cultures without laboratory science; and it allows that indigenous knowledge also contributes to understanding the world. Hence, as stated earlier, it is not necessary to validate indigenous knowledge in and of itself, only in comparison with Western science and the dominance its results and applications exert over indigenous cultures (Bielawski, 1996: 219).

The realist view contrasts with relativist interpretations more commonly invoked in describing and validating the indigenous knowledge of oral societies. On indigenous knowledge, the relativist view would hold that the products of inquiry (research results) were applicable only within the indigenous culture; that is, they are culture-specific. Science and its products are generalizable to all cultures. The realist view of indigenous knowledge, however, allows that the products of inquiry following methods and under frameworks different from science may hold research possibilities, even results, in some of the realms of inquiry where science leaves us wanting.

Further, this approach makes us question what happens to cultures without science, and to their indigenous knowledge, when science and its products are rapidly imported (see Jarvie, 1986: 162–171). On this question, philosophy of science has been largely silent.

Differences perceived between knowledge and science are important in comparing indigenous knowledge with science. Science is a special form of knowledge, and it is easy to forget that not all western knowledge is science. It is not. Nor is science necessarily of intrinsically greater value in problem solving. Andrew Sayer argues that forms of knowledge exist outside of the "intellectualist bias" of western science. Knowledge is 1) gained through activity as well as through contemplation and observation; 2) a skill of doing and making as opposed to writing and saying; 3) a social activity; and 4) available in many forms of which science is the highest. Sayer further argues that science ignores two important contexts of knowledge, those of labor (the manipulation of matter for human purposes) and communication (1984: 16–40). Sayer's description matches Inuit and Dene knowledge well. What people know is as much in what they do and how they solve problems as in what they say. Their collective knowledge and individual knowledge resided in oral traditions and an enduring way of life based on labor, communication, and problem solving in an unforgiving environment, as well as on appropriate, morally grounded behavior in nature.

Paul Feyerabend argues throughout *Against Method* that knowledge from different cultural traditions has not been given the attention it deserves. He writes, "There is no idea, however ancient or absurd, that is not capable of improving our knowledge. The whole history of thought is absorbed into science and is used for improving every single theory" (1988: 33).

DENE VIEWS OF NATURE AND ENVIRONMENT

Nature is the overarching context of Dene life. People are part of nature, not separate from "it" and certainly not in charge of "it". Nature in its broadest sense, usually called "the land", is sacred. From people's place on the land, or within nature, derives Dene morality – that is, expectations and rules for appropriate behavior. Many prescriptions and proscriptions require human behavior that is deeply respectful of animals, especially caribou, and the power of the land, wind and water. Aspects of nature that are alive, thus imbued with spirit, include weather, water, land, rocks and minerals, sky, stars and aurora, as well as plants, fish, birds, animals and humans.

Contemporary Dene educators, drawing on the teachings of Dene elders, state that the Dene perspective is made up of four fundamental tenets:

"1. In our relationship with the land we should strive for respect and a sense of humility.
2. Our relationship with the spiritual world is based upon acceptance of things that can neither be seen nor touched.
3. Our relationship with other people is based upon cooperation and consensus and the welfare of the group.
4. And finally, our relationship with ourselves is one which requires continual self-evaluation and growth while accepting our inherent self-worth" (Dene Kede, 1993: xxv).

In the Dene worldview,

> ... People are the last to be made ... When Dene were created, they were the only people that relied upon everyone else for their survival. They were the weakest of all creatures: hence, the Dene perspective is that survival would be difficult and people, in their relationship to the land, would have to be humble and respectful. As each of the animal people was being defined, a special spiritual relationship between these animal people and all others was defined. For example, as the spider was given its special identity, it was decided that this creature would be the most powerful of people. It would have powers that transcended the earth. Its webs would create beautiful rainbows and be able to capture rain in the heavens enabling the Dene to survive: hence, the Dene perspective that the small, the unseen and the seemingly most insignificant all possess power and thus deserve respect (Dene Kede, 1993: xxiii).

In taking their elders' teachings into modern classrooms for young Dene, Dene teachers guide students through both general topics – such as Water and Rivers – and specific topics – such as Fire or Northern Lights or Fish – along four paths that reflect the tenets above: the Spiritual World, The Land, The People and The Self. In the Dene perspective, then, "the land", or nature, is both a set of objects, creatures and phenomena, and a context within which all living things relate to one another and to the land as a whole (Dene Kede, 1993).

Dene elder Jimmy B. Rabesca states, "All the world has laws. There are many thousands of different animals on this earth and they all have their own laws. When we walk in the bush we think about all of them. This is how we learn the way of all life and the things we don't know" (Dene Kede, 1993: xxiii).

Among the Koyukon Dene, while people are clearly part of the fabric of nature, humans and animals are nevertheless "clearly and qualitatively separated. Only the human possesses a soul ... which people say is different from animals' spirits ... The distinction between animals and people is less sharply drawn than in Western thought – the human organism, after all, was created by the animal's power" (Nelson, 1983: 20).

The land or environment also includes the expression of power, as Sharp details so well:

> The Chipewyan (Dene) are keen observers of their environment and are acute natural historians.

Their knowledge is of a very different order than that of Western ecology and biology but it is often more than the equal of the best science has to offer. Their way of knowledge encompasses not just knowledge of their environment but also includes knowledge of how their social system operates in that environment (Stanner, 1963, 1964, 1965b). Natural historical knowledge is not the basis upon which they situate themselves in their land. They locate themselves upon the basis of the current state of their relationships and then utilize their natural historical knowledge of their environment to wrest their subsistence wherever they have chosen to go. The rhetoric they use to explain where they go sounds as if it is based upon natural history, particularly upon caribou behavior and ecology, but it is only rhetoric. Any attempt to understand how they utilize and relate to their environment by beginning with the physical environment rather than the social environment is ultimately doomed to failure (Sharp, 1988: 128).

Numerous other sources convey much more of the wealth of relationships among the land, the people, the spiritual world and the self in relationship to what western thought calls nature and the environment.[2]

INUIT VIEWS OF NATURE AND ENVIRONMENT

As discussed above, classic ethnography is largely silent on Inuit epistemology (Bielawski, 1992: 61). Treatises titled "intellectual culture" usually include cosmology, religious activities, spirits, poems and songs, and so on. Consequently, it is in the oral histories of Inuit elders, and in the relatively few studies that demonstrate the extreme patience, attention and grasp of language required to articulate Inuit knowledge in English, that views of nature and environment are recorded (Burch, 1971; Nelson, 1969; Briggs, 1970; many of the sources used for the *Handbook of North American Indians*, Volume 5, *Arctic*, 1984, as well as the sources cited below).

Inuit do not separate themselves from nature, nor do they place themselves on a superior footing to other species. Inuit knowledge differs from science in other ways as well.

Western science assumes an inherent differentiation between humans and animals and focuses on the explanation of the relationship between originally independent parts: the Yup'ik Eskimos stand this basic assumption on its head and assume, for instance, that men and animals are analogically related as human and nonhuman persons ... The focus of explanations shifts to the creation of difference out of an original unity (Fienup-Riordan, 1990a: 4; see also 1990b: 167–91).

Yup'ik scholar Anayaqoq Oscar Kawagley provides a clear and detailed exposition of this worldview. "The Yupiaq world view is based on an alliance and alignment of all elements ... there must be constant communication between [the] three constituent realms [human, natural, spiritual] to maintain this delicate balance" among the human, natural and spirit worlds (Kawagley, 1995: 16).

What are the conditions in which this world view works with efficiency, economy and purpose? As young children the traditional Yupiaq people were given specially ground lenses through which to view their world. The resulting cultural map was contained in their language, myths, legends and stories, science and technology, and role models from the community. This oral orientation and learning by observation worked to their advantage (Kawagley, 1995: 17).

Further, Kawagley (1995: 15) writes,

To understand the Yupiaq world view it is necessary to understand the multiple meanings of a word that epitomizes Yupiaq philosophy ... *ella*, which is a base word that can be modified to change its meaning by adding a suffix or suffixes. Examples are *Qaill'ella auqa? How's the weather?*; *Qaill'ellan auqu? How are you feeling?*; *Ellam nunii, the world's land; Ellagpiim yua, Spirit of the Universe; Ellapak, universe; Ella amigligtuq, the sky is cloudy.* Variations of this one word can be made to refer to weather, awareness, world, creative force or god, universe and sky. The key word is awareness, or consciousness ... The human being must possess consciousness to be able to make sense out of values and traditions that would enable them to maintain and sustain their ecological world view (1995: 15).

Thus, nature is one realm, equal with the human realm and the spiritual realm, but intimately, constantly interconnected with humanity and spirituality (Kawagley, 1995).

Inuit oral histories describe in great detail various domains of specific knowledge, akin to scientific disciplines. But Inuit knowledge contrasts with science in that "pure knowledge is never separated from moral or practical knowledge" (Overing, 1985: 17). Inuit Elders describe much practical knowledge in their recorded oral histories. These texts show clearly the integration of nature, humanity and spirituality, as well as the way moral knowledge is embedded in the practical. Stories include environmental and geographical knowledge, midwifery, astronomy, meteorology, and medicine. For example,

If the ice is not supported at the bottom it can be very dangerous ... If the weather was very stormy you won't be able to tell which part of the ice is dangerous. If it's nice weather you can tell by looking at the snow and ice ... The place where the ice or snow has no support is mainly beside the mountains, there is a few on land dangerous too but not as dangerous as the ones on ice and snow ... (Tuniq, 1985: 4).

Tuniq also describes different ways of knowing.

- She heard about it. Describing dangerous ice and snow conditions, she says "I know about it since I've heard about it."
- She saw it (the toxicity of kidney fat). "The reason why I know about it is because I watched a man die from it."
- Learning something from her father. "He used to teach me", and
- Learning by doing something, in this case hunting, again with her father. She describes empirical observation and validation and says, "Everything what my father taught are true up to now".

Salomonie Alayco describes learning astronomy that would be put to use in navigation. "Every single morning, I was asked to check if the constellation Quturtuuk was in view before the constellation Attuuk appeared. My father asked me to go and check whether the constellations Quturtuuk Attuuk were in sight at dawn before the first light of the day. That's how we used stars to tell time in the morning. I was also asked to check from which direction the wind was from and whether there were any clouds" (Alayco, 1985).

Burch (1971) distinguishes between the empirical and non-empirical natural environment of Arctic Alaskan Eskimos. The empirical environment is one scientists recognize: it includes phenomena whose existence and characteristics can be ascertained by means of techniques that *anyone* employing those techniques can replicate; and which utilize exclusively data apprehended through sensory perception or instruments to record those. "Nonempirical phenomena are defined residually relative to the concept of empirical phenomena". Further, "Reality, as perceived by the traditional Arctic Alaskan Eskimos, included an extensive domain which, given the definitions used here, can only be described as 'nonempirical'."

Inuit nature included various beings that populated the landscape along with themselves and animals. Their names and roles vary somewhat across Inuit cultures, but they are often associated with the danger of being alone and unprepared on the land, sea and ice. Some are smaller than "real people", others are beings of great strength. Some ethnographers interpret references to some of these beings as indications of prehistoric contact with peoples living in the central and eastern Arctic at the time ancestors of today's Inuit arrived there about 1000 years ago. But most historic references to them place them in the landscape at the same time as the Inuit who describe them. To the Inuit, nature is both alive and lively.

INUIT AND DENE KNOWLEDGE IN WESTERN ENVIRONMENTAL MANAGEMENT

In the latter part of the 20th century, coupled with the political resurgence of Inuit and Dene and the demise of anthropology's role as objective observer and describer/constructor of Inuit and Dene cultures, Inuit and Dene knowledge once again moved to the fore of western research in the Canadian Arctic and Subarctic. I write "once again", because Samuel Hearne with the Dene, and Charles Francis Hall and Knud Rasmussen with the Inuit, demonstrated conclusively that indigenous knowledge beat western knowledge by far when it came to living in the Arctic and Subarctic. These convincing lessons were submerged, however, in the 20th century race to remove the nonrenewable resources of the Arctic and Subarctic from under the feet of the people who lived, and live, on the land.

Prior to the 1970s, indigenous people and people of European descent could be said to have lived almost in parallel worlds regarding nature in the Subarctic and Arctic. To "white" people (in Alaska, many of these outsiders were actually American Blacks who came north with the building of the Alcan Highway during World War II) the north had been and remained a wilderness from which to remove riches. Even homesteaders looked to trapping, small-scale mining, and wage employment as supplements to making some living from the land.

Alaska Natives' political activism towards a land claims settlement reached fruition with the Alaska Native Claims Settlement Act (ANCSA) in 1971. In the late sixties and early seventies, the U.S. oil supply had been threatened by political instability in the Middle East. The consequent pressure to exploit oil from the North Slope of Alaska opened the way for ANCSA, which recognized for the first time, albeit in a limited way, Alaska Natives' indigenous ownership of the land now called the State of Alaska. Not long afterwards in northern Canada, the Berger Inquiry (1974–1977) into a proposed pipeline from the Arctic Coast to fossil fuel markets in southern Canada concluded that no such industrial development could take place until the environmental and social concerns of the Inuit and Dene, including land claims or fulfillment of treaty promises, were addressed. The Berger Commission recommended a 10-year moratorium on development (Berger, 1988). Both the trans-Alaska oil pipeline and the proposed MacKenzie Valley pipeline neatly transected Inuit and Dene lands and cultures. The trans-Alaska pipeline was completed in 1974; a Mackenzie Valley pipeline is again being planned, 26 years after the Berger Inquiry. The oil supply for the southern industrial market is still "threatened", so much so that oil drilling may commence in the United States' Alaska National Wildlife Refuge, home of the calving grounds for one of the last remaining caribou herds that Inuit and Dene in Alaska and Canada rely on.

Thirty years ago, with Alaska Native political momentum towards ANCSA, and Canada's moratorium on northern industrial development, non-indigenous people of the north, and outsiders, collectively recognized for the first time that indigenous peoples' view of the land, of nature, of the environment contrasted sharply with a western, rationalist perspective. The rise of environmental law that requires industrial developers to minimize their impact on the land dovetailed with renewed western respect for the knowledge that Inuit and Dene carry about their northern lands and waters. The renewed respect followed a long period when governing powers in Alaska and Canada (as well as Greenland) assumed that Inuit and Dene desired to, and would, assimilate into western culture. American and Canadian governments both assumed that native assimilation was preferable because, in their view, Inuit and Dene lives were marginal, even primitive, when compared with southern civilization. In addition, disease, dislocation and other downsides of colonization reduced Dene and Inuit population and cultures throughout the first half of the 20th century. These events, politically significant and directly affecting the modern economies of the north in Alaska and Canada, actually put pressure on the scientific community (where the empirical basis for management strategies are researched) to consider indigenous knowledge – at first empirically and only later, theoretically (Bielawski, 1996 and elsewhere).

However, despite the academic communities' and indigenous groups' demands for recognition of indigenous knowledge in research and planning, little substantial Inuit and Dene knowledge was brought to bear on contemporary concerns in the north and its natural as well as social environment. Indigenous knowledge made no impact on scientific theory or even on the method of western science as it was practiced on Inuit and Dene land.

It was 1996 before Canada's federal government, on the basis of recommendations from a national Environmental Assessment Review Panel, required a northern developer to incorporate indigenous knowledge into its environmental impact statement. The case was diamond mining; both Dene and Inuit used the territory, land and fresh water. Both groups also used the affected watersheds.

Unfortunately, the effort to use indigenous knowledge in first assessing, then planning for, the impacts of the diamond mine was less than successful. At fault were the time frame allowed for incorporating indigenous knowledge into the environmental assessment process and the attempt to use indigenous knowledge towards development before addressing the issues of intellectual ownership of that knowledge. Government regulators who set the terms requiring indigenous knowledge in the developer's proposal had no idea of the research schedule nor funding required to record, much less translate and interpret, indigenous knowledge into terms comprehensible to environmental managers and developers. The individuals and communities holding the knowledge lacked the capacity to record and use their knowledge appropriately in a regulatory process. They were justifiably concerned that if they provided their indigenous knowledge to the proponent, so that the proponent could meet the requirements of the Environmental Impact Statement, then they – the indigenous people – would have given away the very information they required themselves to stop, alter, or monitor the development impact.

Ryan and Robinson, 1990, outline an overview of the methods necessary for good indigenous knowledge research, including intellectual property issues.[3]

Both human concerns (mental, physical, and nutritional health, education, and justice) and the need for environmental management drive current efforts to recognize, preserve, and even apply the body of Dene and Inuit knowledge to environmental and social problems.

All of these aspects have led to attempts – either voluntary or required by law – to incorporate Dene and Inuit views of nature – through what is commonly referred to as "traditional knowledge" – into western environmental planning and impact assessment. [Editor's note: See the chapter on Indigenous Knowledge and Traditional Ecological Knowledge by Dudgeon and Berkes in this volume.] Until recently, industrial development has taken place without regard to indigenous views of nature. In part through ignoring indigenous perceptions of the land, development has also marginalized indigenous people from its benefits. Recent trends, however, have seen a global recognition of the negative impact of developments, especially those that are short-lived and do irreparable damage to the environment. Also, in both northern Canada and Alaska, indigenous people have regained political power with which they may stop development or direct it through participation. For these reasons, incorporating traditional Dene and Inuit environmental knowledge into contemporary northern planning is seen as necessary. What we still lack are proven means to do this effectively, with integrity.

Northern researchers are also interested in indigenous knowledge of climate. Indigenous perceptions of climate "change" or "variability" or perhaps even "global warming" are a much sought-after subset in northern environmental research at present. It is known that the Arctic and Subarctic are seeing the effects of both warming trends (dramatically reduced ice cover) and industrial pollution, which collects in the Arctic at a faster rate than in more temperate areas. Thus, at the turn from the twentieth century to the twenty-first, it is impossible to discuss Inuit and Dene views of nature and environment without touching upon the broader focus of global research on climate change. Scientists are interested in indigenous perceptions of past climate variability as well as of current perceptions; Dene and Inuit are interested in what is happening now and in what will happen.

A last thread in indigenous views of nature and environment concerns perceptions of the effects of industrial pollution on nature and the environment. Large developments receive most of the scrutiny, but indigenous peoples' views are also being shaped by their own use of fossil fuels and plastics on the land and in permanent settlements, in transportation, food preparation, homes, and everywhere else.

INDIGENOUS KNOWLEDGE AND SCIENCE

In the last two decades of the 20th century, while attempts to recognize and utilize indigenous knowledge have grown in northern research, three impediments have emerged. Over and over again in studies attempting to incorporate traditional knowledge, research has faltered, and the quality of research results suffered, because of these problems:

- The problem of extracting empirical bits of indigenous knowledge and attempting to fit them into western scientific categories;
- The problem of attempting to integrate indigenous knowledge and scientific knowledge; and
- The problem of the sacred.

These three problems mean that research drawing from both indigenous and scientific sources on nature and the environment has largely failed to yield substantial advances in knowledge.

INDIGENOUS AND SCIENTIFIC CATEGORIES

Northern scientists first looked to indigenous knowledge to fill gaps in data required to answer scientists' questions and test scientists' hypotheses. In some disciplines – wildlife biology and management, archaeology of the late prehistoric and historic north, meteorology, hydrology, and especially the social sciences (bearing in mind that many physical scientists do not think of social sciences as scientific at all) – indigenous knowledge proved a useful supplement to the scientific method. Usher argues that it is both appropriate and cost effective to extract indigenous knowledge from context and include it in environmental impact assessment and related work (2000), but this approach does not sustain the subtlety and breadth of indigenous knowledge. For research questions that are not limited by the narrow requirements of environmental management reports, scientists attempting to explore indigenous knowledge have often been stymied by the presentation of indigenous knowledge in context, embedded in extremely long narratives, including seemingly nonsensical responses to scientists' specific questions. MacDonald provides a recent, pithy example:

A compelling example of the gulf between mythic and "rational" views of nature occurred in Igloolik a number of years ago. Government biologists, intent on introducing legislation to limit the number of polar bears killed in the area met with community representatives. One old hunter strenuously objected to their proposals, pointing out that polar bears had intelligence matching or exceeding that of humans and, as such, could be 'taken' only when they wanted to 'give themselves.' By way of explanation the hunter told of a time he followed fresh bear tracks

across the island of Qaggiujaq in the Northern Foxe Basin. The tracks suddenly ended, and
there, on the tundra, was a rectangular block of ice. Clearly, the polar bear, not wanting to be
taken, had transformed itself into ice. The government biologists were bemused at this explanation, whereupon the old hunter told them that if they did not, or could not believe him, then
they knew nothing about polar bears (MacDonald, 2000: 18).

In short, while in the last two decades much scientific interest has turned to
mining Dene and Inuit knowledge for nuggets that science in the North has
missed, the amount of effort devoted to the task yielded few research results
of originality and good quality beyond common sense. It proved too difficult,
and seemed counter productive to many scientists, to extract specific bits of
empirical data from indigenous knowledge narratives and to record these in
the data cells that scientists were attempting to fill.

The conceptual problem we are presented with is to derive categories for
data collection that match the aboriginal and scientific worldviews. The empirically oriented scientist does not know what categories of Inuit and Dene
knowledge match the categories of western science; hence the approach must
begin with deduction rather than induction.

ATTEMPTS TO INTEGRATE

The first problem exacerbated the second: the attempt to do what politicians
called for, which was to integrate indigenous knowledge with western science.
This proved impossible as long as western science failed to recognize the
systematic aspects of indigenous knowledge and its theory and method. Without
taking into account indigenous worldviews, or looking for data categories
through Inuit and Dene methods of observation, scientists were left with the
approach described above, and integration proved impossible.

As collaborative attempts utilizing both western science and indigenous
knowledge have extended over years, and as these have given researchers
committed to a long-term research project the first substantial results, the
subtlety of questions asked and data recognized that applies to the questions
is increasing.

THE PROBLEM OF THE SACRED

The only seemingly irresolvable contradiction between western science and
Dene and Inuit indigenous knowledge is the presence of the sacred in the
indigenous worldview regarding nature and the environment. It is fundamental
to western science, and has been since Galileo's time, reiterated in the days of
Bishop Usher, Darwin, Wallace and Huxley, that science and its results are
free of religious, spiritual, and "otherworldly" phenomena, that scientific inquiry
follows a strict method and produces replicable results.

Dene and Inuit views of nature and environment inculcate the sacred in
both inquiry and interpretation of phenomena, observed or "sensed". This
presents a concrete, immovable and impenetrable barrier to most western-
trained scientists when they are presented with Dene and Inuit interpretations
of nature and environment. Even more so are western scientists skeptical, if

not scornful, of Dene and Inuit methods of "inquiry" into natural and environmental phenomena. Very few scholars have attempted even to approach this question, except for Fikret Berkes (1999), but his treatment is limited primarily to environmental ethics. Perhaps with new scholarship showing that western science is in itself a cultural artifact, not free of value or interpretation, some of this will change.

WHAT LIES AHEAD

In 1985, Anne Salmond wrote, "The process of opening Western knowledge to traditional rationalities has hardly yet begun" (p. 260). We know more now about the time it takes to do this work well and how strongly opinions on how to do it diverge. But Dene and their ancestors, Inuit and their ancestors, hold knowledge accumulated over thousands of years. Their environments demanded specialization to temperature extremes and relatively few species. Their respective adaptations hold a good deal of technical knowledge for contemporary life in the North, of empirical data on climate variability and perhaps climate change, and possibly on what we may call theoretical approaches to examining nature and environment. Their views of nature and environment are not artifacts of the past nor written in stone, but living, breathing, changing knowledge informed both by their natural history, their worldviews, and their present lives.

Consequently, to honor, preserve and especially to apply Dene and Inuit views of nature and environment in the contemporary world, we need creative research strategies for western scholars to examine Inuit and Dene knowledge. At the same time, we need creative research strategies for indigenous scholars to examine western science; taken together, these will enrich the body of human knowledge. When such knowledge is applied to questions about nature and environment, human inquiry – questions, methods, theory, practice – can only be enriched. I am convinced we are capable of designing a research methodology for working with both western scientific observations and indigenous observations, without precluding open inquiry through emphasizing either the western scientific method or indigenous methods of observation and inquiry. Contemporary changes in nature and environment might be the catalyst for such balanced studies.

We have left behind the millennium that saw first the development of science and then its domination over every culture on the planet. We are moving through a time where we are risking the planet that supports *Homo sapiens sapiens* and all the species we depend on. This is surely a time when the diversity of indigenous knowledge can only enhance human inquiry and action.

We are beginning to understand that these [Dene] perspectives, which have been at the root of all Dene teachings since time immemorial, have a timeless quality which can be applied to any situation, any place, any people. We understand that we cannot simply talk about Dene survival. In order for us to survive as a people, we recognize the need for the survival of all people and for the survival of the earth (Dene Kede, 1993).

NOTES

[1] The problem of interpretation and translation in all attempts to explain different systems of knowing in the language of another requires more time and careful attention than the average research funding grant allows.
[2] See especially Brody, 1981; Coutu and Hoffman, 1999; Cruikshank, 1998; Ridington, 1988; Sioui, 1992; as well as many of the sources listed in the *Handbook of North American Indians*, Volume 6, 1981.
[3] See also Parlee, 1998 and McDonald, 1997.

BIBLIOGRAPHY

Berger, Thomas. *Northern Frontier, Northern Homeland*. Vancouver: Douglas and McIntryre, 1988.
Berkes, Fikret. *Sacred Ecology: Traditional Ecological Knowledge and Resource Management*. Philadelphia: Taylor and Francis, 1999.
Bielawski, E. 'Inuit indigenous knowledge and science in the Arctic.' In *Naked Science: Anthropological Inquiry into Boundaries, Power, and Knowledge*, Laura Nader, ed. New York: Routledge, 1996, pp. 216–227.
Bielawski, E. *The Desecration of Nanula kue: Impact of Taltson Hydroelectric Development on Dene Sonline*. Ottawa, Canada: Royal Commission on Aboriginal Peoples, 1993.
Bielawski, E. 'Cross-cultural epistemology: cultural readaptation through the pursuit of knowledge.' In *Looking to the Future*, M-J. Dufour and F. Therien, eds. Quebec City: Université Laval, 1992, pp. 59–69.
Briggs, Jean. *Never in Anger: Portrait of An Eskimo Family*. Cambridge, Massachusetts: Harvard University Press, 1970.
Brody, Hugh. *Maps and Dreams*. Vancouver: Douglas and McIntyre, 1981.
Burch, E. 'The non-empirical environment of the Arctic Alaska Eskimos.' *Southwestern Journal of Anthropology* 27(2): 148–165, 1971.
Burch, E. 'Indians and Eskimos in North Alaska, 1816–1977: a study in changing ethnic relations.' *Arctic Anthropology* 16(2): 123–151, 1979.
Burch, E. and T. Correll. 'Alliance and conflict: inter-regional relations in North Alaska.' In *Alliance in Eskimo Society*, Lee Guemple, ed. Proceedings of the American Ethnological Society, Supplement. Seattle: University of Washington Press, 1972, pp. 17–39.
Coutu, Phillip and Lorraine Hoffman-Mercredi. *Inkonze: The Stones of Traditional Knowledge*. Edmonton, Canada: Thunderwoman Ethnographies, 1999.
Cruikshank, Julie. *The Social Life of Stories: Narrative Knowledge in the Yukon Territory*. Vancouver: University of British Columbia Press, 1998.
Dene Kede. *Education: A Dene Perspective*. Yellowknife: Government of the Northwest Territories, 1993.
Feyerabend, Paul. *Against Method*. London: Verso, 1988.
Fienup-Riordan, Anne. 'A problem of differentiation: boundaries and passages in Eskimo ideology and action.' Paper presented at the *Seventh Inuit Studies Conference*, Fairbanks, Alaska, 1990.
Handbook of North American Indians, Volume 5: *Arctic*. Washington, DC: Smithsonian Institution, 1984.
Handbook of North American Indians, Volume 6: *Subarctic*. Washingon, DC: Smithsonian Institution, 1981.
Honigmann, John. *Culture and Ethos of Kaska Society*. New Haven, Connecticut: Yale University Press, 1949.
Jarvie, Ian. *Thinking About Society*. Boston: D. Reidel Publishing Company, 1986.
Kawagley, A. Oscar. *A Yupiaq Worldview: A Pathway to Ecology and Spirit*. Prospect Heights, Illinois: Waveland Press, 1995.
MacDonald, John. *The Arctic Sky: Inuit Astronomy, Star Lore, and Legend*. Iqaluit: Nunavut Research Institute, 2000.
McDonald, Miriam, Lucassie Arragutainaq and Zack Novolinga, eds. *Voices from the Bay*. Ottawa: Canadian Arctic Resources Committee, 1997.

Nelson, R. *Hunters of the Northern Ice*. Chicago: University of Chicago Press, 1969.

Nelson, R. *Hunters of the Northern Forest*. Chicago: University of Chicago Press, 1973.

Nelson, R. *Make Prayers to the Raven: A Koyukon View of the Northern Forest*. Chicago: University of Chicago Press, 1983.

Overing, Joanna, ed. *Reason and Morality*. New York: Tavistock Publications, 1985.

Parlee, Brenda. *A Guide to Community-Based Monitoring for Northern Communities*. Northern Minerals Program Working Paper No. 5. Ottawa: Canadian Arctic Resources Committee, 1999.

Petrone, Penny, ed. *Northern Voices: Inuit Writing in English*. Toronto: University of Toronto Press, 1988.

Ridington, R. *Trail to Heaven: Knowledge and Narrative in a Northern Native Community*. Iowa City: University of Iowa Press, 1988.

Ryan, J. and Mike Robinson. 'Implementing participatory action research in the Canadian North: a case study of the Gwich'in Language and Cultural Project.' *Culture* 10(2): 57–71, 1990.

Salmond, Anne. 'Maori epistemologies.' In *Reason and Morality*, Joanna Overing, ed. New York: Tavistock Publications, 1985, pp. 240–263.

Sayer, Andrew. *Method in Social Science: A Realist Approach*. London: Hutchinson, 1984.

Sharp, Henry S. *The Transformation of Bigfoot: Maleness, Power, and Belief Among the Chipewyan*. Washington, DC: Smithsonian Institution Press, 1988.

Sioui, Georges. *For an Amerindian Autohistory*. Montreal and Kingston: McGill-Queen's University Press, 1992.

Usher, Peter. 'Traditional ecological knowledge in environmental assessment and management.' *Arctic* 53(2): 183–193, 2000.

ANNIE L. BOOTH

WE ARE THE LAND: NATIVE AMERICAN VIEWS OF NATURE

We are the land ... that is the fundamental idea embedded in Native American life ... the Earth is the mind of the people as we are the mind of the earth. The land is not really the place (separate from ourselves) where we act out the drama of our isolate destinies. It is not a means of survival, a setting for our affairs ... It is rather a part of our being, dynamic, significant, real. It is our self ...

It is not a matter of being 'close to nature' ... The Earth is, in a very real sense, the same as our self (or selves) ... That knowledge, though perfect, does not have associated with it the exalted romance of the sentimental 'nature lovers', nor does it have, at base, any self-conscious 'appreciation' of the land ... It is a matter of fact, one known equably from infancy, remembered and honoured at levels of awareness that go beyond consciousness, and that extend long roots into primary levels of mind, language, perception and all the basic aspects of being ...

Paula Gunn Allen, Laguna Pueblo (1979: 191–192)

This is how one Native American presents her interpretation of the indigenous understanding of nature. As we will see in this article, many Native Americans present similar understandings. Their reciprocal relationships with nature permeated every aspect of life from spirituality to making a living and led to a different way of seeing the world, what they might call a more "environmental" way of seeing the world. But is this a true picture? Increasingly there has been debate over the nature of the Native American's relationship to the land, both past and present. This article will examine this debate and the way in which Native Americans view nature.

CAVEATS

Discussing any aspect of Native Americans requires caution, caution which has not always been employed when discussing Native American relationships with nature. Clear distinctions need to be drawn.

The first necessary distinction is that much is disguised in the concept of "Native American" or "Indian". The peoples who ranged, then and now, from the Arctic Ocean to Mexico and between two oceans encompassed a vast diversity of languages, customs, practices and beliefs. They evolved to take advantage of different natural resources, to deal with dramatically different landscapes and climates and evolved again as they moved to different landscapes. To somehow lump this enormous complexity into a singular term is to fool ourselves into thinking we understand something of which we have no clear grasp. It is to misunderstand the nature of people we are interested in

329

H. Selin (ed.), *Nature Across Cultures: Views of Nature and the Environment in Non-Western Cultures*, 329–349.
© 2003 *Kluwer Academic Publishers. Printed in Great Britain.*

understanding. Yet we do that lumping all the time. It leads to difficulty as when we misunderstand why two native tribes may choose different, conflicting courses of action. As for Scotland and England, proximity and the lumping in as "British" does not necessarily result in shared goals. Indeed the title of Native American is most equivalent to the idea of "European". There *is* something distinctly European when compared with North American, but there are limits to the understanding we can draw from the term.

Yet there may well be a "Native American" perspective when it comes to understanding the relationship with nature, and Native Americans as well as non-native writers frequently make this argument. This article will present the case for a sense shared across different native cultures, but I would urge the reader to use caution when thinking about "Native Americans".

The second necessary distinction is between the modern and the historical Native American. When many writers discuss the Native perspective of nature there is a tendency to lump together Native Americans from across the recorded historical period. The problems with this approach should be obvious: none of us lives like our ancestors did 200 years ago. Modern Native Americans face different challenges than did their ancestors, and many of those challenges affect how they can now relate to the land. Further, there is always a question of how well "traditional" philosophy and spirituality manage to translate into the modern world. Care must be taken not to conflate what is known about the beliefs and behaviours of the historical Native American and those of today's Native, although both might be instructive. While they inherit their past, they must live in the present.

Finally, I think it is worthwhile making the distinction between listening to what Native Americans say for themselves and what others write. Both bring different agendas to the table and it is useful to recognise that agenda when assessing the information presented. The Native speaks with the voice of cultural experience, an intimate understanding of native reality an outsider can never hope to achieve. Often they are privy to information or perspectives an outsider could not easily obtain. However, people from within cultures often have difficulty examining those cultures critically and are sometimes constrained in raising controversial issues. Non-natives do not have similar constraints and can be as critical as they wish, but they will almost always lack that essential grasp, that innate understanding of a group of people and the way they think that a cultural resident will have. As such, it can be easy to misinterpret what is going on in a culture as modern critiques of anthropological findings are demonstrating. In terms of trying to grasp the nature of the Native American relation to nature, both past and present, it is most useful to explore both native and non-native writers, keeping in mind the perspectives of both.

THE NATIVE AMERICAN RELATIONSHIP WITH NATURE

Although they varied significantly among different cultures, Native American relationships with the natural world tended to preserve ecological integrity and appear to have done so over a significant period of time. These cultures engaged in relationships of mutual respect, reciprocity, and caring with an Earth and fellow beings as alive and self-conscious as human beings. Such relationships were reflected and perpetuated by cultural elements including religious belief and ceremonial ritual.

In the songs and legends of Native American cultures it is apparent that the land and her creatures are perceived as truly beautiful things. There is a sense of great wonder and of something which sparks a deep sensation of joyful celebration. Above all else, Native Americans were, and are, life affirming; they respected and took pleasure in the life to be found around them, in all its diversity, inconsistency, or inconvenience. Hughes (1983) points out that only the newly arrived Europeans considered the land to be a "wilderness", barren and desolate. To Native Americans, it was a bountiful community of living beings, of whom the humans were only one part. It was a place of great sacredness, in which the workings of the Great Spirit, or Great Mystery, could always be felt.

Standing Bear (1933), a Lakota, wrote that Native Americans felt a special joy and wonder for all the elements and changes of season which characterized the land. They felt that they held the spirit of the land within themselves, and so they met and experienced the elements and seasons rather than retreating from them. For Standing Bear and the Lakota, the Earth was so full of life and beings that they never actually felt alone.

> There was no such thing as emptiness in the world. Even in the sky there were no vacant places. Everywhere there was life, visible and invisible ... Even without human companionship one was never alone (Standing Bear, 1933: 14).

This statement echoes a central belief consistent across many Native American cultures – that the Earth is a living, conscious being. The Koyukon of central Alaska, for example, see the Earth as something alive and powerful, and therefore, something which must be treated with respect.

> For traditional Koyukon people, the environment is both a natural and supernatural realm. All that exists in nature is imbued with awareness and power ... all actions towards nature are mediated by consideration of its consciousness and sensitivity. The interchange between humans and environment is based on all elaborate code of respect and morality, without which survival would be jeopardized (Nelson, 1983: 240).

The belief in a conscious, living nature is not simply an intellectual concept for Native American cultures. For most, perception of the landscape is important in determining perception of self. Native American cultures and histories are based in the land, and their lives are inseparably intertwined with it. In a most real sense, it *is* their life. This interconnection between person and land is not merely a thing of historical significance. Present-day Native Americans continue to acknowledge their ties to the land. Utes in the Southwest, faced with the question of mining on their lands, are deeply troubled, for the land is more than a mere resource, as several individuals have tried to explain.

> The land is a living body with spirit and power, which contains tribal genealogy. It is necessary for the people to remain in the place in which they have always been, as guardians, and as an inseparable part of that place and space.

> The tribe doesn't want to diminish the land, but not because of money issues. But because you diminish *us* when the land is eaten away (emphasis in original) (Romeo, 1985: 160–61).

At the Tellico Dam congressional hearing in 1978, Jimmie Durham, a western Cherokee, tried to express what his people felt for a land they could no longer even live upon, but wished to preserve nonetheless.

> In the language of my people ... there is a word for land: Eloheh. This same word also means history, culture and religion. We cannot separate our place on earth from our lives on the earth nor from our vision nor our meaning as a people ... So when we speak of land, we are not speaking of property, territory, or even a piece of ground upon which our houses sit and our crops are grown. We are speaking of something truly sacred (Matthiessen, 1984: 119).

In Native American relationships with a living, conscious world, reciprocity and balance were required from both sides. Balance was vital: the world exists as an intricate balance of parts, and it was important that humans recognized this balance and strove to maintain and stay within it. All hunting and gathering had to be done in such a way as to preserve the balance. Human populations had to fit within the balance. For everything that was taken, something had to offered in return, and the permanent loss of something, such as in the destruction of a species, irreparably tore at the balance of the world. Thus, offerings were not so much sacrifices, as non-Natives were inclined to interpret them, but an appropriate acknowledgment of a great gift. In this way, the idea of reciprocity emerges. From the Native American perspective, as Hughes describes it, "mankind depends on the other beings for life, and they depend on mankind to maintain the proper balance" (Hughes, 1983: 17).

Momaday (1976: 80), a Kiowa writer and teacher, describes the necessary relationship as an act of reciprocal approbation, "approbations in which man invests himself in the landscape, and at the same time incorporates the landscape into his own most fundamental experience." The respect and approval is two-way: humans both give and receive value and self-worth from the natural world. Part of the idea of reciprocity is the necessity and importance of interaction. Participation in reciprocity is vital; a failure to interact, or a breakdown in interaction, leads to disease and calamity. Thus, everything that is used in everyday life is used for its part in that interaction; it becomes a symbol of sacred interaction and relationship between the people, the plants, the animals, and the land. Rituals such as those used for healing are not designed to ward off illness or directly cure the ill person. Rather, they are designed to remind the ill person of a frame of mind which is in proper relationship with the rest of the world, a frame of mind which is essential to the maintenance of good health.

An old Keres (Pueblo) song goes like this:

> I add my breath to your breath
> That our days may be long on earth
> That the days of our people may be long
> That we may be one person
> That we may finish our roads together
> May our mother bless you with life
> May our Life Paths be fulfilled (Allen, 1986: 56).

The reciprocal relationship embodied in this song – the sharing of breath is the sharing of life – is one of the central insights into Native American worldviews. One does not act in isolation, but with respect towards the others with whom one shares the universe. Involved with a *living* universe, the Native American is engaged in a constant dialogue with a network of relations, human and non-human, natural and supernatural. Allen explains the Indian sense of relatedness as follows:

> [M]an's intelligence arises out of the very nature of being, which is, of necessity, intelligent in and of itself, as an attribute of being ... this idea probably stems from the Indian conception of a circular, dynamic universe: where all things are related, are of one family, then what attributes man possesses are naturally going to be attributes of all beings (Allen, 1979b: 225).

Sam Gill explains that Native Americans hold a "person to person" relationship with the environment as the power of life itself is personified and inextricably linked with, and identical to, the natural world (Gill, 1989: 30). This sense of relationship is explicit in every aspect of life. Consider Ojibwa Winona LaDuke's (1990: 16) comments on the ordinary, mundane activity of hunting and gathering:

> Whether it is wild rice, whether it is fish, whether it is deer or turtles, when you go and take something from the land, you pray before you take it. You offer tobacco, you offer a prayer to that spirit and to the creation of a part of that. *You take those things because you have a relationship with all the other parts of the creation. That is why you are allowed to take those things.* You take that and you give something back as a reciprocal arrangement, because that is how you maintain your relationship (emphasis added).

As countless sacred rituals, ceremonies, songs and teaching stories make clear, maintaining the relationship between human life and non-human life is the heart of Native American spirituality. The Sun Dance of the Plains Indians, which has attracted considerable interest for its more gruesome aspects (including a long government ban on its performance), is not about the ability to endure pain. It is about humanity's willingness to offer blood and spirit towards the maintenance of the balance of life.

> The Sun Dance ... is not a celebration of man for man; it is an honoring of all life and the source of all life, that life may go on, that the circle be a cycle, that all the world and man may continue on the path of the cycle of giving, receiving, bearing, being born in suffering, growing, becoming, giving back to earth that which has been given, and so finally to be born again (Brown, 1978: 12).

A sense of embeddedness in the rest of the world has profound implications for how one chooses to live and interact with others. It is also one reason why the displacement of Native Americans from their lands, and the subsequent damage to the land, was and is so socially and psychically devastating. As Allen points out, the despair that appears in many writings by Native Americans is the despair of having lost "that perfect peace of being together with all that surrounds one," a loss that is irreplaceable (Allen, 1979a: 192). Peter Matthiessen agrees that this understanding is found consistently across a wide diversity of cultures.

It is not a matter of 'worshiping nature,' as anthropologists suggest: to worship nature, one must stand apart from it and call it 'nature' or 'the human habitat' or 'the environment.' For the Indian, there is no separation. Man is an aspect of nature ... (Matthiessen, 1984: 9).

Or consider part of a sacred Navajo chant, designed to remind the person every day of their connections with life:

The mountains, I become part of it ...
The herbs, the fir tree, I become part of it.
The morning mists, the clouds, the gathering waters,
I become part of it.
The wilderness, the dew drops, the pollen ...
I become part of it (Brown, 1989: 20).

Momaday describes living apart from the land with horror: "Such isolation is unimaginable" (Momaday, 1979: 166). All poetry, ceremony, song, and story remind Indians of their part in a living, evolving whole by virtue of their willing participation (Allen, 1979b: 226).

Part of Native relations with nature is encapsulated in traditional spirituality. Traditional Native American spirituality has been the target of attack and suppression by the European invaders since 1492, and it continues to be a significant political, legal and emotional minefield. After 500 years of suppression, some have legitimately questioned whether there is such a thing as "traditional" Indian spirituality. Much knowledge and practice was lost through genocide and through the devastating plagues that swept the New World after the arrival of the Europeans. Many Native Americans were forced, or chose, to convert to European religions, and their descendants remain faithful practitioners of Catholicism or some form of Protestantism. In both the United States and Canada what "traditional" practices remained were outlawed and remained illegal until recently. Yet in spite of everything, Native Americans themselves state that unique, viable spiritual practices continue to exist and flourish.

One key question is, what are spiritual practices intended to do? For many Native Americans, spiritual practices were part of an ongoing dialogue with the world. Lakota Vine Deloria (1973: 102) describes it thus: "The task of the tribal religion, if such a religion can be said to have a task, is to determine the proper relationship that the people of the tribe must have with other living beings." Spiritual practices become a way of learning how to live well, and, in this case, to live well with the natural world. They are methodologies for reaching an appropriate level of consciousness, or mindfulness, of the world. Further, these practices, which interlink behaviours, feelings and ethics, are completely identified with the very survival of the people and the land.

Native Americans saw that all that existed shared in sanctity, as they were all fragments of God. While Native Americans did not worship nature, they recognized that all life around them was sacred, for "each form in the world around them bears such a host of precise values and meanings that taken all together they constitute what one would call their 'doctrine'" (Brown, 1985: 37). When Native Americans saw themselves in terms of community, their definition of community included the natural community. Spirituality requires that humanity and "the rest of creation [be] cooperative and respectful of the task set for them by the Great Spirit" (Deloria, 1973: 96). That task is dwelling with balance and with harmony.

> Respect in the American Indian context does not mean the worship of other forms of life but involves two attitudes. One attitude is the acceptance of self-discipline by humans and their communities to act responsibly towards other forms of life. The other attitude is to seek to establish communication and covenants with other forms of life on a mutually agreeable basis (Deloria, 1999: 51).

Vine Deloria (1973) has given a particularly thorough consideration to the linkage between place and spirituality, contrasting it with Christian practice. Christianity, he notes, although originating in a particular land (the Holy Land), can and does exist almost anywhere on the face of the earth without extensive modification of the central creed, i.e. the Christian god is "portable". Traditional Native American spiritual practices, however, cannot without damage be moved from a given landscape. Rather, they have evolved to incorporate and reflect particular elements of the chosen land. Where would the Navajos be without the Four Mountains, or the Lakota without the Black Hills? When poet Paula Gunn Allen writes "We are the land ... ," her statement, to her, is literal truth. It is who she is as a Laguna (Allen, 1979a). The same sense of identification is true for Kiowa N. Scott Momaday, who writes of giving himself to the "remembered earth" (1979: 164–165) and for Lakotas Black Elk and Standing Bear, who speak of the devastating loss of the Black Hills to gold miners (Standing Bear, 1933; Neihardt, 1932, 1975). The land is who a Native American is and it affects his/her responses to the world. Consider Ronald Goodman's (1990: 1) interpretation of what, to the American government, was merely a question of land expropriation, but to the Lakota was an attack on their spiritual integrity:

> Traditional Lakota believed that ceremonies done by them on earth were also being performed simultaneously in the spirit world. When what is happening in the stellar world is also being done on earth in the same way at the corresponding place at the same time, a hierophany can occur; sacred power can be drawn down; attunement to the will of Waken Tanka can be achieved.

> Our study of Lakota constellations and related matters has helped us appreciate that the need which the Lakota felt to move freely on the plains was primarily religious (Goodman, 1990: 1).

Such an indelible bond with the land, however, is clearly a deliberate construction by the tribes. Indian groups have been migrating across the North American continent for perhaps the last 40,000 years. Some are very recent arrivals; the Navajo only migrated into the American Southwest from Alaska and the Yukon about 600 years ago (Dickason, 1992). They are not "native" to the area. They have, however, reconstructed their religion to reflect the new land in which they dwell, in part by borrowing from earlier residents such as the Hopi. From this, Deloria concludes that it is possible for a cultural group to "consecrate" a particular landscape if it is capable of seeing itself in terms of that landscape (Deloria, 1973: 295).

For the Native American, all living is spiritual practice. Thus, material possessions are symbolic of sacred relationships. Moccasins remind the wearer of his connection with the earth, with the plants and animals on it, and so of the nature of reality itself. A pipe becomes a teaching tool, a history of the people, and a reminder of how all in the world are related. When smoked in a sacred manner, it is offered to "all my relations" (Lame Deer and Erdoes, 1972: 12; 150–253). Activity which maintains life is also part of the ongoing sacred process; for a Native American the acts of hunting or weaving or building a house are also religious activities. Native Americans rarely distinguish between their religious life and their secular life (Toelken, 1976). Instead there is nothing in life that is *not* religious. Everything from hunting to healing is a recognition and affirmation of the sanctity of life. In the weaving of a basket is the creation of the whole world. In a proper life there is never a sense of disconnectedness from the Earth.

> ... [T]he whole universe is sacred, man is the whole universe, and the religious ceremony is life itself, the miraculous common acts of every day. Respect for nature is respect for oneself; to revere it is self-respecting, since man and nature, though not the same thing, are not different ... (Mattheissen, 1981: 12).

In all traditional Native American spiritual practice the sense of the earth as alive was part of a sacred understanding of life and human obligations to life.

> Coming last [in creation] human beings were the "younger brothers" of the other life forms and therefore had to learn everything from these creatures. Thus human activities resembled bird and animal behaviors in many ways and brought the unity of conscious life to an objective consistency (Deloria, 1999: 50).

This idea recurs in contemporary Native Americans as well. Allen sees the distinguishing characteristic between Western and Native American thinking as the "magicalness" in Indian perception. This is not a childish magicalness but a perception of everything in the world as alive, viable, and subject to the need to grow and change. An Indian, she says, is one who

> assumes that the earth is alive in the same sense that he is alive. He sees this aliveness in nonphysical terms, in terms that are familiar to the mystic or the psychic, and this gives rise to a mystical sense of reality that is an ineradicable part of his being (Allen, 1979b: 233).

The vital importance of maintaining essential relationships has serious implications for how Native Americans deal with their fur-clad relatives, the animals. The idea which appears over and over is "kinship" with other living beings.

> All are seen to be brothers or relatives (and in tribal systems relationship is central), all are offspring of the Great Mystery, children of our mother, and necessary parts of an ordered, balanced and living whole (Allen, 1979b: 225).

Brown (1985) comments that non-humans are the links between humans and the Great Mystery. To realize the self, kinship with all beings must be realized. To gain knowledge, humans must humble themselves before all creation, down to and including the lowliest ant. Nature is a mirror which reflects all things, including that which it is important to learn about, understand and value throughout life. Further, many tribes acknowledge that humans found a world to come into and a way of making a living as the result of conscious, caring gifts from the animals. The creation myths of tribes such as the Arapaho or the Iroquois tell of how the Creator created the world with the help of animals, from mud brought up from beneath endless waters by a muskrat, or resting on the back of a willing turtle. Sacred objects are donated, as when eagle grants the use of his feathers, and food is found because holy animals offer their flesh as a gift (Harrod, 1987: 51).

However, some anthropologists, such as Howard Harrod (1987) or Äke Hultkrantz (1981), see a tension inherent in the problem of having to kill animals with whom one had a deep, intimate relationship. Looking at the Plains Indians, Harrod observes that such tensions are apparent in creation myths. The Blackfoot, for example, tell of a time when humans were hunted by the buffalo. To change this, the god taught the Indians to make bows and arrows and then to hunt. In compensation for becoming food, the buffalo and other animals became spiritual helpers which the Blackfoot were to obey. In this fashion, and through other rituals, the tension inherent in eating social equivalents is partially resolved (Harrod, 1987: 44–45, 53–54).

Most Native American legends speak of other species as beings that could shed their fur masks and look human. They once shared a common language with humans and continued to understand humans after the humans had lost their ability to speak to the animals. The animals partook of the sacredness of life, and were often the descendants of the powerful beings who had lived on the earth before humans. An animal was something more than its furry, four-footed shape, more than what was seen by human eyes. It was, says Nelson (1983: 31), a personage and a personality with a lineage far older than humanity's.

> It is to be remembered that the pre-human Immortals (the gods, if you will) were Animals Who Were People. These Forerunners, these Ancient Ones whose bodies shimmered as it were between animal and human forms, these denizens of the elder dream-world, have long since taken their final departure; yet they remain as the visible animals of this everyday world ... Mythic Coyote, supertrickster and pattern-setter for mankind, is not [coyote], raiding fields and howling on the hills before dawn – and yet in a certain mystical sense, he is (Laird, 1976: 110).

Even into the present, the Native Americans respect their relationship with animals and believe that the respect and caring go both ways. Thus, communication is possible, as Yukon Indian Irene Isaacs explains:

> You can tell them and they do it, that wolf. You tell them, "Kill something for me!" And then the wolf will. Then you come, and they are going to kill a moose and you're going to find it and have moose meat. They do that, wolves. So they understand Indians. And other animals, that's what I tell my son about that too. You've got to treat animals good (McClellan, 1987: 280).

Martin, discussing subarctic bands, also notes that a sympathy builds up between the hunter and the animal persons who are hunted, a sympathy which pervades human life. At all times, there is a mutual obligation felt, an obligation to be courteous. An animal is not killed unless the hunter obtains its consent in the spiritual world; the animal must be willing to surrender itself to the hunter. Hunting gives meaning to the hunters' lives; it gives a sense of identity. It is the animal which grants this sense of identity (Martin, 1980).

According to Nelson, the Alaskan Koyukon believe that animals and humans are distinct beings, their souls being quite different, but that animals are powerful beings in their own right (Nelson, 1983 and 1982). Consequently a complex collection of rules, respectful activities, and taboos surround everyday life and assist humans in remaining within the moral codes that bind all life. Hunting, therefore, is conducted with respect and with ritual from the moment the hunt is conceptualized until the animal's remains are properly disposed of. Animals are not offended at being killed for use, but killing must be done humanely, and there should be no suggestion of waste. Nor can the body be mistreated: irreverent, insulting or wasteful behaviour could result in the future loss of the species, which would no longer make itself available for killing. Even gathering must be done respectfully. Yukon women gathering roots are pleased by the find of caches of roots already collected by mice. However, they are careful to leave some for the mice. Otherwise the mice are likely to come raiding Indian supplies during the winter (McClellan, 1987: 139–140).

The practical consequences of such relationships with animals are profound. Species were not endangered or exterminated, for exterminating a species would have meant eliminating not only an essential life but a kindred being. Hughes (1983) believes that the "ecological consciousness" of the Native American was in part due to their sense of kinship with the rest of the world.

The need to both respect *and* use animals continues today, although it can engender controversy. Whaling, for example, continues to be a controversial subject. When a west coast tribe, the Makah, decided to return to their traditional practice of whaling in 1999 it provoked outrage on the part of environmentalists and even ordinary people; it prompted immense television coverage, the attempted intervention of the Sea Shepherd Society and required the protection of the United States Coast Guard for the hunters (Blow, 1998). However as Tom Mexsis Happynook, chair of the World Council of Whalers, argued, whaling represents a modern practice of responsibilities which integrate the native culture into the environment, which "maintain the balance within the environment and ecosystems."

> For the Nuu-chah-nulth hunters the taking of any life was looked upon as the most sacred responsibility that we have had bestowed upon us, because of this my grandfather taught me the hunters had to pay dearly for the honour of taking a life. This especially applied to the whaling chiefs because they hunted the greatest and largest mammal on earth.
>
> People have a very important role to play in the environment which is to help maintain the balance through our relationship with the ecosystem; that one of the most important tools we have at our disposal to meet this obligation is respectful, responsible and sustainable utilization of the resources (the gray whale population on our coast is a perfect example of that responsibility) (Happynook, 1999: 3).

Non-natives see an intelligent, sympathetic endangered species. Whaling Natives see the same but also a creature with which they have had a profound relationship for millennia, the hunting of which underlies who they are as a culture.

It is worth recognizing that the sense of relationship, of reciprocity and of participation in the world encapsulated in Native spirituality, and as reflected in how they interacted with animals, had a very practical ecological link for many Native Americans.

The Indian principle of interpretation/observation is simplicity itself: "We are all relatives." Most Indians hear this phrase thousands of times a year as they attend or perform ceremonies ... this phrase is very important as a practical methodological tool for investigating the natural world and drawing conclusions about it that can serve as guides for understanding nature and living comfortably within it. "We are all relatives" when taken as a methodological tool for obtaining knowledge means that we observe the natural world by looking for relationships between various things in it. That is to say, everything in the natural world has relationships with every other thing and the total set of relationships makes up the natural world as we experience it (Deloria, 1999: 34).

This sounds remarkably like the modern science of ecology. But were Native Americans ecologists?

THE ECOLOGICAL INDIAN

In the last two decades the academic discussion has moved slightly in focus from what view of nature Native Americans had in the past or present to how this might translate into action. The key question appears to have become, were Native Americans truly the first environmentalists? Were they "ecological" Indians? Or were they, as Berkes terms it, "intruding wastrels" destructive of their ecosystems, surviving only because they moved around and had fortuitously small populations (Berkes, 1999: 145). Or are they "fallen angels", former noble savages whose ancestors lived in harmony with nature but whose descendants threaten fragile wildlife populations and at-risk wilderness?

This interest in Native American relationships with nature has an old history. Influential members of the early American conservation movement were deeply impressed by Native Americans and their relations with the natural world (Cornell, 1985). This question however came under intense consideration with the 1960s environmental movement. Former Secretary of the Interior Stewart Udall (1973) articulated this best in his influential article, "Indians: First Americans, First Ecologists." Interest burgeoned, and works such as *Black Elk Speaks* (Neihardt, 1932, 1975), *Rolling Thunder* (Boyd, 1974) and the now controversial speech of Chief Seattle ("The earth is our mother") became cultural icons. The story of Chief Seattle's speech is somewhat instructive as to how eagerly and uncritically the public embraced the idea of the Native environmentalist. While Seattle, a Dwarmish chief, did indeed make a highly articulate speech to the United States government on the appropriation of his tribe's land, what has become know as Seattle's speech was actually a highly liberal adaptation written by a non-native as narration for a 1970s film (Kaiser, 1985).

Native Americans were not unwilling to have this part of their culture articulated. Many writers of the 1970s and 1980s expressing the Native American and nature relationship were Native Americans such as Vine Deloria, Jr., Paula Gunn Allen, John Lame Deer, Winona LaDuke, and N. Scott Momaday.

By the 1980s, however, the issue came under academic scrutiny, and the picture was no longer clear cut. Anthropologists and historians marshalled on both sides. People like Hughes, Nelson, Cronon, Gill, Toelken, Brown and

others presented plausible cases for cultures which, while not gentle nature
lovers, nonetheless maintained functioning ecosystems through cultures which
integrated nature into spiritual and living practices. Many were examining
historical cultures, although Gill, Brown, Toelken and Nelson were working
with modern cultures. Making the case that Native Americans were not model
ecological citizens were scholars such as Callicott, Martin and Krech.

The debate has not as of this writing been resolved. Both sides have over
time made some persuasive cases. The "good ecological" native arguments are,
simplified, based on two threads. The first thread is the ecological knowledge
and "philosophy" contained in Native culture past and present. This has been
discussed in the section on Native views of nature. The second thread is the
ecological state of North America when European explorers discovered it.
While research has demonstrated that the landscape was not unmanipulated
by human residents, nonetheless after an occupation of between 10,000 and
perhaps 40,000 years (Dickason, 1992) North America looked like an Eden to
European arrivals. Forest cover was extensive. Wildlife, including large carni-
vores, appeared in astonishing numbers. It was a sharp contrast to the human-
ized, ecologically impoverished landscape that European explorers had left
behind. This is the crux of the argument of those who argue that Native
Americans offer models of human-nature relationships well worth admiration
and study.

Those who argue against the concept of the "ecological Indian" also draw
on evidence from the pre-European era. Shepard Krech III is now one of the
leaders in the move to debunk the myth of the ecological Indian, which he
does in his most recent book (Krech, 1999) based largely on historical activities.
He defines the truly ecological Indian clearly as one who did *not* change,
damage or irrevocably alter his ecosystem. By doing so, he clearly defines what
we (or at least the scholars debating the ecological Indian myth) think of as
ecological or environmental behaviour: the opposite. By this definition, as
Krech successfully argues, the historical Native American was not ecological
or an environmentalist. There is considerable archaeological evidence, for example, that indicates many cultures did indeed damage their ecosystem through
overuse of resources. Krech cites the mid-western Cahokia culture which flourished between the 12th and 14th centuries. They apparently denuded their
landscape so severely they needed to import wood. The earlier southwestern
civilization of the so-called Anasazi may have also collapsed as a consequence
of overuse of resources (Krech, 1999: 76–77). Other cultures constantly altered
the landscape deliberately. Cultures in the eastern woodland and the prairies
set fire to habitats to alter vegetation to attract different browsers, to force
animals into more easily hunted areas (Krech, 1999: 105), or to produce berries.
While these fires were prompted by considerable ecological knowledge of fire
succession stages, they often got out of control and were highly destructive.
Along the east coast there are still areas burned down to rock (Patterson,
1988). Krech also cites the hunting of bison through massive drives into pens
or over cliffs, with subsequent waste of considerable meat. As Krech does point
out, there is equal evidence of careful use of all resources as well. What is clear

in all his careful scholarship is that Native Americans used, altered and in some cases damaged nature. By non-native definition they are unecological, non-environmental. Krech confirms earlier arguments put forward by scholars such as Callicott and Martin.

Both sides make careful sophisticated arguments that interested readers should approach in their entirety and draw their own conclusions. I remain uncertain as to utility of the debate itself (not that I disagree with debate). I have done a lot of work with a Native Canadian band (the politically correct Canadian term is First Nations) in north central British Columbia, who in turn have given me a nickname: the Blonde Woman. It is said with liking. I am grateful for the name. It means people are willing to talk to me when I show up at their door; the name came with trust. What I find fascinating however is that it reflects the most visible feature by which I am defined as *non-Indian*. I have never met a blonde Carrier native; my blondeness defines me as completely the outsider. By its measurement I am most clearly not native. That definition of who is and isn't is important, even crucial to people. But the band I work with does not make the usual corollary mistake – they do not stereotype me as *the other*. Their focus on my blondeness reflects that too; it is the least important non-native characteristic they could focus on. Instead they let me prove myself as an individual, good or bad, trustworthy or not. They do not stereotype me. It is not a courtesy we tend to return.

To talk about the "ecological" or "non-ecological" Native American is to talk of stereotype. Part of the stereotype is to assume that all Natives are alike and can be characterised as part of that larger group. One bad (or non-ecological) Indian therefore manages to spoil the bunch. To be fair we do this to other groups as well – one bad poor black inner-city dweller casts a shadow on all – but this does not make it a good practice. If we wish to assess a native group's ecological goodness we should assess only that group within its ecosystem, history and circumstance. However, this glosses over a larger problem.

When we are assessing an "ecological Indian" status we are doing what my band does *not* do to me, blonde hair not withstanding: we hold each Native group up against our concept of Indianness and, usually, judge them lacking. Even those disputing the ecological Indian concept are rebelling against an ingrained stereotype of the Native as nature child at home in the wilderness. Those images are as old as Christopher Columbus (Berkhofer, 1979; Francis, 1992; Berkes, 1999). There are two key problems with this stereotype. The first is that there is something like an essential Indian or Indianness. The second is the imposition of a thoroughly non-Native concept of ecological or "good" nature-human relationships.

The first difficulty is part of the problem I identified early in this article, that we take for granted there is something that is a "Native American". I once sat in a university class taught by a Cherokee. After a discussion on blood quantum, a United States government measurement of whether an individual could legally qualify as an American Indian, a student asked the professor, "How much of an Indian are you?" It seems to me that we spend most of our time requiring Natives always to be answering that question. We continually measure

Natives by whether they meet our definition of Indian, whether blood quantum or environmental mysticism, lifestyle or economic choices, rather than allowing Natives to be people, individuals, members of a community. Our ecological Indian is a stereotype of our expectations of an Indian; he is not a real human with real, situationally unique problems, opportunities and choices.

The second problem is that when we allocate goodness as an assessment of ecological or environmentally correct behaviour it is based on a western definition of ecological or environmental goodness. As Paula Gunn Allen and Vine Deloria, Jr., among others, point out, this definition derives more from "the exalted romance of the sentimental 'nature lovers' [or] self-conscious 'appreciation' of the land" (Allen, 1979: 192).

The non-native concept of nature, and consequently good environmental behaviour, is based upon a relationship that is largely separate from a day-to-day intimate interaction with that nature. It is something outside of our everyday lives, in the sense that when we think about nature it is not the bits of grass and single struggling trees along the sidewalks in our cities, or the squirrels and cockroaches that survive with us that we think about. We imagine *Sierra* magazine vistas, forests, steams and charismatic wildlife which we may manage actually to experience once or twice a year at best. As a result of separateness, and of the ability to make a living that is removed from the land, nature becomes a sacred, remote concept. Nature is something we are deprived of, often yearn for, do not understand or know as anything other than a desire. Worse, the undeniable ecological crisis of modern culture creates an image, a truthful image, of nature as fragile, threatened, in need of protection (although this does not mean most of us are willing either to change our lifestyles or voluntarily stay out of natural areas). So the day-to-day use of nature by many Natives is seen as unecological, particularly if such use includes making a living through exploiting resources through mining, trapping, logging, damming, or hunting whales. However, as Berkes (1999: 154) notes, wilderness is a questionable ecological concept; there are few landscapes on earth that were not manipulated by human cultures. Native American perceptions of nature include an intimate understanding of making a living from that nature. Part of living means making a living. We all make a living from the land; some of us can hide that fact from ourselves better than others. If nothing else, a native perception of nature and the human relation with nature is honest. But it does mean that it is often difficult to call Natives environmentalists or even ecological, if by that we mean they did not live with, change and use nature.

Using people to demonstrate a point often involves forgetting that the examples are flesh and blood. Beliefs do not translate perfectly into practice. And the definition of what *is* an ecologically sound practice has changed with the circumstances. Much depends on the natural context: population densities, existing ecological damage, and outside conditions. Today there is disagreement over whether "traditional, real" Indians should drive pickup trucks, snowmobiles, powerboats and use assorted firearms to hunt. There are bitter disagreements over "traditional" harvests, as, for example, the Inuit harvest of the seriously endangered bowhead whale using modern mechanical killing devices.

Time also plays a factor, particularly when looking at Native American practices. What was ecologically acceptable 600 years ago, such as stampeding herds of the incredibly numerous bison off cliffs, is not acceptable now, when bison are endangered. A less dramatic example might be the hunting of eagles for ceremonial feathers, claws, etc. Native Americans are occasionally prosecuted under endangered species laws for these spiritually necessary practices. Yet while eagle populations are on the upswing, they are far from secure, as even Native Americans acknowledge. Who is the more "ecologically sound" citizen, the spiritually motivated Native American or the pragmatic U.S. Fish and Wildlife Service officer who arrests him?

Finally, there is the question of which Indian you chose as a "model": the safely dead "traditional" Indian or the living descendant. The dead no longer have a voice to protest over how they are used. The living are facing very difficult questions about existence in the modern world. Environmentalists often seem to prefer the dead Indian; modern Indians too often make difficult partners. On some issues, Native Americans have welcomed the help of resource rich environmental organizations. On other issues, for example oil and mineral developments on the southwestern or Alaskan reservations, Native American tribes have been on opposite sides from environmentalists. Or consider the bitter divisions that resulted after Greenpeace successfully lobbied against the trade in harp seal skins. Many northern tribes lost much of their yearly income with the loss of the sealskin market, and they are angry in their condemnation of Greenpeace's cultural insensitivity. A similar issue has erupted as animal rights activists are increasingly successful in making the wearing of fur "politically incorrect". While some furbearers are ranch-raised, many are still trapped by northern Indian communities who stand to lose a significant source of income. White middle-class environmentalists are often slow to recognize the imperative of feeding children on reservations with 60% unemployment and chronic poverty. They also raise legitimate questions on the use of animals.

Having demonstrated how difficult it is to use Native Americans as ecological models, I still wish to consider how they have adapted to the land and whether the adaptations could be considered ecologically successful. While most Native Americans, I think, would have considerable difficulty in fitting themselves into a category as shallow and narrow as "environmentalist", I believe that non-Natives have something to learn about living with the land.

It must be remembered that for most Native Americans, the land and its residents have many more shades of meaning than non-Native societies are willing to consider. The land is a source of sustenance and was exploited for that purpose. However, the land also holds other meanings for past and present Indian cultures. Nelson remarks of the Alaskan Koyukon, principally hunters and gatherers, that their land

> is permeated with different levels of meaning – personal, historical, and spiritual. It is known in its finest details, each place unique, each endowed with that rich further dimension that emerges from the Koyukon mind (Nelson, 1983: 245).

To picture Native Americans as people who passed through the land leaving no trace, as some environmentalists are apt to do, is to deny an ecological reality, and, in passing, to damn the Native American with faint praise (with the implication that they were incapable of making such a mark). As Hughes remarks of the Northeastern Indians,

> Like all human cultures, the forest Indians were agents of change in nature. For at least 10,000 years, and perhaps much longer, they had lived within the forest ecosystems, hunting, fishing and gathering with skill and experience. Their land was not a wilderness, but a woodland park which had known expert hunters for millennia (Hughes, 1977: 7).

That the lands remained productive was in part due to Native American lifestyles that were more or less mobile (even the farmers moved to new areas when the soil was exhausted), as Cronon (1983), Merchant (1989) and Krech (1999) document. It may have also been the result, in some cases, of a sense of stewardship. Many tribes deliberately took (and still take) steps to ensure that human-induced changes did not damage their source of livelihood. Among the Yukon Indians, these steps took the form of a belief in ownership of certain areas.

> This kind of ownership meant that a headman and the people who traveled with him had the first right to use the products of the local land and water, but they also had the duty of taking care of these resources ... It was part of their job to see that people did not overhunt, overtrap or overfish the area, as well as to see that as far as possible everybody got enough to eat ... (McClellan, 1987: 151–152).

Among the Cree, efforts to conserve resources took the form of dividing the land into regions, with a hunting camp in the center. The camp was occupied until both animal populations and resources such as firewood and green bedding boughs were reduced. Old campsites were cleaned prior to leaving to avoid offending the spirits, who would send away the animals if the land was not cared for (Tanner, 1979: 74–75). Such concerns did not mitigate the ecological impacts; the game population and the camp site recovered because the Indians moved elsewhere. However, the concern for the spirits' good will, in combination with a mobile population, may have limited irreparable damage, at least where overall human population numbers were low.

Of great concern to Native Americans, for obvious reasons, was the conservation of game animals. Harrod notes the presence of several Trickster stories amongst the Plains Indians in which the Trickster character exploits trusting animals to satisfy an insatiable appetite. In this pursuit he is foolish enough to almost eliminate all the animals, usually missing just one.

> That Trickster is foolish and even a dangerous hunter appears in the theme of the "last animal" which runs through the narratives. As a consequence of escape, often brought on by a momentary lapse in Trickster's character, an animal survives to continue the species. Clearly the hunting techniques of Trickster are not affirmed; rather, the foolishness and potentially disastrous consequences of such activities are underlined in these stories (Harrod, 1987: 63).

There appears to have been a real concern, expressed through such stories, that the animal populations not be decimated. Recent anthropological work among the twentieth century Cree and Koyukon, among others, has uncovered what researchers feel are deliberate conservation practices. Tanner (1979) and

Berkes (1999) observe that the Cree are constantly assessing plant and animal population levels. Long- and short-term changes in populations are understood both in spiritual terms (the activities of the Animal Masters) and in terms of environmental factors (food supplies, water, weather patterns, forest fires, and hunter activities). The Cree govern their hunting activity on the basis of these observations.

Among the Koyukon, conservation activities are also based on their keen observation of the ecological dynamics of their lands. The people regulate their harvests in a number of ways to ensure that plant and animal populations remain healthy. The Koyukon may consciously avoid taking more individuals than they believe can be naturally replaced, or they may take special measures that they hope will enhance the productivity of a species. They avoid killing female waterfowl, bears and moose in the spring when they are breeding. Hunting activities are usually spread over as wide an area as possible. Young plants and animals are usually not harvested but are allowed to mature. Trappers are very cautious about where and how many animals are harvested (Nelson, 1983: 221–223). For example, trappers are careful when trapping the sedentary beaver. [Editor's note: please see Bielawski's chapter on the Arctic Dene and Inuit people.]

While such practices are noteworthy, Nelson is careful to point out that there is no objective data to prove that these intentional limitations actually achieve their conservation goals. He does, however, feel that the Koyukon could be considered to be practising a conservation ethic. Tanner is also unable to document the impacts of the Cree's conservation practices on resource populations. Berkes, however, makes an extensive case for a conservation ethic of the modern Cree, although he notes it was not without many mistakes and misapprehensions.

A final issue to consider is the modern need for Native cultures to make a living from the land upon which they live. This often involves necessary, and not always satisfactory, compromise between traditional beliefs regarding nature and resource extraction. Forestry is a common industry on many Native lands in both Canada and the United States. One well-known, long-term industrial forest operation has been undertaken by the Menominee of Wisconsin. The Menominee have been logging on their reservation for over a century and still have a robust and healthy forest ecosystem. Their forester, Marshall Pecore, argues it is a unique balancing act.

> It is said of the Menominee people that the sacredness of the land is their very body, the values of the culture are their very soul, and the water is their very blood. It is obvious, then, that the forest and its living creatures can be viewed as food for their existence ... Their story is one of successful equilibrium between harvesting and using only what the land can provide, and maximising the jobs and other economic benefits that flow from a sustained-yield harvest (cited in Davis, 2000: 55–56).

However it is also not always easily achieved. The band I work with, Tl'azt'en Nation, also runs a commercial forestry operation and has done so for 20 years. They feel they have been less successful in integrating traditional values and, correctly my analysis suggests, blame it on the restrictions surrounding forestry in British Columbia.

> There's a conflict right there. That one is really a conflict that we're conscious of – trying to look at a bottom line and manage in a traditional way – and we haven't really been able to marry those two successfully and kind of our teachings ... they're still there but we run ... you know there's a real major conflict in thinking and ideology right there along with our teachings, you know when you go into the bush to take a plant for medicine, you return something back. And then in logging you go in there and clearcut an area, you don't put anything back except new trees ... I have a concern about that, and how we do that I'm not sure right now except to try to build those traditional teachings and principles into your management plans, push those, work those. Some of those don't necessarily, are not necessarily acceptable by the Ministry of Forests, so it's an ongoing struggle there too because you have different thinking about how that forest should be managed (Booth, 2000: 33).

However, Natives need to make their own compromises:

> Conservation and development-policy-making and -planning often seems to assume that we, the aboriginal peoples, have only two options for the future: to return to our ancient ways of life or abandon subsistence altogether and become assimilated into the dominant society. Neither option is reasonable. We should have a third option: to modify our subsistence way of life, combining the old and the new in ways that maintain and enhance our identity while allowing our society and economy to evolve (Erasmus, 1989: 229)

*   *   *

Are Native Americans ecologists? Environmentalists? Or are they peoples with a unique understanding of the natural world that might or might not have translated into a better way of living with the earth? Readers will need to draw their own conclusions. However, I agree with Berkes (1999: 182) when he notes that, "A fundamental lesson of traditional ecological knowledge is that worldviews [and beliefs] do matter."

> The challenge is to cultivate a kind of ecology that rejects the materialist tradition and questions the Newtonian, machinelike view of ecosystems ... The indigenous knowledge systems of diverse groups, from the Dene of the North American subarctic to the Fijians of the South Pacific, provide an alternative view of ecosystems. This is a view of an ecosystem pulsating with life and spirit, incorporating people who *belong* to that land and who have a relationship of peaceful coexistence with other beings.

Native American views of nature, as articulated in this article, attract a great deal of interest as well as controversy. There is a fascinating psychological question as to why, as a society, we are attracted to them, even if they are not part of our cultural heritage. Perhaps we need to believe there are ways of seeing the world that might lead to the preservation of functioning ecosystems. Perhaps this is the same basis from which others are driven to criticise the concept of the ecological Indian: we do not have to feel as guilty for our behaviour if we are all equally, inherently, destructive of nature. The problem I see is not that we are able to be inspired by Native Americans, or that they drive us to justify our actions against ecosystems, but that in doing either of these we lose sight of the people themselves and the complex nature of the challenges they face. We need to allow ourselves that inspiration, the critical question as well, but we always need to remember the *people* who articulate a reality that perhaps all humans can understand.

> Once in his life a man ought to concentrate his mind upon the remembered earth, I believe. He ought to give himself up to a particular landscape in his experience, to look upon it from as many angles as he can, to wonder about it, to dwell upon it. He ought to imagine that he touches it with his hands at every season and listens to the sounds that are made upon it. He ought to imagine the creatures that are there and all the faintest motions in the wind. He ought to recollect the glare of noon and all the colours of the dawn and dusk.

> I am interested in the way that a man looks at a given landscape and takes possession of it in his blood and brain. For this happens, I am certain, in the ordinary motion of life. *None of us lives apart from the land entirely; such isolation is unimaginable. We have sooner or later to come to terms with the world around us – and I mean especially the physical world;* not only as it is revealed to us immediately through our senses, but also as it is perceived more truly in the long turn of seasons and of years (emphasis added) (Momaday, 1979: 164, 166).

BIBLIOGRAPHY

Alcorn, J.B. 'Noble savage or noble state? Northern myths and southern realities in biodiversity conservation.' *Ethnoecologica* 2(3): 7–19, 1994.

Allen, Paula Gunn. 'Where I come from, God is a woman.' *Whole Earth Review* 74: 44–46, 1992.

Allen, Paula Gunn. *Grandmothers of the Light*. Boston: Beacon Press, 1991.

Allen, Paula Gunn. 'The woman I love is a planet; the planet I love is a tree.' In *Reweaving the World: The Emergence of Ecofeminism*, Irene Diamond and Gloria Feman Orenstein, eds. San Francisco: Sierra Club Books, 1990, pp. 52–57.

Allen, Paula Gunn. *The Sacred Hoop: Recovering the Feminine in American Indian Traditions*. Boston: Beacon Press, 1986.

Allen, Paula Gunn. 'Iyani: It goes this way.' In *The Remembered Earth*, Geary Hobson, ed. Albuquerque: Red Earth Press, 1979a, pp. 191–193.

Allen, Paula Gunn. 'The sacred hoop: A contemporary Indian perspective on American literature.' In *The Remembered Earth*, Geary Hobson, ed. Albuquerque: Red Earth Press, 1979b, pp. 222–239.

Berkes, Fikret. *Sacred Ecology*. Philadelphia: Taylor and Francis, 1999.

Berkes, Fikret. 'Environmental philosophy of the Cree People of James Bay.' In *Traditional Knowledge and Renewable Resource Management in Northern Regions*, M.M.R. Freeman and L. Carbyn, eds. Edmonton: Boreal Institute, University of Alberta, 1988, pp. 7–21.

Berkhofer, Robert. *The White Man's Indian: Images of the American Indian From Columbus to the Present*. New York: Vintage Books, 1979.

Blackburn, T.C. and K. Anderson, eds. *Before Wilderness: Environmental Management by Native Californians*. Menlo Park: Ballena Press, 1993.

Blow, Richard. 'The great American whale hunt.' *Mother Jones* 23(5): 49–53, 86–87, 1998.

Booth, A. *A Case Study of Community Forestry in the Tl'azt'en Nation*. Prince George: University of Northern British Columbia, 1999.

Booth, A. 'Putting "forestry" and "community" into First Nations' resource management.' *Forestry Chronicle* 74(3): 347–352, 1998.

Booth, A. 'Learning from others: What ecophilosophy can learn from traditional Native American women's lives.' *Environmental Ethics* 20(1): 81–99, 1998.

Booth, Annie L. and Harvey M. Jacobs. 'Ties that bind: Native American beliefs as a foundation for environmental consciousness.' *Environmental Ethics* 12(1): 27–43, 1990.

Brightman, Robert A. *Grateful Prey: Rock Cree Human-Animal Relationships*. Berkeley: University of California Press, 1993.

Brown, Joseph Epes. 'Becoming part of it.' In *I Become Part of It: Sacred Dimensions in Native American Life*, D.M. Dooling and Paul Jordan-Smith, eds. New York: Parabola Books, 1989, pp. 9–20.

Brown, Joseph Epes. *The Spiritual Legacy of the American Indian*. New York: Crossroad Publishing Co., 1985.

Brown, Joseph Epes. 'Sun dance.' *Parabola* 8(4): 12, 15, 1978.

Brown, Joseph Epes. 'Modes of contemplation through action: North American Indians.' *Main Currents in Modern Thought* 30(2): 192–197, 1973.

Callicott, J. Baird. *Earth's Insights: A Survey of Ecological Ethics from the Mediterranean Basin to the Australian Outback*. Berkeley: University of California Press, 1994.

Callicott, J. Baird. 'Traditional American Indian and Western European attitudes towards nature: an overview.' *Environmental Ethics* 4: 293–318, 1982.

Capps, Walter H., ed. *Seeing With a Native Eye*. New York: Harper and Row, 1976.

Churchill, Ward. *Struggle for the Land: Indigenous Resistance to Genocide, Ecocide and Expropriation in Contemporary North America*. Toronto: Between the Lines, 1992.

Cordova, V.F. 'Ecoindian: a response to J. Baird Callicott.' *Ayaangwaamizin* 1: 31–34, 1997.

Cornell, George L. 'The influence of Native Americans on modern conservationists.' *Environmental Review* 9: 105–17, 1985.

Cronon, William. *Changes in the Land: Indians, Colonists and the Ecology of New England*. New York: Hill and Wang, 1983.

Davis, Thomas. *Sustaining the Forest, the People, and the Spirit*. Albany: State University of New York Press, 2000.

Deloria, Barbara, Kristen Foehner and Sam Scinta, eds. *Spirit and Reason: The Vine Deloria, Jr. Reader*. Golden: Fulcrum Publishing, 1999.

Deloria, Vine, Jr. *God Is Red*. New York: Grosset & Dunlap, 1973.

Dickason, Olive Patricia. *Canada's First Nations: A History of Founding Peoples from Earliest Times*. Oxford: Oxford University Press, 1992.

Erasmus, George. 'A Native viewpoint.' In *Endangered Spaces: The Future for Canada's Wilderness*, Monte Hummel, ed. Toronto: Key Porter Books, 1989, pp. 92–98.

Francis, Daniel. *The Imaginary Indian: The Image of the Indian in Canadian Culture*. Vancouver: Arsenal Pulp Press, 1992.

Gill, Sam. 'The trees stood deep rooted.' In *I Become Part of It: Sacred Dimensions in Native American Life*, D.M. Dooling and Paul Jordan-Smith, eds. New York: Parabola Books, 1989, pp. 21–31.

Gill, Sam. *Mother Earth: An American Story*. Chicago: University of Chicago Press, 1987.

Goodman, Ronald. *Lakota Star Knowledge: Studies in Lakota Stellar Theology*. Rosebud Sioux Indian Reservation: Sinte Gleska College, 1990.

Gottesfeld, Leslie M. 'Conservation, territory and traditional beliefs: an analysis of Gitksan and Wet'suwet'en subsistence, Northwest British Columbia, Canada.' *Human Ecology* 22: 443–465, 1994.

Happynook, Tom Mexsis. 'Traditional rights versus environmental protection of a species.' Paper presented at the *Conference on Environmental Law and Canada's First Nations*. Pacific Business and Law Institute, Vancouver, BC, Canada, November 18–19, 1999.

Harrod, Howard L. *Renewing The World: Plains Indian Religion and Morality*. Tucson: University of Arizona Press, 1987.

Hester, L., D. McPherson, A. Booth, and J. Cheney. 'Indigenous worlds meet post(?)modern evolutionary-ecological environmental ethics.' *Environmental Ethics* 22(3): 273–290, 2000.

Hughes, J. Donald. *American Indian Ecology*. El Paso: Texas University Press, 1983.

Hughes, J. Donald. 'Forest Indians: the holy occupation.' *Environmental Review* 2: 2–13, 1977.

Hultkrantz, Åke. *Belief and Worship in Native North America*. Syracuse, New York: Syracuse University Press, 1981.

Kaiser, Rudolf. 'A fifth gospel almost: Chief Seattle's speech(es).' In *Indians and Europe*, Christian Feest, ed. Aachen: Rader Verlag, 1985, pp. 505–526.

Krech III, Shepard. *The Ecological Indian: Myth and History*. New York: W.W. Norton, 1999.

Krech III, Shepard, ed. *Indians, Animals and the Fur Trade*. Athens: University of Georgia Press, 1981.

The People of 'Ksan. *Gathering What Great Nature Provided*. Vancouver: Douglas and McIntyre, 1980.

LaDuke, Winona. 'Environmentalism, racism, and the New Age Movement: the expropriation of indigenous cultures.' Text of a speech. *Left Green Notes* 4 (September/October): 15–18, 32–34, 1990.

Laird, Carobeth. *The Chemehuevis*. Banning: Malki Museum Press, 1976.

Lame Deer, John (Fire) and Richard Erdoes. *Lame Deer: Seeker of Visions*. New York: Simon and Schuster, 1972.

Martin, Calvin. *Keepers of the Game: Indian-Animal Relationships and the Fur Trade*. Berkeley: University of California Press, 1978.

Martin, Paul S. 'Prehistoric overkill.' In *Man's Impact on the Environment*, Thomas R. Detwyler, ed. New York: McGraw-Hill, 1971, pp. 612–624.

Matthiessen, Peter. *Indian Country*. New York: The Viking Press, 1984.

Matthiessen, Peter. 'Native earth.' *Parabola* 6(1): 6–17, 1981.

McClellan, Catherine. *Part of the Land, Part of the Water*. Vancouver: Douglas and McIntyre, 1987.

McLuhan, T.C. *Touch the Earth: A Self-Portrait of Indian Existence*. New York: Outerbridge & Lazard, Inc., 1971.

Merchant, Carolyn. *Ecological Revolutions: Nature, Gender, and Science in New England*. Durham: University of North Carolina Press, 1989.

Momaday, N. Scott. 'The man made of words.' In *The Remembered Earth*, Geary Hobson, ed. Albuquerque: Red Earth Press, 1979, pp. 162–173.

Momaday, N. Scott. 'Native American attitudes to the environment.' In *Seeing With a Native Eye*, Walter H. Capps, ed. New York: Harper and Row, 1976, pp. 79–85.

Moran, Bridget. *Stoney Creek Woman*. Vancouver: Tillacum Library, 1988.

Neihardt, John G. *Black Elk Speaks: Being the Life Story of a Holy Man of the Oglala Sioux*. New York: W. Morrow, 1932; reprinted New York: Pocket Books, 1975.

Nelson, Richard. *Make Prayers to the Raven: A Koyukon View of the Northen Forest*. Chicago: University of Chicago Press, 1983.

Nelson, Richard. 'A conservation ethic and environment: the Koyukon of Alaska.' In *Resource Managers: North American and Australian Hunter Gatherers*, Nancy M. Williams and Eugene S. Hunn, eds. AAAS Selected Symposium No. 67. Boulder: Westview Press, 1982, pp. 211–228.

Patterson, William A., III and Kenneth E. Sassaman. 'Indian fires in the prehistory of New England.' In *Holocene Human Ecology in Northeastern North America*, George P. Nicholas, ed. New York: Plenum Press, 1988, pp. 107–135.

Regan, Tom. 'Environmental ethics and the ambiguity of the Native American's relationship with nature.' In *All That Dwell Therein: Animal Rights and Environmental Ethics*, Tom Regan, ed. Los Angeles: University of California Press, 1982, pp. 206–239.

Romeo, Stephanie. 'Concepts of nature and power: environmental ethics of the Northern Ute.' *Environmental Review* 9(2): 150–170, 1985.

Standing Bear, Luther. *Land of the Spotted Eagle*. Lincoln: University of Nebraska Press, 1933.

Tanner, Adrian. *Bringing Home Animals: Religious Ideology and Mode of Production of the Mistassini Cree Hunters*. New York: St Martin's Press, 1979.

Toelken, Barre. 'The demands of harmony.' In *I Become Part of It: Sacred Dimensions in Native American Life*, D.M. Dooling and Paul Jordan-Smith, eds. New York: Parabola Books, 1989, pp. 59–71.

Toelken, Barre. 'Seeing with a native eye: how many sheep will it hold?' In *Seeing With a Native Eye*, Walter H. Capps, ed. New York: Harper and Row, 1976, pp. 9–24.

Trosper, Robert L. 'Land tenure and ecosystem management in Indian country.' In *Social Conflict Over Property Rights: Who Owns America?*, Harvey M. Jacobs, ed. Madison: University of Wisconsin Press, 1998.

Udall, Stewart L. 'Indians: first Americans, first ecologists.' In *Readings in American History – 73/74*. Guilford, Connecticut: Dushkin Publishing Group, 1973.

Vecsey, Christopher and Robert W. Venables. *American Indian Environments: Ecological Issues in Native American History*. New York: Syracuse University Press, 1980.

LESLIE E. SPONSEL AND PORANEE NATADECHA-SPONSEL

BUDDHIST VIEWS OF NATURE AND THE ENVIRONMENT

> *Buddham saranam gacchami.*
> *Dhammam saranam gacchami.*
> *Sangham saranam gacchami.*
> In the Buddha I take refuge.
> In his teaching I take refuge.
> In the monastic community I take refuge.

The Three Refuges chant that usually begins Buddhist ceremonies reflects the three ultimate components of Buddhism. One becomes a Buddhist by accepting and pursuing the three. The route to enlightenment commences with accepting the *Dhamma* [*dharma*] (teachings), starting with the Four Noble Truths, and then following the Noble Eightfold Path (explained below).[1]

Beyond this core, Buddhism is an enormous and complex subject, with many variations on its basic themes manifested in at least 18 schools and their various sects. However, there are three major schools: Theravada prevails in Sri Lanka, Thailand, and Burma (officially called Myanmar); Mahayana in Japan, North Korea, South Korea, parts of China, and Taiwan; and Vajrayana or Tantric in Tibet, Bhutan, Mongolia and Ladakh. Cambodia, Laos, and Vietnam are a transition zone from Theravada to Mahayana from west to east. In Asia there is tremendous diversity – geographical, ecological, historical, cultural, linguistic and religious, and this has contributed to local variation in the expression of Buddhism. Individuals often adhere to more than one religion or follow a mixture of elements from different religions (syncretism) (Lewis, 1997: 345). Another variable is that, as in any religion, in Buddhism there are differences between ideals and practice, text and context, scholar and laity, and urban and rural (Calkowski, 2000). Lewis (1997: 320) even refers to the "domestication" of Buddhism in local contexts.

Different perceptions and conceptions regarding nature and environment constitute yet another variable. Ultimately nature is the sum total of reality – all beings and all things. In practice nature often means the biosphere of planet Earth. Environment is nature manifested on a local spatial scale or a particular ecosystem. Environments range between the extremes of wilderness as supposedly untouched nature and culturally constructed nature such as Zen gardens.

H. Selin (ed.), Nature Across Cultures: Views of Nature and the Environment in Non-Western Cultures, 351–371.
© 2003 *Kluwer Academic Publishers. Printed in Great Britain.*

Ecology is the natural science of environmental interactions. Environmentalism refers to initiatives to promote the survival and health of relatively natural environments.

Ecology and environmentalism developed in the West, mainly since WWII, and it was inevitable that some who are Buddhologists and/or Buddhists would inquire into the relationship between Buddhism and nature. Most view this as discovering (or re-discovering) green thinking in Buddhism, but a few consider this as a perverse imposition of green thinking on Buddhism in response to contemporary environmental concerns.

What are the relationships between Buddhism and nature? This is our central question here. Our answer is distinguished by pursuing it mainly within the framework of the Triple Refuge – Buddha, *Dhamma*, and *Sangha* (monastic community). Specific examples are drawn from Theravada in Thailand with which we are most familiar (see Rajavaramuni, 1984). Also we emphasize two themes: Buddhism has a long history of mutualistic relationships with trees and forests (Sponsel and Natadecha, 1988: 309), and it has endured so long in such diverse contexts because of the continuing relevance of its core principles.[2]

BUDDHA

It is not easy to distinguish between fact, legend, and myth in the early accounts of Buddha's life. Although some biographical information is preserved in the Pali scriptures, it was some 200 years after his death before even partial accounts appeared (Roscoe, 1994: 38). Nevertheless, the basic facts are clear: his birth as a human being, his renunciation, studies with Hindu spiritual teachers (*gurus*), long period of asceticism and intense meditation, enlightenment, forty five years of teaching followers, establishing the *sangha*, and discourses. Also the core principles of the Buddha's teachings, such as the Four Noble Truths and the Noble Eightfold Path, can be taken as factual.[3]

Buddha is an honorific title meaning an enlightened or awakened being. His personal name was Siddhattha Gotama. The Buddha's conventional dates are 566–486 B.C.E. He was born under a Sal tree (*Shorea robusta*) in a grove called Lumbini Park near Kapilavastu (now Madeira), about 130 miles north of Benares in present-day Nepal. While in his youth Siddhattha's father, King Suddhodana, sheltered him from suffering beyond the palace, but eventually he witnessed sickness, old age, and a corpse. These impressed upon him the suffering and impermanence of human existence. Then he observed a religious mendicant. That inspired him to search for a spiritual solution to the problems of the human condition. At the age of 29, shortly after the birth of his son Rahula, Siddhattha made the Great Renunciation, a radical decision to leave his family, wealth, and future as a king to become a wanderer in a religious vision quest.

Nature was the context and source of Siddhattha's search for and eventual achievement of enlightenment (Ryan, 1998: 63–76). For six years he pursued spiritual awakening with various gurus and through rigorous asceticism in the forest. In the process, among many other things, Siddhattha realized that moderation (the middle way) was the best path, instead of the extremes of either asceticism or hedonism.

In six years of study gurus took Siddhattha only so far. Eventually he pursued contemplation and meditation on his own during seven days at the base of a bodhi fig tree (*Ficus religiosa*) near the town of Bodhgaya. There Siddhattha finally attained enlightenment (Buddhahood) at age 35. Today the bodhi at Bodhgaya, supposedly the same tree under which the Buddha was enlightened, is considered by some to be the oldest historical tree in the world (Altman, 1994: 163).

In ancient times the Buddha was represented by an image of a bodhi tree rather than in human form. The word "bodhi" means awakening, and the bodhi is considered the "tree of knowledge". Buddhists revere it everywhere as the chosen sacred symbol of Buddha and enlightenment. In Asia and beyond, wherever the bodhi can grow Buddhists have planted it, especially in temple compounds (Kabilsingh, 1998: 55). After his enlightenment, the Buddha subsequently spent a week meditating under each of several other trees: Nigrodha or Indian fig (*Ficus indica*), Mucalinda (*Barringtonia acutangula*), and Rajayatana or kingstead (*Buchanania latifolia*) (Kabilsingh, 1998: 56).

For the remaining 45 years of his life the Buddha was a wandering teacher over much of northern and eastern India. He died at the age of 80 while reclining between two Sal trees in a grove outside the small town of Kusinara and in the company of many of his followers. When death was near he declined to appoint a successor. Instead he said the Dhamma and monastic rules (*vinaya*) should guide followers. In addition, in his sermons he repeatedly advised that every individual should think on his own and test his teachings against their own reason and experience: "Be ye lamps unto yourselves" (Roscoe, 1994: 73). Accordingly, Buddhism lacks any single centralized authority, although different schools and sects may have recognized leaders, such as the Dalai Lama for Tibetan Buddhism.

Reflecting the associations between the Buddha and trees, groves, and forests, Buddhists in general hold them in reverence, particularly ancient giant trees (Ryan, 1998: 52–56). In what many consider to be the most important text in the Pali Canon (*Tipitaka*), the *Mahasatipatthana Sutta*, a discourse on the foundations of mindfulness, the Buddha advocates meditating in a forest or at the base of a tree (Walshe, 1995: 335). The banyan tree (*Ficus bengalensis*), because of its peculiar ability to lower roots to the ground from its spreading branches, also became a symbol of the spread of the Dhamma (Ryan, 1998: 54).

The concern for nature in the Buddha's life and teachings may reflect the historical ecology of his homeland (Gadgil and Guha, 1992: 82, 87–90). The Buddha lived in the Gangetic plain which includes what is now a small portion of southernmost Nepal and in India the southeastern third of the state of Uttar Pradesh and the northern half of the state of Bihar. At the time much of this plain was densely forested, and forest products were still being harvested there into the 19th century. However, as farming and settlements expanded, growing pressure on the land and resource base led to deforestation in many areas (Erdosy, 1998; Ling, 1973: 37–49). The Buddhist "creation myth" from the

Agganna Sutta even describes society and government as arising from the need to control competition and disputes over food and other resources. According to Gadgil and Guha (1992: 88), "The best-known ancient state-sponsored conservation campaign was undertaken by the Mauryan emperor Ashoka, following his conversion to Buddhism. The Ashokan edicts advocate both restraint in the killing of animals and the planting and protection of trees." Kabilsingh (1990: 12) asserts that the early Buddhist literature even reflects a concern about tigers and other species being endangered.[4]

The relevance of the Buddha's life and teachings to nature is clear in at least three respects. First, major events in his life were associated with trees, groves, and forests. Second, he often drew on parables about animals and other aspects of nature to illustrate moral and other principles (see *Jatakas* below). Third, during his lifetime it appears that population and economic pressures were leading to deforestation and resource depletion in some areas. If nature was so relevant to the Buddha, then surely nature is relevant to Buddhism and vice versa. Indeed, one meaning of Dhamma is nature.

DHAMMA

The Buddha's teachings were memorized and transmitted orally by monks chanting them through successive generations over four centuries. The earliest scripture preserved intact is the Pali canon which was written down around the middle of the first century BCE in Sri Lanka (Table 1). Printed editions only began to appear as recently as the end of the 19th century (Mills, 1999: 187–192). Over the ages, as various schools of Buddhism evolved, so did their versions of sacred texts, each considered authentic – the original Buddhism of the Buddha.[5]

Most laity do not study Buddhist texts, but learn about them through monks in their community and popular media. Most commonly known is the *Jatakas*, a collection of 547 parables about the Buddha's previous lives. In these he is often represented as an animal that sacrifices its life to save others. They illustrate how the Buddha cultivated the core virtues of wisdom, nonviolence, compassion, loving-kindness, generosity, and so on. They also illustrate the interrelationships and interdependencies between beings. Furthermore, they imply that animals have a moral sense and are capable of ethical behavior, a point recognized for some species by modern science (de Waal, 1996). The great popularity of these stories probably influenced positive attitudes of people towards the environment.[6]

Table 1   Pali Canon (*Tipitaka*)

1.	*Sutta Pitaka*	Discourses or sermons (*sutta*) of the Buddha and some of his immediate disciples
2.	*Vinaya Pitaka*	Monastic Code or rules of conduct for monks and nuns
3.	*Abhidhamma Pitaka*	Scholastic treatises and commentaries that codify and explain the first two

Dhamma is a complex concept with many related meanings. Buddhadasa (1998: 74) succinctly explains them:

> Everything arising out of Dhamma, everything born from Dhamma, is what we mean by "nature." This is what is absolute and has the highest power in itself. Nature has at least four aspects: nature itself; the law of nature; the duty that human beings must carry out toward nature; and the result that comes with performing this duty according to the law of nature.

Like Buddhadasa, many others who are renowned as Buddhist practitioners as well as scholars emphasize Dhamma as nature.[7] Buddhism views the universe as including a physical aspect and beings that inhabit it. World systems, essentially galaxies, arise, decline, and disintegrate over billions of years. Humans and natural systems influence one another. Thus, humans also shape their environment, for better or worse.

In Buddhism the main characteristics of existence are suffering, impermanence, interdependence, and no-self. The practice of Buddhism involves progressing to a deeper understanding of these and related matters. Knowledge, understanding, and wisdom begin with the Four Noble Truths (Table 2) (Dhamma, 1997). As Ryan (1998: 2) asserts, "The fundamental postulate of Buddhism is that all beings are united in distress, in *dukkha*." The Buddha repeatedly stated that he taught only two things, the causes and the end of suffering (Ryan, 1998: 9). Suffering covers a wide diversity of conditions including dissatisfaction, discontent, disharmony, discomfort, irritation, friction, pain, illness, dying, and death.

Ending both suffering and rebirth is the ultimate aim of Buddhism. Endless wandering (*samsara*) is the cycle of repeated rebirth that all living creatures experience, unless they attain enlightenment or Buddhahood (*nibbana*[*nirvana*]). Enlightenment means achieving a peaceful and supramundane state as well as liberation from the endless cycle of rebirth and suffering. With intelligence and free will humans can understand their situation and improve it. Moral decisions and actions have consequences and determine future rebirths (*kamma* [*karma*]). As a natural law, kamma does not depend on rewards or punishments from any supernatural being(s). Thus, the life and destiny of the human individual is the result of self-determination through individual actions, although some of the consequences may not occur until future reincarnations. Other forces or agents including accidents also influence an individual's life and rebirth. In all of this Buddhism stresses individual responsibility in making moral decisions and actions to develop the human capacity for wisdom, virtuous living, and happiness. This is accomplished, first, through the recognition of the Noble Truths, and second, through following the Noble Path (Table 3).

It could be argued that each of the eight components of the Noble Path is

Table 2 Four Noble Truths

1. All existence is suffering (*dukkha*).
2. Suffering is caused by ignorance and desire.
3. Suffering can end.
4. The way to end suffering is the Noble Eightfold Path.

Table 3 Noble Eightfold Path (see Dhamma, 1997).

Widsom (*panna*)	1. **Right understanding**: individual acceptance and confirmation of the Noble Truths and other components of the Dhamma through reason and experience; it includes letting go, abandoning egoism and selfishness
	2. **Right resolve**: a commitment to developing right attitudes including peacefulness and goodwill
Morality (*sila*)	3. **Right speech**: speaking the truth with thoughtfulness and sensitivity for harmonious interactions with others
	4. **Right action**: avoiding immoral behavior such as killing, stealing, and sensual pleasures
	5. **Right livelihood**: avoiding any occupation that might harm other beings
Meditation (*samadhi*)	6. **Right effort**: eliminating evil impulses, controlling thoughts, and nurturing a positive mental attitude
	7. **Right mindfulness**: cultivating constant awareness and seeing the laws of nature such as interdependence
	8. **Right meditation**: developing deep levels of mental calm and inner peace as well as avoiding any distractions from the pursuit of the Dhamma and nibbana

relevant to nature to the degree that it is correlated with the pivotal principle of extending nonviolence, compassion, and loving-kindness to all beings and things. Indeed, Skolimowski (1990: 29) writes that only after he first published his book, *Eco-Philosophy: Designing New Tactics for Living*, in 1981 did he realize that embedded within it was the Noble Path. Right livelihood, for example, would include any occupations or activities which do not harm any beings and things. Of course, it is impossible to live without causing some harm; even vegetarians depend on plants for food. However, harm can be minimized as much as possible.

Buddhism makes the fundamental distinction between need and greed. The Middle Way would modestly satisfy the four basic needs that the Buddha recognized – food, medicine, clothing, and shelter – and thereby minimize harm to other beings and reduce pressure on the environment including pollution that accompanies resource exploitation. Voluntary simplicity is an important ingredient for developing a sustainable and green society. Ideally Buddhists should be concerned with the contemplation of nature instead of its consumption. Thus, Buddhism aims at the extinction of egotism and selfishness.

Impermanence (*anicca*) is one of the central concepts in Buddhism. Indeed, the Buddha's last words were, "Decay is inherent in all things: be sure to strive with clarity of mind (for nirvana)" (Keown, 1996: 30). The Buddha viewed everything as a constantly changing collection of aggregates. Thus, a human is simply a temporary manifestation of natural processes. Ultimately, from its conception a human life is only a temporary compromise with death and decay.

The Buddha realized that the notion that an individual exists as an isolated or independent entity is an illusion. All beings and things are interrelated, interconnected, and interdependent, the actions of each affecting others. Interdependence (*paticcasamuppada*) refers to the principle that everything conditions and is conditioned by others. Nhat Hanh (1987: 45–46) illustrates this. "If you are a poet, you will see that there is a cloud in this sheet of paper. Without a cloud there will be no water; without water, the trees cannot grow; and without trees, you can not make paper." All of nature is a complex system of conditions, causes, and effects. Kaza (2000: 168) observes, "When one sees one's self as part of a mutually causal web, it becomes obvious that there is no such thing as an action without effect."

Ideally, Buddhists pursue liberation from suffering in life by studying nature, living a virtuous life, and aspiring to oneness with their natural environment. In this non-dualistic thinking humans and nature are a single reality. Humans are an integral part of nature, rather than apart from it, a realization fundamental to some views of ecology and environmentalism. Thus, as Kaza (1990: 25) says, "An environmental ethic is not something we apply outside ourselves; there is no outside ourselves. We are the environment, and it is [sic] us."

The unity of human and nature, as well as the unity of mind and body, is realized through meditation. Kraft (1994: 165) says,

> Meditation can serve as a vehicle for advancing several ends prized by environmentalists: it is supposed to reduce egoism, deepen appreciation of one's surroundings, foster empathy with other beings, clarify intention, prevent what is now called burnout, and ultimately lead to a profound sense of oneness with the entire universe.

For example, an environmental meditation from Nhat Hanh (1990: 195) could be considered whenever one uses water.

> Water flows from the high mountains.
> Water runs deep in the Earth.
> Miraculously, water comes to us,
> and sustains all life.

The key to Buddhist practice and ethics is the primacy of the individual's mind (Saddhatissa, 1970: 28). Ultimately this has profound importance and numerous ramifications in terms of recognizing the causes and solutions of environmental problems. Although important, government, science, technology, industry, and/or business are neither the cause nor the solution. Instead, the cause and the solution are found in the collective and accumulative consequences of the behavior of the individuals who compose humanity, although some may be more responsible (or irresponsible) than others. Environmental health is then not so much a matter of the scientific and technological management of land, resources, and pollution as it is the spiritual management of ourselves. People must be informed and motivated to limit reproduction, consumption, and waste in order to minimize harm to other beings and things (Gross, 1997).

Buddhists are supposed to pursue the moral life by following the Dhamma. By far the most important are the first negative precept – to abstain from killing, or, more broadly, to do no harm to any living being (ahimsa); and the first positive precept – compassion (karuna) and loving-kindness (metta) toward all life (Saddhatissa, 1970). Nonviolence is the minimal moral requirement for all monks and laypersons, as regularly stressed in the early texts of Buddhism.

In Buddhism sentience refers to the capacity of something to be aware of its surroundings and to act with intention. However, there is disagreement on what is sentient. In ancient India, animals, plants, seeds, water, and earth were all considered sentient (Schmithausen, 1991a: 5–7). In other times and places, Buddhists have limited the attribute of sentience to animals. Today some Western researchers claim that plants may respond to external physical stimuli, including even human influences like conversation and music. One reason for specifying sentient beings is that if they feel pain and suffering as well as cling to life and are afraid of death, then nonviolence and universal compassion and loving-kindness must be applied to them as well. Some would extend this principle beyond animals to plants, and some even to rocks and other things, particularly in Mahayana.[8]

Mahayana emphasizes dedicating one's life to the wellbeing of the world. A *bodhisatta* is someone who, because of his/her great compassion, forgoes nibbana and takes a vow to work through unlimited lifetimes to lead others to nibbana. Compassion and loving-kindness is extended to all beings in recognition of their suffering and to try to reduce or eliminate it (Keown, 1996: 60–61).

Buddha-nature in Mahayana refers to the capacity of all beings to become enlightened (Eckel, 1997: 332). Buddha-nature is extended to all beings and things in Chinese and Japanese Buddhism, including trees and rocks. Furthermore, it is assumed that all beings strive for and are eventually destined to achieve enlightenment (see King, 1991 and La Fleur, 2000).

Buddhism emphasizes the intrinsic value of humans and nature. In his classic book, *Buddhist Ethics*, Saddhatissa (1970: 88) states, "A Buddhist does not sacrifice living beings for worship or food, but sacrifices instead his own selfish motives." Regarding the Buddha and his followers, Kabilsingh (1998: 59) writes, "Their attitude to forests and trees cannot be interpreted otherwise than as an appreciation of their spiritual worth and the desire by believers to conserve them."

If it is understood that the first negative precept and the first positive precept apply to *all* beings, not just to humanity or some sector thereof, then the ecological implications are immediately obvious and undeniable. There is an eco-logic here: it is possible to detect elements of ecology in Buddhism, and elements of Buddhism in ecology, although obviously one cannot call ancient Buddhists ecologists or environmentalists as we define them today. Both Buddhism and ecology pursue a monistic rather than dualistic worldview. Instead of dichotomizing organism and environment or human and nature, they consider that all life is subject to natural laws. They adopt holistic and systems approaches regarding the unity, interrelatedness, and interdependence of the components of nature and teach respect and even reverence for nature, including the intrinsic as well as extrinsic values of other beings (Natadecha-Sponsel, 1988; Robinson, 1972).

As Kraft (1994: 164) recognizes, Buddhism has the potential to transform not only the individual, but also the individual's relationship with all other beings and things, and many are realizing that only such a radical change of human-environment relationships will resolve the environmental crisis and avert catastrophe. Kaza (2000: 174–175) writes,

> At the heart of the Buddha's path is reflective inquiry into the nature of reality. Applying this practice in today's environmental context, eco-activists undertake rigorous examination of conditioned beliefs and thought patterns regarding the natural world. This may involve deconstructing the objectification of plants and animals, the stereotyping of environmentalists, dualistic thinking of enemy-ism, the impacts of materialism, and environmental racism.

Such considerations have led some to characterize Buddhism as a religious ecology (Batchelor and Brown, 1992: viii).

Bodhi (1987: vii) summarizes much of the foregoing on the relevance of Dhamma to nature:

> With its philosophical insight into the interconnectedness and thoroughgoing interdependence of all conditioned things, with its thesis that happiness is to be found through the restraint of desire in a life of contentment rather than through the proliferation of desire, with its goal of enlightenment through renunciation and contemplation and its ethic of non-injury and boundless loving-kindness for all beings, Buddhism provides all the essential elements for a relationship to the natural world characterized by respect, care, and compassion.

SANGHA

The Buddha provided a framework for the relationship between humans and their environment as well as spiritual inspiration, and so do monks through teaching and by example. For instance, monks lead a simple life, and this reminds laity to avoid overindulgence, or materialism and consumerism, which are, of course, major causes of the ecocrisis.

Many of the more than 200 rules for monks (*vinaya*) are relevant to nature. Several aim to prevent monks from knowingly harming any living being. They can only eat fruit without seeds or if already damaged by someone else. *Bhutagama*, the Pali term for a living plant, means the home of a being. For a monk to intentionally cut, burn, or kill a living plant is an offense. It might endanger some animal's habitat, among other things. Harming even small animals like ants should also be avoided, although harming large animals is much more serious. Harming an organism like a worm by digging in the ground is forbidden and could lead to expulsion from the sangha. Also a monk can not deliberately have someone else kill an animal for him. Monks refrain from traveling during the rainy season to avoid trampling on small animals and young crops that are abundant then (Rains Retreat or *pansa*). Monks are supposed to check water or strain it before using it for drinking or other purposes in order to avoid knowingly harming any visible organisms in it, even mosquito larvae. They are prohibited from polluting water in any way as well. A monk is even prohibited from self-defense if it injures another being. In these and many other ways monks are supposed to protect and respect all life (Thanissaro, 1994: 294–300, 317–319, 420–424).

Many if not most of the monks and nuns who followed the Buddha spent much of their time in parks and forests rather than in a town, temple, or monastery. They meditated at the base of a large tree, a tradition of holy people that long antedated Buddhism which was practiced and advocated by the Buddha as well.[9]

There have been forest monks in Thailand since the first kingdom, Sukhothai, in the middle of the 13th century CE (Kabilsingh, 1998: 136). In recent times in Thailand an exemplar of a forest monk is Buddhadasa Bhikku (*Phutthathat* in Thai) (1906–1993), a very influential Thai monk, whose name means "Servant of the Buddha". This was a fitting name since Buddhadasa consistently focused on the basic principles of Buddhism. He felt that the practice of Buddhism in Thailand had diverged from this focus, and his "radical conservatism" emphasized a return to fundamentals.

Buddhadasa integrated some of the different concerns and practices of two types of monks. Forest monks are usually solitary and wandering followers of the Dhamma, emphasizing meditation. City monks live in communities in a temple or monastery in or adjacent to settlements where they provide services to laity such as conducting ceremonies and/or pursue scholarship (e.g., Lewis, 1997: 324). Buddhadasa sought to integrate practice, service, and study as a forest monk in a balanced middle way (Santikaro, 1996: 153).

As a child, Buddhadasa was strongly influenced by nature while grazing cattle for his father and while collecting herbs from the forest for the local temple. Later, as a monk, he continued to study animals and plants to gain insight into the Dhamma (Santikaro, 1996: 149). Buddhadasa pursued monastic studies in Bangkok, but soon returned home to establish a monastery in a natural setting that he believed would be more conducive to practicing the Dhamma (Santikaro, 1996: 152). The forest monastery he established in 1932 is called Suan Mokkh (The Garden of Liberation). Suan Mokkh is about 33 miles from Surat Thani in southern Thailand. It occupies about 120 acres of forest. Individuals and groups are encouraged to meditate, study, and discuss Buddhism among the trees whenever weather permits (Santikaro, 2001: 3).

When the present authors interviewed Buddhadasa at Suan Mokkh in 1986, he mentioned the connections between the behavior of people and environmental changes in the case of local deforestation with agricultural expansion and the decline of water resources. Only Suan Mokkh, by virtue of being a monastery, remained an island of forest in a sea of farmland. Buddhadasa also told us that he thought that only when people realized the negative consequences of some of their behavior on the environment, and their own suffering as a result, would they eventually change for the better. Beyond ignorance and greed Buddhadasa also attributed environmental deterioration to weakening adherence to Buddhism and related Thai traditions with Westernization. Buddhadasa strongly influenced many other monks who became activists in environmental conservation and sustainable economic and community development, as well as prominent lay individuals who became engaged Buddhists, including Sulak Sivaraksa and Chatsumarn Kabilsingh.[10]

Other forest monks have become environmental activists. They view deforestation as sacrilegious, a threat to the forest monk tradition, since the forest is their sacred habitat. They understand the suffering that deforestation causes to local humans and other beings (e.g., Usher, 1994). Out of compassion for the suffering which diverse beings experience as a result of deforestation, numerous monks have initiated environmental education programs, sustainable economic development projects, and rituals to protect remaining forests (Darlington 1998, 2000). [Editor's note: See Darlington's article on Conservation in Buddhist Thailand in this volume.]

Naturalists Graham and Round (1994: 71), who worked for many years in Thailand, recognize monks as custodians of nature. We argue three propositions concerning the environmental relevance of the sangha. First, ideally a monastic community in any religion exhibits attributes that are similar, and in some instances even identical, to many of the characteristics of a green society. Among these is a small-scale community based on nonviolence, moderation, cooperation, and reciprocity in satisfying basic needs. Second, monks have extraordinary status and power to transform Thailand into a more ecologically appropriate society because of their unique social and moral roles. By contrast monks hold a mirror to society, for example, given their vow of poverty. Spiritual development is their goal, rather than Western style economic development through materialism and consumerism. Third, by drawing on the ecological wisdom of the Dhamma, the sangha has significant potential to contribute to the environmental ethics and education of the populace and, as a consequence, to help create a greener society (Sponsel and Natadecha-Sponsel, 1993, 1995, 1997).

There is another aspect of the sangha that deserves mention. Together with two Thai biologists we argue that collectively and accumulatively over time temples and monasteries may help promote both the *ex situ* and *in situ* conservation of biodiversity in Thailand (Sponsel, *et al.*, 1998). Many temple complexes can be viewed as sacred ecosystems with groves of bodhi, banyan, and other trees and associated animals. People are prohibited from disturbing plants and animals in and near a temple complex. Planting trees, especially in a temple yard or its vicinity, is one way to gain merit as advocated by the Buddha.

During our field research in southern Thailand we found small but healthy forests associated with some temples, entire mountain forests protected by temples or shrines, sections of community forests donated by villagers to their local temple to conserve them for future generations, and areas of forest restoration associated with initiatives of monks. We have hypothesized that particular temples may serve one or more of the following conservation functions: forest reserves, botanical gardens, germplasm banks, medicinal plant collections, zoological gardens, wildlife sanctuaries, restoration ecology, model of a green society, environmental education, and environmental action (Sponsel, et al., 1998).

LAITY

To the extent that the laity follows the Buddha's example and teachings and those of the sangha, they are also relevant to nature. The laity and the sangha are interdependent, the laity providing for the material needs of the monks and the monks providing for the spiritual needs of the laity. However, whereas monks can not harm other beings, the laity must do so to obtain food and income from farming, fishing, and so on. Ryan (1998: 102) asserts that this mutuality between sangha and laity can also serve as a model for the mutuality between nature and humans.

For Buddhists the ultimate pilgrimage is the spiritual journey through meditation to discover the Buddha-nature within oneself (Barber, 1991: 103), a vision quest that reveals the oneness of the individual and nature. However, since ancient times Buddhists have also gone on pilgrimages to sacred sites associated with the Buddha and other Buddhist dignitaries, as well as to temples, shrines, and other sacred places, many on mountains and/or in forests (Barber, 1991: 105). Many of the natural sacred places associated with Buddhism involve certain prescriptions and proscriptions such as not killing any animals (Barber, 1991: 115). Thus, in effect, even if inadvertently, these sites often function as wildlife sanctuaries.

In several places in the *Tipitaka* when the issue of vegetarianism arose the Buddha explicitly refused to prohibit monks from eating meat, mainly because they subsist on alms food volunteered by local laity. Vegetarianism was an option, and today, although widely admired, it is not common in Theravada countries and Japan. However, most monasteries in China and Korea, and Zen monks, are strictly vegetarian. Vegetarianism is one way to reduce the human impact on the environment (see Kapleau, 1982).

Engaged Buddhism belies the myth or stereotype that Buddhism is necessarily a detached or escapist religion. A phrase coined by Nhat Hanh, engaged Buddhism is simply the active application of Buddhism for the benefit of others; that is, putting compassion, loving-kindness, and other principles into practice. This is simply a matter of emulating the Buddha's life and following his teachings. Today, among the concerns of engaged Buddhists are nonviolence, peace, human rights including women's and workers' rights, participatory democracy, social work, alleviating poverty, appropriate technology such as organic farming and recycling, sustainable economic development, community development, and environmentalism. The more prominent engaged Buddhists today include the Dalai Lama, Sulak Sivaraksa, and Thich Nhat Hanh. The Buddhist Peace Fellowship and the International Network of Engaged Buddhists are among the more prominent organizations dedicated to engaged Buddhism.[11]

An outstanding example of engaged Buddhism focused on environmentalism is the Buddhist Perception of Nature Project. It aims to enhance awareness, attitudes, and actions regarding nature through applying Buddhism to develop a new perspective for environmental education and ethics. This was also the first attempt to compile the environmentally relevant literature of Buddhism (Davies, 1987; Kabilsingh, 1987). In turn such projects provided a stimulus and model for similar initiatives in neighboring Cambodia (Buddhist Institute, 1999).[12]

WEST

Buddhism originated in northern India and spread into other parts of Asia millennia ago. By the mid-19th century Buddhism started to spread into North America, Europe, and beyond. Now it is becoming global as evidenced, for example, by Buddhist contributions to the development of the United Nations Earth Charter (Morgante, 1997).

Buddhists comprise only 6% of humanity. There are at least 353,794,000 Buddhists worldwide in 86 countries, although more than 95% live in Asia (*Encyclopedia Britannica*, 1999). It is estimated that there are 3–4 million Buddhists in the United States, 1,139,100 in Europe, 1 million in Russia, 140,000 in Australia, and 5,000 in South Africa (Baumann, 1997: 2–3).

East and West are not completely separate. For centuries they have influenced each other in numerous and sometimes profound ways. For instance, Henry David Thoreau, a student of Asian philosophy as well an icon for environmentalists, first translated into English from the French the Lotus Sutta, and his life at Walden Pond was somewhat reminiscent of a forest monk. In the other direction, American Henry Steel Olcott and Ukrainian Helena Petrovna Blavatsky helped revive Buddhism in Ceylon in the late 19th century after it had declined under waves of European colonialism.

PROBLEMS

The problems and limitations of the relevance of Buddhism and Buddhists to nature appear to be basically an issue of ideal vs. actual behavior. The ideals of Buddhism seem to be environmentally friendly, yet in practice Buddhists are often environmentally unfriendly. This is evidenced in natural resource depletion and environmental degradation in countries that are predominantly Buddhist. Nevertheless, this does not invalidate the idea that Buddhism is environmentally friendly, at least in principle. One must be careful to avoid confusing Buddhism and Buddhists. There are internal contradictions and discrepancies in every religion. An important part of the solution is to educate people about the negative consequences of their behavior so that they can minimize them. Also Buddhists need to adhere more closely to the Noble Path as it applies to ecology and environmentalism.[13]

By far the strongest critic of the relevance of Buddhism to nature is Ian Harris (1991, 1994, 1995a,b, 1997, 2000). The gist of his critique of what he terms ecoBuddhism is that: first, this is a recent American imposition on Buddhism and not an authentic interpretation of the texts; second, in Buddhism there is neither a concept of nature nor of the inherent value of other beings comparable to the West; and, third, Buddhism is escapist and not concerned with practical problems like the environment (cf. Suzuki, 1956).

Actually, Buddhist ecology and environmentalism have much broader geographical, historical, cultural, and national foundations than Harris allows. Whereas Harris sees modern Western environmentalism as an imposition on Buddhism, we argue that Buddhism contributed significantly to the development of Western environmentalism in the first place.

As should be clear from the tremendous variation and variability in Buddhism, it is problematic to pronounce one thing authentic and another not. Puritanical, literalist, or fundamentalist criticisms of the application of Buddhism to contemporary concerns do not afford sufficient attention to the actual practice of Buddhism by Buddhists where text and rigid doctrine are not considered so important (Reynolds and Carbine, 2000). Moreover, the Buddha repeatedly emphasized that individuals should test his teaching against their own reason and experience, instead of blindly accepting authority, tradition, or dogma. If a practicing Buddhist, such as a monk of many years who also happens to be a Buddhist scholar, as are Buddhadasa, Nhat Hanh, and the Dalai Lama, thinks that Buddhism has significant environmental relevance, then it does. (These individuals represent each of the three major schools of Buddhism).

Harris (1997: 378) accuses those who espouse the environmental relevance of Buddhism of being selective. However, Swearer (1997: 39) replies, "His position is founded on too narrow a construction of the Buddhist view of nature and animals based on selective reading of particular texts and traditions." Kraft (1994: 175) observes, "It is clear that ecologically sensitive Buddhism exhibits significant continuities with traditional Buddhism, continuities that can be demonstrated textually, doctrinally, historically, and by other means." Also, contrary to Harris, some Asian scholars, like Kabilsingh (1998), have identified numerous textual sources that are relevant to contemporary environmentalism.

Beyond skeptics who question the relevance of Buddhism and nature, there are many other problems. For instance, Kraft (1997) identifies these dilemmas facing Buddhist environmental activists:

- There are gaps between the Buddha's teaching and contemporary sociopolitical realities.
- Buddhism emphasizes individual morality and action, but many environmental problems demand collective responsibility and action.
- There is competition between meditation and action.
- Is it better to be identified as Buddhist or just blend in with other activists?
- Are spiritual practices and rituals environmentally effective?

At the same time, in our opinion, as long as Buddhists are faithful to the core principles of the Dhamma, then compassionate actions on behalf of all beings are still Buddhist practices no matter how innovative.[14]

Ultimately perhaps the strongest counter to any criticisms or backlash against Buddhist ecology and environmentalism is the mere fact that, as this essay and its bibliography reveal, there is tremendous concern and action on behalf of nature by many Buddhist individuals and organizations throughout the world, and momentum has been growing markedly in the 1990s.

FUTURE

Kaza (2000: 177–178) wonders if Buddhist environmentalism may be overwhelmed by a general backlash against environmentalism, resistance to engaged

Buddhism, or environmental catastrophes with consequent economic and social collapse. If the worst scenarios do not transpire, then Buddhists in the West, by virtue of being only a fraction of the population, may not have much impact (Kraft, 1994: 178). However, we think that this underestimates the potential force of Buddhist ideas themselves, which may be far greater than that imagined from mere numbers of practitioners.

It is unlikely that people millennia or even centuries ago ever conceived of possibilities like anthropogenic mass species extinction or global warming. However, inevitably new problems and issues arise that require new interpretations and approaches by Buddhists, and this must have happened throughout the 2,500 years of the history of Buddhism, considering the diversity of contexts in which it flourished (Lopez, 2001). That Buddhism has endured for so long is a clear demonstration of its continuing adaptability, responsiveness, and relevance. Now the most important challenge for Buddhists is to cultivate social, political, and environmental actions that more closely approximate the ideals of Buddhism, especially the Noble Path and basic precepts.

In a benchmark symposium of the Siam Society in 1987, an underlying theme was the connection between the environmental crisis in Thailand and decline of adherence to Buddhism with Westernization. In the conclusion, Kunstadter (1989: 548–550) records that the participants, mostly Thai, subscribed to a "theory of a moral collapse" as the cause of growing ecological disequilibrium.

If more people were aware of the negative environmental consequences which are the collective and cumulative results of their behavior, then many might change for the better and thereby improve the environmental situation. When reason is combined with adequate knowledge of the ecocrisis, then this may lead to wisdom in improving how humans relate to nature.

Undoubtedly the primary concern of the Buddha was suffering, its causes and alleviation. The world is bound to suffer even more during the 21st century as a result of population and economic growth. Already humanity consumes about 40% of the gross biological production on the planet. Yet the world population will nearly double within the next 50 years, 95% of the growth in the poorest countries (Gelbard, 2001; Vitousek, et al., 1986). This alone will place tremendous pressure on the land and resource base of societies, and in turn on nature (Sponsel, 2001b). In this case, Buddhism may well prove even more relevant. Buddhists must apply critical and radical thought in examining contemporary problems and issues, something needed in our time as much as in the time of the Buddha and probably more so (see Ryan, 1998: 24–25). Being a Buddhist remains a geopolitical act (Timmerman, 1992: 66, 75).[15]

The task ahead is to explore deeper and wider into the relevance of Buddhism, especially in each of the different schools and sects in both text and context, and for those who are practicing Buddhists to apply the Dhamma as much as possible. Buddhists owe no less to the Buddha, the sangha, other beings, and themselves. It is a matter of survival for Buddhism as well as nature and humanity.

NOTES

[1] In this essay all Buddhist terms are in Pali, unless in quotes or personal or place names. Pali is the ancient literary language for Theravada Buddhism and remains its ecclesiastical language. For introductions to Buddhism read Keown, 1996 or Lopez, 2001; for reference works consult Fischer-Schreiber, *et al.*, 1991, Powers, 2000, and Snelling, 1998.

[2] For early statements on Buddhism and nature see Bloom, 1970; Barash, 1973; Robinson, 1972; de Silva, 1987; and Suzuki, 1956. More recent explorations include Eckel, 1997; Kabilsingh, 1998; Ryan, 1998; and Schmithausen, 1997. Since the late 1980s there has been a flurry of books on Buddhist ecology and environmentalism. Monographs include Habito, 1993; Kabilsingh, 1998; Macy, 1991; Nisker, 1998; Ryan, 1998; Schmithausen, 1991a,b; de Silva, 1998; and Titmuss, 1995. Edited books include Badiner, 1990; Batchelor and Brown, 1992; Kaza and Kraft, 2000; Martin, 1997; Sandell, 1987; and Tucker and Williams, 1997. Sadakata, 1997 and Schmithausen, 1991a most directly address Buddhist views of nature. This is part of a broader movement called spiritual ecology (Sponsel, 2001a).

[3] For biographies of the Buddha see Nelson, 1996 and Saddhatissa, 1998.

[4] See the *Aggana Sutta* in Walshe, 1995:407–415; Ryan, 1998:77–93, 95–103; and Sadakata, 1997.

[5] For an inventory of Buddhist scriptures see Mills, 1999: 187–192, and for a selection of the scriptures Conze, 1959. Also see Bodhi, 1995; Khantipalo, 1966; Stryk, 1968; Walshe, 1995; and Watson, 1993.

[6] On Buddhism and animals see Chapple, 1996, 1997; Cowell, 1905; and Rhys Davids, 1989.

[7] Kabilsingh, 1998; Ryan, 1998; and Schmithausen, 1991a,b, 1997 are especially useful analyses of the early Buddhist literature on Buddhism and nature. Also see Brown, 1992: 99; Nyanasobhano, 1998; Sadakata, 1997; and Suzuki, 1956: 250, 255. See Sadakata, 1997, especially, for more detail.

[8] See Ford, 1999 on communication in plants and animals; Schmithausen, 1991b on sentience of plants; and Kaza, 1996.

[9] On forest monks see Darlington, 1998, 2000; Taylor, 1993; Tiyavanich, 1997.

[10] For more on Buddhadasa see Buddhadasa, 1994, 1998; Jackson, 1988; Sivaraksa *et al.*, 1990; and Swearer, 1997.

[11] On engaged Buddhism see Queen, 2000, and Queen and King, 1996.

[12] For another area, Ladakh, see Norberg-Hodge, 1991, 2000.

[13] Callicott and Ames, 1989: 279–289; Tuan, 1968; and Ward, 1993 discuss discrepancies.

[14] Among other critics are Eckel, 1997 and Schmithaussen, 1991a,b, 1997.

[15] Metzner, 1999; Sivaraksa, 1992; 1999; Sivaraksa *et al.*, 1993; Sponsel and Natadecha-Sponsel, 2000; and Titmuss, 1995.

BIBLIOGRAPHY

Allendorf, Fred W. and Bruce A. Byers. 'Salmon in the net of Indra: a Buddhist view of nature and communities.' *Worldviews: Environment, Culture, Religion* 2(1): 37–52.

Altman, Nathaniel. *Sacred Trees*. San Francisco: Sierra Club Books, 1994.

Badiner, Allan Hunt, ed. *Dharma Gaia: A Harvest of Essays in Buddhism and Ecology*. Berkeley: Parallax Press, 1990.

Barash, David P. 'The ecologist as Zen master.' *American Midland Naturalist* 89: 214–217, 1973.

Barber, Richard. *Pilgrimages*. Woodbridge, Suffolk: Boydell and Brewer, Ltd., 1991.

Batchelor, Stephen. *The Awakening of the West*. London: Aquarian, 1994.

Batchelor, Martine and Kerry Brown, eds. *Buddhism and Ecology*. London: Cassell Publishers, 1992.

Bauman, Martin. 'The Dharma has come West: a survey of recent studies and sources.' *Journal of Buddhist Ethics* 4: 1–10, 1997.

Bloom, Alfred. 'Buddhism, nature and the environment.' *The Eastern Buddhist* 5(1): 115–129, 1972.

Bodhi, Bhikkhu. 'Foreword.' In *Buddhist Perspectives on the Ecocrsis*, Klas Sandell, ed. Kandy, Sri Lanka: Buddhist Publications Society, 1987, pp. v–viii.

Bodhi, Bihikkhu. *The Middle Length Discourses of the Buddha: A New Translation of the Majjhima Nikaya*. Boston: Wisdom Publications, 1995.

Brown, Kerry. 'In the water there were fish and the field were full of rice: reawakening the lost harmony of Thailand.' In *Buddhism and Ecology*, Martine Batchelor and Kerry Brown, ed., 1992, pp. 87–99.

Buddhadasa Bhikkhu. *Heartwood of the Bodhi Tree*. Boston: Wisdom Publications, 1994.

Buddhadasa Bhikkhu. 'Conserving the inner ecology.' *Tricycle* 8(2): 73–75, 1998.

Buddhist Institute. *Cry from the Forest: A Buddhism and Ecology Community Learning Tool*. Phnom Penh: Buddhist Institute, 1999.

Calkowski, Marcia. 'Buddhism.' In *Religion and Culture: An Anthropological Focus*, Raymond Scupin, ed. Upper Saddle River, New Jersey: Prentice Hall, 2000, pp. 249–274.

Callicott, J. Baird, and Roger T. Ames, eds. *Nature in Asian Traditions of Thought: Essays in Environmental Philosophy*. Albany: State University of New York Press, 1989.

Chapple, Christopher Key. *Nonviolence to Animals, Earth, and Self in Asian Traditions*. Albany: State University of New York Press, 1996.

Chapple, Christopher Key. 'Animals and environment in the Buddhist birth stories.' In *Buddhism and Ecology: The Interconnection of Dharma and Deeds*, Mary Evelyn Tucker and Duncan Ryūken Williams, eds. Cambridge, Massachusetts: Harvard University Press, 1997, pp. 131–148.

Clarke, J.J. *Oriental Enlightenment: The Encounter Between Asian and Western Thought*. New York: Routledge, 1997.

Conze, Edward. *Buddhist Scriptures*. New York: Penguin, 1959.

Cowell, E.B., ed. *Jataka Stories*. London: Pali Text Society, volumes I–VI, 1895–1905.

Darlington, Susan Marie. 'The ordination of a tree: the Buddhist ecology movement in Thailand.' *Ethnology* 37(1): 1–15, 1998.

Darlington, Susan Marie. 'Rethinking Buddhism and development: the emergence of environmental monks in Thailand.' *Journal of Buddhist Ethics* 7: 1–14, 2000.

Davies, Shann, ed. *Tree of Life: Buddhism and Protection of Nature*. Geneva, Switzerland: Buddhist Perception of Nature, 1987.

Dhamma, Rewata. *The First Discourse of the Buddha*. Boston: Wisdom Publications, 1997.

Eckel, Malcolm David. 'Is there a Buddhist philosophy of nature?' In *Buddhism and Ecology: The Interconnection of Dharma and Deeds*, Mary Evelyn Tucker and Duncan Ryūken Williams, eds. Cambridge, Massachusetts: Harvard University Press, 1997, pp. 327–349.

Encyclopedia Britannica Book of the Year. Chicago: Encyclopedia Britannica, 1999.

Erdosy, George. 'Deforestation in pre- and protohistoric South Asia.' In *Nature and the Orient: The Environmental History of South and Southeast Asia*. Delhi: Oxford University Press, pp. 51–69.

Fischer-Schreiber, Franz-Karl Ehrhard, and Michael S. Diener. *The Shambhala Dictionary of Buddhism and Zen*. Boston: Shambhala, 1991.

Ford, Brian J. *Sensitive Souls: Senses and Communication in Plants, Animals and Microbes*. London: Little, Brown, and Company, 1999.

Gadgil, Madhav and Ramachandra Guha. *This Fissured Land: An Ecological History of India*. Berkeley: University of California Press, 1992.

Gelbard, Alene. 'Human population stabilization.' In *Encyclopedia of Biodiversity*, Simon Asher Levin, ed. San Diego, California: Academic Press 2001, vol. 4, pp. 799–810.

Graham, Mark, and Philip Round. *Thailand's Vanishing Flora and Fauna*. Bangkok: Finance One Public, 1994.

Gross, Rita M. 'Buddhist resources for issues of population, consumption, and the environment.' In *Buddhism and Ecology: The Interconnection of Dharma and Deeds*, Mary Evelyn Tucker and Duncan Ryūken Williams, eds. Cambridge, Massachusetts: Harvard University Press, 1997, pp. 291–311.

Habito, Ruben L.F. *Healing Breath: Zen Spirituality for a Wounded Earth*. Maryknoll, New York: Orbis Books, 1996.

Harris, Ian. 'How environmentalist is Buddhism?' *Religion* 21: 101–114, 1991.

Harris, Ian. 'Buddhism.' In *Attitudes to Nature*, Jean Holm and John Bowker, eds. New York: Pinter Publishers, 1994, pp. 8–27.

Harris, Ian. 'Buddhist environmental ethics and detraditionalization: the case of ecoBuddhism.' *Religion* 25: 199–211, 1995a.

Harris, Ian. 'Getting to grips with Buddhist environmentalism: a provisional typology.' *Journal of Buddhist Ethics* 2: 173–190, 1995b.

Harris, Ian. 'Buddhism and the discourse of environmental concern: some methodological problems.' In *Buddhism and Ecology: The Interconnection of Dharma and Deeds*, Mary Evelyn Tucker and Duncan Ryūken Williams, eds. Cambridge, Massachusetts: Harvard University Press, 1997, pp. 377–402.

Harris, Ian. 'Buddhism and ecology.' In *Contemporary Buddhist Ethics*, Damien Keown, ed. London: Curzon Press, 2000, pp. 113–116.

Holy Places of the Buddha. Berkeley: Dharma Publishing Co., 1994.

Jackson, Peter. *Buddhadasa: A Buddhist Thinker for the Modern World*. Bangkok: The Siam Society, 1988.

Johnson, Wendy and Stepahnie Kaza. 'Earth Day at Green Gulch.' *Journal of the Buddhist Peace Fellowship*, Summer 1990: 30–33.

Kabilsingh, Chatsumarn. *A Cry from the Forest: Buddhist Perception of Nature, A New Perspective for Conservation Education*. Bangkok: Wildlife Fund Thailand, 1987.

Kabilsingh, Chatsumarn. 'Early Buddhist views on nature.' In *Dharma Gaia: A Harvest of Essays in Buddhism and Ecology*, Allan Hunt Badiner, ed. Berkeley: Parallax Press, 1990, pp. 8–13.

Kabilsingh, Chatsumarn. *Buddhism and Nature Conservation*. Bangkok: Thammasat University Press, 1998.

Kapleau, Philip. *To Cherish All Life: A Buddhist Case for Becoming Vegetarian*. San Francisco: Harper and Row, 1982.

Kaza, Stephanie. 'Towards a Buddhist environmental ethic.' *Buddhism at the Crossroads*, Fall 1990: 22–25.

Kaza, Stephanie. *The Attentive Heart: Conversations with Trees*. Boston: Shambhala Press, 1996.

Kaza, Stephanie. 'American Buddhist response to the land: ecological practice at two West Coast retreat centers.' In *Buddhism and Ecology: The Interconnection of Dharma and Deeds*, Mary Evelyn Tucker and Duncan Ryūken Williams, eds. Cambridge, Massachusetts: Harvard University Press, 1997, pp. 219–248.

Kaza, Stephanie. 'To save all beings: Buddhist environmental activism.' In *Engaged Buddhism in the West*, Christopher S. Queen, ed. Boston: Wisdom Publications, 2000, pp. 159–183.

Kaza, Stephanie, and Kenneth Kraft, eds. *Dharma Rain: Sources of Buddhist Environmentalism*. Boston: Shambhala Publications, 2000.

Keown, Damien. *Buddhism: A Very Short Introduction*. New York: Oxford University Press, 1996.

Ketudat, Sippanondha, and Robert B. Textor. *The Middle Path for the Future of Thailand: Technology in Harmony with Culture and Environment*. Honolulu: East-West Center and Chiang Mai, Thailand: Chiang Mai University, 1990.

Khantipalo, Bhikkhu. *Dhammapada: Growing the Bodhi Tree*. Bangkok: Buddhist Association of Thailand, 1966.

King, Sallie B. *Buddha Nature*. Albany: State University of New York Press, 1991.

Kraft, Kenneth. 'The greening of Buddhist practice.' *Cross Currents* 44(2): 163–179, 1994.

Kraft, Kenneth. 'Nuclear ecology as engaged Buddhism.' In *Buddhism and Ecology: The Interconnection of Dharma and Deeds*, Mary Evelyn Tucker and Duncan Ryūken Williams, eds. Cambridge, Massachusetts: Harvard University Press, 1997, pp. 269–290.

Kunstadter, Peter. 'The end of the frontier: culture and environment interactions in Thailand.' In *Culture and Environment in Thailand*, Siam Society, ed. Bangkok: Siam Society, 1989, pp. 543–552.

LaFleur, William R. 'Enlightenment for plants and trees.' In *Dharma Rain: Sources of Buddhist Environmentalism*, Stephanie Kaza and Kenneth Kraft, eds. Boston: Shambhala Publications, 2000, pp. 109–116.

Lewis, Todd T. 'Buddhist communities: historical precedents and ethnographic paradigms.' In *Anthropology of Religion: A Handbook*, Stephen D. Glazier, ed. Westport, Connecticut: Praeger, 1997, pp. 319–368.

Ling, Trevor. *The Buddha: Buddhist Civilization in India and Ceylon*. New York: Charles Scribner's Sons, 1973.

Loori, John Daido. 'The precepts and the environment.' In *Buddhism and Ecology: The*

Interconnection of Dharma and Deeds, Mary Evelyn Tucker and Duncan Ryūken Williams, eds. Cambridge, Massachusetts: Harvard University Press, 1997, pp. 177–184.

Lopez, Donald S., Jr. *The Story of Buddhism: A Concise Guide to its History and Teachings.* New York: Harper Collins Publishers, 2001.

Lorie, Peter, and Julie Foakes. *The Buddhist Directory.* Rutland, Vermont: Charles E. Tuttle Co., 1997.

Macy, Joanna. *World As Lover, World As Self.* Berkeley: Parallax Press, 1991.

Martin, Julia, ed. *Ecological Responsibility: A Dialogue With Buddhism.* Delhi: Tibet House, 1997.

Metzner, Ralph. *Green Psychology: Transforming our Relationship to the Earth.* Rochester: Park Street Press, 1999.

Mills, Laurence-Khantipalo. *Buddhism Explained.* Chiang Mai, Thailand: Silkworm Books, 1999.

Morgante, Amy. *Buddhist Perspectives on the Earth Charter.* Boston: Boston Research Center for the 21st Century, 1997. (Also see http://www.brc21.org/bec-toc.html).

Morreale, Don, ed. *The Complete Guide to Buddhist America.* Boston: Shambhala, 1998.

Natadecha-Sponsel, Poranee. 'Buddhist religion and scientific ecology as convergent perceptions of nature.' In *Essays on Perceiving Nature,* Diana MacIntyre Deluca, ed. Honolulu: University of Hawaii Perceiving Nature Conference Committee, 1988, pp. 113–118.

Nelson, Walter Henry. *Buddha: His Life and His Teaching.* New York: Penguin Putnam, 1996.

Nhat Hahn, Thich. *Being Peace.* Berkeley: Parallax Press, 1987.

Nhat Hahn, Thich. 'Earth Gathas.' In *Dharma Gaia: A Harvest of Essays in Buddhism and Ecology,* Allan Hunt Badiner, ed. Berkeley: Parallax Press, 1990, pp. 195–196.

Nisker, Wes. *Buddha's Nature: A Practical Guide to Discovering Your Place in the Cosmos.* New York: Bantam Books, 1998.

Norberg-Hodge, Helena. *Ancient Futures: Learning from Ladakh,* San Francisco: Sierra Club Books, 1991.

Norberg-Hodge, Helena. 'Economics, engagement, and exploitation in Ladakh.' *Tricycle* 10(2): 77–79, 114–117, 2000.

Bhikkhu Nyanasobhano. *Landscapes of Wonder: Discovering Buddhist Dhamma in the World Around Us.* Boston: Wisdom Publications, 1998.

Payutto, P.A. *Buddhist Economics: A Middle Way for the Market Place.* Bangkok: Buddhadhamma Foundation, 1994.

Powers, John. *A Concise Encyclopedia of Buddhism.* Oxford: One World Publications, 2000.

Queen, Christopher S., ed. *Engaged Buddhism in the West.* Boston: Wisdom Publications, 2000.

Queen, Christopher S. and Sallie B. King, eds. *Engaged Buddhism: Buddhist Liberation Movements in Asia.* Albany: State University of New York Press, 1996.

Phra Rajavarmuni. *Thai Buddhism in the Buddhist World: A Survey of the Buddhist Situation Against a Historical Background.* Bangkok: Mahachulalongkorn Buddhist University Wat Mahadhatu, 1984.

Reynolds, Frank E., and Jason A. Carbine, eds. *The Life of Buddhism.* Berkeley: University of California Press, 2000.

Rhys Davids, Caroline A.F. *The Stories of the Buddha: Being Selections from the Jataka.* New York: Dover Publications, 1989.

Robinson, Peter. 'Some thoughts on Buddhism and the ethics of ecology.' *Proceedings of the New Mexico-West Texas Philosophical Society* 7: 71–78, 1972.

Roscoe, Gerald. *The Triple Gem: An Introduction to Buddhism.* Chiang Mai, Thailand: Silkworm Books, 1994.

Ryan, P.D. *Buddhism and the Natural World.* Birmingham: Windhorse Publications, 1998.

Sadakata, Akira. *Buddhist Cosmology: Philosophy and Origins.* Tokyo: Kosei Publishing Co., 1997.

Saddhatissa, Hammalawa. *Buddhist Ethics.* New York: G. Braziller, 1970.

Saddhatissa, Hammalawa. *Before He Was Buddha: The Life of Siddhartha.* Berkeley: Ulysses Press/Seastone, 1998.

Sandell, Klas, ed. *Buddhist Perspectives on the Ecocrisis.* Kandy, Sri Lanka: Buddhist Publication Society, 1987.

Santikaro Bhikkhu. 'Buddhadasa Bhikkhu: life and society through the natural eyes of voidness.' In

Engaged Buddhism: Buddhist Liberation Movements in Asia, Christopher S. Queen and Sallie B. King, eds. Albany: State University of New York Press, 1996, pp. 147–193.

Santikaro Bhikkhu. *The Garden of Liberation*. http://ksc.goldsite.com/Suanmokkh/archive/garden1a.htm, pp. 1–15, 2001.

Schmithausen, Lambert. *Buddhism and Nature*. Tokyo: International Institute for Buddhist Studies, 1991a.

Schmithausen, Lambert. *The Problem of the Sentience of Plants in Earliest Buddhism*. Tokyo: The International Institute for Buddhist Studies, 1991b.

Schmithausen, Lambert. 'The early Buddhist tradition and ecological ethics.' *Journal of Buddhist Ethics* 4: 1–42, 1997.

Seager, Richard Hughes. *Buddhism in America*. New York: Columbia University Press, 1999.

Silva, Lily de. 'The Buddhist attitude towards nature.' In *Buddhist Perspectives on the Ecocrisis*, Klas Sandell, ed. Kandy, Sri Lanka: Buddhist Publication Society, 1987, pp. 9–29.

Silva, Padmasiri de. *Environmental Philosophy and Ethics in Buddhism*. New York: St. Martin's Press, 1998.

Sivaraksa, Sulak. *Seeds of Peace: A Buddhist Vision for Renewing Society*. Berkeley: Parallax Press, 1992.

Sivaraksa, Sulak. *Global Healing: Essays and Interviews on Structural Violence, Social Development, and Spiritual Transformation*. Bangkok: Thai Inter-Religious Commission for Development and Sathirakoses-Nagapradipa Foundation, 1999.

Sivaraksa, Sulak, Uthai Dulayakasem, Anant Viriyaphinij, Niphon Chamduang, and Jonathan Watts, eds. *Buddhist Perception for Desirable Societies in the Future*. Bangkok: Thai Inter-Religious Commission for Development and Sathirakoses-Nagapradipa Foundation, 1993.

Sivaraksa, Sulak, Pracha Hutanuvatra, Nibhond Chaemduang, Santisukh Sobhanasiri, and Nicholas P. Kholer, eds. *Radical Conservatism: Buddhism in the Contemporary World*. Bangkok: The Sathirakoses-Nagapradipa Foundation, 1990.

Skolimowski, Henryk. *Eco-Philosophy: Designing New Tactics for Living*. Boston: M. Boyars, 1981.

Skolimowski, Henryk. 'Eco-philosophy and Buddhism.' *Buddhism at the Crossroads*, Fall 1990: 26–29.

Snelling, John. *The Buddhist Handbook*. Rochester, New York: Inner Traditions, 1998.

Sponsel, Leslie E. 'The historical ecology of Thailand: increasing thresholds of human environmental impact from prehistory to the present.' In *Advances in Historical Ecology*, William Balée, ed. New York: Columbia University Press, 1998, pp. 376–404.

Sponsel, Leslie E. 'Do anthropologists need religion, and vice versa? Adventures and dangers in spiritual ecology.' In *New Directions in Anthropology and Environment: Intersections*, Carole L. Crumley, ed. Walnut Creek, California: AltaMira Press, 2001a, pp. 177–200.

Sponsel, Leslie E. 'Human impact on biodiversity, overview.' In *Encyclopedia of Biodiversity*, Simon Asher Levin, Editor-in-Chief. Vol. 3, pp. 395–409. San Diego: Academic Press, 2001b.

Sponsel, Leslie E. and Poranee Natadecha-Sponsel. 'Buddhism, ecology and forests in Thailand: past, present and future.' In *Changing Tropical Forests: Historical Perspectives on Today's Challenges in Asia, Australasia and Oceania*, John Dargavel, Kay Dixon, and Noel Semple, eds. Canberra: Centre for Resource and Environmental Studies, 1988, pp. 305–325.

Sponsel, Leslie E. and Poranee Natadecha-Sponsel. 'The potential contribution of Buddhism in developing an environmental ethic for conservation of biodiversity.' In *Ethics, Religion and Biodiversity: Relations Between Conservation and Cultural Values*, Lawrence S. Hamilton, ed. Cambridge, U.K.: The White Horse Press, 1993, pp. 75–97.

Sponsel, Leslie E. and Poranee Natadecha-Sponsel. 'The role of Buddhism in creating a more sustainable society in Thailand.' In *Counting the Costs: Economic Growth and Environmental Change in Thailand*, Jonathan Rigg, ed. Singapore: Institute of Southeast Asian Studies, 1995, pp. 27–46.

Sponsel, Leslie E. and Poranee Natadecha-Sponsel. 'A theoretical analysis of the potential contribution of the monastic community in promoting a green society in Thailand.' In *Buddhism and Ecology: The Interconnection of Dharma and Deeds*, Mary Evelyn Tucker and Duncan Ryūken Williams, eds. Cambridge, Massachusetts: Harvard University Press, 1997, pp. 41–68.

Sponsel, Leslie E., Poranee Natadecha-Sponsel, Nukul Ruttanadakul, and Somporn Juntadach.

'Sacred and/or secular approaches to biodiversity conservation in Thailand.' *Worldviews: Environment, Culture, Religion* 2(2): 155–167, 1998.

Sponsel, Leslie E. and Poranee Natadecha-Sponsel. 'Does Buddhism have any future? some thoughts on the possibilities of Buddhist responses to the 21st century.' *Seeds of Peace* 16(1): 36–39, 2000.

Stryk, Lucien. *World of the Buddha: A Reader.* Garden City, New York: Doubleday and Company, 1968.

Suzuki, Daisetz Teitaro. 'The role of nature in Zen Buddhism.' In *Zen Buddhism: Selected Writings of D.T. Suzuki*, William Barrett, ed. New York: Anchor Books, 1956, pp. 229–258.

Swearer, Donald K. 'The hermeneutics of Buddhist ecology in contemporary Thailand: Buddhadasa and Dhammapitaka.' In *Buddhism and Ecology: The Interconnection of Dharma and Deeds*, Mary Evelyn Tucker and Duncan Ryūken Williams, eds. Cambridge, Massachusetts: Harvard University Press, 1997, pp. 21–44.

Taylor, J.L. *Forest Monks and the Nation-State: An Anthropological and Historical Study in Northeastern Thailand.* Singapore: Institute of Southeast Asian Studies, 1993.

Thanissaro Bhikkhu. *The Buddhist Monastic Code.* Valley Center, California: Metta Forest Monastery, 1994.

Thanissaro Bhikkhu. 'Metta Forest Monastery.' In *The Complete Guide to Buddhist America*, Don Morreale, ed. Boston: Shambhala, 1998. pp. 26–28.

Timmerman, Peter. 'It is dark outside: Western Buddhism from the Enlightenment to the global crisis.' In *Buddhism and Ecology*, Martine Batchelor and Kerry Brown, eds. London: Cassell Publishers, 1992, pp. 65–76.

Titmus, Christopher. *The Green Budhha.* London: Wisdom Publications, 1995.

Tiyavanich, Kamala. *Forest Recollections: Wandering Monks in Twentieth-Century Thailand.* Honolulu: University of Hawaii Press, 1997.

Tuan, Yi-Fu. 'Discrepancies between environmental attitude and behavior: examples from Europe and China.' *Canadian Geographer* 12(3): 176–191, 1968.

Tucker, Mary Evelyn, and Duncan Ryūken Williams, eds. *Buddhism and Ecology: The Interconnection of Dharma and Deeds.* Cambridge, Massachusetts: Harvard University Press, 1997.

Usher, Ann Danaiya. 'After the forest: AIDS as ecological collapse in Thailand.' *Thai Development Newsletter* 26: 20–32, 1994.

Vitousek, P.M., P.R. Ehrlich, A.H. Ehrlich, and P.A. Matson. 'Human appropriation of the products of photosynthesis.' *BioScience* 36(6): 368–373, 1986.

Waal, Frans de. *Good Natured: The Origins of Right and Wrong in Humans and Other Animals.* Cambridge, Massachusetts: Harvard University Press, 1996.

Walsche, Maurice. *The Long Discourses of the Buddha: A Translation of the Digha Nikaya.* Boston: Wisdom Publications, 1995.

Ward, Tim. *What the Buddha Never Taught.* Berkeley: Celestial Arts, 1993.

Watson, Burton. *The Lotus Sutra.* New York: Columbia University Press, 1993.

Yamauchi, Jeff. 'The greening of Zen Mountain Center: a case study.' In *Buddhism and Ecology: The Interconnection of Dharma and Deeds*, Mary Evelyn Tucker and Duncan Ryūken Williams, eds. Cambridge, Massachusetts: Harvard University Press, 1997, pp. 249–265.

JOHN BERTHRONG

CONFUCIAN VIEWS OF NATURE

Just as there are many diverse forms of Confucianism, there are divergent Confucian views of nature. The tradition called Confucianism in the West has a long and developmentally complex history in East Asia (Schwartz, 1985; Graham, 1989). Although Confucianism or the teaching of the Ru (scholars) has Chinese origins, Confucian teachings spread and flourished in Korea, Japan and Vietnam (Berthrong, 1998). In order to understand characteristic Confucian reflections on nature, two questions concerning definitions must be addressed. What is Confucianism as a self-reflective tradition or set of traditions? And what do Confucians make of nature?

The terms Confucian and Confucianism were invented and promoted by early Jesuit scholar-missionaries to China in the 16th and 17th centuries (Berthrong, 1998, 2000; Jensen, 1997). The Jesuits assumed that the various peoples of Asia possessed what early modern Europeans defined as religions and that religions had founders (Jensen, 1997). The early modern Western models for religions were Christianity, Judaism and Islam. Based on their examination of Chinese intellectual and religious history, the Jesuits decided that the literati scholars of late imperial China did indeed have a religion; it was equally obvious that the founder of the religion of the Chinese educated elite was Kongzi or Master Kong. The religion of Master Kong, or Kongfuzi as the Jesuits honored him, then became Confucianism.

There is actually no precise pre-modern Chinese equivalent term for Confucianism, although modern Chinese intellectuals easily recognize the general outlines of the tradition that the Jesuits were trying to define during their encounter with Chinese literati scholars during the Ming dynasty within the larger matrix of Chinese intellectual and religious life. For most of its history, the common Chinese term for a Confucian has been *ru* or scholar. A scholar did not necessarily have to be a Confucian, but most literate Chinese *ru* for the last two thousand years were probably Confucians. Because of this fact, some modern scholars now use the term Ruist for the older term Confucian. The other main religio-philosophic choices were the indigenous Daoism or Buddhism after it arrived in China via Central Asia sometime early in the Common Era (Fung, 1952–53; de Bary, 2000).

373

H. Selin (ed.), Nature Across Cultures: Views of Nature and the Environment in Non-Western Cultures, 373–392.
© 2003 *Kluwer Academic Publishers. Printed in Great Britain.*

The immense history of Chinese science and technology begun by Joseph Needham demonstrated that the Chinese had a robust concern for the study of the natural world (Needham, 1954-). Of course, not all the naturalists, scientists, doctors and philosophers discussed in *Science and Civilisation in China* were of the Ruist persuasion; adherents of many other schools of Chinese thought were keen naturalists as well (Bodde, 1991).

The Confucians had many ways of looking at nature (Li *et al.*, 1982). They were interested in natural phenomena and human nature and wondered about the relationship of human beings to the broader world around them (Henderson 1984; Graham 1989; Schwartz, 1985). Moreover, not all Confucian scholars were philosophers, proto-scientists or naturalists. In fact, many modern historians of science have made the counter argument that the state-sponsored version of Confucian thought that sustained the imperial state apparatus as its formal ideology was not supportive of the study of nature (Qian, 1985). However, such claims about Confucian hostility to the study of nature are overdrawn. Some Confucians at some times were passionately concerned with the study of nature based on their interpretation of the Ruist tradition; other Ruists at other places and times had no interest at all. The reasons for these diverse views will be explained below.

Given the complexity of the Confucian heritage in East Asia, one pragmatic way to appraise representative Confucian views of nature is to review the broad historical development of Ruist thought over the centuries. Although focused Confucian interest in nature is not inextricably linked to the historical development of the tradition, there are enough congeries of commonplace views that arise with the flow of Confucian intellectual history to make a historical approach a very useful way to explain Confucian views of nature.

THE SIX ERAS OF CONFUCIAN HISTORY

During its spread throughout East Asia, there have been six paradigmatic historical transformations in the Confucian tradition (Berthrong, 1998):

1. The rise of the classical tradition in Shang and Zhou China (*ca.* 1700–221 BCE);
2. The Commentary synthesis of the Han Dynasty (206 BCE–220);
3. The defense of the Confucian Way: the challenge of Neo-Daoism and Buddhism – from the Wei-Jin to the Tang Dynasty (220–907);
4. The renaissance of the Song and the flowering of the Ming Dynasties (960–1644) and Confucianism's spread to Korea and Japan;
5. The Evidential Research of the Qing Dynasty and Korean and Japanese appropriations of the tradition (1644–1911); and
6. New Confucianism in the modern world: variations on an East Asian theme (1911 to the present).

At first glance this chart of the historical development of Ruist thought appears somewhat overcomplicated – at least when compared and contrasted to the more usual way of presenting the history of Chinese thought. In its most simple form, the argument is made that in China the rise of Chinese culture occurred in the great classical period (*ca.* 551–221 BCE); after the flowering of the Hundred Schools of Thought, the rest of Chinese intellectual history is one of commentary on the foundations provided by the great masters of the classical age (Fung, 1952–53). There are arguments like this about Western thought as well. Alfred North Whitehead once quipped that the entire history of Western philosophy was a series of footnotes to Plato. But it is hardly the case that Whitehead meant to dismiss the work of Western philosophy from Aristotle to the modern world by this comment. Unfortunately, many Western intellectuals still labor under the outmoded assumption that there was no real intellectual development in China and Sinitic East Asia (or India for that matter) after the classical age. Hegel's theories about Asian intellectual history die slowly and painfully, if they die at all (Clarke, 1997).

THE CLASSICAL AGE

From the traditional Ruist perspective, the classical age stretches far back into the pre-history of China. Traditional Ruists scholars, including Master Kong (551–479 BCE), believed that they were the heirs of a long series of dynastic achievements that began with the civilizing acts of the legendary sage kings of high antiquity. Master Kong defined this extended intellectual heritage as "this culture of ours" (Berthrong and Berthrong, 2000; Bol, 1992).

In terms of the early classical texts there is nothing like a sustained reflection on the natural world as a special object of concern. The early texts, such as the *Analects*, relentlessly worry about the cultivation, reform and practice of human conduct both personally and within the larger society. However, this does not mean that the earliest texts of the Ruist canon did not contain material about the natural world. For instance, Master Kong remarked that the great early collection of poetry, the *Shijing* (*Classic of Poetry, Book of Songs*), listed many names for material objects and plants. A scholar was expected to have a detailed knowledge of such naturalist information embedded in the *Classic of Poetry*. In other classics there were also descriptions of natural phenomena, geographical details and sources for ores, plants and foodstuffs, both commonplace and exotic. It is fair to say that these lists of material items were probably produced more for purposes such as tax collection than interest in the flora and fauna of ancient China, although this scholarly focus cannot be discounted either (Puett, 2000).

The main focus of Ruist philosophic concern was neither in proto-science nor descriptions of the natural world. Above all else, Confucians sought to cultivate ethical life and reform society (Hall and Ames, 2001). Master Kong sought to reform the customs of his time by attention to restoring the glories of the great founders of his beloved Zhou dynasty. But if Master Kong and his immediate disciples can be characterized as philosophers with a strong bent towards social ethics, it is also important to note that there was nothing anti-natural about their philosophic visions (Ho, 1985; Smith, 1993). According to the early classical Ruist worldview, human beings were embedded in a world that began in the family, expanded into society, the state, and finally into the entire world or cosmos (Ames and Hall, 2001; Jullien, 1995; Smith, 1993).

Other thinkers and schools were developing a rich vocabulary designed to describe the natural world. These included terms such as *qi* or vital force, *yin-yang* (a pair of correlative terms for the passive and active forces of nature that have passed into common English usage) and *wuxing* or the five phases.[1] As we shall see, towards the end of the classical period and with the transition to the Han dynasty, all of these terms and concepts were incorporated into Ruist thought. But again, it was not the case that the Ruist thinkers were unaware of the development of concepts such as the five phases, but rather that this was not the kind of philosophic reflection that interested them in the earliest phases of the Ruist tradition (Berthrong, 1998).

The second sage of the emerging Ruist tradition is Master Meng (371–289 BCE), known in English as Mencius. Master Meng, like his revered teacher Master Kong, was not interested in nature per se, although he inclined to use metaphors drawn from the natural world in order to defend Ruist thought from the many philosophic attacks of his day. One of Master Meng's most arresting parables is about the fate of Ox Mountain.

> Mencius said, 'There was a time when the trees were luxuriant on the Ox Mountain, but as it was on the outskirts of a great metropolis, the trees were constantly lopped by axes. It is any wonder that they are no longer fine? With the respite they get in the day and in the night, and moistening by the rain and dew, there is certainly no lack of new shoots coming out, but then the cattle and sheep come to graze upon the mountain. That is why it is as bald as it is. People, seeing only its baldness, tend to think that it never had any trees. But can this possibly be the nature of a mountain?' (Lau, 1984, 2: 231).

This has become one of the most quoted stories in the Confucian repertoire about Ruist thought about ecology. However, Master Meng, true to the sensibilities of classical Confucian thought, is using a stunningly effective rhetorical story drawn from the observation of nature for human moral instruction. Mengzi goes on to liken the common state of human beings to the sad condition of Ox Mountain.

Mengzi's point is that neither the mountain nor human nature need be left in this sorry condition. What is needed is intelligent cultivation of human nature and sensible conservation on Ox Mountain. In fact, the Ruist tradition's interpretation of what we now label ecology is a consistent conservationist program. The population of the classical Chinese world had obviously grown large enough to be a threat to the bounty of the North China plain and the mountains and rivers that surrounded, bisected and ran through it. The Ruist scholars believed that there could be a balance between human action and the environment. Just like human nature, nature itself needed careful tending. Unlike their Daoist rivals, Ruists believed that intelligent management of the natural world would allow for the flourishing both of nature and human beings (Tucker and Berthrong, 1998).

Master Meng's favorite metaphors for human renewal often made use of plant imagery. Human beings are like the plants of the world: they are rooted in the soil of the family, society and state. If carefully nurtured, they will become abundant. But if there is lack of husbandry, then the plants will wither and die. Mencius is often condemned for having an excessively melioristic view of human beings. However, he took great pains to point out that human beings have only the seeds of goodness; the rest is up to sensible, enlightened self-cultivation. One of the clear lessons that the growing Ruist tradition took from Master Kong and Master Meng was that human beings are rooted in the natural order. There is no other order to which human beings can escape, and therefore it is essential to deal intelligently with the self and with nature. Ruist thinkers were never tempted to conceive of some heaven to which human beings could flee or appeal. This world, this Ox Mountain, was and is the home of humanity.

The third of the great classical Confucians, Xunzi (310–221 BCE), is the most systematic and most problematic of the early Ruist masters (Graham, 1989; Puett, 2001). The problem is not with the quality of his thought but rather with his sharp disagreement with Master Meng. In short, Xunzi believed that Master Meng was wrong in stating that human nature was good. Master Xun held, to the contrary, that human nature was evil and whatever good human beings could claim came from self-cultivation and artifice. To hold that human nature is evil struck later Ruists as completely wrong, and for his disagreement with the Second Sage, Mengzi, Xunzi was not considered to be orthodox at all after the great Ruist revival in the Song period.

Whatever later philosophers made of Xunzi, in his own time he was a towering intellectual force (Knoblock, 1989.1: 1–128). Xunzi gave voice to a strain in Ruist thought that has become almost second nature to the tradition. For this service, Xunzi is considered the grand rationalizer or naturalizer of the emerging Ruist worldview. The locus of the debate revolved around the ancient term *tian*. This term is most often translated as "heaven". It indeed meant the sky above, but was part of the ancient Zhou religion as well. It meant Heaven as the high god of the Zhou religious tradition along with its more mundane meaning as the vault of heaven. Both Master Kong and Master Meng would have understood tian in its dual modalities of Heaven and as the heavens.

Xunzi was the classical Confucian thinker who naturalized tian. He did so in two interlocking ways. First, he denied any theo-volitional qualities to tian. For Master Xun, tian no longer was the high god of the Zhou pantheon; tian was not conscious and it most certainly did not play a volitional role in the governance of the natural order or in human destiny. It was natural in the sense that it was the order structure of the world. Because of this view of heaven, Xunzi tried as best he could to give a rational explanation of the world. Second, this rational explanation of the world and of human beings within it caused Xunzi to be interpreted as the great naturalist among the classical Confucians. For Xunzi it rained on the good and bad because of the nature of the world and not because some god willed it as a reward or punishment for proper conduct or perverse acts.

On the one hand, to paraphrase Xunzi, tian has a constant way; earth has constant categories. Although he did not spend a great deal of time describing them, Xunzi believed that the world, including human beings, emerged out of the interaction of the vital force/*qi* and was governed by the agency of yin-yang. But on the other hand, this did not mean that he was a mechanistic determinist. Human flourishing arises from the conscious efforts of human beings to understand their world and to act according to this knowledge. For instance, unsettled social conditions were not the fault of heaven's decrees but of decline in agricultural production, mistaken governmental interference in the economy, and social disharmony.

The root problem, as Xunzi saw it, was that things were limited and human desire was infinite. This imbalance led to social disharmony; the image is rather like Thomas Hobbes' belief that the natural condition of humans is a state of perpetual war of each against all, where no morality exists, and everyone lives in constant fear. Unless sage rituals carefully govern human society, there will be incessant conflict caused by the struggle for everyone to try to take more of the limited material resources of the world. Xunzi's solution to the problem of limited resources was to order the world in a hierarchical fashion. However, to make this a true civilization rather than just a hierarchy of power relations, social ritual was a key element in his plan for the reform of society. For Xunzi, as for Plato, civilization is the victory of (persuasion) ritual over violence, of deference over conflict.

Xunzi also assumed more and more of the natural knowledge of the late classical age even if he did not incorporate it systematically into his philosophy. As we mentioned before, the synthesis of classical Ruist philosophy and the growing body of sophisticated naturalist lore was the work of the next generation of Ruist thinkers, the Han scholars.

THE HAN SYNTHESIS

The story of Han Confucian speculation on the natural order begins considerably before the Han itself. Along with all other kinds of philosophic speculation, there was a growing body of naturalist lore that became the pan-Sinitic basis for all future East Asian views of nature. Rather like the pre-Socratic cosmologists, Chinese thinkers formulated ever more complex explanations of nature. The very first of such cosmologies has been called *Taiyi shengshui* or *The Great One Gave Birth to Water* based on the opening line of the short text.

> The Great One gave birth to water. Water returned and assisted Taiyi, in this way developing heaven. Heaven returned and assisted Taiyi, in this way developing the earth. Heaven and earth [repeatedly assisted each other], in this way developing the 'gods above and below.' The 'gods above and below' repeatedly assisted each other, in this way developing Yin and Yang. Yin and Yang repeatedly assisted each other, in this way developing the four seasons. The four seasons repeatedly assisted each other, in this way developing cold and hot. Cold and hot repeatedly assisted each other, in this way developing moist and dry. Moist and dry repeatedly assisted each other; they developed the year, and the process came to an end (Henricks, 2000: 123).

Of course the process did not really come to an end. The *Taiyi shengshui* itself then goes on to show how the cycle of natural and perhaps not so natural causes or actions outlined above started yet again and became "the mother of the ten thousand things" (see also Allan, 1997).

During the Han dynasty two great streams of thought were fused together, providing the basic materials of the Confucian view of nature. The first stream is the fundamental Confucian stress on social ethics or what the great New Confucian Mou Zongsan (1909–1995) called profound concern consciousness (de Bary, 2000, 2: 558–561). Mou's definition of concern consciousness is helpful at the point when Han dynasty Confucian scholars such as Dong Zongshu (179–104 BCE) were creating their grand interpretation of the cosmos. While social ethics are directed towards human society, Mou's idea of the foundational aspect of concern consciousness can be expanded to include more aspects of human life.

For instance, the Han Ruist scholars were convinced that their philosophic vision included the natural world as well as the various theories of human self-cultivation and social formation that they inherited from Master Kong, Master Meng and Master Xun. The most representative of the early Han Confucian masters was Dong Zongshu (Berthrong, 1998). Dong is lauded for fusing Ruist social ethics, via an enhanced understanding of concern consciousness, with the pan-Sinitic naturalist cosmology of his day. Dong carried out his philosophic task with the best of Confucian aims in mind, namely to show how even the cosmos itself was moral and hence exerted a constraint on the actions of the emperor. In other words, Dong conjoined social ethics with cosmology. It was this encompassing vision of heaven, earth and humanity that formed the foundations of the imperial Confucian ideology of the Han dynasty. It was also a total schema qua political ideology that proved highly resilient. It was not formally replaced in China until the fall of the Qing dynasty in 1911.

The Han Ruist cosmology was based, as noted above, on an emerging body of naturalist speculation that was the common property of all the various Chinese philosophic schools. But commonality did not stop the Confucians from claiming the pan-Sinitic cosmology as their own. In fact, they argued that as a cosmology it would obviously have to be part of the teachings of the ancient sage kings and hence a fit object of study and veneration for serious Ruist scholars. Han Ruist scholars such as Dong Zongshu simply deemed that they were giving a full account of all of the teachings of the sages and that any such full teaching included a robust theory of the natural world.

This pan-Sinitic naturalist view contained all the terms that are now considered commonplace when traditional Confucian philosophers talked about the natural world. The enhanced and expanded cosmology of Dong Zongshu provided the shared language for future Ruist views of nature. This new vocabulary was only newly formalized within Ruist discourse and had been developing for centuries. The list includes terms such as *yin-yang*, *qi*, *wuxing* and the *taiji* or the Supreme Ultimate as the incipient focal point of the processive unfolding of the cosmos. The pivotal text that Confucians could claim was an essay appended to the classic *Yijing* (*I Ching*, or *Book of Changes*). The *Great Appendix*, as this essay was known, showed how the emergent late classical and early Han cosmology could be given an impeccable Confucian pedigree and how Ruist thinkers could now make use of the most up-to-date cosmological and naturalist thinking available at the beginning of the Han dynasty.

After Dong Zongshu grafted the emerging naturalist lexicon as recorded in the appendices of the *Yijing* onto the body of Ruist social/ethical thinking, Confucians would then be equipped with a subtle language for the description of the natural world. This Han synthesis is often called a correlative philosophy or cosmology. What this means is that all agents, objects, events, powers, influences, causes and forces are related inextricably to each other within the ever-fertile matrix of the Dao.[2] The Supreme Polarity or Ultimate of the Dao gives rise to the primal forces of yin and yang, and yin and yang are manifested via the five phases of the material world. The correlative synthesis of the Han thinkers proved to be a rich matrix for Ruist speculation about the natural order.

FROM THE FALL OF THE HAN TO THE GLORIES OF THE TANG

The great Han dynasty fell from effective power by the end of the second century of the Common Era. The collapse of the Han in China was somewhat like the fall of the Roman Empire at the other end of Eurasia, although China's period of disunity was not nearly as culturally disruptive as what happened in the post-Roman West. For instance, the great Han Confucian synthesis held firm and became part of the Confucian patrimony. However, it is likewise fair to say that Ruist thought was no longer the center of Chinese intellectual history.

During the third and fourth centuries there was a revival of interest in other schools of Chinese thought, especially various forms of Daoism and slightly later, Buddhism. Along with the flourishing of new religious schools of Daoism, there was a renaissance of philosophic speculation on the earlier Daoist classics. One of the most fascinating outcomes of this revival of interest in texts such as the *Laozi* and the *Zhuangzi* was the emergence, for the first time in any important way, of philosophic reflection on what in Western terms can be called ontology and cosmology strictly defined.

While there had been a great deal of self-reflection and observation of the material world, there had been little interest in asking the fundamental ontological question: Why is there something rather than nothing at all? The grand cosmological question then followed: If there is something rather than nothing, what is the nature of this something and how do things relate to each other? The great Neo-Daoist and Confucian Wei-Jin thinkers asked these fundamental questions about the being or non-being of the world and whether or not the things of world could be known through the analysis of causality.

At that point, something else began transforming the Chinese intellectual landscape, and that was the arrival of Buddhism via Central Asia from its original home in India. No one will ever know just when the first Buddhists journeyed into China, but by the second century their presence was being noticed. Initially the Chinese were not sure what to make of the new religion. Some thought that it was a version of Daoism that had somehow gone west and now was returning to its Chinese home; others were convinced that Buddhism was a completely new worldview.

It is important to remember that Buddhism was the first religiously and philosophically sophisticated non-Sinitic religion, indeed cultural system, which the Chinese had ever encountered. Buddhism also arrived at a propitious time: the collapse of the Han dynasty shook the Chinese world. In short, the Chinese were as ready as they ever would be for a new religion. Buddhism provided the Chinese with a new explanation of how to achieve liberation from the bondage of the pain and suffering of the world. Moreover, along with proffering a new religious worldview, Buddhism came with a vast panoply of spiritual teachings, meditations, scriptures, chants, music and visual arts. The Chinese were captivated by Buddhism and thus began the great Buddhist period in East Asia.

Buddhism did not replace Daoism or Confucianism, but it became the third of the great Chinese religious traditions. By the rise of the great Tang dynasty in the 7th century, Buddhism had become completely Chinese. After a phase of learning from Indian and Central Asian masters and texts, the Chinese created their own distinctive schools such as Chan, Pure Land, Huayan and Tiantai (de Bary, 2000). Along with their religion, Buddhists introduced new conceptions of the natural world into China, including major contributions to astronomy, medicine and mathematics. However, it was in the religious and philosophic dimensions of Chinese culture that Buddhism had its greatest impact.

In terms of Confucian views of nature it was the Daoist tradition that contributed as much or more than the newly Buddhist establishment from the 2nd to the 10th centuries. The organized Daoist religious communities that had emerged at the end of the Han period grew and proliferated during these centuries, and they articulated new forms of practice that had lasting impacts on Ruist naturalist sensibilities (Kohn, 2001). For instance, Daoists of this period cultivated all kinds of meditation technologies. One of the chief aims of these meditations was to extend life through what the Chinese call inner and outer alchemy (*neidan* and *waidan*). Both forms of alchemy sought to transform the human person to become one with the Dao or at least extend the lifespan towards infinity (Kohn, 2001).

Daoist outer alchemical practice (based on taking various drugs that were supposed to offer immortality) led to a vast expansion of the traditional Chinese pharmacopoeia, as the adepts experimented with various minerals and plants in order to find the elixir of immortality. Although the Daoist alchemists never found the key to eternal life, they did find all kinds of useful new medicines and other forms of bodily self-cultivation that proved to be a positive contribution to Chinese medical practice. Because Confucian scholars were often as interested in medicine as their Daoist and Buddhist colleagues, these new Daoist methods and materials became part of the Ruist world as well (Porkert, 1974).

The Confucians found that they could include new Daoist and Buddhist ideas and practices in their own cosmology. The typical Confucian appreciation of nature was strengthened during this period. But it was probably the religious dimensions of the shared Chinese cultural world that was most affected by the sinification of Buddhism and the flourishing of Daoism during these long centuries between the fall of the Han and the demise of the Tang dynasties.

THE SONG CONFUCIAN REVIVAL

Confucians, stimulated by Buddhist and Daoist examples, mounted a major renewal of their tradition in the Northern Song period (960–1127). In the West this revival gained its own name, Neo-Confucianism. Neo-Confucianism became the second period of Confucian thought, rivaling the classical period itself (Bol, 1992). The Neo-Confucians also reflected on nature. In fact, at least for gentry-literati culture, the renewed Confucian views of nature dominated the Sinitic intellectual world until the recent impact of modernity with the Western shock of the 19th century.

The Northern Song Confucian revival was a complex affair and made up of more than just one school. One particular school, later systematized by Zhu Xi (1130–1200), came to dominate the philosophic scene, but this group of thinkers, who called themselves *daoxue* or the Study of the Way School, was only one group of Ruist scholars who created a new cultural world in the Northern Song (Bol, 1992). There have been endless debates about the level and intensity of the borrowing of the new Ruist scholars from Daoist and Buddhist sources. Suffice it to say, in some cases there was clear borrowing of texts and ideas; beyond explicit borrowing there was the obvious stimulation of having to respond to complex and sophisticated alternative cosmological worldviews, religious practice and institutions and social sensibilities.

In order to explain this new naturalist cosmology we will explore the interconnected thought of Zhou Dunyi (1017–1073), Shao Yong (1011–1077), Zhang Zai (1020–1077) and Zhu Xi (1130–1200). This is the philosophic lineage that most carefully explicated the renewed Ruist vision of the natural world (de Bary, 2000).

The manifesto of the renewed Ruist philosophy was Zhou Dunyi's *Taijitu shuo* (*Explanation of the Diagram of the Supreme Polarity*). In this very short commentary on a cosmological diagram, Zhou outlined a vision of the natural order that became dominant in East Asia until the 19th century. It is a creative restatement of the classical natural worldview of the Han with a fundamental reaffirmation of the centrality of Ruist concern consciousness. The opening paragraph sums up the essential features of this vision of the cosmos.

> Non-Polar (*wuji*) and yet Supreme Polarity (*taiji*)! The Supreme Polarity in activity generates yang; yet at the limit of activity it is still. In stillness it generates yin; yet at the limit of stillness it is also active. Activity and stillness alternate; each is the basis of the other. In distinguishing yin and yang, the Two Modes are thereby established (de Bary, 1999, 1: 673).

The text then goes on to show how all the myriad things are generated out of the ceaseless creativity of the Supreme Polarity. Zhou shows how qi, wuxing, the modality of the yin and yang and the four seasons give rise to the cosmos.

Within this potentially harmonious balancing of various polar forces, human beings have a central place and role to play.

> Only humans receive the finest and most spiritually efficacious [*qi*]. Once formed, they are born; when spirit [*shen*] is manifested, they have intelligence; when their fivefold natures are stimulated into activity, god and evil are distinguished and the myriad affairs ensue (de Bary, 1999, 1: 675).

Zhou sets the human sage squarely in the midst of the natural order. Just as with the classical, Han and Tang Confucians, there is no other world – and it is just this Ruist conviction of the role of nature as the only reality that differentiates the Neo-Confucians from Daoists and Buddhists.

As we have noted before, although Zhu Xi is given credit for standardizing the emerging Dao Learning (from the Chinese term for Zhu's school, *daoxue*) worldview, Zhu's role, albeit creative, made use of material from his beloved Northern Song masters. Later Ruist views of nature, therefore, are a collectively woven tapestry, skillfully expanded and sometimes unraveled by generations of Song, Yuan, Ming and Qing Ruist scholars. This being said, Zhu Xi remains paramount as the most influential cosmologist of early modern China.

According to Zhu Xi, the natural world is a manifestation of the constant conjunction of principle/*li* and vital energy/*qi*. Li provides the pattern for everything that is or can be. In fact, for anything to be, that is, to be concrete in the sense taken by the Song Dao Learning scholars, it must have its own li. Moreover, each thing was also qi; there is nothing without qi. Zhu Xi would be asked often if it were possible for something to be without the union of li and qi and he always answered that this would be a philosophic impossibility (Berthrong, 1998).

How does this union of principle and vital force happen? Zhu Xi, reflecting the ancient social and moral commitments of the Ruist tradition, answered this question in terms of the human person. Zhu believed that the human person was an icon for the whole of reality. There was a macro level, with the heavens and the vast expanse of the world itself as good examples of this cosmic level; there was the mesocosmic level of human beings; and there could even be a microcosmic level of the forces of the five phases and yin-yang modalities as the circulation of qi within the human body. Because humans were placed, as the classics taught, between heaven and earth, there was no reason not to draw lessons about nature from analyzing them.

Each person had a heavenly mandated nature or *xing*. This nature was the principle allotted by heaven for the specific person. A person also had an allotment of qi given to him or her by his or her parents. The qi is identified with the emotional, passionate side of human life. The question that most intrigued Zhu was how the li and the qi could achieve anything substantial in terms of human flourishing. His answer was that the mind-heart or *xin* unified the principle as nature with the qi as the passions and specific bodily allotment. In the famous formula Zhu borrowed from Zhang Zai, "the mind-heart unifies the nature and the emotions."

When pressed to give a philosophic account of what this fusion of nature, emotions and the mind-heart meant, Zhu provided the following schematic.

Nature is really li, the emotions are qi, and the xin functions as the Supreme Polarity or raison d'être of the emerging person. The principle functions in two ways. First, it is the pattern of what a person should be and, modeled on Zhou's Supreme Polarity, it is also a plan for how a person ought to be. Zhu's version of Ruist philosophy is resolutely axiological in that facts and values, principles and vital forces, are part of a unified pattern of existence and can never be separated. It is important to note, and we will return to this point later, that all Ruist philosophers agreed with Zhu's analysis of the human being as part of the natural world.

The next question is: how then does the xin play such a pivotal role in the process? Zhu's short version of the answer was that the xin was the most refined part of the qi and was in fact that part of qi that can recognize li. The xin, therefore, when properly cultivated, could recognize the specific li in it, its own human nature, and then extend this recognition to a knowledge of the other patterns and vital forces of the world. What about the rest of a person's qi? Zhu answered that each of us received a unique allotment of qi. If we are lucky, it is a very clear portion; if we are unlucky it is a turbid or stolid allotment. However, any human being has the potential, through proper self-cultivation, of making the qi clear and of following the heavenly-endowed principle of human nature.

Zhu then goes on to show how the yin-yang modalities and the five phases play roles in the cultivation of the human person, and indeed, of the whole of creation. This grand cosmic vision allowed Zhu to be in touch with the best science of his day, because to be a human was to be part of nature itself. For instance, Zhu recognized the nature of fossils (Kim, 2000). From this empirical observation, Zhu hypothesized that the world must be immeasurably old and that at one time there must have been an ocean where the mountain rocks containing the bodies of the ancient animals were now to be found.

Zhu Xi also provided an explanation of how a person goes about investigating the material world. His favorite method was, again borrowing terminology from the ancient classics, the method of *gewu* or the examination of things. Zhu believed that a person needed to examine the world in order to find out the linkage of li and qi in everything. Of course, Zhu was not a modern empirical scientist. What concerned him the most was the examination of the human person per se, and the moral nature of the person more specifically. However, the point still remains that more naturalistically inclined Ruist philosophers could appeal to Zhu's doctrine of the examination of things in order to create the impressive body of material studied by Needham and his colleagues.

Two other Northern Song masters should be mentioned as contributors to Zhu Xi's grand view of nature. The first is Zhang Zai. Zhang is remembered as the patron saint of qi theory. Zhang provided Zhu with a sophisticated and extended meditation on the role of qi in a Neo-Confucian perspective, and later thinkers constantly returned to Zhang's theory of qi in the late imperial developments of Ruist views of nature. The second figure is Shao Yong. Shao has always been something of a controversial figure in Ruist studies because he was, perhaps unjustly, believed to be too closely identified with the more magical aspects of numerological lore and even Daoist influences. Nonetheless, Shao gave pride of place to the role of number as pattern in Ruist philosophy and thus made sure that mathematics remained an area of speculative reflection for later Ruist thought (Chan, 1963).

LATE IMPERIAL CHINA AND EAST ASIAN DEVELOPMENTS

Confucian discourse originated in China but it did not remain just a Chinese worldview. Ruist thought was exported to Vietnam, Korea and Japan and played a vital role in the development of the worldviews of these distinctive East Asian cultures.

The Southern Song dynasty was finally conquered by the Mongol Yuan dynasty in 1279. By 1369 a great Chinese revolt threw the Mongols out and the indigenous Ming dynasty was founded. The Ming dynasty and the Qing dynasty that followed it in 1644 are called the era of later imperial China. This long period did not end until the overthrow of the Qing dynasty in 1911. The Ming and Qing Ruist thinkers continued patterns of thought inherited from the Song and Yuan periods, but also contributed new visions of the natural order. In some cases Qing thinkers were harsh critics of Zhu's grand cosmological worldview and offered revised Ruist interpretations of nature.

Until now we have been stressing the formal view of nature found in the evolving Ruist discourse. However, the Ruist view of nature was not just the result of philosophic speculations about the natural world. Ruist scholars were also artists of the highest order. For instance, Chinese painting and poetry have been considered part of the ecumenical treasures of the entire civilized world. All aspects of nature fascinated Chinese painters. The great landscape paintings of the Song and Yuan masters are considered masterpieces of the universal human spirit. These landscapes show the ceaseless flow and power of the vital forces of the natural world. There is a reciprocal relationship between the refined Ruist worldview and way human beings find their place in the complex web of mountains, rivers, lakes, winds and plants. Chinese painters were also famous for their depictions of plants, birds and animals. Painting was common to Daoist and Buddhist-inspired artists as well, and Confucians could not claim a monopoly of this artistic form.

Other arts also served as a venue for Confucian thoughts about nature. The great tradition of poetry often took the natural world as its focus. The Chinese poetic world is full of rich and vibrant images of nature. Here too it is often hard to say whether the poet is Ruist, Daoist or Buddhist. The Ruist, in fact pan-Chinese, vocabulary of principle, heaven, earth, humanity, yin and yang, the five phases, the forces of the storm and the shimmering of rivers and the majesty of mountains infuses Chinese lyric poetry. Along with painters and calligraphers, the Chinese deem their lyric poets their most sublime artistic figures.

Even when Confucians were not great artists, they still loved to describe their travels through the natural world and were avid nature writers. There was even a genre devoted to travel narratives; Zhu Xi, polymath that he was, contributed short descriptive pieces about the scenery he observed on his travels (Strassberg, 1994). Closer to home, any accomplished scholar-official prized his private studio. Gardens were integral elements of a studio complex, and scholars lavished a great deal of time and effort on creating a beautiful environment for study and reflection. And where there was not space for a large garden, the Chinese invented what the Japanese call bonsai or miniature plants and gardens. There was even a whole art form for collecting strange and beautiful rocks to be placed in the scholar's garden. Water, plants, refined architecture and rocks were all arranged to manifest the inseparable linkage of heaven, earth and humanity (Clunas, 1991, 1996).

Ruist scholars also made use of popular arts such as *fengshui* (geomancy; the art of siting and aligning a building correctly based on traditional lore) for designing their houses and other buildings. Even today Chinese and other East Asians will call upon the services of a fengshui master before they begin construction of any edifice. Nor is fengshui used only for the construction of the house itself; fengshui considerations govern how furniture is placed and even how a room should be used for maximum benefit. The complicated fengshui calculations are governed by the same general view of nature, such as the flow of the yin-yang forces and the dynamics of the five phases. [Editor's note: see Graham Parkes' chapter on Fengshui in this volume.]

In terms of the evolution of Ruist views of nature, there were two major developments in the Ming-Qing periods, both in China and internationally in Korea and Japan. The first was continued speculation on the essential role of qi in Ruist cosmologies. The heart of the matter revolved around how to interpret Zhu Xi's teaching about the relationship of li and qi (Black, 1989). Zhu was clear, at least to himself, that you could never separate principle and vital energy. Other scholars were not so sure. They argued that Zhu, notwithstanding the master's protestations, had what they called a "two origin theory". By this they meant that Zhu had two sources of things and events, namely li and qi.

Later Confucians scholars in China, Korea and Japan were not convinced that this was an appropriate Ruist strategy. These scholars, honoring the teachings of Zhang Zai, suggested that there was only one source of the things and events of the world, namely qi. They considered that proper Confucian views of nature should be a "one source theory" and that the axis of this one source was always and everywhere qi (Tucker, 1989). These scholars did not deny the role of principle in the manifestation of objects and events, but stressed that principle was to be understood, without exception, as pattern or principle within the living, creative matrix of vital energy.

The critique of Zhu Xi's "two source theory" arose so vigorously because Ruist scholars were highly sensitive to the claim that their post-Song naturalism was too heavily indebted to Daoist or Buddhist influences. In order to protect Ruist philosophy from such claims about possible Daoist or Buddhist intellectual contamination, these Chinese, Korean and Japanese scholars believed it more prudent to locate all Ruist views of nature within the natural domain of qi. In terms of Western philosophy, late imperial Ruist thought tended towards a robust but non-reductionist naturalism. The Ruists wanted to differentiate their teachings from Daoist and Buddhist talk about the void and emptiness as being the hallmarks of objects and events. The affirmation of the centrality of qi was deemed sufficient to separate Ruist teachings from Daoist and Buddhist speculations. In the hands of Wang Fuzhi (1619–1692), Dai Zhen (1724–1777), Yi Yulgok (1536–1584), or Kaibara Ekken (1630–1714), Qi philosophy became a highly sophisticated meditation on the social, historical and natural worlds.

The other great Chinese reaction against Zhu Xi style moral metaphysics was a movement labeled the Evidential Research School (*kaozheng*) (Elman, 1984). This school was even more critical in its complaints against earlier forms of Song and Ming Ruist orthodoxy than were the revivers of qi theory. The Evidential Research scholars argued that all Song and Ming moral philosophy was hopelessly compromised because it was all tainted by Daoist and Buddhism idealist influences. They contended that even the great Zhu Xi had borrowed too much from the Buddhists and Daoists. What real Confucian scholars needed to do was adjure the moral metaphysics of the Song and Ming masters and return to a pristine form of Confucian scholarship.

According to the Evidential Research scholars, there was no way to redeem Song-Ming moral metaphysics; the whole enterprise needed to be abandoned. One of the slogans of this radical new school was the appeal to skip over the commentarial traditions of Tang, Song and Ming Ruist scholars and return to the more solid learning of the Han scholars. The idea was that if there were to be a reform of Ruist thought, then scholars needed to return to genuine Ruist teachings before they were contaminated by the great medieval schools of Daoist and Buddhist thought. Another name of the Evidential Research School was the School of Han Learning.

The scholars of the Han Learning and the Evidential Research School developed new research agendas. One of their favorite arguments was that any kind of abstract moral metaphysics was not in line with the true teachings of the classical Zhou Ruist masters. A typical slogan was to try to find the truth in the concrete facts of the situation. The Han Learning scholars vociferously argued that no one could ever find out the true nature of the world merely from *a priori* metaphysical principles. It was all well and good for Zhu Xi to speculate on the relationship of human nature to li and qi, but where was his proof in the empirical or historical world?

The Han Learning scholars sought the truth in the facts of the world as they construed the world. For instance, they were great local historians, geographers, astronomers, hydrologists and philologists. In order to understand their society, they demanded that any such speculation be based on a careful reading of the actual history of a country, region, province or social institution. Metaphysics was of no help in figuring out what a solid tax policy should be. Along with a demand for historical and empirical research, the Han and Evidential Learning scholars desired that scholarship be practical. The ultimate aim was still very much Ruist: human flourishing was the goal of true Confucian scholarship and must support the formulation of sound economic, educational and social policies.

One of the claims of these scholars was that Ruists must be tangible in their studies. Even today modern Chinese social, political and intellectual historians depend on the local histories and philologically informed textual studies of these scholars. It was this commitment to the concrete nature of scholarship that should, they contended, differentiate real Ruist learning from that of the Song and Ming moral philosophers and their Daoist and Buddhist cousins.

This renewal of interest in the real details of the natural world bore fruit in Japan in the 17th and 18th centuries. For instance, the great Ruist philosopher Kaibara Ekken (1630–1714) was also a keen naturalist (Tucker, 1989). Along with works on Confucian social ethics and epistemology, he wrote numerous naturalist treatises on the flora and fauna of Japan (Tucker, 1989). He is particularly famous for his writings on plants and birds; his work is full of careful illustrations. He wrote geographies and appreciations of flowers and was interested in medicine. In short, Ekken was a naturalist Confucian philosopher who carried out his naturalist agenda while remaining firmly within the Song-Ming traditions of moral philosophy. Empirical studies were not just the prerogatives of the Qing Evidential Research and Han Learning scholars in China. [Editor's note: See John Tucker's chapter on Japanese Views of Nature and the Environment in this volume.]

Another fascinating naturalist philosopher in Japan was Ando Shoeki (1703–1762). In his *Hosei Monogatari* (Tales of the World of Law), Ando had various animals discourse on the natural order of things. No Ruist-inspired thinker ever developed a more sensitive ecological view of nature. In "A Method for Making the Thieving and Violent World of Law Tally with the World of the Self-Acting Living Truth", Ando paralleled the Way of Heaven, the way things ought to be, with "Thieving Heaven's Way", the way that things are when human beings misuse the natural world for greedy purposes (Yasunaga, 1992: 356). In a typical Confucian stance, Ando notes that ecological violations are much moral misdeeds as are any transgressions of traditional Ruist ritual codes. As with the classical thinker Xunzi, Ando persisted in seeing the world as an ecological matrix of forces in which human beings needed to pay heed to the natural order if there were to be real human flourishing.

Although it belies the popular modern view – both Asian and Western – that Ruist philosophy was hopelessly moribund in later periods, at least some Ruist thinkers in China, Korea and Japan came to a robust naturalist philosophic vision just before the intrusion of the Western powers in East Asia. The classical, middle and late imperial periods of Ruist learning proffer a wide-ranging set of naturalist philosophies and social policies that are still to be mined as East Asia moves into its own version of modernity.

MODERN TIMES

At the beginning of the 20th century many critics believed that Confucianism was a doomed tradition. In China, the great imperial civil service examinations, based on Ruist texts, were abolished in 1905. Ruist scholars in short order lost control not only of their dominant role in the imperial civil service but also of the educational institutions that had trained Chinese youth for hundreds of years for the all-important examinations. Moreover, many of the best Western-educated young Chinese scholars themselves became harsh critics of the Confucian Way. More and more ordinary Chinese people soon began to reject the age-old views of the family and ritual that had been one of main glues that held Chinese society together. With the victory of the Chinese Communist Party in 1949, the final demise of the Ruist world seemed in sight.

However, the Confucian tradition has not passed into the museum of East Asian cultural history. Like Buddhism, Daoism and Shinto, Confucianism is making a comeback in modern East Asia. Prophecy about just what form the Confucian revival will take is hazardous; nonetheless, there has been a remarkable renewal of interest in Confucian thought.

The question that is raised over and over again is what can Confucianism contribute to the formation of a modern East Asian world? (de Bary, 2000). Many Confucians believe that a reformed Ruist tradition has a great deal to contribute to the human flourishing of the modern epoch. For instance, contemporary Confucians, often called New Confucians in China in order to distinguish them from the Confucians of the classical and imperial periods, have called for the modernization of their worldviews. Scholars in Korea and Japan, who are also committed to the reformation and renewal of the Confucian Way, join these modern Chinese intellectuals. Among other things that contemporary Ruist scholars need to do is to transform the traditional views of nature in light of modern physical and social science teachings about the world.

The pan-East Asian New Confucians assert that there is nothing essential in the traditional Confucian worldview that contradicts the teachings of modern ecumenical science. The New Confucians point out that the persistent naturalist tendencies of the classical and imperial schools of Ruist scholarship throughout East Asia make accommodation with modern science one of the easiest parts of the modern Confucian reformation. Confucians are also eager to engage in conversation with all other social movements, religions and philosophic traditions about the ecological crisis. As many scholars have pointed out, the ecological future of the world might well depend on decisions made about industrialization in India and China in the 21st century. The Confucians remember Mengzi's fable of Ox Mountain all too well and are seeking ways to provide a balanced view of the place of humankind in the natural order. From the Confucian perspective, ecology and social ethics must be linked for either to have any effect whatsoever.

New Confucians are confident that the inherent naturalism of their classical traditions will provide them with a platform for the articulation of a renewed Confucian vision of the natural order. Ruists have everywhere and always realized that human flourishing can only take place within the larger matrix of nature.

NOTES

[1] "In Chinese thinking, yin and yang are the most basic concepts. Originally they were certainly connected with darkness and light, but gradually their meanings were greatly extended. Yin came to represent things or natures that are cold, dark, female, negative, passive and so on; while yang represented things or natures that are warm, light, male, positive, active, and so on. According to the Yin-yang school, ... the universe was balanced with the yin and yang forces, and all phenomena are the result of the interplay of these two opposite yet reciprocal forces. *Wu-xing* – earth, wood, metal, fire and water – at the very beginning apparently referred to material substances. They were later conceived in terms of their functional attributes ... The Five Phases conception was developed, again by the Yin-yang school of Zhou Yan, to be the basis of a comprehensive theory for explaining change in the cosmos. The Five Phases were arranged in particular sequences of succession, either by "mutual conquest" or by "mutual production". Thus by mutual production wood produces fire, fire produces earth, earth produces metal, metal produces water, and water produces wood. By mutual conquest wood is conquered by metal, metal reduced by fire, fire extinguished by water, water blocked up by earth, and earth manipulated by implements made of wood.

Cosmologists of the late Zhou and Han eras used these sequences of the Five Phases to explain cosmic and historical evolution, as well as various seasonal, diurnal, and medical changes and rhythms. A good example is the explanation for the change of the seasons. Spring corresponds to wood, summer to fire, the third month of the summer to earth, autumn to metal, and winter to water. Thus the seasons change in the sequences of mutual production. ...

During the Han, the yin-yang and Five Phases theories were developed further. Dong Zhongshu of the Former Han incorporated the theory into the exposition of Confucian classics, thus extended the yin-yang and Five Phases mode of thinking into the realm of politics and social relations. By that time Han cosmologists could use this theory to build a complicated framework of correspondences between man, the state and the universe. They worked up a long list of things grouped by five and associated them with the *wu-xing*, including five planets, five seasons, five directions, five colors, five musical tones, five wise emperors, five viscera, five orifices, five animals, five conducts, five punishments, and still others. And all these were balanced with yin and yang." (Sun, Xiaochun, "Crossing the boundaries between heaven and man: astronomy in ancient China." In *Astronomy Across Cultures: The History of Non-Western Astronomy*, Helaine Selin, ed. Dordrecht, The Netherlands: Kluwer Academic Publishers, 2000, pp. 426–427).

[2] The Dao, conventionally translated as The Way, is explained in great detail, as is correlative thinking, in James Miller's article on Daoism and Nature in this collection.

BIBLIOGRAPHY

Allan, Sarah. *The Way of Water and the Sprouts of Virtue*. Albany: State University of New York Press, 1997.

Ames, Roger T. and David L. Hall. *Focusing the Familiar: A Translation and Philosophical Interpretations of the Zhongyong*. Honolulu: University of Hawaii Press, 2001.

Berthrong, John H. *Transformation of the Confucian Way*. Boulder, Colorado: Westview Press, 1998.

Berthrong, John H. and Evelyn Nagai Berthrong. *Confucianism: A Short Introduction*. Oxford: Oneworld Publications, 2000.

Black, Alison Harley. *Man and Nature in the Philosophical Thought of Wang Fu-chih*. Seattle: University of Washington Press, 1989.

Bodde, Derk. *Chinese Thought, Society, and Science: The Intellectual and Social Background of Science and Technology in Pre-modern China*. Honolulu: University of Hawaii Press, 1991.

Bol, Peter K. *This Culture of Ours: Intellectual Transition in T'ang and Sung China*. Stanford, California: Stanford University Press, 1992.

Chan, Wing-tsit. *A Source Book in Chinese Philosophy*. Princeton, New Jersey: Princeton University Press, 1963.

Chu Hsi and Lü Tsu-ch'ien. *Reflections on Things at Hand: The Neo-Confucian Anthology*, Wing-tsit Chan, trans. New York: Columbia University Press, 1967.

Clarke, J.J. *Oriental Enlightenment: The Encounter Between Asian and Western Thought*. London and New York: Routledge, 1997.

Clunas, Craig. *Superfluous Things: Material Culture and Social Status in Early Modern China*. Urbana: University of Illinois Press, 1991.

Clunas, Craig. *Fruitful Sites: Garden Culture in Ming Dynasty China*. Durham, North Carolina: Duke University Press, 1996.

de Bary, William, Irene Bloom Theodore, and Richard Lufrano, eds. *Sources of Chinese Tradition*. 2 vols. New York: Columbia University Press, 2000.

Elman, Benjamin A. *From Philosophy to Philology: Intellectual and Social Aspects of Change in Late Imperial China*. Cambridge, Massachusetts: Harvard University Press, 1984.

Feng, Yu-lan. *A History of Chinese Philosophy*. 2 vols., Derk Bodde, trans. Princeton, New Jersey: Princeton University Press, 1952–53.

Graham, A. C. *Disputers of the Tao: Philosophical Argument in Ancient China*. La Salle, Illinois: Open Court, 1989.

Henderson, John B. *The Development and Decline of Chinese Cosmology*. New York: Columbia University Press, 1984.

Hendricks, Robert C. *Lao Tzu's Tao Te Ching: A Translation of the Startling New Documents Found at Guodian*. New York: Columbia University Press, 2000.

Ho Peng Yoke. *Li, Qi and Shu: An Introduction to Science and Civilization in China*. Hong Kong: Hong Kong University Press, 1985.

Ivanhoe, Philip J. *Confucian Moral Self Cultivation*. 2nd ed. Indianapolis and Cambridge: Hackett Publishing Company, Inc., 2000.

Jensen, Lionel M. *Manufacturing Confucianism: Chinese Traditions and Universal Civilization*. Durham, North Carolina: Duke University Press, 1997.

Jullien, François. *The Propensity of Things: Towards a History of Efficacy in China*, Janet Lloyd, trans. New York: Zone Books, 1995.

Kim, Yung Sik. *The Natural Philosophy of Chultsi, 1130–1200*. Philadelphia, Pennsylvania: Memoirs of the American Philosophic Society, 2000.

Knoblock, John. *Xunzi: A Translation and Study of the Complete Works*. 3 vols. Stanford, California: Stanford University Press, 1988–1994.

Kohn, Livia. *Daoism in Chinese Culture*. Cambridge, Massachusetts: Three Pines Press, 2001.

Kuriyama, Shigehisa. *The Expressiveness of the Body and the Divergence of Greek and Chinese Medicine*. New York: Zone Books, 1999.

Lau, D.C. *Mencius*. 2 vols. Hong Kong: The Chinese University Press, 1984.

Lee, Peter H. *Sourcebook of Korean Civilization*. 2 vols. New York: Columbia University Press, 1993–1996.

Li, Guohao, Zhang Mengwen, and Cao Tianqin, eds. *Explorations in the History of Science and Technology in China*. Shanghai: Shanghai Chinese Classics Publishing House, 1982.

Lloyd, G.E.R. *Adversaries and Authorities: Investigations into Ancient Greek and Chinese Science*. Cambridge: Cambridge University Press, 1996.

Najita, Tetsuo, ed. *Tokugawa Political Writings*. Cambridge: Cambridge University Press, 1998

Nakayama, Shigeru. *Academic and Scientific Traditions in China, Japan, and the West*, Jerry Dusenbury, trans. Tokyo: University of Tokyo Press, 1984.

Needham, Joseph, *et al. Science and Civilisation in China*. Cambridge: Cambridge University Press, 1954–.

Porkert, Manfred. *The Theoretical Foundations of Chinese Medicine: Systems of Correspondence*. Cambridge, Massachusetts: The MIT Press, 1974.

Puett, Michael J. *The Ambivalence of Creation: Debates Concerning Innovation and Artifice in Early China*. Stanford, California: Stanford University Press, 2000.

Qian, Wen-yuan. *The Great Inertia: Scientific Stagnation in Traditional China*. London and Sydney: Croom Helm, 1985.

Schwartz, Benjamin I. *The World of Thought in Ancient China*. Cambridge, Massachusetts: The Belknap Press of Harvard University, 1985.

Smith, Richard J. and D.W.Y. Kwok, eds. *Cosmology, Ontology, and Human Efficacy: Essays in Chinese Thought*. Honolulu: University of Hawaii Press, 1993.

Strassberg, Richard E. *Inscribed Landscapes: Travel Writing from Imperial China*. Berkeley, California: University of California Press, 1994.

Sung, Ying-hsing. *T'ien-kung K'ai-wu: Chinese Technology in the Seventeenth Century*, E-Tu Zen
    Sun and Shiou-Chuan Sun, trans. University Park, Pennsylvania: The Pennsylvania State
    University Press, 1966.
Tsunoda, Ryusaku, ed. *Sources of the Japanese Tradition*. New York: Columbia University Press,
    1958.
Tucker, Mary Evelyn. *Moral and Spiritual Cultivation in Japanese Neo-Confucianism: The Life and
    Thought of Kaibara Ekken (1630–1714)*. Albany: State University of New York Press, 1989.
Tucker, Mary Evelyn and John Berthrong, eds. *Confucianism and Ecology: The Interrelation of
    Heaven, Earth, and Humans*. Cambridge, Massachusetts: Harvard University Center for the
    Study of World Religions, 1998.
Wang, Aihe. *Cosmology and Political Culture in Early China*. Cambridge: Cambridge University
    Press, 2000.
Yasunaga, Toshinobu. *Ando Shoeki: Social and Ecological Philosopher of Eighteenth-Century Japan*.
    New York and Tokyo: Weatherhill, 1992.

JAMES MILLER

DAOISM AND NATURE

DEFINITIONS, ORIGINS, CONTEXTS

Daoism is an English word that covers a wide variety of religious and philosophical traditions. These stem from an understanding of the human condition
as being inextricably enfolded in a matrix of cosmic power and creativity
known as Dao 道, conventionally translated by the term "Way" (and also
transliterated, under an older romanization system, as "Tao"). The concept of
Dao as it is used today originated in China during the period of political
disunity that saw the gradual dissolution of the Zhou 周 empire
(1122–256 BCE). This disintegration began with the Spring and Autumn
(Chunqiu 春秋) period (770–476 BCE) and continued with the Warring States
(Zhanguo 戰國) period (475–221 BCE). It was accompanied by the flourishing
of an intellectual culture whose main figures we know today by their Latinised
names (e.g. Confucius, Mencius), all of whom were concerned with one fundamental (and therefore pragmatic) question: Where is the Way? This penetrating
question was answered on a wide variety of levels – ethical relationships,
political organization, moral self-cultivation, and ritual order – that were seen
as being mutually related. Indeed the notion of the unity and relationship of
all dimensions of life was held as axiomatic. There was, however, one seminal
text that penetrated right to the heart of this deeply human, and deeply humane,
question by stating that the Way is to be sought in the very vitality of nature,
in the wholly natural and wholly spontaneous transformation and flourishing
of the world. The *Daode jing* (*Scripture of Way and Power*, 4th century BCE)
or the *Laozi* (after its mythical author) is the key text around which the variety
of Daoist traditions continue to construct themselves by means of written
commentaries, ritual recitation and meditation. The worldview of this text
implies a certain redundancy to the title of this present essay: there can be no
"Dao and Nature" as though these were two discrete categories of being (*cf.*
Creator and the created in Christian thought) that must be brought by some
means or other into relation (*cf.* an absolute and singular act of creation *ex
nihilo*). Dao is no more – and no less – than the flourishing of nature itself.

This natural spontaneity translates a Chinese term *ziran* 自然 (lit. self-so) that

393

H. Selin (ed.), Nature Across Cultures: Views of Nature and the Environment in Non-Western Cultures, 393–409.
© 2003 *Kluwer Academic Publishers. Printed in Great Britain.*

is the basis of the Modern Standard Chinese term for nature (*ziranjie* 自然界).
In *Daode jing* 25 we read:

> Humans model Earth.
> Earth models Heaven.
> Heaven models Dao.
> Dao models natural spontaneity (*ziran*).

The three basic dimensions of existence (human, earthly, and heavenly) are
thus folded into the natural evolution of the Way, which proceeds without
reference to any wholly external power or transcendent force. It is important
to bear in mind, however, that the ancient Chinese term may not simply be
equated with the English word "nature", for the natural operation of the Dao
is not limited to one dimension of life or being. Indeed it lies at the root of all
activity, whether human, celestial, political, animal, or vegetal.

The earliest Daoists, if we may permit for the sake of convenience such a
cross-cultural, cross-historical designation, were hardly armchair philosophers
idly speculating as to the metaphysical foundations of the universe. They sought
a practical experience of nature's creative power, manifested in the spontaneous
arising and decaying of things in an ceaseless flow of activity (*yang* 陽) and
receptivity (*yin* 陰) within the energetic field (*qi* 氣) that constitutes the material
of the universe. And they sought to model their lives after this natural spontaneity, making *ziran*, or naturalness, the core value of their philosophy (Liu, 2001).
In the *Daode jing* this core value entails a strategy of non-(artificial) action
(*wuwei* 無為) as a means to achieve the optimal state of harmonic integration
between the various dimensions of life. For an authoritarian monarch in feudal
China, such a strategy was absolutely necessary for survival.

> The more prohibitions and rules,
> The poorer people become.
> The sharper people's weapons,
> The more they riot.
> The more skilled their techniques,
> The more grotesque their works.
> The more elaborate the laws,
> The more they commit crimes.
>
> Therefore the Sage says:
> I do nothing [*wuwei*]
> And people transform themselves.
> I enjoy serenity
> And people govern themselves.
> I cultivate emptiness
> And people become prosperous.
> I have no desires
> And people simplify themselves (*Daode jing* 57, trans. Addiss and Lombardo, 1993).

While this chapter clearly refers to the application of Daoist strategy in the
political realm, the broad teaching of the *Daode jing* – cultivating emptiness,
enjoying tranquility, becoming desireless and *wuwei* – came to be extended and
applied to all realms of life. The *Zhuangzi* (3rd century BCE) speaks of the
Dao in strongly metaphysical tones.

The Way has its reality and its signs but is without action or form. You can hand it down but you cannot receive it; you can get it but you cannot see it. It is its own source, its own root. Before Heaven and earth existed it was there, firm from ancient times. It gave spirituality to the spirits and to God; it gave birth to Heaven and to earth. It exists beyond the highest point and yet you cannot call it lofty; it exists beneath the limit of the six directions, and yet you cannot call it deep. It was born before Heaven and earth, and yet you cannot say it has been there for long; it is earlier than the earliest times, and yet you cannot call it old (Zhuangzi, 1964: 77).

The Xiang'er 想爾 commentary on the *Daode jing* (*ca.* 200 CE, ascribed to Zhang Daoling 張道陵) views the text in terms of personal self-cultivation in the context of the Daoist religious community and stresses values such as tranquility and non-aggression, while deprecating strong emotions such as jealousy and hatred.[1]

Modern environmentalists, who see in it the principles of conservation, organic harmony and respect for nature that they themselves espouse, have adopted the *Daode jing*, together with selected parts of the tradition, enthusiastically. Although it is foolish, not to say blatantly anachronistic, to suggest that early Daoists were environmentalists, nonetheless it is easy to see why the text and its traditions should be so appealing in the present climate of continuing environmental degradation. Typical of the modern environmentalist appropriation of Daoism is Doris LaChapelle's "extended rhapsody", *Sacred Land Sacred Sex – Rapture of the Deep* (1992).

Now after all these years of gradual, deepening understanding of the Taoist way, I can state categorically that all these frantic last-minute efforts of our Western world to latch on to some "new idea" for saving the earth are unnecessary. It's been done for us already – thousands of years ago – by the Taoists. We can drop all that frantic effort and begin following the way of Lao Tzu [Laozi] and Chuang Tzu [Zhuangzi] (LaChapelle, 1992: 90, cited in Paper, 2001).

There is much support for the view that early Chinese thought is deeply embedded in the natural world. Sarah Allan (1997) persuasively explains how the root metaphors of Chinese culture are derived from images of nature or explained in terms of natural phenomena. The term Dao, for instance, is analogous to the flowing of water. It provides irrigation-life for the ten thousand things (*Daode jing* 62), and like water it is soft, weak, pliable, yielding, and ultimately unstoppable. The analogy of Dao with natural phenomena is extended in Confucian discourse to the question of the moral nature of humanity. In the argument between Mencius and Gaozi over the moral status of human nature, both "accept as a premise that human nature (*xing*) and water are like phenomena".

Gaozi said, 'Nature is like a bubbling spring. If you make a channel for it to the east, then it will flow eastward. If you make a channel for it to the west, then it will flow westward. Human nature is not biased toward good or bad; it is like water which is not biased toward east or west'.

Mencius said, "Water certainly is not partial to east or west, but is it not partial to above and below? Human nature being good is like water going downwards. Among people, there are none who have [as their nature] not being good; of water there is none which does not descend (Cited in Allen, 1997:42).

On the other hand, the attitude that seemingly equates Daoism (or at least the classical, textual Daoism) with an environmentally friendly naturalism, has not been universally shared amongst modern commentators, whose views are expertly summarized and reviewed by J. J. Clarke. Clarke (2000) notes that the simple equation of all that is natural with all that is Daoism leads to a peculiar ontological problem. Roger Ames (1987: 342) puts it this way: "If all is *tao* and *tao* is natural, what is the source, the nature, and the ontological status of unnatural activity?" Although it is impossible to think Daoism without thinking nature, the Daoist *problematik* calls into question the nature of nature itself and offers many seemingly conflicting and paradoxical answers to the question of the human relationship with, and our being embedded within, the natural world. This is entirely appropriate as the root understanding of Dao as the spontaneous and creative unfolding of nature is, for Daoists, fundamentally a concept that is dark, obscure and mysterious (*xuan* 玄) and does not readily disclose the normative patterns (*li* 理) for cultured conduct (*li* 禮) that Confucian moral philosophers would ideally like.

THE PRACTICE OF CORRELATIVITY

It was during the former Han dynasty (206 BCE to 22 CE) that a basic cosmological system was established that remains at the heart of the traditional Chinese worldview. This basic microcosm/macrocosm relationship can be seen in such early Han dynasty cosmological texts as the *Huainanzi*:

> The roundness of the head is an image of heaven, the squareness of earth is the pattern of earth. Heaven has four seasons, five phases, nine directions and 360 days. Human beings have accordingly four limbs, five inner organs, nine orifices, and 360 joints. Heaven has wind, rain, cold, and heat. Human beings have accordingly the actions of giving, taking, joy, and anger. The gall bladder corresponds to the clouds, the lungs to the breath, the liver to the wind, the kidneys to the rain, and the spleen to the thunder (*Huainanzi* 7, cited in Kohn, 2001: 52).

As traditional Chinese medicine developed, a more mature and technical understanding of the relationship between the various cosmic-human dimensions of life emerged. This cosmology is based on the concept of a universe of multiple, interrelated dimensions of *qi*-energy that resonate synchronically with each other and diachronically in a sequence of five phases. The standard "generative" sequence is denominated in terms of natural elements: Wood > Fire > Earth > Metal > Liquid. There is also a converse controlling or "ruling" sequence: Earth > Water > Fire > Metal > Wood. The five phases are also correlated synchronically across potentially infinite dimensions of life. In Table 1 the correlations are mapped in terms of the generative sequence.

This way of correlative thinking is most well known as the basis for traditional Chinese medicine in which qi flows through the body along twelve meridians, each correlated to one of six major and six minor orbs or energy systems. One of the earliest medical texts, the *Yellow Emperor's Classic of Internal Medicine, Simple Questions* (*Huangdi neijing suwen*) demonstrates the correlations this way:

Table 1   Five Phases

Agent	Direction	Colour	Season	Orb	Emotion	Sense	Flavours
Wood	East	Green	Spring	Liver	Anger	Eyes	Sour
Fire	South	Red	Summer	Heart	Joy	Tongue	Bitter
Earth	Centre	Yellow	Late summer	Spleen	Worry	Lips	Sweet
Metal	West	White	Fall	Lungs	Sadness	Nose	Pungent
Water	North	Black	Winter	Kidney	Fear	Ears	Salty

The orb of the heart includes the pulse. Its splendour shows in the complexion. It rules over the kidneys.

The orb of the lungs includes the skin. Its splendour shows in the body hair. It rules over the heart.

The orb of the liver includes the muscles. Its splendour shows in the nails. It rules over the lungs.

The orb of the spleen includes the flesh. Its splendour shows in the lips. It rules over the liver.

The orb of the kidneys includes the bones. Its splendour shows in the hair on the head. It rules over the spleen.

If one eats too salty, the pulse hardens, tears appear, and the complexion changes.

If one eats too bitter, the skin withers and the body hair falls out.

If one eats too sour, the flesh first hardens, then wrinkles and the lips become slack.

If one eats too sweet, the bones ache and the hair on the head falls out.

These are the injuries caused by the five flavours (Kohn, 1993:166).

In this extract, the diachronic correlations are based on the blocking or "ruling" sequence. The first line of each of the two paragraphs describe how an excess in the water phase causes problems for the fire phase, which regulates it according to the ruling sequence. Thus an excess of salt (water phase, synchronic correspondence = kidneys) would create problems diachronically for the fire phase (synchronic correspondence = heart) that regulates it. A disturbance in the fire phase (heart) means problems for the pulse and the complexion.

All of the problems of the body must, however, take into account the rotation of the seasons. Thus in winter (water phase) it is appropriate that there should be more salty foods than in the other seasons, but in the spring one should be more careful about eating salty food.

Although there is nothing specifically Daoist about this correlative thinking, it lies at the heart of all traditional Chinese thinking. Moreover, since the body is the pre-eminent field or domain in which Daoist practices take place, it is important to understand how the functioning of the body is located within, and synchronically affected by, the constantly transforming phases of the natural world. In fact, in the fully realized or perfected (zhen 真) Daoist, the boundaries between self and world are completely porous; it is as though one is fully transparent to the cosmic location in which one is situated. In psychological terms, there is no assertion of ego over and against what is non-ego.

This holistic or correlative way of thinking is also emblematic of Daoist texts and practices, in which words and gestures signify and actualise objects of a wholly other dimension. For instance, in Highest Clarity (Shangqing) Daoist texts, meditation practices developed that sought to visualize astral deities dressed in certain coloured clothes, inhabiting certain organs of the body at certain times of the year. Here the energy cycles of the body are fully aligned with the seasons, the stars, and the colours in an elaborate and highly technical exercise of meditative harmonization.

It is important to appreciate how this multi-dimensional all-pervasive way of thinking threads itself throughout the Daoist worldview. In this way heavens, the earth and the body are woven together into the seamless fabric of the spontaneously self-creating dao. Below I examine these three basic categories of nature (astrology, geography, physiology) separately, but I do not wish to give the impression that they represent wholly distinct domains of nature that are somehow disconnected from each other, as in the traditional specializations of the natural sciences of the modern Western university. Rather they are to be understood as participating in an "ecological" relationship of reciprocity and mutual influence.

THE SKY

With the overthrow of the ancient Shang Dynasty by the Zhou (1100 BCE) came the substitution of the high lord (Shangdi 上帝) by the sky or heaven (Tian 天) as the supreme principle of religion. Tian, not a personal deity but an impersonal cosmic force, ruled the seasons and the stars and, by extension, determined the calendar and the festivals. In later thought, heaven was associated with the yang aspect of the transformation of the Dao, in contrast to the earth, which was associated with yin. By means of interaction between heaven and earth, yang and yin, the ten thousand things come into existence. Since Tian was essentially a generative force within the natural world, it does not quite have the same meaning as "heaven" and is perhaps better translated simply as "sky" (Allan, 1997: 21–22).

Astrology thus became an important divinatory art throughout Chinese culture and within Daoism in particular. Since the early Han dynasty there were almanacs based on a variety of different schemes that sought to bring order to the apparently unrelated phenomena that occur at the same time and to advise on the most propitious times for beginning projects. The most important divinatory schemes include the 64 hexagrams of the Yijing (I Ching, Book of Changes), allegedly composed by Confucius. This binary system of 64 (2^6) combinations of yin (represented by a broken line) and yang (represented by a solid line) is one of the world's most pervasive and enduring methods of divination (Yoshinobu, 2000). Aside from the Yijing we find various numerological schemes based on the 60 day cycle that combines the 12 Chinese zodiac signs with the ancient 10 day week cycle; these are also combined and correlated with the five phase and yin-yang systems. The heavenly (and therefore natural) world was thus perceived as a complex of interlocking cycles, knowledge of which was important when undertaking any important activity. Thus arose the art of hemerology, calculating auspicious days, the original application for which was in military strategy (Yoshinobu, 2000: 548).

The cyclical sequences that were observed in the natural world were from pre-Han times applied to what we would call the political realm. The doctrine of the Mandate of Heaven (*tianming* 天命) held that the power to govern was granted, and revoked, by Tian. Whereas Confucian philosophers such as Mencius viewed this in moral terms, others viewed it in naturalistic terms. The doctrine of the five "virtues" (*de* 德) that was developed and systematized by Zou Yan 騶衍 (*ca.* 350–270 BCE) held that the cycle of natural conquest applies to politics as much as to nature (as if those two concepts could properly be separated). Simply put, Zou applied or extended[2] the fundamental natural philosophy of the five virtues to the political sphere, holding that the sequence of dynastic mandates mirrored the natural sequence of the five phases or virtues.

> Each of the Five Virtues is followed by the one it cannot conquer. The dynasty of Shun ruled by the virtue of Earth, the Xia ruled by the virtue of Wood, the Shang dynasty ruled by the virtue of Metal, and the Zhou dynasty ruled by the virtue of Fire (Wen Xian 59,9b; adapted from Needham, 1956: 238).

Thus the virtue or power to govern is not only symbolised by the cyclical patterns of nature, but was directly encoded in the revolutions – or revelations – of heaven. By paying attention to the night sky, it was possible to give predictions and warnings about political matters. The ramifications of such a naturalistic political theory are complex. Liu Xin 劉歆 (d. 23 CE), for example, sought

> in Joseph Needham's words, 'to reconcile the irreconcilable' though discovering the regular interrelations of the sexagenary cycle, the lunar cycle, tropical years, eclipse periods, the 'year-star' Jupiter's synodic revolutions, and conjunctions of the five naked-eye planets – all in accord with the celestial and terrestrial movements of the Five Phases (Bokenkamp, 1994 and Sivin, 1969).

The result of such calculations is the determination of times of "apocalypse" that occur at the conjunction of all the cycles, once every 23,639,040 years, and the corresponding conclusion that the world undergoes periodic radical transformation (Bokenkamp, 1994: 65).

To help with similarly complex calculations, a diviner's board or compass (*shi*) was invented in the Han period. This assisted in correlating the 60-day cycle, the 28 days of the lunar cycle, and the 24 periods of the solar year (Kalinowski, 1986: 12). To the modern western imagination such an approach seems wholly superstitious, but in fact it conveys a deep-seated organic naturalism that pervades Chinese culture. Because of the regularity of the stellar cycles, fate, as such, is ultimately explainable. In Daoism the stars came to be understood as powerful cosmic forces and celestial residences for astral deities. Though they are remote, our lives are bound up with them. Fate (*ming* – in Heideggerian terms the givenness of our situation) is perhaps the most intractable feature of the human condition. Yet in China fate was not seen as wholly capricious, but as ultimately fathomable and explicable. Looked at this way, Chinese astrology is comparable to a popular view of genetics, in which the ultimate destiny of human beings is seen as capable of being symbolized in a complex sequence of the four letters ACGT. Both instantiate the view of Paracelsus (or perhaps fulfil his prophecy) that nature itself is a text that we must study and translate for ourselves (Hahn, 1993: 688 and Blumenberg, 1986).

Daoist religion takes the view that since nature is in a state of constant flux, our lives have not been fixed forever in genetic tablets of stone. On the contrary, human beings are in a privileged position to shape their own destinies. A classic Daoist maxim says, "My destiny lies with myself and not with anyone else". For many Daoists, then, dealing with destiny involves dealing with the stars and negotiating with the deities they represent. This negotiation can be understood as symbolizing on a religious-imaginative plane the continuous interaction between human beings and the natural conditions of their existence. Nature thus exists as the sum total of finite conditions in which a human life is embedded and with which it has to do. Human life is not only empowered by its location in the natural environment, but also circumscribed and limited by it.

One of the earliest Chinese religious traditions that was absorbed into the Daoist mainstream was the shamanic tradition of voyages to the stars. We know of this tradition through ancient poetic texts such as the *Songs of Chu* (*Chuci*; ca. 200 BCE; trans. Hawkes, 1967) that relate (generally woeful) encounters with goddesses as a result of astral voyages. One such poem, "Faraway Journey" (*Yuanyou*) relates in splendid dramatic imagery a voyage to the stars and the transformation of the human voyager:

My face, like jade, is flushed with radiant colour;
My pure essence is starting to grow strong,
My solid body dissolving into softness,
My spirit's ever subtler and more unrestrained.
...
Holding to my sparkling soul, I climb to the empyrean;
Clinging to the floating clouds, I ride up further high.

As these shamanic journeys became integrated into the Daoist mainstream, attention focused on the stars of the Big Dipper (*beidou* 北斗, Plough, *Ursa Major*) located close to the central ridgepole (*taiji* 太極; the "supreme ultimate" of Neo-Confucian philosophy) from which the sacred canopy of the cosmos is suspended. Since it was thought that the whole sky revolved around it, it was correlated with the actions of the Emperor.

When the filial conduct of the kingly one spills over, the Dipper lets its germinal essence fall: the seven stars of the Northern Dipper are bright and luminous as if about to fall (*Ruiying tu* 瑞應圖 3a; trans. Schafer, 1977: 49).

The seven stars of the constellation had traditional Chinese names, to which Daoist traditions appended their own designations and correlations. Thus in the texts of the Highest Clarity (*Shangqing* 上清) tradition, the fourth star was known as Occult Tenebrity (*Xuanming* 玄冥), correlated in the *Scripture of the Inner Radiances of the Yellow Court* (*Huangting neijing jing* 黃庭內景經; ca. 3rd century CE) with the spirit that that rules over the kidneys. The fifth star was known as Cinnabar Prime (*Danyuan* 丹元), the spirit of the heart. In the case of this Daoist tradition, the spirits in the body are the corresponding inner radiances (*neijing* 內景) of the constellated effulgences of the Big Dipper. Indeed the fully realized (*zhen* 真) spiritual body may be described as being as deeply obscure and as fully radiant as the night sky itself. The celestial being, Lady

Wei Huacun, who revealed to the medium Yang Xi much of the original Shangqing corpus in the early 360s CE, is described in her biography as a radiant being from outer space:

> Empyreal phosphor, glistening high;
> Round eye-lenses doubly lit;
> Phoenix frame and dragon bone;
> Brain coloured as jewel-planetoids;
> Five viscera of purple webbings;
> Heart holding feathered scripts (*Sandong zhu'nang* 8. 22b: 230).

The most advanced adepts of the Shangqing tradition, moreover, feasted on pure light energy rather than the diet of grains that were thought to bring about death in ordinary mortals. The practice of absorbing the light (*fuguang* 服光) caused the adept to become a transparent, luminous being. The *Esoteric Biography of Purple Perfected Yang* (*Ziyang zhenren neizhuan* 紫陽真人內傳 4th century CE) relates that as a young man Zhou Ziyang would soak up the dawn light, and after five years became so luminous and transparent that one could see his internal organs (Porkert, 1979). To be a fully realized Daoist being, then, is to have appropriated for oneself the glistening jewels of the natural world and to embody in oneself the radiant light of the stars.

THE EARTH

The category of sacred space is of supreme importance in the comparative study of religions (Eliade, 1961), and, in common with many religions, Daoism accords pre-eminent status to the mountain. In this way the earth is viewed, like the sky (and hence the calendar), as containing naturally occurring points or configurations that particularly lend themselves to the spiritual transformation of human beings. In these spaces, the natural world is pregnant with mystery and numinous power. From the Daoist point of view, however, there is nothing particularly magical or otherworldly about this, since in Daoism the sacred is located within the ordinary world. It is simply that the contours of energy-matter in the universe are not distributed with the random evenness of Brownian motion, but configured and constellated in particular forms and spaces. In fact the categories of "sacred" and "profane" do not really hold up to intense scrutiny in the Daoist tradition, since both are no more and no less than the evolutionary outworking of the Dao. The radical otherness or discontinuity that this Eliadean dichotomy implies does not feature in a metaphysics of correlation and mutual implication.

China possesses five sacred mountains or marchmounts (*wuyue* 五嶽) that were designated as such under the Han dynasty to protect the five directions of the empire (the four cardinal points, plus the centre) and were objects of the imperial cult (Landt, 1994 and Geil, 1926) (see Table 2).

In addition to the five imperially-sponsored sacred mountains, there are numerous other mountains that are of particular importance to Daoism, including, for example, Qingchengshan 青城山 in Sichuan Province where there is still

Table 2 Sacred mountains

Direction	Marchmount	Location
Northern Peak (Beiyue 北嶽)	Hengshan 恒山	Shanxi
Southern Peak (Nanyue 南嶽)	Hengshan 衡山	Hunan
Eastern peak (Dongyue 東嶽)	Taishan 泰山	Shandong
Western Peak (Xiyue 西嶽)	Huashan 華山	Shaanxi
Central peak (Zhongyue 中嶽)	Songshan 嵩山	Henan

a significant Daoist presence today. Many of these mountains are the locations for Daoist temples (*guan* 觀) or altars (*tan* 壇), or serve as homes for recluses and hermits. As James Robson has pointed out, no one sectarian lineage or religion can claim a monopoly on Chinese sacred mountain space; each tradition added its own layers of significations and interpretation to the steadily accreting wealth of religious meaning (Robson, 1995 and Naquin and Yü, 1992). Nevertheless, mountains hold particular significance for Daoists, and feature prominently in Daoist hagiography. Zhang Daoling, the founder of *Tianshi dao* 天師道 (Way of the Celestial Masters, the first formalized Daoist religious movement) received his first Daoist revelation on Mt. Heming 鶴鳴山 in 142 CE.

> He heard that the people of Shu were very pure and generous and could easily be taught; moreover there were many famous mountains in Shu [present day Sichuan province]. So he entered Shu with his disciples and dwelt on Mt. Heming where he composed twenty-four volumes of Daoist writings. He then concentrated his spirit and refined his will and suddenly there were heavenly beings descending with a thousand chariots, ten thousand riders and golden carriages with feathery canopies drawn by countless dragons on the outside and tigers on the inside. He called himself the Archivist; at other times he called himself the child of the Eastern sea. They then gave Ling the newly emerged Way of the Covenant of Orthodox Unity.[3]

From this revelation on Mt. Heming began the organised religious movement that we know as Daoism today.

Mountains are also significant because they are the home to grottoes. Grottoes or caverns (*dong* 洞) form a means of communication (*tong* 通) or pathway between the Daoist adept and the mysterious workings of the Dao, whether symbolized as the starry heavens above or as the internal functioning of the body.

> Despite a singular solidity, their physical permeability in terms of air- and water-flow reflects the inner workings of the human body. Blood equals water; air equals breath. Spermatic liquids form pools; walls constitute shapes like inner organs or viscera. Their resident, left windowless and in an enclosed void, experiences the dignity of complete independence and autarky (Hahn, 2000: 695).

This correlation of external physical space and internal physiological space is typical of the Daoist approach to nature in which natural images are replicated and reformulated across a multitude of dimensions and categories of life. There is, in the end, nothing discrete about the Daoist mental universe, for it is capable of reaching across worlds and mapping layers of meaning over and against and on top of each other. All this is with the aim of comprehending the polymorphic, transfigurative character of the natural world, a world of continuous transformation (*bianhua* 變化) in which a pupa becomes a chrysalis and a chrysalis becomes a butterfly, and next we dream we are a butterfly, or are we a butterfly dreaming of us? (*Zhuangzi*, 1964: 45).

An example of the multivalent character of natural imagery in the Daoist imagination can be seen in the biography of Zhou Ziyang (*q.v.*). Zhou's spiritual progress takes place by means of a pilgrimage to China's famous mountains in search of the way to become a fully perfected being. His journey begins at Songshan, the central peak, where he meets the deity known as the Central Huanglao Lord (*Zhongyang Huanglao jun* 中央黄老君). Zhou travels from mountain to mountain throughout China amassing a treasury of Daoist revealed scriptures, recipes and talismans, and finally ascends the Empty Mountain (*Kongshan* 空山) and in a grotto there has a vision of the Huanglao Lord flanked by the Lord of Infinite Lustre and the White Prime Lord. The Huanglao Lord tells Zhou he "should look back in his own grotto chamber". At this point Zhou closes his eyes and for the first time practices internal vision (*neishi* 内視) and receives a vision of the White Prime Lord and the Lord of Infinite Lustre in the inner grotto chamber of his own imagination. The Huanglao Lord congratulates him on his subtle powers of mental concentration and advises him to return to the Everlasting Mountain (*Changshan* 常山) where he will be initiated into the highest levels of perfection: "This is the way to ascend to the heavens in broad daylight". Eventually Zhou achieves the level of perfected being and is officially granted the title Perfected Purple Yang.

It is impossible to interpret such a story in an unequivocal, literal manner. The mountains referred to are both physical mountains and metaphors for moments of meditative transformation. Similarly the grotto chamber refers both to a physical space in which meditation is practiced and also the inner-body space in which Zhou's internal vision takes place. "The way to ascend to the heavens in broad daylight" is paradoxically attained through the practice of internal meditation in darkness and shadow. The Daoist religious imagination draws upon a wealth of natural imagery, and in so doing reveals the deep mystery of nature's self-transformative power. In this way nature is like a text whose mysteries are waiting to be revealed to the properly initiated.

The textuality of nature is made concrete and explicit in the association of mountain grottoes with repositories of Daoist texts. In the biography above, Zhou Ziyang was able to amass a large quantity of sacred texts and talismans by visiting mountains. The Daoist literatus Ge Hong (287–347) explains,

> All noted mountains and the Five Marchmounts harbor books of this sort, but they are hidden in stone chambers and inaccessible places. When one who is fit to receive the Dao enters the mountain and meditates on them with utmost sincerity, the mountain spirits will respond by opening the mountain, allowing him to see them (*Baopuzi neipian* 19/336, cited in Campany, 2001: 134).

A grotto thus permits communication between the earthly and celestial realms. It is dark and mysterious, but paradoxically a repository of revelations and enlightenments. Grottoes thus form the medium of communication with the central realms of light in the heavens; they are libraries of sacred texts that permit the communion of heaven, earth and humans.

The association between mountain grottoes and sacred libraries was formalised when the Daoist scriptures first came to be compiled by Lu Xiujing (406–77). Liu arranged them into three subdivisions that he termed grottoes: the Cavern of Mystery, the Cavern of Perfection and the Cavern of Spirit. These grottoes were the repositories of the original texts, presided over by deities who had transmitted them to earth at various times and places. These three grottoes moreover corresponded to the three major heavens, Jade Clarity, Highest Clarity and Great Clarity, each divided into twelve sub-heavens, making thirty-six in total (*Daojiao sandong zongyan*, trans. Kohn, 1993: 65–71). The grottoes, as libraries, thus represent the chief means of communication between the celestial and earthly realms in Daoism. Some of these methods of communication are publicly available to us today as the revealed texts and talismans contained within the Daoist Canon (*Daozang* 道藏). Others are secretly handed down from master to initiate in lineages of transmission and ordination that continue to this day.

It is no surprise, therefore, that the most important text of the Shangqing (Highest Clarity) revelations was the scripture of the "great cavern" *Dadong jing* 大洞經 (*Scripture of Great Profundity*), several versions of which exist in the Daoist canon. The term *dadong* is glossed by a commentator as *taiji* 太極, the "supreme ridge-pole" of the cosmos from which the cosmic canopy is suspended (see Robinet, 1993: 97–119). In this way, the dark earthly caverns and the radiant heavens above are seen as equal aspects of the same cosmic fabric from which the revealed texts are woven. In the Daoist vocabulary, therefore, earth, heaven and text are bound up together in the fabric of the Dao.

THE BODY

It is impossible to treat any aspect of Daoism without referring at length to the idea of the Daoist body (Schipper, 1993). The body is the pre-eminent terrain for the Daoist religious imagination. It is within the body that the cosmic landscape is imaged and upon which the subtle alchemical processes of internal meditation take place. The body is seen as co-extensive with the natural world and in microcosmic sympathy with the macrocosm.

The most important body for the Way of the Celestial Masters (*Tianshi dao* 天師道) is the body of the community. We have some insight into how, historically, this communal movement functioned in its earliest days through a text the Celestial Masters adopted and transmitted, known as the *One Hundred and Eighty Precepts* (*Yibaibashi jie* 一百八十戒). In his study of this text, Kristofer Schipper notes, "not less than twenty [of the precepts] are directly concerned with the preservation of the natural environment, and many others indirectly:

14. You should not burn [the vegetation] of uncultivated or cultivated fields, nor of mountains and forests.
18. You should not wantonly fell trees.

19. You should not wantonly pick herbs or flowers.
36. You should not throw poisonous substances into lakes, rivers, and seas.
47. You should not wantonly dig holes in the ground and thereby destroy the earth.
53. You should not dry up wet marshes.
79. You should not fish or hunt and thereby harm and kill living beings.
95. You should not in winter dig up hibernating animals and insects.
97. You should not wantonly climb in trees to look for nests and destroy eggs.
98. You should not use cages to trap birds and [other] animals.
100. You should not throw dirty things in wells.
101. You should not seal off pools and wells.
109. You should not light fires in the plains.
116. You should not defecate or urinate on living plants or in water that people will drink.
121. You should not wantonly or lightly take baths in rivers or seas.
125. You should not fabricate poisons and keep them in vessels.
132. You should not disturb birds and [other] animals.
134. You should not wantonly make lakes (Schipper, 2001: 82–3).

In answer to the question of why the earliest Daoist communities were concerned with the state of the natural environment, Schipper draws the conclusion (p. 83) that the natural environment functioned as a kind of sanctuary, in the sense of a sacred space and in the sense of a place of refuge from the human world. There is a more fundamental point at stake here, which is evident in the language used: the precepts are directed at members of the community, and in fact we know that they were adopted as the code of practice for the heads of the Celestial Masters community, known as libationers (*jijiu* 祭酒). The implication of the imperative "you should not" is that the libationer himself, and by extension the community as a whole, will suffer the consequences of failing to abide by the precepts. Lu Xiujing wrote, "When a Daoist master has not received Laojun's *One Hundred and Eighty Precepts*, then his body will have no virtue, and he cannot be deemed to be a Daoist master and receive the homage of the people, nor can he rally and administrate the gods and ancestors" (*Lu xiansheng daomen kelüe* DZ1127: 16b. Cited in Schipper, 2001: 91). Thus the purpose of Daoist priests' observing this code of conduct is that they will be able to perform their religious tasks correctly and be exemplary leaders to their congregations.

Although the *Precepts* have been replaced by other texts as the code for Daoist priests today, their influence is very much in evidence. At a 1998 conference on Daoism and Ecology at the Harvard University Center for the Study of World Religions, Liu Ming, a contemporary North American Daoist priest, said,

> As a Daoist, I do not share the notion that life is a series of problems to be solved. I do not
> see human beings at odds with nature. Not in the past, not now, and not in the future. I see
> no crisis. What I do see are human beings, who, having the option to be at odds with themselves,
> choose to practice greed, wastefulness, and fear (Kohn, 2001: 378).

The similarity between this statement and the *Precepts* is that the discussion of the Daoist approach to nature is couched in terms of the person-within-the-world, not in terms of "nature" or "environment" as though these terms referred to some external entity or object with which we have to do. In this light the respect for the environment indicated in the *One Hundred and Eighty Precepts* indicates a respect for the life of the communal body and for the life of the individual.

From the earliest days of the Daoist tradition the practice of nourishing the vitality of the body has been a central concern (Roth, 1999), a concern that was elaborately developed in the Daoist tradition of inner alchemy (*neidan* 內丹) later known as the golden elixir (*jindan* 金丹). The foundations of internal alchemy practices can be found in the cosmogony set forth in chapter 42 of the *Daode jing*: "Dao gives birth to One; One gives birth to Two; Two gives birth to Three; Three gives birth to the ten thousand things". This cosmogony accounts for the gradual decay and dissipation of energy within the cosmos. The aim of the alchemist is to reverse this dissipation by reverting or countering (*ni* 逆) the cosmogonic process, a process further elaborated in terms of the cyclical mutation of yin into yang and yang into yin, as symbolized in the sixty-four hexagrams of the *Yijing*. Briefly, the aim is to arrive, through a series of purifications, at the decoction of undifferentiated yang- and yin-energy (the "Two" of the cosmogonic sequence) and to fuse these two primal energies into the undifferentiated Oneness of the original Dao.

Alchemists such as Ge Hong had sought to arrive at this stage through an "external", operative or laboratory alchemy (*waidan* 外丹) based on the use of mercury sulphide or cinnabar (HgS; *dan* 丹). In one version, cinnabar was heated seven or nine times to produce a pure form of mercury, representing pure yin energy. In a second version, mercury (pure yin) was extracted from cinnabar, and lead (pure yang) from native lead. These two were then fused together, thus reversing the cosmogonic direction of One into Two.[4] Many Daoist traditions, however, viewed laboratory alchemy as distinctly inferior to internal forms of alchemy and meditation, reflecting the overall priority of the inner landscape as the field of operations for Daoist practices.

Within the landscape of the body, pure yin (mercury) is imaged as the pure energy of the kidneys (corresponding to water), and pure yang (lead) as the pure energy of the heart (corresponding to fire). The elixir is decocted in the three "cinnabar fields" (*dantian* 丹田) of the body: the *niwan* 泥丸; (probably derived from a Chinese transliteration of the Sanskrit term *nirvana*) located in the head; the Purple Palace (*zigong* 紫宮), located in the chest; and the lower cinnabar field, also known as the *dantian*, located in the abdomen, an inch below the navel.

By correctly directing the essence (*jing* 精) and the qi of the body through a series of internal meditations, the adept produces an "immortal embryo", the pure distillation of the primal energy from which the adept was created. Having been appropriately nurtured and nourished, the embryo is birthed through the head in a complete inversion of the physical birth of the adept.

*   *   *

It is not easy to sum up the Daoist view of nature and environment, because the tradition is so rich and complex in terms of the appropriation of nature by adepts and the metaphysical imagings of the operation of nature. There are, however, three key principles that tend to distinguish the Daoist view of nature and which commend further study: the surprising recursivity of nature, the practical correlativity of all dimensions of life, and the textuality of the fabric of the Dao.

The recursivity of nature means that nature is evolving in a way that continuously folds back on itself and, as it were, gathers itself up in its hands. Nature is always pregnant with itself in an irrepressible superfluity of vitality and power. This is the theoretical explanation for the fractal-like identity of microcosm and macrocosm in which the overarching patterns of creation, transformation and decay are imaged in both the tiniest and the grandest processes of the cosmos.

The practice of correlativity is thus the way in which the human mind fits together the many different dimensions of life so as accurately to reflect the interlocked and interconnected trajectories of evolution (dao) that are woven together into the fabric of time. This fabric, symbolised spatially as a rotating canopy that is suspended from the central ridgepole of the cosmos, contains many different dimensions of being, but they are all made from the same qi. This vibrates in particular forms and frequencies to configure the various arrays of matter and energy in the universe, all of which resonate sympathetically with each other.

The Daoist tradition, moreover, holds that the deep mysteries of nature are available to the properly initiated adept in the form of texts and talismans that decode the very nature of the Dao for us. These texts are the symbolic revelation of the root processes of the Dao, processes that are ordinarily veiled from our understanding, and which have the appearance of magic to the uneducated. In fact, however, the whole of nature itself may be understood as an ongoing activity of communication or dao – whose alternate meaning is "to speak". In this drama whose script – the script of nature – is continuously evolving, we are both privileged actors and mere fragments of self-consciousness. But this is no Hegelian drama that is aiming towards a final purpose of absolute self-communication. It is a drama that is spontaneously rewriting itself in unpredictable, marvellous and deeply mesmerizing ways.

NOTES

[1] For a translation and study of the Xiang'er commentary, see Bokenkamp, 1999.

[2] Sima Qian's description of Zou's method of *tui* 推 (extension) is as follows: "First he had to examine small objects, and from these he drew conclusions [*tui* 推] about large ones, until he reached what was without limit. First he spoke about modern times, and from this went back to the time of Huang [Di]. ... Moreover he followed the great events in the rise and fall of ages, and by means of their omens and [an examination into] their systems, extended [*tui*] his survey [still further] backwards to the time when the heavens and the earth had not yet been born, [in fact] to what was profound and abstruse and impossible to investigate" (*Shi ji* 74: 3a. In Needham, 1956: 233).

[3] There is a biography of Zhang Daoling in Ge Hong 葛洪, *Shenxian zhuan* (*Biographies of Divine Immortals*).

[4] See the works of Ho Peng Yoke [aka Ho Ping-Yü] especially, *Li, Qi and Shu: An Introduction to Science and Civilization in China* (Hong Kong: Hong Kong University Press, 1985). See also the works of Fabrizio Pregadio. A good starting point is his "Chinese Alchemy: An Annotated

Bibliography of Works in Western Languages," *Monumenta Serica* 44 (1996): 439–76, and his overview "Elixirs and Alchemy" in Livia Kohn, ed., *Daoism Handbook* (2000), pp. 165–195.

BIBLIOGRAPHY

Allen, Sarah. *The Way of Water and Sprouts of Virtue.* Albany: State University of New York Press, 1997.

Ames, Roger T. *Thinking Through Confucius.* Albany: State University of New York Press, 1986.

Blumenberg, Hans. *Die Lesbarkeit der Welt.* Frankfurt: Suhrkamp, 1986 [1981].

Bokenkamp, Stephen. *Early Daoist Scriptures.* Berkeley: University of California Press, 1999.

Bokenkamp, Stephen. 'Time after time: Taoist apocalyptic history and the founding of the T'ang Dynasty.' *Asia Major,* 3rd series, 7: 59–88, 1994.

Campany, Robert Ford. 'Ingesting the marvellous.' In *Daoism and Ecology,* N.J. Girardot, James Miller and Liu Xiaogan, eds. Cambridge, Massachusetts: Harvard University Center for the Study of World Religions/Harvard University Press, 2001, pp. 125–147.

Campany, Robert Ford. *To Live as Long as Heaven and Earth.* Berkeley: University of California Press, 2002.

Clarke, J.J. *The Tao of the West.* New York: Routledge, 2000.

Daode jing 57 Tao Te Ching, Stephen Addiss and Stanley Lombardo, trans. Indianapolis, Indiana: Hackett Publishing Company, 1993.

Eliade, Mircea. *The Sacred and the Profane,* Willard A. Trask, trans. New York: Harper and Row, 1961.

Geil, William E. *The Sacred Five of China.* London: Houghton Mifflin, 1926.

Hahn, Thomas H. 'Daoist sacred sites.' In *Daoist Experience,* Livia Kohn, ed. Albany: State University of New York Press, 1993, pp. 683–708.

Ho Peng Yoke [aka Ho Ping-Yü]. *Li, Qi and Shu: An Introduction to Science and Civilization in China.* Hong Kong: Hong Kong University Press, 1985.

Kalinowski, Marc. 'Les traités de Shuihuidi et l'hémérologie chinois à la fin des Royaumes Combattantes.' *T'oung Pao* 72: 175–228, 1986.

Kohn, Livia. *The Taoist Experience.* Albany: State University of New York Press, 1993.

Kohn, Livia. 'Change starts small: Daoist practice and the ecology of individual lives.' In *Daoism and Ecology,* N. J. Girardot, James Miller and Liu Xiaogan, eds. Cambridge, Massachusetts: Harvard University Center for the Study of World Religions/Harvard University Press, 2001, pp. 373–390.

Kohn, Livia. *Daoism and Chinese Culture.* Cambridge, Massachusetts: Three Pines Press, 2001.

LaChapelle, Doris. *Sacred Land Sacred Sex – Rapture of the Deep.* Asheville, North Carolina: Kivaki Press, 1992.

Landt, Frank A. *Die fünf heiligen Berge Chinas. Ihre Bedeutung und Bewertung in der Ch'ing-Dynastie.* Berlin: Koster, 1994.

Liu Xiaogan. 'Non-action (*Wuwei*) and the environment today: a conceptual and applied study of Laozi's philosophy.' In *Daoism and Ecology,* N.J. Girardot, James Miller and Liu Xiaogan, eds. Cambridge, Massachusetts: Harvard University Center for the Study of World Religions/ Harvard University Press, 2001, pp. 315–339.

Naquin, Susan and Chun-Fang Yü, eds. *Pilgrims and Sacred Sites in China.* Berkeley: University of California Press, 1992.

Needham, Joseph. *Science and Civilisation in China.* Vol. 2. Cambridge: Cambridge University Press, 1956.

Paper, Jordan. 'Daoism and deep ecology: fantasy and potentiality.' In *Daoism and Ecology,* N.J. Girardot, James Miller and Liu Xiaogan, eds. Cambridge, Massachusetts: Harvard University Center for the Study of World Religions/Harvard University Press, 2001, pp. 3–21.

Porkert, Manfred. *Biographie d'un Taoïste Légendaire: Tcheou Tseu-yang,* Mémoires de l'Institut des Hautes Études Chinoises, Vol. 10. Paris: Collège de France, 1979.

Pregadio, Fabrizio. 'Chinese alchemy: an annotated bibliography of works in western languages.' *Monumenta Serica* 44: 439–476, 1996.

Pregadio, Fabrizio. 'Elixirs and alchemy.' In *Daoism Handbook*, Livia Kohn, ed. Leiden: Brill 2000, pp. 165–195.

Robson, James. 'The polymorphous space of the southern marchmount.' *Cahiers d'Extrème-Asie* 8: 221–264, 1995.

Roth, Harold. *Original Tao.* New York: Columbia University Press, 1999.

Schafer, Edward. *Pacing the Void: Tang Approaches to the Stars.* Berkeley: University of California Press, 1977.

Schipper, Kristofer. *The Taoist Body*, Karen C. Duval, trans. Berkeley: University of California Press, 1993.

Schipper, Kristofer. 'Daoist ecology: the inner transformation. a study of the precepts of the early Daoist ecclesia.' In *Daoism and Ecology*, N.J. Girardot, James Miller and Liu Xiaogan, eds. Cambridge, Massachusetts: Harvard University Center for the Study of World Religions/Harvard University Press, 2001, pp. 79–93.

Sivin, Nathan. *Cosmos and Computation in Early Chinese Mathematical Astronomy.* Leiden: E.J. Brill, 1969.

Sun, Xiaochun. *The Chinese Sky During the Han.* Leiden: Brill, 1997.

Yoshinobu, Sakade. 'Divination as Daoist practice.' In *Daoism Handbook*, Livia Kohn, ed. Leiden: Brill, 2000, pp. 541–566.

Zhuangzi. 'Discussion on making all things equal.' In *Chuang-Tzu: Basic Writings*, Burton Watson, trans. New York: Columbia University Press, 1964, pp. 31–45.

HAROLD COWARD

HINDU VIEWS OF NATURE AND THE ENVIRONMENT

In contrast to some attitudes toward nature as an "It" that is separate from humans, Hindus see the surrounding world as a "Thou" of which they are an interdependent part. Humans and their society are imbedded in nature and dependent upon cosmic forces. Individual human life is experienced as a microcosm of the universe. Human life is in continuity with the cosmos. Hindu religion has a strong ethical direction aimed at keeping this relational continuity in balance. This approach has much in common with traditional Chinese and Aboriginal Australian views and practices. For the Hindu the universe is God's body, of which we humans, along with everything else in nature, are but a part. The essence of earth, air, water, the tree, cow, you and me is the same divine spirit manifesting in different forms. Therefore it is natural that the ethic of radical non-violence (ahiṁsā) to all forms of human, animal and plant life should have originated in India. To harm another (person, animal or plant) is to harm God's cosmic body of which one is a part. It is tantamount to harming oneself which one would not want to do on the grounds of logic, self-interest or, at the highest level, respect for the divine.

But if one visits India today, the idealistic Hindu ecotheology described above seems to be little in evidence. In the cities, pollution of earth, air and water, together with the pressure of India's exploding population, is overwhelming. Modern industrialization, with its attitude that nature is a vast store of resources to be exploited for economic profit, has overtaken the more environmentally sound cultures that existed in pre-modern India. However, these cultures and their traditional approaches to nature may still be found intact in many rural villages, which foreign visitors seldom see. It is from these sources that present-day Hindu environmental movements frequently arise – more often than not from the so-called "tribals" (adivasis) whose existence depends much more intimately on the land. But these movements seldom speak directly to the experience of mega-cities like Mumbai (Bombay) and Calcutta which continue to expand at an alarming rate. The ethical challenges posed by large-scale industry and urbanization are not part of previous Hindu experience, and so one does not find ethical guidelines in the tradition to use in regulating modern development (Klostermeier, 1995: 143). Hindu scholars are just now

411

H. Selin (ed.), Nature Across Cultures: Views of Nature and the Environment in Non-Western Cultures, 411–419.
© 2003 Kluwer Academic Publishers. Printed in Great Britain.

searching their scripture and philosophy for principles that could be employed
to limit and direct modernization. Eventually such scholarly Hindu ecotheology
will link up with the grassroots movements already active in India on behalf
of the protection of the environment. Then there will be a better chance that
Hindu ecotheology – which, as we shall see, has rich resources – can influence
government regulations, corporate decision-making and the excessively con-
sumerist lifestyle of the 250 million or so well-off upper class Indians, especially
those living in the rapidly growing cities.

HINDU SCRIPTURES AND THE ENVIRONMENT

Ancient Hindu myth is founded on what we would now call a profoundly
ecological vision. The human/nature relationship was at the core of that vision
and permeated the biological, physical and spiritual dimensions of life. Kapila
Vatsyanan summarizes this ancient wisdom:

> Man's life depends upon and is conditioned by all that surrounds him and sustains him, namely,
> inanimate, mineral and animate, aquatic, vegetative, animal and gaseous life. It is, therefore,
> Man's duty to constantly remind himself – in individual and collective life – of the environment
> and the ecology (Vatsyanan, 1992: 160).

All of this is encapsulated in one of the most ancient Hindu scriptures, *Rgveda*
10.90, a story which tells of the creation of the cosmos from the sacrifice of a
giant person, Purusa (Radhakrishnan and Moore, 1957: 19). In the beginning
the gods gathered to create the world by sacrificing Purusa. From his feet came
the earth, from his mid-section the sky, from his eye the sun, from his mind
the moon, from his mouth the gods, and from his breath the winds. Animals
and people (divided by caste) also came from different body parts. This is only
one of many different creation myths told in the Vedas.[1] In all of them, the
universe is described as an interconnected whole in which each part is interdependent with every other part. It is a living organism or ecosystem. In later
Hindu thought this universe is understood to be God's body (*Bhagavad-Gita*
11.7). Consistent with this is the idea that the earth and the rivers of India are
sacred. In the Hindu scriptures the earth is referred to as a great goddess who
nourishes and sustains all creatures. In the early texts she is called Pṛthivi
(*Rgveda* 5.84), and as the earth goddess she requires that people treat the earth
well. Although issues of overuse, pollution or desertification are obviously not
specified in the ancient texts, it is clear that humans must act in ethical ways
or risk the wrath of the earth herself. The earth feeds and sustains us as humans,
and in return we are to behave in ways that do not damage the earth. Rivers
too are seen to be sacred, especially the Ganges. Hindu mythology describes
the Ganges as a great goddess that originates in heaven and flows down to
earth giving both food and purification. In Northern India, Hindus bathe in
the Ganges and cremate their dead on its banks. Its waters are carried around
the world by Hindus to use in ritual worship. The Ganges and other rivers,
together with the land, are suffused with sacred power and evoke a response
of awe, gratitude and respect. Sadly this attitude has not been embodied in
environmental regulations to control modern industry, forestry and agriculture.

As a result even the sacred Ganges is today badly polluted with industrial and other wastes. The Vedas also speak of air as the breath of God, yet air and noise pollution are serious problems in the larger cities.

Since God is in everything, anything in nature has the potential to function as a symbol of the divine. In addition to rivers, trees are taken to be manifestations through which God may be revered. Among animals it is the cow especially that serves as a sacred symbol of God's preserving and sustaining power. Spending time in India rapidly convinces one that this is not a superstition but a true reflection of reality. For the average village family, the cow provides milk, yogurt, dung for fuel and transport of produce to market – as well as being the family pet. Without the cow, the family would not survive; therefore, the cow is a fitting symbol for God's sustaining power. Vegetarianism as practiced by Hindus is not a mere superstition but is founded upon two key principles: non-violence (*ahimsā*) and rebirth (*samsara*). Rigorous following of the teaching of non-violence means that we must not kill animals – for food or any other reason. The concept of rebirth is the idea that animals are beings in different karmic forms and that as they move up the ladder of existence over repeated rebirths, they will eventually be born as humans. Therefore to kill and eat an animal is tantamount to cannibalism. In another tradition in India, Jainism, this idea is extended down to include the plants and even the molecules of matter. From this perspective, everything is continuously interconnected. It is also the case that people must survive, so that any ancient proscription against killing plants could not be followed.

HINDU PHILOSOPHY AND THE ENVIRONMENT

Indian philosophers developed the above worldview affirming the interconnectedness of humans, animals, plants and even matter into what is called Karma Theory. Unlike Christian views which subject nature to human domination, Karma Theory rejects the dualism of nature and humans, and maintains, especially from the Jaina perspective, that there is no radical separation between humans and other forms of beings (animals, plants, air, water, atoms of matter). Instead, a radical continuity is proposed. While Hindu thinkers do not extend the continuity as far as the Jains – usually restricting themselves to a rebirth relationship with the animals – it is the principle of continuity between the different forms that compose God's body that is important. All are individual souls (*jivas*) trapped in different stages of karmic bondage. Karma that weighs down the jivas is the sum of the memory traces of good or bad actions done in this and previous lives. These karmic memory traces (*samskāras*) are stored up in the unconscious (to use a concept from modern psychology) and carried with one from one life to the next. They are created by one's freely chosen actions and are colored by their moral quality. Good actions that foster the interconnected health of the whole of God's body (the cosmos) are light colored and foster one's progress toward salvation or release from rebirth; bad actions are selfish acts that do the reverse (Coward, 1995: 48–54). Good actions move us up the ladder of rebirth to final release and union with God; bad actions take us in the opposite direction. According to Karma Theory we are each authors of and morally responsible for our own destiny. Note that in the above theory the karmic memory trace in the unconscious does not cause anything; it is not mechanistic in nature. Rather, it simply acts like an impulse which predisposes one to think a thought or do an action. It is all up to one's own free choice, which means that the moral responsibility for one's choices rests squarely upon one's own shoulders. Even the initial impulse arising from the *saṁskāra* or memory trace in the unconscious got there in the first place by a freely chosen action in one's past (including this and all previous lives) and therefore is an impulse for which one is morally responsible.

The application of Karma Theory to environmental concerns is straightforward and powerful. The impulses I am now feeling in the way I behave toward the animals, plants, earth, air and water are a direct result of the way I have freely chosen to behave in past lives. If my arising karmic impulses are suggesting irresponsible behaviour towards the environment, it is because I have acted in immoral ways toward nature in this and previous lives. And since I freely chose to behave in those ways, I created for myself the impulses now arising from my own unconscious. If I find myself wanting to cut down the forest, foul the water, pollute the air, and selfishly over-consume the earth's resources, I cannot blame these impulses on God, the devil, my parents, the multinational market economy, etc. They are coming into my mind at this time because I laid them down as memory traces from actions I chose to do in this or past lives, so I alone am responsible for the environmental impulses, be they good or bad, that I am now experiencing. But I am not trapped by my past environmental karma. While I may have habitually fouled the air or water in past actions (in this and previous lives), I can now use my free choice, when the situation arises, to act in ecologically responsible ways. Thus, through repeated good choices, environmentally destructive karmic patterns can be removed from my unconscious and replaced by good ones. From the perspective of Karma Theory, I am totally responsible both for my impulses toward the environment and the way I choose to act or not act on those impulses. And the way I choose to act today creates the karmic impulses I will experience tomorrow and in my future lives as I interact with nature. However, because I am part of a continuum with all other beings – with all of nature – what I choose to do affects not only my future but also the rest of the cosmos of which I am but an interconnected part. Thus, my ecological responsibility is both individual and cosmic. The way I make my choices conditions not only my future lives but also the future of all other beings – which, in the karma perspective, includes all of nature.

There is another strand in Hindu philosophy which could function to undercut the strong sense of ecological responsibility fostered by Karma Theory. The most famous Hindu philosopher, Sankara (*ca.* 700 CE), argues that there is a radical separation between reality (the divine Brahman) and everything else in experience, including all of nature. Suresvara, Sankara's early follower, put it clearly: "Between the world (Existence) and the rock-firm Self (Reality) there is no connection whatever except that of ignorance" (Deutsch,

1989: 262). The realm of the world, of nature, is called by Sankara *maya* or illusion in the sense that we take it to be ultimately real while in reality it is not. Because of our mistake we become focused on and trapped in the interplay of the world of maya assuming, in our ignorance, that it is real. Sankara's prescription for release from this state of ignorance which causes our continued rebirth is to realize existentially the radical separation of our true self (*atman*) from nature or maya. Then, in our enlightened state, nature or maya simply disappears. The danger in all of this from an environmental perspective is that it could lead one to assume that what happens in the world of nature does not matter. Since maya or the world ultimately does not exist, it is not worth worrying about in ethical terms. While this may be a common reading of Sankara, it is, in my view, not correct when the systematic nature of his thought is fully considered. Before one can realize the non-duality of Brahman (Sankara's goal), one must first have met all the ethical requirements of the Hindu scriptures – including the ecological requirements specified by Karma Theory. Sankara accepts and requires that one meet all the demands of Karma Theory but then suggests that there is yet another level of spiritual realization which lies beyond the ethical – namely a direct union with God (Brahman). And at this final level, the world and its ethical environmental concerns are left behind – in reality they simply cease to exist. A later Hindu philosopher, Ramanuja (*ca.* 1100 CE) rejects this transcendent final experience of Sankara and instead maintains that the ethical experience of the cosmos as God's body is as high as one can go (Lipner, 1986).

HINDU SPIRITUALITY IN ACTION

One important Hindu scripture, the *Bhagavad Gita*, sets forth a strong basis for ecological action. According to the *Gita* 3.25 the enlightened person is one who acts unselfishly to maintain the world. In traditional literature such virtuous action is described as including the digging of wells and the planting of shade and fruit trees along roads for the benefit of all. In addition to those who work with nature in adapting it to human needs are those who act to protect it from pollution or destruction. Hindu holy men have a protective attitude to all living beings including the saving of animals from predatory hunters. In 1973 the Chipko movement was formed to fight the deforestation of the Himalayan foothills. The movement uses a non-violent approach – people join hands around trees to prevent them from being cut. The movement harkens back to 1730 when tribal women in the state of Rajasthan put their arms around trees to keep them from being cut for fuel for the Maharaja's lime kilns. The women, who were cut down with the trees, gave their lives for trees that were an essential part of their lifestyle.

In Hindu India, says Vandana Shiva, women and ecology go together (Shiva, 1992: 205–214). Modernity and its developmental logic lay waste to both women and nature. Her analysis provides a Hindu parallel to Rosemary Radford Ruether's *Gaia and God: An Ecofeminist Theology of Earth Healing*. Both argue that the result of modern development in industry and agriculture has been to exploit women and nature ruthlessly. But Shiva offers an extra element to her analysis: "Diversity is, in many ways, the basis of women's politics and the politics of ecology" (Shiva, 1992: 207). In India women have traditionally been responsible for the production and preparation of plant foods, including seed preparation, planting, climate requirements, plant diseases, pruning, staking, companion planting, growing seasons, harvesting, storage and soil maintenance. The men mainly plowed the land. The women also fed and milked the cows and managed the forests for the production of firewood. In line with the teaching of the *Gita*, they have worked to be in harmony with nature, to maintain the world and to benefit society. The impact of modern development, says Shiva, has been devastating on both women and the biodiversity they have worked to sustain. Mass mechanized agriculture imposes highly productive uniform and homogenous farming which, while it may produce more in the short run, rapidly exhausts the soil and upsets the diversity so essential to a sustainable ecology. It also takes the production and handling of produce out of the hands of women, thus taking away the livelihood essential for their families. The modern approach seems to fail the criteria of the *Gita*, namely, to seek unselfishly to maintain one's family, society and the world in which one lives.

Recent developments offer some hope, and the leadership is coming from the very heart of Hindu religion. Vasudha Narayanan has searched Hindu sources for a new basis to respond to the contemporary ecological challenge (Narayanan, 2000: 111–130). Narayanan notes the close connections between the teachings in the Hindu epics and puranas on *dharma* (righteousness, duty, justice) and the ravaging of the earth. As she puts it, when dharma declines, humans take it out on nature. It is in the dharma rather than the moksa or enlightenment texts that Narayanan finds resources for a Hindu response to contemporary need for positive practices in the face of the problems of ecology, population pressure and excess consumption. Narayanan searches out dharma texts with helpful teachings and matches up positive dharma practices in which present-day Hindus are engaged. She finds many teachings condemning the cutting down of trees and supporting the planting of trees – even to the goddess Parvati's teaching that one tree is equal to ten sons! While governments may be the leaders in the non-dharmic behaviour of harvesting forests, Hindu temples such as the Tirumala-Tirupati temple in South India are showing great initiative in fostering the dharma of tree planting. This famous pilgrimage temple used to give pilgrims an Indian sweet called *laddus* as a *prasada* or material symbol, which when blessed by the deity and eaten gave one divine grace. Around 100,000 *laddus* were given out daily. Now, however, sapling trees are given instead and the pilgrims are instructed to plant them at home or in the temple grounds. As a result, over 2.5 million temple trees have been planted. This practice is more powerful than it may seem, since the Tirumala-Tirupati temple is the richest in India and carries considerable dharmic clout with Hindus at home and in the diaspora communities. That the practice is catching on is evidenced by the fact that the chief minister of Tamilnadu requested that, in lieu of other devotional expressions on his birthday, trees should be planted.

According to reports, over 100,000 were planted. In Hindu texts, trees, like cows, are recognized as preservers and sustainers of life and therefore to be appropriate symbols of God.

Turning from trees to rivers, Narayanan notes that rivers also feature in dharma texts as sacred purifiers of pollution, with the Ganges as a prime example. Sadly observing that the rivers of India are rapidly being dammed and fouled by both industrial and human waste, she points out that it is the women of India who lead the fight against these practices – appropriately perhaps since most rivers are considered to be female. Some successes have been achieved, such as the protest led by Ms. Medha Patkar resulting in the stopping of the Narmada river dam project. Women have also led the way in pressing classical Hindu dance into the service of ecology thus spreading the message through art.

Narayanan also considers India's population problem, pointing out that the dharma texts that emphasize the duty of procreation were formulated during periods when epidemics and famines kept population levels down, child mortality was high and death came early. Now with modern medicine and improved sanitation all of this has changed, and India's population has rapidly increased to levels that are causing serious ecological damage. However, in certain states, such as Kerala, where girls are educated and women are employed at all levels, reproduction is at the replacement rate only. In this case, the patriarchal dharmic ideals espoused in these texts were superseded by sound politics. This points to an interesting dilemma. It is clear that we cannot just take the words of scripture, in whatever religion, and apply it today. We do not want to over romanticize the past or take the words of the texts as truth. At the same time, we can often use the words and thoughts of the ancients to help with environmental and other problems. For example, with regard to consumption, Narayanan notes that the Hindu texts are replete with the dangers and futility of possessions. Yet consumerism is rapidly taking over in India in the name of modernization. Even worse perhaps is the use of the dowry system as a convenient way of fulfilling greed for consumer luxury items. While the Hindu theological response is showing some success in producing positive ecological practices, it seems to be losing ground to the influx into India of the market economy.

CONTRASTS WITH WESTERN ECOTHEOLOGY

Western ecology movements often begin from the assumption that humans and nature are separate – that nature was created by God for the benefit of humans. While an unfair reading of the ecotheology of Genesis or the Qur'ān, this view, together with the impetus of Enlightenment thinkers like Descartes and Kant, has fostered the separation hypothesis. Today it shows up strongly in some of the ecology movements of the modern West and evokes a critical response from India. Ramachanadra Guha recounts how the Western separation principle has been forcefully superimposed upon India by Non-Governmental Organizations (NGOs) like the World Wildlife Fund (WWF) and the

International Union for the Conservation of Nature and Natural Resources (ICUN) (Guha, 1995: 281–289). Guha notes that the Western principle of separation between humans and nature is translated by the Deep Ecology movement into a focus on the preservation of unspoiled wilderness and the restoration of degraded areas to a more pristine condition. When applied as public policy by the WWF and the ICUN this approach requires that people be excluded from nature to keep it wilderness. Guha's critique is based on the Hindu perspective that people are not separate from but an interconnected part of nature. Therefore ecological goals are best served not by separating people from nature but by ensuring that people live in nature intimately in ways that support the diversity and harmony of the whole – which is God's body.

The specific example highlighted by Guha is Project Tiger, a network of parks established in India as tiger reserves which were created by the displacement of existing tribal villages and their livestock. Initial impetus for setting up the parks came from two social groups: first, ex-hunters turned conservationists belonging mostly to the declining Indian feudal elite; and second, representatives of the WWF and ICUN seeking to transplant the American system of national parks onto Indian soil.

Guha notes that nowhere in the process were the needs of the local peasant population taken into account, the result being a direct transfer of resources from the poor to the rich. Also by giving such projects high profile, funds and attention were diverted from environmental problems that impinge more directly on the lives of the poor (e.g. water shortages, soil erosion and air and water pollution). In this example, the action of powerful Western NGOs has perhaps served to heighten rather than reduce ecological tensions in India. The Deep Ecology approach of separating people and nature to preserve nature seems to run counter to the traditional ecology of Hinduism.

Guha's critical example may be somewhat one-sided. With the explosion of human population in India, perhaps there is a need to limit human expansion and create a dedicated space for the tiger – and maybe the need for a balanced approach here could also find support in Hindu ecology. However, Guha's critique raises one very important point, one which has to do with the way we deal with key policy issues in today's pluralistic world. Rather than superimposing a Western basis for ecology upon India, the better approach may be to search for a response to ecological problems from within India's own wisdom. And here, as outlined above, Hinduism has much to offer. The fact that Hindus, when challenged by modern development, have been slow to translate their wisdom into government laws or guidelines makes them no different from those from Judaeo-Christian backgrounds in Europe and Northern America. And, as Narayanan has shown, authentic responses to the contemporary environmental crises are occurring. In India the motto "A tree is equal to ten sons" seems a hard sell indeed. Yet, as Narayanan's evidence indicates, with the strong leadership of the Tirumala-Tirupati Temple, the behaviour of millions of Hindus has taken a radical change for the better.

NOTES

[1] The word *Veda* means knowledge, and the *Vedas* are considered the most sacred scripture of Hinduism.

BIBLIOGRAPHY

Coward, Harold. 'The ecological implications of Karma theory.' In *Purifying the Earthly Body of God: Religion and Ecology in Hindu India*, Lance E. Nelson, ed. Albany: State University of New York Press, 1998, pp. 39–50.

Deutsch, Eliot. 'A metaphysical grounding for natural reverence: East-West.' In *Nature in Asian Traditions of Thought*, J. Baird Callicot and Roger T. Ames, eds. Albany: State University of New York Press, 1989, pp. 259–265.

Guha, Ramachandra. 'Radical environmentalism.' In *Ecology: Key Concepts in Critical Theory*, Carolyn Merchant, ed. New Jersey: Humanities Press, 1994, pp. 281–289.

Klostermaier, Klaus. 'Hinduism, population and the environment.' In *Population, Consumption and the Environment*, Harold Coward, ed. Albany: State University of New York Press, 1995, pp. 137–152.

Lipner, Julius. *The Face of Truth*. London: Macmillan, 1986.

Narayanan, Vasudha. 'One tree is equal to ten sons: Hindu responses to the problems of ecology, population, and consumption.' *Journal of the American Academy of Religion* 65(2): 291–332, 1997.

Potter, Karl, ed. *Advaita Vedanta up to Sankara and his Pupils*. Princeton, New Jersey: Princeton University Press, 1981.

Radhakrishnan, S. and C. Moore, eds. *A Sourcebook in Indian Philosophy*. Princeton: Princeton University Press, 1957.

Ruether, Rosemary Radford. *Gaia and God: An Ecofeminist Theology of Earth Healing*. San Francisco: Harper, 1992.

Shiva, Vandana. 'Women's indigenous knowledge and biodiversity conservation.' In *Indigenous Vision: Peoples of India Attitudes to the Environment*, Geeti Sen, ed. New Delhi: Sage Press, 1992, pp. 205–214.

Vatsyanan, Kapila. 'Ecology and Indian myth.' In *Indigenous Vision: Peoples of India Attitudes to the Environment*, Geeti Sen, ed. New Delhi: Sage Press, 1992, pp. 157–182.

PARVEZ MANZOOR

NATURE AND CULTURE: AN ISLAMIC PERSPECTIVE

Man as a creature, Lewis Mumford (1967: 46) once acutely observed, "is never found in a 'state of nature', for as soon as he becomes recognizable *as* man, he is already in a state of culture". Certainly the dialectic of culture and nature is seminal to any image of Man, be it theological, philosophical or even biological. However, as soon as we speak of humans, we become conscious of the enormous ideational gulf that separates all traditional, pre-modern worldviews from modern ones based on natural sciences. For man, from the vantage point of any philosophical or theological discourse, is a given, a precept rather than a concept; s/he is the measure of everything and it is through her/him that the world – cosmos and nature – acquires its meaning and form. The very raison d'être of modern science, and the incontestable premise of its epistemology, on the other hand, is the rejection of all anthropocentric visions and principles.[1] The Islamic perspective on nature and culture, emanating from the Islamic image of Man, is incontrovertibly anthropocentric.

Islam's anthropological vision devolves from its belief about the ultimate scheme of things, about the totality of being of which God, *Allāh* in the language of the Quranic revelation, is the creator. The "ultimate" in the Islamic worldview is trans-cosmological; it stretches beyond the world of men and stars. "The throne of God extends beyond heavens and the earth" (*Qur'ān* 2: 255). It also follows that the very idea of "nature", which informs every modern discourse, is problematic when examined from Islam's theocentric vantage point.[2] In fact, the sharpest intellectual encounter in classical Islam, that took place between the champions of orthodoxy and those of philosophy, was over the Muslim *falāsifa*'s[3] "naturalism" – their doctrine of the eternity and necessity of the world.[4] No modern attempt to expound the theme of nature and culture in Islam can therefore ignore the inherent tension between the theological and the philosophical (including the scientific) modes of discourse.

All arguments against the existence of a cosmic order that is ruled by the will of an all-powerful and wise God are contingent upon a necessary, sovereign and self-subsistent order of "nature". Nature is the sum total of everything that exists, but it is also the impersonal necessity of the world: everything behaves the way it does because it is in its "nature" to do so. The Greek concept of

H. Selin (ed.), Nature Across Cultures: Views of Nature and the Environment in Non-Western Cultures, 421–432.
© 2003 *Kluwer Academic Publishers. Printed in Great Britain.*

"nature", whether postulating a self-contained, self-sufficient, self-regulating universe, or signifying the intrinsic disposition of a thing to obey immanent laws, is alien to the Quranic worldview. The world exists, according to the *Qur'ān*, not because of any intrinsic necessity but because of the gratuitous act of a transcendent will; it is radically contingent rather than naturally necessary.[5]

And yet, Islamic thought, which does not propose any autonomous and self-regulating order of nature, is not condemned to inhabit a normless and erratic cosmos. It is neither unreceptive to the quest for an intelligible universe, nor dismissive of the logical and moral demands for a predictable, coherent and reliable order of reality. Islam's insights about the regularity, constancy and dependability of the created order, and its own pathways towards a scientific and rational mode of thinking, proceed from certain Quranic concepts, the most salient of which is *Qadr* (or *Qadar*), which acts as a bridge between divine and natural causation.

As a sign of God's power, *Qadr* suggests the orderliness of creation, for the basic idea it conveys is that of measure and norm. If things and beings act as they do, if they have a natural disposition, it is because of the measure they have been allotted to by the transcendental power of God. Nevertheless, the order, measure and finitude of every created thing is the obverse of God's freedom, power and infinitude, and there exists a definite ontological disparity between God and nature, between the Creator and the creation. The Quranic ontology is essentially moral and teleological in significance and is meant to address the problem of man rather than to unravel the riddle of the universe. Or, as Fazlur Rahman (1980: 65) explains, "The Qur'anic cosmogony is minimal". Hence, the Quranic view of the natural world is logically bound up not only with the concept of God but also with that of man. And whereas both nature and man have their pre-ordained roles in the divine scheme of things, it is man who is the principal actor in the drama of creation. In fact, Muslim tradition has always recognized the birth of Adam to be the most significant event in the Quranic account of creation (*Qur'ān*: 2: 30ff.) Therefore, while nature submits to God's command, willingly or unwillingly *(Qur'ān*: 3: 83, 41: 11), man's obedience to God has to be voluntary. He has been endowed with the faculty of choice, the ability to transgress God's commands, and hence he is to be tested within history and tried at the end of his term. Significantly, while judgment in history concerns the accomplishment of a collectivity, a people, a nation or a community, in the Final Judgment it is the total performance of individuals that comes under review.

Rahman (1980: 67) has also suggested that the notion of divine "measuring" has a strong holistic basis in terms of patterns and trends, and entails some form of "holistic determinism", if not a theory of predetermination itself. Within the Muslim culture most of the terms that denote destiny or fate, such as *qadar, taqdīr, muqaddar* etc., are semantically and etymologically related to this Quranic usage, as is the designation of "nature" itself *(Qudra/t)* in most Muslim languages. For the "measured" in its Quranic setting conveys the sense of being limited or finite rather than predetermined or destined.

And the sun – it runs its course to a resting-place – that is the measuring (ordaining) of the All-Mighty, All-Knowing. And as for the moon, we have appointed its stations, till it reappears like a dried palm-branch. It is not for the sun to overtake the moon, nor for the night to outstrip the day, each moving in its own orbit (*Qur'ān* 36: 38–40).[6]

The measure of nature (creation) is thus finitude and limited existence, another sign of its dependency on the transcendent Creator. The mystics of Islam construe the measuring, finitude and limited existence of the world in a different vein. While in Avicenna's philosophy (cf. note #4), *qadr* as an emblem of God's power yields a theory of the radical contingency of the world, for Ibn 'Arabī (1165–1240), it is a sign of God's compassion. For him, and for Sufis in general,[7] God's mercy (*rahma*) is nothing but bestowing on everything existence *qua* existence, an ontological act by dint of which anything becomes recognizable as something. Ibn 'Arabī names this ontological reality, which precedes any moral judgment on the world as fallen or unredeemed, "the Breath of the Compassionate" (*Nafas al-Rahmān*, or *Nafas Rahmānī*).[8] In this ontological capacity, Ibn 'Arabī regarded the Breath of the Compassionate as Nature (*tāb'īa*) (Izutsu, 1984: 131). S. Hussein Nasr (1993: 210, n. 37) has this to say regarding the difference in outlook between the Hellenized philosophers and the mystical Sufis:

> Among the followers of Hellenized cosmology in Islam, Nature is usually the common ground or substance of the four natural qualities and elements and of all the changing forms brought into being by them. In the ultimate sense, Nature for them, as for the Greeks, is the principle of change, while for the Sufis it is the breath of God (*al-nafas al-ilāhī*) or the Breath of the Compassionate (*nafas al-Rahmān*), and the feminine, passive and "motherly" aspect of the Divine act of creation.

For Corbin (1960), *Nafas al-Rahmān*, which he translates as "the Sigh of the Compassionate", reveals "Divine Sadness", the sadness of Godhead for being unknown before the creation. The starting point for Ibn 'Arabī's speculation is the *hadīth qudsī*,[9] according to which God declares: "I was a hidden Treasure and I yearned to be known. I created the world that I may be known by them".

Other related Quranic concepts that have been instrumental in influencing the Muslim attitude towards the created order are *Amr* (command) and *Sultān* (authority). Thus, when God creates a thing (*khalq*), he simultaneously awards it its potentialities and the laws of its behavior (*amr*) whereby it falls into a pattern and becomes a factor in the cosmos (Rahman, 1980: 23). *Sultān*, which connotes the idea of power and authority, on the other hand, is used in the *Qur'ān* in the sense of a divine proof or sign. Rahman (1980: 74) suggests that *Sultān* is best translated as "that which overwhelms without having an alternative", or as "knock-down proof". As with other Arabic terms, it conveys both the concrete sense of power and force and the metaphoric one of authority and reason. The whole of nature, accordingly, is a proof of God's authority and a realm of His power. The malaise of man is that he takes the natural world – God's primary miracle – for granted and yearns for a supernatural sign. Fritjhof Schuon (1972: 20) holds that the attitude of reserve adopted by Islam towards miracles is explained by the predominance of the pole of 'intelligence' over the pole of 'existence': the Islamic outlook is based on what is spiritually evident, on the feeling of the Absolute, in conformity with the very nature of man which in this case is seen as a theomorphic intelligence and not as a will waiting to be seduced in either a good or a bad sense, seduced, that is to say, by miracles or by temptation ... Now spiritual certainty ... is an element to which the devil has no access; he can imitate a miracle but not what is intellectually evident.

From the Quranic perspective, the natural and the commonplace, be it the beginning of life in the embryo or the resurrection of plants after a rainfall, is more of a cause for wonderment and a reason for gratitude. That the world exists at all, that there is the plenitude of being instead of nothingness, is adduced in the *Qur'ān* as the most compelling argument for man's acknowledgement of the lordship of Allāh. Man must take stock of his situation from the fact that he may not have been, that he once was not (*Qur'ān*, 76: 1). The Quranic discourse on nature, in short, is situated in what is essentially God's address to man, and its purpose is nothing but the guidance of man.

If Muslim perception and understanding of nature was largely determined by the monotheism of the *Qur'ān*, the character of the Arabic language was also vital in influencing the study of natural philosophy and science. For, as soon as Muslims came in contact with the heritage of the classical world, they developed a rich indigenous vocabulary of Arabic words for the study of philosophy and its cosmological doctrines (Goichon, 1938). The word "nature" itself, *physis* in Greek and *natura* in Latin, was rendered in Arabic by the term *ṭābʻia*, from the root *ṭabʻ*, to impress, to stamp, but also, in the passive mood, to have a disposition.[10] The earlier translators also occasionally used the word *kiyan* as a synonym for *ṭābʻia* (Massignon, 1946). Though scholastic terminology of *natura naturans* and *natura naturata*[11] may have been derived from their Arabic precedents, the Arabic usage of *ṭābʻia* was different in meaning from the classical languages. Muslim authors generally distinguished between *ṭibāʻ* as the essential attributes of something, or, in other words, the inner disposition of a thing to behave in a certain way, and *tabʻ* as that which gives movement or rest to a thing without possessing a will of its own (Nasr, 1993: 8). Eventually, all such expressions of naturalistic philosophy and metaphysics were rejected by the orthodoxy, whose champion, al-Ghazālī, philosophically defended the monotheistic doctrine of *creatio ex nihilo* (creation out of nothing).

In order to comprehend the Greek idea of "nature", a notion which has no equivalent either in the *Qur'ān* or in the Hebrew Bible and which has been the source of much tension between philosophy and orthodoxy in both these traditions, one must rediscover its philosophical counterpart in the Semitic linguistic consciousness itself. The philosophical equivalent of nature may be found in the concept of "custom" or "way" (*Sunna* in Arabic) (Strauss, 1989: 253ff.).[12] When beings or things behave in a way that is regular and foreseeable, when they do what is customary to them, it is tantamount to saying that it is in their nature to do so. For instance, the fact that fire burns may be expressed either as: a) it is so because of its custom or habit (Semitic); or because it is in its nature to do so (Greek). In other words, for perceiving the predictability of existence, neither the Aristotelian metaphysics of nature nor a theory of causality is indispensable.[13]

Given the normative nature of *Sunna*[14] in Islam, and its ubiquitous presence in the sciences of theology, ethics, and jurisprudence, one may conclude that this concept forms the characteristically Islamic pathway to the consciousness of a regular and predictable order of reality, sustained not by any inner logical necessity, by the natural disposition of things, but by divine custom (*Sunnat Allāh*). Thus, Islamic theology, especially in its Ashʿrite form, which emphasizes the power of God and the insignificance of human beings, holds that God is both the ultimate and the direct cause of all things. It reasons, for instance, that fire does not burn because it is its nature to do so, but because God so willed it. Hence, speaking strictly from within this perspective, there can be no laws of nature, only the custom of God, and miracle is a breach of that custom (Gardet and Anawati, 1981: 430–431).[15]

One must not conclude from this that the Ashʿrites, because of their intimate affiliation with Sunnism, represented the common, consensual Islamic position. The following remark by Ibn Taimiyya (1263–1327) is sufficient to dispel any such misapprehensions. To him, the denial of nature – the self-subsistent and self-regulating principle of the cosmos – by the Ashʿrites goes counter to the opinion of "the majority of reasoning men, both Muslim and non-Muslim, both men of the *Sunna* such as theologians, jurists, traditionists, and mystics, and men not belonging to the *Sunna* such as the Muʿatazlites and others" (Al-Azmeh, 1986: 10). Further, the denial of "nature" by certain theological schools did not concern the regularity of nature's behavior, but its autonomy, or the identity of the ultimate source of this regularity. It was a theoretical contention rather than a practical issue. For instance, in the context of medicine it was both permissible and commonplace to invoke the concept of nature, but it was formally inadmissible in the context of the theory of human action to which certain theologians subscribed.

Notwithstanding all the logical discomforts of superimposing a divinely-ordained order on a natural one, Islam's fundamental attitude towards the world, seen as creation and a divine gift, is one of reverence and gratitude. The most recurring claim of the *Qur'ān* about nature is that it is magnificent, orderly, beautiful and immense. It is the handiwork of God, His prime miracle and, as befits such a sublime artist, faultless.

> He who created seven heavens one on top of another –
> Thou shalt not see in the creation of the All-merciful any imperfection.
> Return thy gaze; seest thou any fissure?
> Then return thy gaze again, and again, and thy gaze comes back to thee dazzled, aweary (67: 3–4; Translation by Arberry).
> Thou shalt see the mountains, that thou supposest fixed, passing by like clouds – God's handiwork, who has created everything very well (27: 88).

The immensity of space and matter and the plenitude of being that comprise the universe are signs (*āya*), or proof, of its Maker; it is the recurring theme of the *Qur'ān*. In fact, the Quranic characterization of nature as a symbol of the Transcendent reality, as an immanence pointing beyond itself, best depicts the characteristically Islamic attitude towards it. For the Muslim mind, the perception of nature as a cipher of a higher truth is further reinforced by the

Quranic claim about the correspondence of natural and scriptural signs. Like natural phenomena, the verses of the *Qur'ān* are signs of God and are similarly named *āyāt*. Indeed, it is the soul (intellect) of man which, as a repository of divine signs, mediates between the natural world and the transcendent truth beyond, and assures man of his ultimate felicity: "We shall show them our signs in the horizons and in their souls, till it becomes clear to them that it is the truth" (41: 53).

God, man and nature in the Islamic perspective form a unity of purpose and goal that articulates the quintessentially Islamic position with regard to nature. Or, as Ismā'īl al Fārūqī (1980: 25) expresses it more lyrically: "In Islam nature is creation and gift: as creation it is teleological, perfect and orderly; as gift it is as an innocent good placed at the disposal of man. Its purpose is to enable man to do the good and achieve felicity. This treble judgment of orderliness, purposiveness and goodness characterizes and sums up the Islamic view of nature".

Muslim scholars have always been struck by the strong parallels that the Scripture itself draws between nature and divine guidance (*huda*, or *hidāya*),[16] between the revelation of the *Qur'ān* and the creation of the universe. God's communication to man, according to this schematic view, is of two kinds: verbal and non-verbal, scriptural and natural, through the verses (*āyāt*) of the *Qur'ān* and through the signs (*āyāt*) of nature. Moreover, the *Qur'ān* and nature are mentioned together so often "not fortuitously but because of the intimate connection (that exists) between the two" (Rahman, 1980: 72). Passages like the following make this insight abundantly clear:

> Indeed, in the creation of the heavens and the earth and the succession of day and night are signs for the people of wisdom – those who remember [i.e. recite the Scripture] standing and sitting and lying on their sides, and who ponder over the creation of the heavens and the earth [exclaiming]: Our Lord, You have not created all this in vain! (3: 190ff. Translation by Fazlur Rahman).

> These are the verses of the Book. That which has been sent down to your Lord is the Truth; yet most people do not believe. It is God who has raised the heavens without any pillars that you can see; then He established Himself on the Throne and subdued the sun and the moon – each running to a destined term (13: 1ff. See also 10: 1–3; 12: 102–105; and 20: 1–6; translation by Fazlur Rahman).

The ideational affinity between nature and revelation is so pronounced in the *Qur'ān* that not only are the two seen as complementary systems of divine guidance, but the Quranic view of nature becomes directly relevant to the truth of the *Qur'ān* itself (Rahman, 1980: 76). In every attempt at the definition of Islam's relationship to nature, nothing less than the most fundamental issues of metaphysics and morality are at stake. In fact, from the Islamic viewpoint, the theme of nature and culture impinges on the problem of transcendence and that of the purpose and goal of human existence. Any Islamic reflection in this regard is therefore obliged to focus on them, no matter how elliptically or meekly, and irrespective of all the strident modernist claims that these are non-issues, impossible to resolve from within the ideational framework of empiricism.

To reflect on the current theme is therefore not to become prisoner to the metaphysical prejudices of our age, as the subject of nature and culture is particularly subversive of, and recalcitrant against, modernism as a worldview. In choosing to put nature and culture under scrutiny, we are also inviting ourselves to revisit unsettled debates and rekindle old controversies. Indeed, we even put a question mark on the whole project of modernity and its proposed solution to the problem of man. Further, any serious reflection on a historical tradition's relationship to nature and culture perforce stretches far beyond the modern concerns of ecology and political economy. Similarly in the case of Islam, the dialectic of nature and culture, or the delineation of a normative vision of man's legitimate role in the world, cannot be exhausted by, or made hostage to, the current jargon about the malaise of global capitalism.

Transcendence has always been recognized as the essence of Islam. No Islamic vision can ever be cogent or authentic if it does not take cognizance of the ineluctable transcendence which is the sine qua non of being human in the world. An Islam without the perception and recognition of a transcendent realm (*'Ālam al-Ghaib*) beyond the borders of the empirical world (*'Ālam al-Shahāda*) would not be Islam at all. And it is here that the epistemological disparity with modern science and the moral discrepancy with scientism or doctrinaire secularism come to light.[17] However, it is also a dispute that does not admit of any solution through arbitration or impartial, empirical judgment; hence, the phenomenological, noncommittal twist of contemporary discourses, which allows one to discuss these issues without deciding upon the question of truth. Speaking in this vein, we may restate the Islamic position by insisting that while transcendence as an attribute of the world (a contradiction in terms!) may or may not be rationally demonstrable, some perception of it certainly is indispensable to being human. Fazlur Rahman (1980: 21) echoes the central conviction of all monotheists that "the removal of God from human consciousness means the removal of meaning and purpose from human life". And it is this commitment to man's humanity that forms the backbone of Islam's attitude towards nature and culture.

Any inquiry into the dialectic of nature and culture in Islam must have its central focus on humans. For it is in them that nature and culture, transcendence and immanence, norms and facts, intersect. Islamic humanism, unfortunately, is a much neglected subject and the true image of *homo islamicus* has been obscured as much by the heartless positivism of modernity as by the mindless literalism of the Islamic tradition itself. The absence of any proper and balanced study and the abundance of polemical and highly politicized tracts make this enterprise quite daunting. What follows here can only be regarded as tentative and exploratory.

Any Muslim attempt to re-examine the Islamic perspective on man runs into formidable difficulties, not the least demanding of which is the conceptualization of transcendence in such a manner that it promotes a neutral, universal conversation without denying the possibility, if not the actuality, of the Islamic revelation. But we have an equally recalcitrant problem of history, as every attempt to grasp the immanence of the human condition, underwritten by modern philosophy's attempt to explain temporality and historicity, terminates in the cul-de-sac of relativity and nihilism. The claim of the meaningfulness of human existence, which can only be derived from postulating a transcendent source, may be bartered for an immanent, temporally contingent and intellectually graspable reality, but such a reality is indifferent to the human quest for meaning. The conundrums of norm and history, existence and meaning, Self and Other afford no transparent view of the human condition, and what goes under the rubric of academic criticism is often nothing more than a veiled apology for the writer's own political constituency.

Even if the *Qur'ān* has very little to say about cosmogony, it does present an image of man that is transcendental without being anti-historical. It "does not appear to endorse the kind of a doctrine of radical mind-body dualism found in Greek philosophy, Christianity, or Hinduism", even if later, under the influence of al-Ghazālī, orthodox Islam did come to accept it (Rahman, 1980: 172). Like every other being and non-being, man is a creation of God. Yet his status is special on two accounts: ontologically, because he has been infused with God's spirit (15: 29; 38: 72: 32: 9), and morally, because he is God's Deputy and the custodian of his creation on earth (2: 30ff; 7: 11ff; 20: 116ff). It is through the story of the birth of Adam that the *Qur'ān* alludes to the most significant act of creation. Adam, from the Quranic account, may be envisaged in both transcendental and immanentist terms; he is both the primordial, eternal man and the individual, historical human being.[18] The transcendence of Adam, which is reflected in his intelligence (*'ql*) and which endows him with rational faculty and moral judgment, however, must be seen in conjunction with his immanence, his mission in history. For Adam has on his own accord accepted the challenge of creating a just moral order on earth, an enterprise described by the *Qur'ān* as "trust" (*amāna*) (33: 72).[19]

The Quranic designation of Adam as the Representative or Vicegerent (*Khalīfa*) of God is, again, moral in scope and purpose. It presents a conceptual scheme that mediates between transcendence and immanence, that bridges the gap between the *de facto* and the *de jure*, the *is* and the *ought*, of the human situation – without invoking the ontological language of incarnation. Although there is no ontological relationship between God and Adam in the manner of the Christian doctrine of incarnation, the Quranic Adam does appear to have some functional resemblance to Jesus in being a bridge between transcendence and immanence. Adam's role, however, can only be conceived in moral terms. (3: 58). In Christian theology too Jesus is referred to as the "Second Adam", redeeming mankind of the sin that the first Adam committed. Apparently, due to the absence of the Original Sin (or, at least, of a strong version of it) in Islam, the first Adam retains the functions which in Christianity are the preserve of the second. Man is denied the attribute of sovereignty but given all the freedom, royal power and responsibility that are the privileges of the Viceroy. In moral terms, it is tantamount to denying man the right to be a norm unto himself and a source of his own values. The Quranic view of Adam's *khilāfa* is a supremely humanistic doctrine, but without the hubris and arrogance of humanism that according to the critics of modernity is its bane and the source of its nihilism (Gray, 1995: 144–184).

Modern civilization, held responsible by some for creating the environmental crisis, has also generated an ecological consciousness which, paradoxically, seems to have reintroduced the sacred into the discourse of science. Ecological awareness however has brought the realization that between the human world and the non-human one, there exists an ineluctable tension. Any vision of man which is sensitive to the ultimate claims of nature *and* culture is therefore obliged to conceive man's relationship with nature in terms of harmony, balance and coexistence rather than those of dominion and subordination.[20] Some radical strands of ecological thought however seek to underrate the significance of the cultural, uniquely human, factors in this relationship. For instance, Atkinson (1991: 94) says, "For a society which has achieved a stable, self-reproducing relationship with nature, there's little purpose in historical awareness". Whatever the attributes of a social world without historical consciousness, it is obvious that such a world would lack scientific inquisitiveness. It would have no need for knowing the secrets of nature and no urge for disrupting its amoral harmony. In the final analysis, it would be a world without the predicament, perhaps even the perception, of any ought, and hence bereft of any moral imperative. From the Islamic point of view, it would also be a sub-human world.

To view man from the perspective of culture is to give legitimacy to his search for order and meaning. It is also to attribute to him some measure of transcendence. For it is by overcoming his natural constraints that man reveals his need for transcendence and demonstrates his capacity for the creation of the symbolic world of culture. If to be human is to mediate between the worlds of nature and culture, then from the Islamic point of view, the gist of this mediation is a moral responsibility. Indeed, so pronounced and pervasive is the theme of the unity of nature and man's moral responsibility in the Islamic tradition that even outsiders have not failed to notice it. For Marshall Hogson, it constitutes the essence of the Islamic ethos, which he expresses as: "The demand for *personal responsibility for the moral ordering of the natural world*" (Hodgson: 1976: 337; emphasis in the original).[21]

NOTES

[1] To modern people, convinced of the import of science for the wellbeing of the human community and appreciative of its role in the progress of civilization, any claim about the non-anthropocentric character of modern science might appear incomprehensible. A few words of explanation are therefore in order.

Modern natural science is a human creation, and as such bears our stamp. Nor can it be gainsaid that we use science, wisely or foolishly, to promote our own interests. As for the fruits of science, these are subservient to human ends. The paradox is that man was able to penetrate the secrets of nature, and acquire prodigious power over it, only by forsaking his propensity for viewing it in human terms. In fact, only when nature ceased to be described in an anthropomorphic language could the progress of science become a reality. According to the anthropocentric vision of medieval physics, the entire world of nature existed for the sake of man and was fully intelligible in his mind. Thus the very categories in terms of which nature was interpreted – substance, essence, matter, form quantity, quality – presupposed that man was active in his acquisition of knowledge and nature passive. At the same time, the knowledge of the physical world, according to modern science, is accessible to us in categories – space, time, mass, energy and the like – that accord no privileged status to the human observer. Not surprisingly, the avoidance of all forms of teleological theories, explanations with a human quotient, is strictly de rigueur in modern science.

E.A. Burtt expresses this central metaphysical contrast between medieval and modern science in terms of man's relation to his natural environment: "For the dominant trend in medieval thought, man occupied a more significant and determinative place in the universe than the realm of physical nature, while for the main current of modern thought nature holds a more independent, more determinative, and more permanent place than man" (Burtt, 1954: 17–18). The elimination of an anthropocentric vision from the study of nature has also revealed a universe that is without any value and produced a science that is self-confessedly value-free. Such a science "is unable to establish its own meaningfulness or to answer the question whether, and in what sense, science is good" (Strauss, 1989: 33). Strauss also contends that, "the assumption that we should act rationally and therefore turn to science for reliable information – this assumption is wholly outside the purview and interest of science proper." Our reverence for science then is a cultural trait that does not ensue from its self-acclaimed value neutrality or non-anthropocentric cosmology.

² Islam shares this essentially trans-cosmological vision with other monotheistic traditions, especially the Biblical ones.

³ The Arabic term, denoting a collectivity, refers to the Hellenized Muslim theorists whose thought was based on Greek cosmology.

⁴ A critical account of this debate, perceptively revealing the underlying issues and sensibilities involved, is found in Rahman, 1979.

⁵ The Muslim philosopher Ibn Sīnā (Avicenna) (980–1037) produced a synthesis between the Greek, Aristotelian, metaphysics of being as an eternal given and the monotheistic, creationist account of the radically contingent world. Nevertheless, for his orthodox critics like al-Ghazālī (1058–1111), such a philosophical account of being was heretical and rationally incoherent. For an introduction to Avicenna's thought, see Goodman, 1992. S. Hussein Nasr (1993) also provides, from a traditionally Islamic vantage point, a critical perspective on Avicenna's natural philosophy. Al-Ghazālī's work is available in a lucid English translation by Michel E. Marmura. His anti-philosophical arguments, further, have been disputed by the Muslim philosopher and jurist Ibn Rushd (1126–1198), repeated by the Jewish sage, Mūsā b. Maimon, better-known as Maimonides (1135–1204), and critiqued by Orientalist scholars (almost everyone).

⁶ For the revolutionary change which the Qur'ān wrought in the worldview of the jahili (literally, ignorant; figuratively, non-Muslim) Arab community, replacing its fatalistic and tragic outlook on human existence with a faith in the power of an All-Wise and All-Just God, see Izutsu, 1964. Daud Rahbar (1960) perceptively analyzes the problem of Quranic determinism.

⁷ Sufism is generally understood to be the mystical dimension of Islam.

⁸ For some modern attempts to expound this aspect of Ibn 'Arabī's thought, see Izutsu, 1984: 116–140, and Corbin, 1960: 115–116; 297–300, and passim.

⁹ Ḥadīth qutsī is a prophetic saying, not part of the Quranic revelation, in which God speaks in the First person.

¹⁰ The Hebrew language, which also lacks the concept of "nature" in its biblical vocabulary, has likewise acquired a modern term teva that corresponds to the Arabic ṭabʿ or ṭābʿīa.

,¹¹ Natura naturata or created nature, describes the passive reality of our daily experience. Natura naturans, or creating nature, is the active power that directs and governs life.

¹² Although Strauss presents his insight with a ring of original discovery, for students of Islam and Arabic, both medieval and modern, the logical correspondence between "nature" and "custom" has never been part of any esoteric wisdom.

¹³ This is the gist of al-Ghazālī's counter-argument against philosophy as a worldview. His critique of causality was later echoed in modern thought, including in Jung's observations on synchronicity. The connection between cause and effect, he claims, cannot be established, for observation shows only concomitance and never any necessary causal connection (al-Ghazālī, 1997).

¹⁴ The Sunna may simply be regarded as the Islamic theory of practice.

¹⁵ "Il n'est que de loi positive revélée et les 'loi de la nature' sont simples 'coutumes de Dieu', que

l'expérience et la répétition des actes nous font connaître, à tire, en quelque sorte de contestation provisoire. Le miracle en conséquence se definra comme une 'rupture d'habitude'". Cf. also Nasr, 1993: 9.

[16] This Quranic concept, which may be construed as the phenomenological equivalent of Christianity's "grace", also provides the rationale for divine intervention in human affairs, disclosing, to continue speaking in Christian terms, God's plan for the salvation of humanity. Some of this discussion is also found in Izutsu (1987: 139–150).

[17] Scientific knowledge is based on sense perception; it discards all empirically non-verifiable data. Religious faith, on the other hand, affirms the realm of transcendence, of the unknown and the unknowable, even if claims that this realm is in part cognitively accessible through trans-sensory channels of intuition, inspiration and revelation. The mystery of humans, accordingly, is impenetrable; it is beyond the ken of empirical science. Secularism has many names. In contemporary literature it is presented, either humbly, as a rejection of the ecclesiastical authority, as a model for pluralism, a theory of society, or a system of governance; or augustly, as a philosophy of history, a creed of atheism, an epistemology of humanism; or as a metaphysics that corresponds to the ultimate scheme of things. Within academic discourse, it is also customary to distinguish its various manifestations as a process of history (secularization), a state of mind and culture (secularity) and a theory of truth (secularism). Religious faith in general, and Islam in particular, can coexist with both science and secularity. The incompatibility becomes an issue when they are transformed into doctrines, into scientism and secularism, that have pretense to explaining everything and providing answers to the ultimate questions of the meaning of human existence.

[18] One must not confound this transcendental perspective with the biological one of modern science and construe Adam as an emblem of *homo sapiens* (in the manner of Lucy!), or reduce his being to atoms and genes. It is licit to speak of man in biological terms, as the *Qur'ān* itself employs biological images and metaphors (23: 12–14; cf. also 32: 8), as well as to refer to humanity as Adam's progeny, but it is always within the grand paradigm of transcendence that Adam finds his habitat and name.

[19] There are some parallels between the Quranic conception of Adam's vicegerency (deputized by God to exercise his authority in government and religious matters) and the Hegelian philosophy postulating the march of history through the self-realization of the Spirit. However, on account of his free will, the Quranic Adam is vulnerable and not assured of a happy ending.

[20] A summary of the debate that followed the publication of Lynn White Jr's thesis on the complicity of Biblical "dominion ethics" and contemporary environmental malaise is to be found in Spring and Spring, 1974.

[21] Marshall G. Hodgson, professor of history at the University of Chicago, was a sympathetic student of Islam who authored a monumental, 3-volume work, *The Venture of Islam*. From his Quaker vision and sensitivities, he similarly views Christian calling as "The demand *for personal responsiveness to redemptive love in a corrupted world*" (vol. 2, p. 337; emphasis supplied).

BIBLIOGRAPHY

Al-Azmeh, Aziz. *Arabic Thought and Islamic Societies*. London: Croon Helm, 1986.

Al-Faruqi, Ismāʿīl R. *Islam and Culture*. Kuala Lumpur: ABIM, 1980.

Al-Ghazālī. *The Incoherence of the Philosophers. Tahāfut al-falāsifah*: a parallel English-Arabic text, translated, introduced, and annotated by Michael E. Marmura. Provo, Utah: Brigham Young University Press, 1997.

Arberry, A.J. *Koran Interpreted*. 2 vols. London: George Allen & Unwin, 1955.

Atkinson, A. *Principles of Political Ecology*. London: Belhaven Press, 1991.

Burtt, E.A. *The Metaphysical Foundations of Modern Science*. New York: Doubleday Anchor Books, 1954. First published in 1924; second revised edition, 1932.

Corbin, Henry. *Creative Imagination in the Sufism of Ibn Arabi*. Princeton, New Jersey: Princeton University Press, 1960.

Gardet, Louis and M.M. Anawati. *Introduction à la théologie musulmane; essai de théologie comparée*. Paris: Vrin, 1981.

Goichon, A.M. *Lexique de la langue philosophique d'Ibn Sina (Avicenne)*. Paris: Desclée de Brouwer, 1938.

Goodman, L.E. *Avicenna*. London and New York: Routledge, 1992.

Gray, John. *Enlightenment's Wake*. London: Routledge, 1995.

Hodgson, Marshall G. *The Venture of Islam*. 3 vols. Chicago: University of Chicago Press, 1974.

Ibn Rushd (Averroës). *The Incoherence of the Incoherence. Tahāfut al- Tahāfut*. Simon Van der Bergh, trans. London: Luzac, 1954.

Izutsu, Toshihiko. *God and Man in the Koran*. Tokyo: Keio Institute of Cultural and Linguistics Studies, 1964; reprint Salem, New Hampshire: Ayer, 1987.

Izutsu, Toshihiko. *Sufism and Taoism: A Comparative Study of Key Philosophical Concepts*. Berkeley: University of California Press, 1984.

Maimonides, Moses. *A Guide for the Perplexed*. Chicago: University of Chicago Press, 1963.

Massignon, Louis. 'La nature dans la pensée islamique.' In *Eranos-Jahrbuch* 14, 1946, pp. 144–148.

Mumford, Lewis. *Technics and Human Development (The Myth of the Machine, vol. I)*. New York: Harcourt, Brace, and World, 1967.

Nasr, Seyyed Hussein. *An Introduction to Islamic Cosmological Doctrines*. Cambridge, Massachusetts: Harvard University Press, 1964; reprint: Albany: State University of New York Press, 1993.

Rahbar, Daud. *God of Justice*. Leiden: Brill, 1960.

Rahman, Fazlur. *Major Themes of the Qur'ān*. Chicago: Bibliotheca Islamica, 1980.

Rahman, Fazlur. *Prophecy in Islam: Philosophy and Orthodoxy*. London: Allen and Unwin, 1958; reprint, Chicago: University of Chicago Press, 1979.

Schuon, Fritjhof. *Understanding Islam*. Baltimore: Penguin Books, 1972.

Spring, David and Eileen, eds. *Ecology and Religion in History*. New York: Harper and Row, 1974.

Strauss, Leo. 'Progress or return.' In *The Rebirth of Classical Political Rationalism: Essays and Lectures by Leo Strauss*, Thomas L. Pangle, ed. Chicago: University of Chicago Press, 1989, pp. 227–270.

JEANNE KAY GUELKE

JUDAISM, ISRAEL, AND NATURAL RESOURCES: MODELS AND PRACTICES

ISSUES OF DEFINITION

A summary of Judaism and the environment in a volume on the history of non-western science is a more problematical undertaking than may appear at first glance. Well-informed Jews today debate the issue of defining a Jew. Is identification based on ethnicity, matrilineal descent, a shared sense of history and culture, or adherence to a specific set of religious practices? In the latter case, are all of Judaism's denominations equally valid? What about secular Jews, such as the majority of the Israeli population today? Orthodox Jews, defined by their fundamentalist belief in the binding obligations of ancient Jewish law (*Talmud, halakhah*), reject the competing truth-claims of liberal denominations and of Jewish groups like the Falashas of Ethiopia who became geographically isolated before the laws were encoded (Robinson, 2000: 459–450, 474–477).

How does one define the environment: through the science of ecology, natural resource management strategies, or as a set of beliefs and ideologies about nature in the abstract? Is commercially oriented range management, for example, in or out? What about vegetarianism or good feelings about nature?

To the extent that the Near East may be defined as non-western, Jews and Judaism fit the concept of non-Western, through their attachments to the land of Israel and through residence in predominantly Islamic countries. Yet the Hebrew Bible became the Christian Old Testament, and as such, it remains one of the pillars of western civilization. The United States, moreover, is arguably the major center of Jewish activity outside of the state of Israel.

Most types of scholarship experience periodic paradigm shifts and ideological battles between academic factions. Since ancient times, believers have debated whether the Bible should be understood literally or metaphorically and esoterically. Since the Enlightenment, researchers in biblical "criticism" have linguistically inferred different authors of the Bible and rejected the notion of a unified text (Friedman, 1987). Their perspective, however, has not attracted scholars who interpret the Bible more holistically as a work of literature (Alter and

H. Selin (ed.), *Nature Across Cultures: Views of Nature and the Environment in Non-Western Cultures*, 433–456.
© 2003 *Kluwer Academic Publishers. Printed in Great Britain.*

Kermode, 1987). Today Biblical literary and archaeological studies are undergoing a period of relative turmoil over the extent to which the Bible can be legitimately read as a work of history, especially as evidence supporting Jewish claims to an ancient homeland within the present-day state of Israel and the West Bank. This debate stems from the intense political situation of the Near East (see Dever, 1998) and ambiguous or contested archaeological evidence about ancient Israelites' real identity as a discrete ethnic group. The entire concept of ethnicity is open to debate today, with increasing criticisms of once widely accepted definitions and parameters.

Finally, is science an appropriate concept to apply at all to a population or belief system defined on the basis of theology? (see Tirosh-Samuelson, 2002). The basic core of Judaism and Jewishness emerged during the Iron Age among a profoundly religious population. The Bible indicates that some of the ancient Hebrews were back-sliders, but they principally slid into the pagan idol-worship practices of their neighbors, rather than into analytical, empirical modes of secular inquiry or "science" as developed by Greek scholars of the classical period. Ironically, natural scientists positioned Darwin's theory of natural selection, a cornerstone of contemporary scientific ecology, in opposition to the Hebrew Bible's principal environmental message, the creation narrative of the book of Genesis.

This is tricky scholarly terrain, for which no traverse is likely to be fully successful. The strategy taken here is to follow a few key lines of inquiry, with the understanding that alternative approaches are also possible and desirable.

The environment will be addressed through a suite of themes key both to ancient and modern Near Eastern experiences as well as standard textbooks in environmental studies: water, agriculture, grazing lands, and heritage preservation. The concept of natural heritage recognizes that wild plants, animals, and the physical environment have cultural meaning and value to people independent of their economic worth (see Leopold, 1949; Thompson, 2000). As examples of the latter, the cedars of Lebanon emerge in the Bible as religious symbols of pre-eminence and authority, but sometimes also of excessive pride. Restoration efforts in Israel's parks today are underway to show visitors cedars and other plants mentioned in the Bible in their natural habitats (see www.neot-kedumim.org.il).

My primary focus will be on biblical and ancient post-biblical Judaism as it emerged in and near Israel (Palestine), today within the modern political borders of the contested West Bank and Israeli state; and with a concluding summary of twentieth century environmental issues within Israel. Jews and Judaism will be primarily identified through a set of belief systems identified in ancient texts or practices inferred from archaeological evidence, although not exclusively those meeting the Orthodox criterion of *halakhah*, or observance of Jewish law (see Gottwald, 1985; Alter and Kermode, 1987). Obviously a religion as long-lived as Judaism experienced significant change over time, with geographical diversity and competing authorities during its long history, which this brief chapter unfortunately must conflate.

The founding text of Jewish belief is the Hebrew Bible or *Tanakh*, codified

perhaps around 250 BCE. The first five books, or *Torah*, form the core of
Jewish belief and liturgy (Robinson, 2000: 257–264). Christians subsequently
incorporated the Jewish Bible into their own sacred texts, calling it the Old
Testament and adding a separate New Testament, which Judaism does not
recognize. After the Hebrew Bible was completed, Jewish rabbis (teachers,
religious leaders) in Palestine and Babylon (modern Iraq) began the task of
interpreting the Torah's laws in a body of rulings known as the *Mishnah* and
further commenting upon those interpretations. Their various analyses became
codified into a second multi-part Jewish scripture, the *Talmud*, by about 500
CE. For Orthodox Jews it has sacred canonical status and remains the basis
upon which the Bible is to be understood and its precepts enacted (Werblowsky
and Wigoder, 1966: 373–375; Neusner, 1984; Steinsaltz, 1976). Additional rabbinical commentaries and Jewish lore on biblical themes were incorporated
into another extensive collection called *Midrash*, redacted by 600 CE (see
Genzburg, 1967–69; Freedman and Simons, 1971; Bialik and Ravnitsky, 1992).
Also dating from Roman times are descriptions of Jews penned by classical
authors (Stern, 1980) and a few preserved religious volumes (apocrypha, pseudepigrapha) of non-scriptural or ambivalent status, ranging from the story of
the Maccabees and the origin of Chanukah to more explicitly Christian works
(Charles, 1963–64; Goodspeed, 1989). These offer additional information on
Jewish life in ancient times. A rich and extensive reference work on the above
and on most things Jewish is the *Encyclopaedia Judaica* (1971–72), abbreviated
below as *EJ*.

Although Jewish texts like the Bible are explicitly religious, they also reveal
some mental habits in common with science as it is defined today. These
include keen observation of natural phenomena, inquiry about their underlying
causes or explanations, bodies of information or evidence assembled in response
to one's inquiry, as well as interpretation of natural phenomena through models,
propositions, and simple declarative statements. Today, scientists call the latter
hypotheses, theories, or laws; religiously inspired writers may call them truths
or articles of faith. The methods and criteria that each side uses to obtain their
models may differ, but they can serve the same function in a society seeking
to explain natural events. Judaism, like modern science, postulates a universe
that is fundamentally orderly and predictable. Though punctuated by unfore-
seen disasters like earthquakes or floods, these anomalies are themselves understandable through inquiry into fundamental principles.

Judaism differs from modern science, however, in postulating a universe
whose laws are essentially moral principles (*hochma*, or wisdom) established by
a single all-powerful creator God (*Proverbs* 8: 22–36). Once His moral principles
are grasped, however, cause and effect in natural phenomena may often be
discerned. Accordingly, a God who created the entire universe is necessarily so
superior to human beings that some explanations for His creation lie beyond
human understanding (*Isaiah* 55: 8–10; *Job* 38–41). From the perspective of
contemporary scientific discourse, Judaism does present an elegant or simple
explanation for all natural phenomena: a divine mind who fashions the universe

through wisdom and encourages human humility in the face of the universe's vastness and complexity.

There is an applied or practical side to both religious and western scientific environmental evidence and models, moreover, in the form of management theories and practices. Both religion and resource management address just what one should *do* to ensure a supportive environment. Wise practices may then be applied to promote productive agriculture, an adequate water supply, or wildlife preservation. But because modern environmental science and Judaism are based upon some fundamentally different premises about causality, they may promote radically different types of practices.

BIBLICAL CONCEPTS OF WATER RESOURCES

The significance of water to ancient Jews, specifically in the form of rainfall and managed run-off, cannot be overemphasized. Without the major rivers for flood plain irrigation that characterized the older and larger "hydraulic" societies of Egypt and Mesopotamia, ancient Hebrews depended on rainfall and minor streams to water their crops (Smith, 1966: 62–76; Hillel, 1982; Ron, 1985). Israel straddles the boundary between the seasonally arid Mediterranean climate region and the low-latitude hot desert. Mediterranean-type climates are known for their cool, moist winters and rainless, sunny summers. As one traverses towards the Judaean desert east of Jerusalem and south along the seacoast towards the Sinai Peninsula, precipitation declines both in quantity and predictability. Within the deserts to the east and south, precipitation may be abundant enough to support grass for grazing in one year but be nearly absent the next year. The timing of the onset of the rainy season and the sequencing of dry and wet years are also unpredictable, particularly as one moves towards the true deserts. In both environmental zones the occasional oasis, spring, or well may be used to water livestock and to irrigate orchards, but the grass itself parches away in the hot summer. Precipitation amounts may increase with elevation in the hill regions, sufficient to produce forests and pasturage, but farming on hill slopes may require labor-intensive construction and maintenance of terraces, or the slopes may simply be too steep or rocky to farm profitably.

These are highly challenging environments to cultivate or even to graze (Frick, 1989). For ancient peoples attempting to make a living from such places, how could they understand the variability of rainfall in their region? Can humans influence rainfall amounts? How best to understand the puzzling emergence of springs and streams from beneath the surface of desert landscapes?

The Bible begins in its very first book (*Genesis*) and chapter to provide a model for the environment. An all-powerful, autonomous, and pre-existing God (*Elohim*) creates in the beginning from a chaotic void a division of sky, earth, and surface water, over which He retains dominion. He subsequently creates all living things, over which He gives humanity preeminence and conditional control. God pronounces the whole of his creation to be good – the basis on which Jewish and Christian creation theologians today argue for environmental protection. The Genesis interpretation of nature is clearly anthropocentric and utilitarian, but ultimate control over nature rests with God rather than man.

The book of Genesis also provides an ancient sketch of the hydrologic cycle (Stadelmann, 1970). The problem of how precipitation comes from the sky is explained by an invisible celestial water body, usually translated in English Bibles as the *firmament*, which is the source of rainfall (*Genesis* 1: 6–8; *Psalms* 148: 4; Josephus, *Antiquities* bk. I ch. 1, sec. 1; *Wisdom of Sirach* 43: 1, 14–15 in Goodspeed, 1989). Elsewhere the firmament is described as the heavens in general or more specifically as God's storehouse or treasury of rain, snow, and hail (Bialik and Ravnitzky, 1992: 765–767). Why the sea does not overflow from constant run-off (*Ecclesiastes* 1: 7), or how springs emerge from desert ground remain something of a mystery, known only to God, in the absence of an empirical understanding of evaporation, water vapor, and ground water.

Although contemporary Jews seldom conceptualize Him in these terms, the Bible clearly establishes God, whatever else He may be from a theological perspective, as a powerful weather deity (*Deuteronomy* 28: 22–24; *Jeremiah* 14: 1–6; *Job* 37: 1–18, 38: 22–38; *Psalms* 18: 9–13, 147: 16–18, 148: 8). Although God may be located anywhere, His special abode is the *shmayim*, the Hebrew word meaning both "sky" and "heaven". From this atmospheric vantage point, God controls rainfall and drought. The earlier books of the Bible generally attribute God's dispensation of these critical phenomena to his reward or punishment for good and evil human conduct – a transaction so central to the Torah that it became the foundational covenant or contract between God and the Jews, as transmitted by the leader Moses on Mount Sinai (*Deuteronomy* 11: 8–17). The biblical books of *Leviticus* and *Deuteronomy* enumerate in detail a set of laws that the Jews are to observe, citing drought, famine, military defeat, and plagues as punishment for disobedience. A rainy day in the Near East does not therefore signify "bad weather" or a spoiled picnic: rather, it is a material form of God's reward and blessing that offers hope for good crops and full stomachs. Only rarely do especially righteous individuals have the ability to invoke rain themselves as God's deputies, as for example the prophets Samuel and Elijah (*I Samuel* 12: 16–18; *I Kings* 18).

From the Biblical concept of God as the pre-eminent weather deity who dispenses or withholds rain for crops and human survival, a number of corollaries follow within the Jewish belief system. First is the absolute abhorrence of worshipping man-made idols, god-kings, or spirits of animals, plants, or natural features. Ancient Jewish texts offer plenty of evidence that lapsed Jews did indeed worship such alternatives. But normative Judaism's moral messages are that idols, as the handiwork of artisans, cannot possibly be considered autonomous creators of nature (see *Wisdom of Solomon* 13: 10–19 in Goodspeed, 1989); and worshipping them may anger God and cause Him to withhold rain in retribution, as is His entitlement under the terms of His covenant with the Jews. To worship an ancient tree or tree spirit is therefore highly misplaced in Judaism, as the tree itself depends on God's rainfall for its survival. Nor should humans trust in their own wits or wealth accumulation to sustain them through hard times. God can easily retract these through natural or military disasters.

This reasoning might almost be cast in terms of science's independent and dependent variables. The pre-eminence of water in the survival of all living creatures, and their dependence upon an autonomous, causal Creator who controls it, seemed self-evident to the ancient scribes.

During the early centuries CE, a few rabbis living in Babylon attempted to calibrate this relationship more closely (Bialik and Ravnitzky, 1992: 767). They developed weather forecasts based on observation of the types of cloud formations that preceded rain showers, recorded annual variations in rainfall, and devised rigorous definitions of drought (*EJ* 13: 1520). They implemented an elaborate schedule of voluntary prayers and fasts in which drought-stricken congregations were supposed to participate; their intensity was based on the length of delays in the expected onset of the rainy reason or on the volume of rainfall they received (*Ta'anit* tractate of the *Mishnah*, *EJ* 15: 675).

Rainfall was less of a concern for the large Jewish population who lived in Babylon in Talmudic times, where irrigation water was more predictable (Lasker and Lasker, 1984). In Palestine, however, the Phoenician and Canaanite rain-god Baal and goddesses Asherah and Anath (Astarte), the "Queen of Heaven", were serious competitors for the Jews' allegiance, as their neighbors held that that these deities controlled the life-giving rains and the earth's fecundity, instead of the God of the Jews (*I Kings* 18: 19–40, *II Kings* 11: 15–20, *Jeremiah* 44: 15–19; Patai, 1990: 62–66; Frymer-Kensky, 1992: 89–95, 156–7).

How, then, should one understand permanently arid environments, according to Hebrew beliefs? Here, too, the model applies. Where the scribes had tangible evidence of abandoned former settlements in the desert, such as Sodom and Gemorrah on the Dead Sea, they attributed their demise to the wickedness of their old inhabitants. But the waterless desert, as the place where human survival is most in jeopardy, is also the environment where the righteous individual who faces death in dire necessity may attain spiritual grace and unexpected discoveries of water and shade. An alternative interpretation, figuring most prominently in the book of *Exodus*, is that the dreaded deserts are a place where God tests the faithful through a series of trials and obstacles. If equal to the challenge, their reward for righteous conduct will surely follow.

The ancient Hebrew model for the earth extends far beyond the popular modern concept of Gaia as one complex planetary organism or of the ecosphere as the proximate concentric spheres of earth, life, surface water, and atmosphere. God also creates and rules the sun, moon, and the stars "for signs and for seasons", plus heavenly realms and creatures invisible to all but a few prophets. Along with the other ancient civilizations, the ancient Jews probably observed the rising and setting of constellations and key stars as a celestial calendar; synagogue mosaics and rabbinical discussions reveal at least a keen interest in the zodiac as a link between the distant heavens and human events (*Wisdom of Sirach* 43: 6–10; Josephus, *Antiquities* bk. I, ch. 1, sec. 1; L. Levine, 2000: 572–575). Judaism's religious calendar is lunar, with an interpolated thirteenth month every few years, and with holy days timed to new moons and full moons of particular months.

Key Jewish festivals and liturgy also follow from Mediterranean precipitation cycles and the agricultural systems adapted to them in the Near East. Most of the holy days outlined below originated in biblical times or in the few centuries thereafter (see Robinson, 200: 92–127). They address the questions of what observant Jews should do to encourage God to bless them with abundant rainfall and crops and to minimize the possibility of punishing droughts (see *Jeremiah* 14: 1–6). Prayer, repentance, fasting, and – during the existence of the temple in Jerusalem – animal and plant-food sacrifices were believed to propitiate God and to restore a supportive balance between Israel, nature, and God (James, 1961: 92–120). Thus the period just before the expected onset of autumn rains became a crucial period in which to rectify one's wrongdoing and to petition God for rainfall.

The Jewish New Year begins in early autumn with the festival of *Rosh Hashanah*. It is a time of celebration, but it also initiates a profound ten-day period of repentance and prayers for mercy as God judges Jews' conduct over the preceding year. *Yom Kippur*, the most solemn day in the Jewish calendar, is spent in fasting in the synagogue (house of worship), and in collective confessions of wrongs committed, in the hope of atonement. At the close of the day, Judaism holds that God decides the course of events for the coming year.

The weeklong festival of *Sukkot* (Tabernacles or Booths in English Bibles) follows five days later. Before the Jerusalem temple was destroyed, *ca.* 70 CE, Jews brought sacrifices to it from the previous harvest. Priests conducted a special water-drawing ceremony. Israelites spent Sukkot living in small temporary huts in their fields. Worship services and family prayers petitioned God for rain, particularly on the final day of the festival. Special ritual plants used in the ceremonies are all species characteristic of riparian or irrigated environments: the willow, myrtle, date palm, and citron. These practices indeed suggest a kind of sympathetic magic or theory of correspondences, in which humans hope to influence natural outcomes by mimicking or ritually using elements from the desired aspects of the environment (Schaffer, 1982). Prayers for rain or dew continue from the last day of Sukkot until Passover, at the beginning of the harvest and dry period.

Shortly after the onset of the autumn rains, farmers planted winter barley and wheat. This schedule enabled them to get the maximum benefit from the rain while it was available, and also to time their harvests to the dry season, as rain might otherwise spoil the ripened crops. Prudent farmers planted their seed grain in different batches from about late October through December to minimize chances that a cloudburst or lack of soil moisture during germination would spoil the entire crop (Frick, 1989; *EJ* 1: 376; *Ecclesiastes* 1: 26). Despite the higher prestige accorded to wheat, barley was probably the more common staple, especially among the poor, because it is the more drought-tolerant crop.

Talmudic passages suggest that Jewish farmers practiced some conservation techniques. Deep plowing and wide spaces between furrows encouraged penetration of rain into the soil. Farmers probably spread livestock manure on their fields to fertilize them. In the absence of information on crop rotation to retain soil fertility, they also let their grain fields lie fallow in alternate years and on a seventh "sabbatical" year. They plowed fallow fields, however, to inhibit weeds. Extensive systems of terraces on hillsides, visible around the city of Jerusalem today, inhibited the rapid run-off of rainfall and diminished soil erosion. Ancient farmers sited their terraces principally on north-facing slopes with less insolation and hence better moisture retention and soil development (Ron, 1966). In the Judaean hills, they also supplemented rain water with irrigation from springs.

Olive trees were widely grown on stony hillsides. Harvesting of olives, principally for their oil, occurred in October and November. Olive oil use extended beyond cooking to lighting, ointment, and ritual lights in the temple, the latter commemorated in the December festival of *Chanukah*.

Although ancient water diversion and storage practices are only tangentially mentioned in the Bible (see *2 Chronicles* 26: 10, *Wisdom of Sirach* 48: 17), archaeologists have located many examples of small check-dams in hillside gullies, deep cisterns and large urban storage tanks hewn out of bedrock, and ingenious schemes for directing surface run-off to crops planted down-slope (see Hillel, 1982; Rogerson and Davies, 1996; Tsuk, 2000). The winter must have been a time when cisterns refilled and crops benefited both from "the rain of heaven" (*Deuteronomy* 11: 10–11) as well as from its intensive management once it reached the ground surface.

Most rainy-season holidays are post-biblical, and would seem to have little to do with climate, with the exception of *Tu b'shevat* in February, when the early spring plants and orchard trees start to bloom. This "new year for trees" originally marked the age of a tree for purposes of consumption and tithing of its produce.

Perhaps because of Israel's lower latitude, solar festivals marking solstices and equinoxes are less apparent in Judaism than in northern European traditions. Although Passover occurs around the spring equinox, Judaism's modified lunar calendar varies the actual date of the festival by as much as several weeks from March 21.

The festival of Passover in the spring at the onset of the dry season, however, has explicitly agricultural connotations. Commemorating the Jews' exodus from Egypt, it was also the beginning of the barley harvest. In ancient times, it was also a pilgrimage festival in Jerusalem, when farmers offered up their first sheaves of barley at the temple. No barley was to be eaten before God received His due. Passover initiates the "counting of the *omer*," of seven weeks to the next pilgrimage festival of *Shavuot* ("Weeks" or Pentecost). This second holiday of "first fruits" coincides with the wheat harvest.

As with barley and wheat, the grape harvest occurred during the dry season (Walsh, 2000). The Jewish covenant with God specifically mentions grain, wine, and oil as rewards for living an observant Jewish life. Summer was thus a time of reckoning: if the grain harvests and vintage proved good, there was plenty to store through the dry season and following winter. If not, famine might ensue. The biblical book of Ruth, read in synagogues during Shavuot, indicates that population migrations were common outcomes of famine, and it also outlines Israelite practices during the grain harvest such as women's binding sheaves or gleaning kernels from the fields behind the male harvesters.

Summer was also the season for warfare, and the covenant cites military invasions as another form of punishment for misconduct. Biblical imagery links the vintage with armed conquest, when it compares the crushing of grapes underfoot with armies invading a helpless populace, or abandoned settlements with a vineyard stripped of its vintage (*Isaiah* 5: 1–16, 6: 31–38, 24: 13; 63: 3–4; *Jeremiah* 25: 30–31; *Micah* 7: 13). The mourning and fast day of *Tisha be Av* (ninth day of the Hebrew month of *Av*) laments various military defeats inflicted upon the Jews, particularly the destruction of the two Jerusalem temples. In contrast, the ancient custom on the fifteenth of *Av* was for marriageable girls to dance in the vineyards (*Mishnah Ta'an.* 48, *Judges* 21: 21).

The annual cycle ends with the month of Elul. Observant Jews spent it in repentance, in preparation for the more intense penitential period between *Rosh ha Shanah* and *Yom Kippur* at the beginning of the Jewish year.

Although this cycle of Jewish holy days may seem well removed from modern science or environmental management, they share some fundamental features. Jews directly observed highly variable annual precipitation amounts and resultant crop yields. Initially they explained the underlying cause through a model of a just Creator who rewarded and punished human conduct. Regardless of how scientists today may evaluate this model, the precept that humans should propitiate the Creator for any transgressions, particularly just before the onset of the autumn rains, and should express special thanks for the key harvests in order to extend His good will, logically follows from it.

As observers of nature, the ancients also eventually began to question this simple management scheme; i.e., to test the thesis they adopted, over the centuries in which the Bible was written down in final form (see Kay, 1989). The scribes of the chronologically more recent books of Ecclesiastes and Job (38: 25–26), for example, noted that rain fell on good and bad men alike – indeed, that rain fell on wild places where no one was living at all. Evil kings might live happy lives and their sons might experience tragedies during their reigns that by rights should have befallen their fathers. Sometimes bad things did happen to innocent people (*Ecclesiastes* 8: 14, 11: 5, 12: 13). Nevertheless, the sages argued, it was still important to follow the commandments as divinely given, in light of such apparent contradictions, because the impenetrable wisdom of the Master of the Universe in giving the laws was necessarily superior to man's understanding.

As Jews immigrated to other countries, they retained the seasonal structure of the Mediterranean climate in their holy day calendar, even where climates were vastly different, and even after theological explanations for their observances replaced the agricultural ones. The ability of Diaspora Jews to remain in farming at all varied considerably between historical periods and places, depending upon government restrictions on the types of work and land tenure permitted to non-Christians in Europe and to non-Muslims within the Islamic world. Often forbidden to farm or own land, and living as outsiders in feudal societies, Diaspora Jews increasingly turned to commerce and urban residence. Their holy days and liturgies, however, continued to commemorate their Mediterranean roots.

GRAZING AND RANGE LANDS

When the book of Genesis moves from a narrative on the origins of humanity and nature to the specific origins of the Jewish people, it describes a clan or extended family of nomadic herders, emigrating from modern-day Iraq, and looking to settle in the "promised land" of modern-day Israel/Palestine. Over the passage of generations, the Jews emigrated to Egypt, then back to the Holy Land, only to face further dispersion and partial return under later foreign military invasions. Subsequent texts, when viewed chronologically, describe the Jews eventually as settling permanently in villages and even cities, and taking up farming, with livestock often fed from mangers in stables (*Isaiah* 11: 7, 30: 24; *Jeremiah* 46: 21). By 70 CE, Israel was apparently one of the more densely settled districts of the Roman Empire, with every scrap of suitable land brought under cultivation (Josephus, *Wars*, bk. 3 ch. 3 sec. 2). By *ca.* 300 CE, Israeli rabbis forbade keeping small livestock like sheep and goats within the confines of settled communities to minimize their destructive impact on plant crops (Alon, 1980: 277–285). Prior to the destruction of the temple *ca.* 70 CE, the demand for livestock used for ritual sacrifices, for cavalry horses, and for donkeys, oxen, and camels as draft or pack animals ensured that the Jews would continue to raise livestock in outlying areas, apart from any dietary considerations.

Ancient Jews similarly adapted their grazing practices to the seasonal exigencies of rainfall and drought. Late winter and spring were times when dormant grasses and herbs resprouted, so that livestock could be grazed in the desert areas. By mid-summer, however, dry weather and any grazing impacts reduced the landscape to sparsely vegetated drylands. Seasonal pasture is, apparently, the original meaning of the Hebrew word *midbar*, translated in most English Bibles as "wilderness" (E. Levine, 2000). The Bible gives ambivalent interpretations of the *midbar*. On the one hand, it is an area beyond the fringe of settlements, where shepherds tend flocks, political refugees may seek hiding-places, and spiritually advanced individuals may experience epiphanies. But ancient Jews also knew the locations of older, abandoned settlements from previous civilizations within the *midbar* where their own livestock grazed. They characteristically attributed the demise of these towns to the inhabitants' wickedness and God's retribution (*Isaiah* 5: 17). As with crop yields, Jews attributed good and poor pasture to God's beneficence (*Psalms* 23) or retribution (*Jeremiah* 25: 34–38).

They adopted systems of livestock raising common in many parts of the world in the past and today, known as transhumance. While most of the family resided in town, a few members of a livestock-owning family would bring grazing animals out to lands too rocky or arid to be cultivated, particularly during the rainy season and spring, when forage was available for them. This seems to have been a job for the sons or sons-in-law of herding families (*Genesis* 32: 2, 12–18; *Exodus* 3: 1; *I Samuel* 16: 11). In some areas they might return animals to the village nightly or seasonally, but the pentateuch model suggests that herders kept livestock in more remote areas for extended periods, moving them according to the availability of forage. The New Testament describes shepherds grazing their sheep around Bethlehem (*Luke* 2: 8), whose name in Hebrew translates literally as the "house of bread", i.e., grain. Although scholars dispute the month of Jesus' actual nativity, the leafing-out of the barley crop in December provided a source of pasturage for livestock (or cut fodder for the manger) at a point in the crop's life cycle before the grain could be damaged from grazing.

The seasonal regime influenced ancient breeding preferences. Sheep were wool bearing and fat-tailed, comparable to breeds still raised in the Near East today (*EJ* 14: 1133–1134). The great fat content of the tails serves as an energy reservoir for the sheep, enabling them to survive times of poor forage, much like the better-known camel's hump. Goats were probably much like those raised today by Bedouin; their long black hair was woven into useful clothes and tents, while affording the animals protection from the winter downpours. Donkeys, the common beasts of burden, require less feed and water than horses. Sheep and goats in particular are fairly prolific, and overgrazing must have been a common concern.

The Bible provides some evidence of range management practices. Extensive grazing was open range. Brothers agreed to move their herds into separate areas to avoid overgrazing; invading pagans are accused of overgrazing the land in addition to their religious deficiencies (*Genesis* 13: 6–13; 36: 6–7; *Judges* 6: 5). Cisterns hewn out of rock provided water for livestock. In one passage, a village well was a common water supply for various families' animals, but the herdsman kept it covered when not actually in use to prevent livestock from trampling it and sullying the water (*Genesis* 29: 2–8). Predation by lions apparently was a problem in the time of King David (*I Samuel* 17: 34–37); as were the more typical concerns about strayed animals and access to water (*Psalms* 23).

The "good shepherd" whose animals have enough to eat and drink and who remain obediently under his watchful care understandably became a powerful biblical metaphor for ideal human and divine conduct. Similarly, the Bible compares lapsed Jews living in a state of distress to parched and arid pastures; sincere repentance, however, will return them to happiness just as the dry lands may green up and ephemeral streams flow when it rains. Such literary tropes perhaps stray from modern science, in which nature-culture dichotomies owe much to their classical Greek heritage. In contrast, the Jewish texts demonstrate a fairly holistic view of humanity and nature, together with many other non-western societies. Although the rabbis taught that humans were superior to other organisms and that God was spatially detached from Creation, nevertheless ancient Hebrew has no separate words for nature and culture, and texts often conflate the earth and the heaven; the Hebrew word *olam* stands in for "everything" or the "universe". In their dependence upon the Creator, humans, plants, and animals are essentially fellow travelers. Ancient scribes were keenly aware of how drought affected humanity through its effects on pasturage, livestock, and grain crops. These further provided the mental images and language with which to reflect upon human conduct and social wellbeing.

JEWISH DIETARY LAWS

Both ancient and contemporary Orthodox Judaism have been extremely concerned with specific dichotomies within groups of people, animals, and crops. Among their extensive list of proscriptions, observant Jews are not to intermarry or even to drink wine produced by non-Jews. Categories of animals, birds, and seafood are considered "unclean" and forbidden as food sources (Klein, 1979; Dresner, 1982; Houston, 1993). Even among permitted food animals, meat is unfit for Jews if slaughtered improperly, blemished, or from a forbidden part of the animal. Certain crops may not be grown mixed together within a field in Israel. Jews may not weave together linen and wool to make cloth. These commandments are first found in the Torah, specifically in *Leviticus* and *Deuteronomy*, but much of the Talmud extends the dietary and agricultural regulations so that Jews might not transgress them even accidentally. For example, a biblical injunction against boiling a young animal (kid) in its mother's milk led to rabbinical prohibitions against mixing dairy products and meat in the same recipe, against serving meat and milk at the same meal, and even against using the same (washed) earthenware for serving dairy and meat foods at different meals. The most conspicuous example of dietary restrictions is the Jewish proscription against eating pork and raising pigs, as one of a class of animals forbidden in the Bible because they do not chew a cud.

The Jewish dietary laws (*kosher, kashrut*) are mentioned here because many Jewish environmentalists today see them, plus a prototype commandment in Genesis, as forming the basis for more humane and environmentally sound food practices. Vegetarianism is indicated to Jewish vegetarians in the verse in which God gave Adam and Eve and their descendants only plant foods to eat (*Genesis* 1: 29). God later voided this ordinance, however, both when He granted to Noah's extended family both plant foods and meat, provided they avoided any blood (*Genesis* 9: 3–4), and later in the larger set of laws specifically transmitted to the Jews by Moses on Mount Sinai (*Leviticus* 11). Many Jewish vegetarians believe that vegetarianism is the original desired state, while others find little support for it in the deposit of Jewish rabbinical tradition (Bleich, 2001; Gerstenfeld, 1998: 136–138). The accumulated body of rabbinical dietary edicts makes the practice of keeping strictly kosher a daunting task for some cooks: vegetarianism greatly simplifies life in a kosher kitchen.

The ecological argument for vegetarianism, apart from religious considerations, is fairly diverse and complex (see Adams, 1998), but one principal point is the significant loss of food energy or calories with each step up the food chain. If, for example, only 10–20% of the energy contained in a food organism is available to the next trophic level on the food chain, then the earth could support more people at a better standard if everyone ate grain directly, rather than feeding grain to, say, cattle and eating the beef. Rangeland overgrazing and deterioration are additional ecological concerns. Some vegetarians cite spiritual values about the sanctity of all life or humane considerations regarding the keeping and slaughtering of livestock under modern agribusiness conditions.

Most kosher individuals eat specially processed kosher meat as their tastes and budget permit, particularly for Sabbaths and festival meals. Environmentalists among them may argue that kosher slaughtering is more humane than normal slaughtering (a point debated by the vegetarians) and that kashrut teaches Jews to limit their appetites and so to respect life. Meat eating is taking a life, but it is to be done with reverence and with the conscious choice to leave alone many life forms that are otherwise edible.

Unfortunately, there is no simple way to link kashrut unambiguously to environmental science. To be sure, even highly conservative environmental practices and religions do evolve over time, and today's students of the essentially modern science of ecology may develop discourses that make kashrut or Jewish vegetarianism environmentally as well as religiously sound dietary strategies. Jewish sages since ancient times through the present, however, have attempted to explain the biblical food prohibitions in logical or theological terms, and their debates have not resolved the problem of rationale: just why did God prohibit specific foods and combinations (*EJ* 6: 42–43)?

Among the earliest of the ecological explanations advanced over the centuries, by such sages as Maimonides (*Guide for the Perplexed* 3: 48) and Nahmanides, is that pigs and the equally forbidden crustaceans are not merely ritually "unclean" but that these are creatures of unhygienic habits and habitats; they live in mud and effluvia. More recent scholars postulated ancient Jewish knowledge that parasites that thrive in poorly cooked pork would flourish in the sub-tropical Mediterranean temperatures. Unfortunately there are either logical or historical problems with most such explanations (Kass, 2001): bottom-dwelling fish are classified as kosher, for example, so long as they have fins and scales. Ancient people probably knew nothing of trichinosis. Pigs, with their rooting habits and love of mud, are not particularly well adapted to hot dry climates or rocky soils, but this does not explain why the Bible prohibits them as "abominations". (So much for the sanctity of all life!) Some scholars suggest that within classes of animals delineated in ancient Jewish ethnobiology, such as birds, fish, and "beasts", permitted animals were really the idealized animals within the type. Carnivorous mammals and carrion-eating birds, for example, were viewed as unsuitable nourishment for a holy people. Yet this explanation also is troubling, because elsewhere in the Bible lions and eagles epitomize strength or royal majesty (*Proverbs* 30: 19, 30).

Some neighboring ancient cultures like the Greeks and Romans used pigs in their own rites and sacrifices. Pigs were sacred to their agricultural goddess Demeter/Ceres; their rooting habits taught early humans how to plow, and sacrificed pigs figured prominently in her worship ceremonies. Other cultures in the Near East apparently did cook kids in goat milk. The Jews no doubt wished to avoid duplicating their pagan practices. Yet mere avoidance of other cult animals or foods is problematical also, as neighboring societies also sacrificed species like cattle that the Mosaic code in fact commanded Jews to sacrifice on appropriate occasions. Some classical authors likewise noted that priests of the ancient Egyptians – the biblical Jews' enemies – also abhorred pork (Stern, 1980 II: 37). The concept of separating out and distinguishing different species is clearly important, as it extends beyond diet to certain crop combinations and mixed fabrics, but elsewhere in the Bible mixtures of various foodstuffs pass without comment as typical menu or ritual fare.

The moral and theological explanations of kashrut probably come closer to the mark for an ancient religious group. The talmudic principle of *Bal Tashchit* – avoiding wanton destruction of fruit trees during warfare and needless cruelty to animals – also derived from the Mosaic code (Schwartz, 1997). Boiling a kid in its mother's milk might have seemed like an act of cruelty towards the mother animal, for example. Nevertheless the question of *why* cows but not pigs cannot yet definitively be answered. An ecologist might also agree with Gerstenfeld (1998: 151) that there is little evidence of explicit environmentalist concern in ancient Jewish scriptures and rabbinical interpretation in any modern sense, although following Orthodox Jewish precepts may nevertheless lead to environmental or humanitarian benefits today. An observant Jew might agree with Ecclesiastes or the concluding arguments of the Book of Job – that sometimes God declines to explain troubling phenomena, but the proper response is to follow the laws anyway as a matter of faith in the Creator.

HERITAGE RESOURCES AND THEIR PRESERVATION

The concept of heritage resources basically combines and sometimes conflates natural resources like ecosystems with cultural resources desired for historic or archaeological preservation. The notion that both are part of a culture's patrimony further incorporates the cultural meaning that biota or particular places have for societies (Leopold, 1949: 177; Graham, Ashworth, and Tunbridge, 2000: 13; Thompson, 2000). The meanings that societies ascribe to nature may not be very scientific, but they surely affect the application of environmental science in the real world. In Israel today, for example, special reserves that reintroduce rare plant and animal species mentioned in the Bible probably owe more to the species' religious-historical associations than they do to biologists' concerns about the species' gene pool or role in the food web (see Hareuveni, 1984). Practical knowledge of wildlife zoology and plant ecology is nevertheless applied to the creation of suitable habitat for the showcased specimens.

In the international preservation scene, both ecological and historical curators increasingly realize that management of heritage resources does not occur in an ethnic or political vacuum. In sites with competing groups of early occupants, just whose narratives about the past and whose symbolic species take precedence? Who manages the protected sites? How are the subordinate groups interpreted?

Despite its comparative novelty in western environmental textbooks, the concept of heritage resources seems particularly appropriate to Jewish environmental thought, both in ancient times and in the present. Paramount in Jewish texts past and present is the concept of Israel/Palestine as the eternal Jewish homeland and patrimony (*Genesis* 17: 8. *Exodus* 3: 8, *Joshua* 1: 1–6; Kunin, 2000), a concept defined as Zionism in the late nineteenth century. The Bible and rabbis recorded the land as itself having a voice or role to play in human history (see *Genesis* 4: 11, *Deuteronomy* 30: 19; Vilnay, 1973), along with special practical issues with environmental repercussions for Jews dwelling in Israel. Although they portrayed plant and animal species as having their own needs and existence (albeit subjected to human mastery), they also valued species for their symbolic or iconographic possibilities, particularly in humans' ability to mimic their noble qualities (see *Proverbs* 30: 24–31).

Although the Bible never describes the land of Israel as vacant or free for the taking, it insists at many points that it is to become or has become the homeland of the Jews. The capital city of Jerusalem on Mount Zion evolves as the absolute spiritual center of this homeland, just as the temple containing the "Holy of Holies" chamber, where the high priest may actually experience God's eminence, becomes the heart of Jerusalem (Kunin, 2000). At the time of initial nation building in the land of Israel, the Jewish people consisted of twelve clans or tribes, each taking up its own portion of the country. The distinctive identities of most of the clans disappeared through assimilation with the passage of time, and the political boundaries of the country changed with the Jews' varying political fortunes and misfortunes (Aharoni, 1979).

Even during centuries when nearly all Jews lived outside its borders, the belief in the land of Israel as a fundamental root of Jewish identity remained strong in Jewish daily prayers, rabbinical teachings, synagogue liturgies, and particularly in the festival of Passover. Jews living outside of Israel were admonished to move, or at least to pray that they might move, to Israel as soon as political, personal, and divine circumstances permitted. (Such ideology is, needless to say, troubling to modern-day Palestinian Arabs and their sympathizers, who operate under very different narratives about Palestinian rights to the land and about Israeli Jews as recent colonists from Europe or other parts of the Near East.)

God is the actual landowner in Jewish thought (Davies, 1989; Kark, 1992). Human access to land is thus conditional. The "gift of the good land", together with abundant food to be harvested from it, is God's principal reward for the stringent demands of an observant Jewish life, just as exile and famine are His principal punishments for evil doing. When the Jews can acquire the Promised Land from its earlier inhabitants only by military force, God assures them that the Canaanites or Philistines have forfeited their prior right to the land through wicked deeds and that Jewish forces will prevail if they obey God's commandments. As with Jewish skepticism about human conduct influencing droughts, Jews were sometimes hard-pressed to blame their exiles on misconduct. In later centuries, life in the *Galut* (Diaspora) might be attributed to God's inscrutability, persecution, or, more prosaically, to individual Jews' decisions to seek better economic opportunities in other lands (Roth, 1970).

The biblical books of Leviticus and Deuteronomy, together with the body of Jewish law that developed to interpret them, place considerable emphasis on proper treatment of the land base. Both within and outside of Israel, Jews are to observe a Sabbath, or weekend day of rest and worship, for themselves as well as for their lands and draft animals (Waskow, 2000). Fields may not be cultivated, animals may not be yoked for plowing, fires necessary for many crafts and even for cooking may not be lit. Many of the festivals carry proscriptions against doing work of any kind, notably farming.

Within Israel, a variety of land tenure restrictions were formulated to apply in recognition of God's prerogatives. Agricultural products were subjected to complex rules about temple offerings and tithing to support the religious institutions. Corners of the grain fields were to be left unharvested for the poor to use. Every seventh year was to be a Sabbatical year, culminating in a Jubilee year every fifty years. During these special years of rest or Sabbath, all arable land was to lie fallow; Jews were to live on previously stored foods. Anything that grew spontaneously from the fallow land was common property, to be shared with the poor, not for commercial use, and exempt from tithing (Blidstein, 1966). Because of profound beliefs in patrilineal inheritance as the most valid means for land conveyance, all land "sold" was to revert to the original owners during the Jubilee year, to minimize loss of families' patrimony. Slaves gained their freedom on the Jubilee year. The Jews did not observe these practices outside of the land of Israel, and some practices fell into disuse within Israel during the Roman era. Yet their fundamental purpose seems to be an interpretation of the soil of Israel as something more than real estate or a natural resource, but rather as sanctified ground.

Concepts of various plants and animals in Jewish thought are somewhat ambiguous. On the one hand, Jewish prophets warned against worshipping pagan gods "under every leafy tree" or worshipping animal effigies, as in the golden calf scene in the book of Exodus (32: 1–6). Rabbis feared that appreciation of nature's beauty would turn scholars away from the more important study of scripture (Benstein, 1995). They often subverted the most "natural" of biblical passages as mere metaphors in esoteric spiritual or ethical teachings, stripping the passages of literal ecological content (see Freedman and Simon, 1961). The proscription against representational art has not been as strong in Judaism as it has been in Islam. Nevertheless, the commandment against making "graven images" as potential objects of idolatry limited Jewish depiction of both human and animal forms.

On the other hand, the Bible accorded souls to animals as well as to humans (*Ecclesiastes* 2: 19–22), and commanded that draft animals be given a Sabbath day of rest. Some passages describe plants, animals, and physical nature as singing praises to God (*Psalms* 148: 3–10; *I Chronicles* 16: 23–33). Nature is at a clear disadvantage in the Mosaic covenant: a drought resulting from human wickedness would necessarily harm innocent creatures. Yet the prophets recognized this problem and postulated a future messianic era when God would offer a new covenant that would enable nature to thrive peacefully. Noah, as he prepared the ark to weather out a forty-day storm that would destroy the rest of the earth, was commanded to take on board representatives of all animal species, including those deemed ritually unclean (*Genesis* 7: 2–3). The knowledge of King Solomon, the wisest man in ancient Jewish lore, included a thorough understanding of the ways and even languages of animals and plants (*I Kings* 4: 33; *Wisdom of Solomon* 7: 17–20). Talmudic sages developed a set of blessings that one should recite on viewing wonders of nature (Robinson, 2000: 20). The Torah is called "a tree of life".

Such ambiguity particularly relates to God's commandment of human dominance over plants and animals in the creation story of Genesis. This mastery appears to depend upon moral fitness, however – a state that fallible humans in the Bible seldom attain (Kay, 1989).

Mainstream Judaism never quite developed the attitude in which this world is merely a painful theater of temptations and hardships through which souls must pass before attaining their permanent afterlife. Nevertheless, the didactic or iconographic possibilities of plants, animals, and physical nature for purposes of moral instruction, with an eye towards the "world to come", underlay much of the Bible's and rabbis' treatment of nature (*Proverbs* 26: 2, 31: 24–28). The ant, for example, is praised for its habits of thrift and industry, which humans should emulate. Eagles and lions become symbols of majesty and pride (*Proverbs* 30: 19, 30; *Jeremiah* 25: 28). The saying from the Bible – "all flesh is grass" – relies upon experiences of pastures that green-up during the rainy season, only to shrivel and parch during the summer dry period. Since grasslands teach people that life is ephemeral, people should not trust in material accumulation, but rather in God. In one of the most impressive biblical passages about nature, God mocks the man Job (38–40) for daring to question Him, as Creator of the universe and of wilderness species that man cannot domesticate or even understand. Thus the unfathomable behavior of the ostrich and wild ass of desert regions teach Job about humility and the limits to human knowledge. Some rabbis taught that each creature in fact had its own special quality for purposes of moral instruction (Ginzberg, 1937 I: 43).

Ancient sages sometimes developed little didactic tales about creatures and landscapes to answer Bible questions. For example, why did God choose Mount Sinai as the location where He delivered the commandments? The answer was that both Mount Carmel and Mount Tabor in Israel vied for the honor, each claiming pre-eminence. God chose Mount Sinai because it was humble and did not boast of its achievements. In another version, God chose Mount Sinai over numerous peaks contending for the honor like a group of clamoring children. The reason was that the other sites were scenes of prior idol worship: only Sinai remained pure (Vilnay, 1978, III: 21–22; cf. *Isaiah* 2: 12–19). One learns nothing about actual landforms from these fables; they merely convey moral instruction about the evils of arrogance and idol worship.

There are several environmentalist implications of such Jewish readings of nature, however. One is that, in common with other ancient pre-scientific people, Jews did not view nature as something separate from the human condition. A wise Creator wants both ants and people to be industrious, both hill tops and people to be humble. One moral code applies across the natural-cultural spectrum, despite humans' superordinate position, blurring some of the dualism so prevalent today in western thought. A corollary of the didactic interpretation of life forms and landforms is that killing creatures unnecessarily or other maltreatment of nature would minimize the earth's capability for moral instruction and thwart God's purpose in creating them.

Environmental conservation was probably very far from the minds of the sages who wrote down the Bible and its subsequent commentaries. They lived in times when nature's dominance over people through floods, droughts, insect infestations, and the like seemed far more powerful than any impact that ancient humans might have upon the environment. Yet their beliefs about nature's "lessons" may be another example of an interpretive model that was not originally intended to have environmental benefits, like *bal tashchit*, but that could do so today if widely put into practice (Schwartz, 1997; Vogel, 2001). It is not inconsistent with the pre-Darwinian natural theology that engendered many taxonomic and natural history investigations in early modern Europe.

FROM ANCIENT TO MODERN TIMES

The Jews of ancient times spent most of their history under the domination of one or another foreign government. They were brought into the sphere of the Greek empire under Alexander the Great and subsequently under Roman domination. As part of their management strategy for their unruly Jewish subjects, the Romans destroyed the second temple, exiled and enslaved many Jews, gave "abandoned" farms to non-Jews, and imposed increasingly heavy burdens of taxes and religious discrimination on the Jews. Many Jews emigrated in hopes of a better life. The Christian Byzantine Empire in 324 CE encompassed Israel as part of the old Roman colonies (Alon, 1984: 752–757). Some aspects of Jewish intellectual life flourished during the early centuries of the Common Era, notably with the production of the Babylonian Talmud, but the centers of Jewish life increasingly were located outside Israel's biblical borders.

With the rise of Islam and Arab conquest in 636 CE, the ideology of Israel as a Jewish homeland became primarily an historic memory and a hope for a messianic future, as the area fell under the control of various Muslim empires governed from abroad, notably the Ottomans in the centuries prior to World War I (Roth, 1970). With the defeat of Turkey, Britain governed Israel until 1947.

Jews continued to live in Israel during the medieval and early modern periods, but they were usually few in number and restricted to specific communities. Jewish communities in the Diaspora, however, expanded throughout the Near East and in Europe, depending on the type of welcome, persecution, or expulsion they received in foreign countries (Roth, 1970). With the opening of the western hemisphere to European colonization, Jews settled in most countries of the world. Large Jewish populations developed in Eastern Europe prior to World War II.

Many Eastern European Jews participated in the new intellectual and political currents of the nineteenth and early twentieth centuries, such as modernity, political liberation movements, and especially socialism. Some Jews became increasingly integrated into their mainstream society's practices and beliefs and abandoned Orthodox religious practices. Still faced with ethnic discrimination, however, some Jewish political leaders began to feel that a return to Israel as a Jewish homeland could and should be a practical reality as much as it had been a millennial vision. Not all Zionists shared the identical ideology, but prominent political goals included the Jews' return to a Jewish homeland as a means of securing self-rule, civil rights, and a Jewish cultural revival. Zionists envisioned the ennoblement of Europe's oppressed urban ghetto Jews through manufacturing labor and farming the land of their forefathers. Although some Orthodox Jews shared the political Zionist ideology, many observant Jews believed that a return of the Jews to Israel was a strictly spiritual matter that must necessarily await God's intervention and the advent of the messiah, and they chose to remain in the *Galut*. The modern origins of the state of Israel thus reflect secular European ideas and experiences as much as ancient religious teachings.

Notwithstanding the agrarian Zionist ideology, perhaps 80% of the European Jews who immigrated to Palestine during the late nineteenth and early twentieth centuries settled in towns and cities (Ben-Artzi, 1997; Divine, 2000; Near, 1992). Those who enacted the agrarian Zionist vision purchased land from Arab owners – often land that the Arabs didn't want to cultivate themselves, such as the Hulah Valley swamp south of the Sea of Galilee in 1934. In keeping with their socialist influences, the Jewish pioneers set up collective farming communities. They proceeded to drain swamps, to replace small-scale Arab cultivation with extensive fields, and to mechanize crop raising in their efforts to build up the land. Israel's Mediterranean climate enabled the early *Halutzim* (settlers) to cultivate and market crops unsuited to temperate Europe, such as citrus fruit grown with irrigation.

Settlers and their Jewish supporters outside of Israel established the Jewish National Fund (JNF) in 1901. A major component of its agenda was to plant forests, typically conifers, on land that is non-arable (though often grazed by Palestinian and Bedouin herders) (*Israeli Yearbook & Almanac*, 1998: 215; Cohen, 1993).

Needless to say, local Palestinian Arabs often stoutly resisted what they saw as a kind of European invasion of their own patrimony in the land of Palestine. As the horrors of the Nazi Holocaust against the Jews became known, international pressures mounted on Britain to allow Israel to become an independent Jewish state. The United Nations recognized the state of Israel in 1947, over the objections of its Arab member nations.

The subsequent history of Israel until today has been checkered by periods of relative calm interspersed with wars and raids pitting Israeli Jews against Arab residents and sometimes against bordering Muslim states. Constantly fearing guerrilla and terrorist attacks if not outright military invasion, the state of Israel placed national defense as its highest priority. In various migration waves and movements, Jews from around the world nevertheless began to immigrate to Israel, many from Russia or Islamic countries. Some emigrated for ideological reasons, such as ethnic or religious solidarity; others fled even more precarious circumstances at home.[1]

Rather than continuing the traditional Palestinian Arab economy at mid-century, the new State of Israel chose to develop as a modern western industrial state. Thus within the boundaries of present-day Israel/Palestine, two cultures and economies uneasily co-evolved: one oriented towards the West, industry, urbanization, and modernity, and one more conservatively oriented towards long-standing agrarian and village traditions. The two economies have connected at many contact points, however. Israeli capital, for example, benefited from the cheaper wages paid to Arab workers in such sectors as tourism, construction, and even agriculture.

Western-style scientific agronomy essentially supplanted the Talmudic precepts in agriculture. During the latter part of the twentieth century, fewer than 20% of Israelis called themselves "religious" or Orthodox, with a much smaller percentage of religiously observant Israelis engaged in agriculture. Israelis maintained the sabbatical year primarily through the fiction of temporarily "selling" the land to non-Jews every seventh year so that cultivation could continue. *Bal taschit* was ignored when Arab olive trees and vines were cleared to make way for Israeli commercial agriculture or for national security strategies, such as the construction of military roads and towns in the West Bank (Benvenisti, 2000).

The above are just some of the events and factors that influenced Jewish environmental practice in the twentieth century. Some of the results were entirely predictable, based generally upon experiences of other industrializing countries, and particularly upon the circumstances of Israel's recent history.

> Israel's first priority in 1948 was survival – not the long-term survival of humankind through clean air and water, but immediate escape from the perils of war and hunger. Pollution prevention, a low priority worldwide, was an irrelevant luxury. After the War of Independence, construction of housing and industry and the development of agriculture were the major objectives; preservation of the environment was not (1998 *Israel Yearbook & Almanac*: 214; see also Tal, 2000; Brooks, 2000).

Draining the swamps expanded agricultural land, but it threatened the survival of wetland species. In the absence of strict emissions regulations, Israeli industries polluted rivers with toxic effluent. Cities with inadequate sewage treatment facilities dumped untreated wastes. Urban and agricultural water demands borrowed heavily from coastal aquifers that began to absorb seawater, together with chemical fertilizers from agricultural applications. Water diversion projects on the Jordan River radically lowered the levels of the Dead Sea and raised neighboring Arab countries' concerns about Israel's "stealing" their share from the Jordan watershed (Lees, 1998). Low water volumes in Israel's rivers exacerbated concentrations of water pollutants. Pollution of the Mediterranean Sea along the Israeli coast raised international environmentalist concerns.

Other environmental impacts related to Israel's concern for national defense. When the army concluded that it needed an installation on the summit of

Mount Meron, for example, the national park there lost land to make space for it.

Yet western-style environmental conservation reached Israel as well (Odenheimer, 1991). Israelis concerned about environmental protection founded the non-government organization, the Society for the Protection of Nature in Israel, in 1953. It publishes the popular environmentalist journal *Israel – Land and Nature* and conducts a series of field trips for the public. During the 1960s, the government established an Environment Ministry and a Nature and National Parks Protection Authority, followed by other agencies dedicated to protecting water quality, abating pollution, and so on. The government established a series of national parks, nature preserves, and heritage resources sites – notably the *Chai Bar* desert wildlife preserve for species mentioned in the Bible and its botanical counterpart *Neot Kedumim* (see Haruveni, 1984). Israel has been drawn into international environmental conservation initiatives, such as water quality and fisheries in the Mediterranean Sea. Israeli universities support a variety of courses, degrees, and research programs in western natural science and environmental conservation.

Israelis have sometimes been conservationists in spite of themselves. The country lacks conventional energy resources, and few petroleum exporting nations will sell oil to Israel. Consequently solar water heaters on apartment rooftops are common sights, just as low indoor temperatures are common winter experiences. The exclusion of motor vehicles from much of the old city of Jerusalem to preserve its characteristic narrow streets, and proscriptions against Sabbath-day driving in ultra-religious neighborhoods equally enhance their air quality.

Within such contrasts, much of the ambivalence that environmentalists feel towards western science may be seen. On the one hand, science and technology contributed to much of modern Israel's pollution and loss of green space. On the other hand, natural sciences and resource management disciplines provide the insights and tools to rectify environmental problems. The role of Judaism in environmental protection in Israel suggests some dialogue between religion and environmental science as well. National parks and nature reserves often protect sites as biblical heritage resources. The implications of new genetically engineered foods for kosher dietary laws are currently under debate. Internationally, Jewish philosophers, environmentalists, and congregations inquire into prospects for linking Judaism more clearly to the environmentalist agenda. With Palestinians' and Jews' competing human ecologies and landscape ideologies, political ecology concerns remain paramount. The prospects for peace and warfare in Israel, and the conflicting aspirations of its Jews and Palestinians, will probably be the overriding issues in the future of Israel's environments.

NOTES

[1]   See *Israel Yearbook & Almanac* in annual editions for summaries and demographic and economic statistical information.

BIBLIOGRAPHY

Adams, Carol J. 'Ecofeminism and the eating of animals.' In *Environmental Ethics: Divergence and Convergence*, 2nd. ed., R.G. Botzler and S.J. Armstrong, eds. Boston: McGraw-Hill, 1998, pp. 505–514.

Aharoni, Yohanan. *The Land of the Bible: A Historical Geography*. London: Burns and Oates, 1979.

Alon, Gedaliah. *The Jews in Their Land in the Talmudic Age (70–640 C.E.)*, vol. 1. Jerusalem: Magnes Press, Hebrew University, 1980.

Alter, Robert and Frank Kermode, eds. *The Literary Guide to the Bible*. Cambridge, Massachusetts: Belknap Press, 1987.

Ben-Artzi, Yossi. *Early Jewish Settlement Patterns in Palestine, 1882–1914*. Jerusalem: Magnus Press, 1997.

Benvenisti, Meron. *Sacred Landscape: The Buried History of the Holy Land Since 1948*. Berkeley: University of California Press, 2000.

Benstein, Jeremy. 'One, walking and studying ... : nature vs. Torah.' *Judaism* 44: 146–169, 1995.

Bialik, Hayim Nahman and Yehoshua Hana Ravnitsky, eds. *The Book of Legends: Sefer Ha-Aggadah, Legends from the Talmud and Midrash*. New York: Schocken Books, 1992.

Bleich, J. David. 'Vegetarianism and Judaism.' In *Judaism and Environmental Ethics: A Reader*, Martin D. Yaffe, ed. Lanham, Maryland: Lexington Books, 2001, pp. 371–383.

Blidstein, Gerald. 'Man and nature in the sabbatical year.' *Tradition: A Journal of Orthodox Thought* 8: 48–55, 1966.

Brooks, David B. 'The struggle for Israel's environment: tougher than ever.' In *Torah of the Earth: Exploring 4,000 Years of Ecology in Jewish Thought*, vol. 2, A. Waskow, ed. Woodstock, Vermont: Jewish Lights Publishing, 2000, pp. 72–98.

Charles, R. H. *The Apocrypha and Pseudepigrapha in English*. Oxford: Clarendon Press, 1963–64.

Cohen, Shaul Ephraim. *The Politics of Planting: Israeli-Palestinian Competition for Control of Land in the Jerusalem Periphery*. Chicago: University of Chicago Press, 1993.

Davies, Eryl W. 'Land: its rights and privileges.' In *The World of Ancient Israel*. R.E. Clements, ed. Cambridge: Cambridge University Press, 1989, pp. 349–370.

Dever, William G. 'Archaeology, ideology, and the quest for an "ancient" or "biblical" Israel.' *Near Eastern Archaeology* 61(1): 39–52, 1998.

Divine, Donna Robinson. 'Zionism and the transformation of Jewish society.' *Modern Judaism* 20: 257–276, 2000.

Dresner, Samuel H. *The Jewish Dietary Laws*. New York: Rabbinical Assembly of America, United Synagogue Commission on Jewish Education, 1982.

Encyclopaedia Judaica. Jerusalem: Encyclopaedia Judaica, 1971–72.

Freedman, H. and M. Simon. *Genesis Rabbah I*. London: Soncino Press, 1961.

Frick, F.S. 'Ecology, agriculture and patterns of settlement.' In *The World of Ancient Israel*, R.E. Clements, ed. Cambridge: Cambridge University Press, 1989, pp. 67–93.

Friedman, Richard E. *Who Wrote the Bible?* Englewood Cliffs, New Jersey: Prentice Hall, 1987.

Frymer-Kensky, Tikva. *In the Wake of Goddesses*. New York: Macmillan, 1992.

Genzburg, Louis. *Legends of the Jews*, 7 vols. Philadelphia: Jewish Publication Society of America, 1967–69.

Gerstenfeld, Manfred. *Judaism, Environmentalism and the Environment: Mapping and Analysis*. Jerusalem: The Jerusalem Institute for Israel Studies and Rubin Mass Ltd., 1998.

Goodspeed, Edgar J. *The Apocrypha: An American Translation*. New York: Vintage Books, 1989.

Gottwald, N.K. *The Hebrew Bible: A Socio-Literary Introduction*. Philadelphia: Fortress Press, 1989.

Graham, Brian, C.J. Ashworth, and J.E. Tunbridge. *A Geography of Heritage: Power, Culture, and Economy*. London: Arnold, 2000.

Hareuveni, Nogah. *Tree and Shrub in our Biblical Heritage*. Kiryat Ono, Israel: Neot Kedumim Ltd., 1984.

Hillel, Daniel. *Land, Water, and Life in a Desert Environment*. New York: Praeger Publishers, 1982.

Houston, Walter. *Purity and Monotheism: Clean and Unclean Animals in Biblical Law*. Sheffield, UK: JSOT Press, 1993.

James, E.O. *Seasonal Feasts and Festivals*. London: Thames and Hudson, 1961.

Kark, Ruth. 'Land – God – man: concepts of land ownership in traditional cultures in Eretz-Israel.' In *Ideology and Landscape in Historical Perspective*, Alan R.H. Baker and Gideon Biger, eds. Cambridge: Cambridge University Press, 1992, pp. 63–82.

Kass, Leon R. 'Sanctified eating.' In *Judaism and Environmental Ethics: A Reader*, Martin D. Yaffe, ed. Lanham, Maryland: Lexington Books, 2001, pp. 384–409.

Kay, Jeanne. 'Human dominion over nature in the Hebrew Bible.' *Annals of the Association of American Geographers* 79: 214–232, 1989.

Klein, Isaac. *A Guide to Jewish Religious Practice*. New York: Jewish Theological Society of America, 1979.

Kunin, Seth D. 'Sacred place.' In *Themes and Issues in Judaism*, S.D. Kunin, ed. London: Cassell, 2000, pp. 22–55.

Lasker, Arnold A. and Daniel J. Lasker. 'The Jewish prayer for rain in Babylonia.' *Journal for the Study of Judaism* 15: 123–144, 1984.

Lees, Susan H. *The Political Ecology of the Water Crisis in Israel*. Lanham, Maryland: University Press of America, 1998.

Leopold, Aldo. 'Wildlife in American culture.' In *A Sand County Almanac and Sketches Here and There*, A. Leopold, ed. Oxford: Oxford University Press, 1949, p. 177.

Levine, Etan. 'The land of milk and honey.' *Journal for the Study of the Old Testament* 87: 43–57, 2000.

Levine, Lee I. *The Ancient Synagogue: The First Thousand Years*. New Haven, Connecticut: Yale University Press, 2000.

Near, Henry. *The Kibbutz Movement: A History*, 2 vols. Oxford: Oxford University Press, 1992.

Neusner, Jacob. *Invitation to the Talmud: A Teaching Book*, rev. ed. New York: Harper and Row, 1984.

Odenheimer, Micha. 'Retrieving the Garden of Eden: environmental organizations in Israel.' *The Melton Journal* 24: 12–14, 20, 1991.

Patai, Raphael. *The Hebrew Goddess*. 3rd ed. Detroit: Wayne State University Press, 1990.

Robinson, George. *Essential Judaism: A Complete Guide to Beliefs, Customs, and Rituals*. New York: Pocket Books, 2000.

Rogerson, John and Philip R. Davies. 'Was the Siloam tunnel built by Hezekiah?' *Biblical Archaeologist* 59 (3): 138–148, 1996.

Ron, Z. 'Agricultural terraces in the Judean Mountains.' *Israel Exploration Journal* 16: 33–49, 111–122, 1966.

Ron, Z. 'Development of irrigation systems in mountain regions of the Holy Land.' *Transactions of the Institute of British Geographers* n. s. 10: 149–169, 1985.

Roth, Cecil. *A History of the Jews, From Earliest Times to the Six-Day War*. New York: Schocken Books, 1970.

Schaffer, Arthur. 'The agricultural and ecological symbolism of the four species of Sukkot.' *Tradition: A Journal of Orthodox Thought* 20: 128–140, 1982.

Schwartz, Eilon. '*Bal taschit*: A Jewish environmental precept.' *Environmental Ethics* 19: 355–374, 1997.

Smith, George Adam. *The Historical Geography of the Holy Land*. London: Fontana Library, 1966.

Stadelmann, L.J. *The Hebrew Conception of the World: A Philological and Literary Study*. Rome: Pontifical Biblical Institute, Analecta Biblica 39, 1970.

Steinsaltz, Adin. *The Essential Talmud*. New York: Basic Books, 1976.

Stern, Menahem, ed. *Greek and Latin Authors on Jews and Judaism*, 3 vols. Jerusalem: The Israeli Academy of Sciences and the Humanities, 1980.

Tal, Alon. 'An imperiled promised land.' In *Torah of the Earth: Exploring 4,000 Years of Ecology in Jewish Thought*, vol. 2, A. Waskow, ed. Woodstock, Vermont: Jewish Lights Publishing, 2000, pp. 42–71.

Thompson, Janna. 'Environment as cultural heritage.' *Environmental Ethics* 22: 241–258, 2000.

Tirosh-Samuelson, Hava, ed. *Judaism and Ecology: Created World and Revealed World*. Cambridge, Massachusetts: Harvard University Press, 2002.

Tsuk, Tsvika. 'Bringing water to Sephoris.' *Biblical Archaeology Review* 26(4): 34–41, 2000.

Vilnay, Zev. *Legends of Jerusalem: The Sacred Land*, vol. 1. Philadelphia: The Jewish Publication Society of America, 1973.

Vilnay, Zev. *Legends of Galilee, Jordan, and Sinai: The Sacred Land*, vol. 3. Philadelphia: The Jewish Publication Society of America, 1978.

Vogel, David. 'How green is Judaism? Exploring Jewish environmental ethics.' *Judaism* 50: 66–91, 2001.

Walsh, Carey Ellen. *The Fruit of the Vine: Viticulture in Ancient Israel*. Winona Lake, Indiana: Eisenbrauns, 2000.

Waskow, Arthur. 'Earth, social justice, and social transformation: the spirals of sabbatical release.' In *Torah of the Earth: Exploring 4,000 Years of Ecology in Jewish Thought*, vol. 1, A. Waskow, ed. Woodstock, Vermont: Jewish Lights Publishing, 2000, pp. 70–83.

Werblowsky, R.J. Zwi and Geoffrey Wigoder. *The Encyclopedia of the Jewish Religion*. New York: Holt, Rinehart and Winston, Inc., 1965.

Whiston, William A.M., trans. *The Works of Josephus*. Peabody, Massachusetts: Hendrickson Publishers, 1987.

Whitekettle, Richard. 'Where the wild things are: primary level taxa in Israelite zoological thought.' *Journal for the Study of the Old Testament* 93: 17–37, 2001.

Index

Science Across Cultures

KLUWER ACADEMIC PUBLISHERS – DORDRECHT / BOSTON / LONDON

Printed in the United States
94879LV00001BA/5-6/A